The Arenaviridae

THE VIRUSES

Series Editors
HEINZ FRAENKEL-CONRAT, *University of California*
Berkeley, California

ROBERT R. WAGNER, *University of Virginia School of Medicine*
Charlottesville, Virginia

THE VIRUSES: Catalogue, Characterization, and Classification
Heinz Fraenkel-Conrat

THE ADENOVIRUSES
Edited by Harold S. Ginsberg

THE ARENAVIRIDAE
Edited by Maria S. Salvato

THE BACTERIOPHAGES
Volumes 1 and 2 • Edited by Richard Calendar

THE HERPESVIRUSES
Volumes 1–3 • Edited by Bernard Roizman
Volume 4 • Edited by Bernard Roizman and Carlos Lopez

THE INFLUENZA VIRUSES
Edited by Robert M. Krug

THE PAPOVAVIRIDAE
Volume 1 • Edited by Norman P. Salzman
Volume 2 • Edited by Norman P. Salzman and Peter M. Howley

THE PARAMYXOVIRUSES
Edited by David W. Kingsbury

THE PARVOVIRUSES
Edited by Kenneth I. Berns

THE PLANT VIRUSES
Volume 1 • Edited by R. I. B. Francki
Volume 2 • Edited by M. H. V. Van Regenmortel and Heinz Fraenkel-Conrat
Volume 3 • Edited by Renate Koenig
Volume 4 • Edited by R. G. Milne

THE REOVIRIDAE
Edited by Wolfgang K. Joklik

THE RETROVIRIDAE
Volumes 1 and 2 • Edited by Jay A. Levy

THE RHABDOVIRUSES
Edited by Robert R. Wagner

THE TOGAVIRIDAE AND FLAVIVIRIDAE
Edited by Sondra Schlesinger and Milton J. Schlesinger

THE VIROIDS
Edited by T. O. Diener

The Arenaviridae

Edited by
MARIA S. SALVATO
University of Wisconsin Medical School
Madison, Wisconsin

PLENUM PRESS • NEW YORK AND LONDON

Library of Congress Cataloging-in-Publication Data

The Arenaviridae / edited by Maria S. Salvato.
p. cm. -- (The Viruses)
Includes bibliographical references and index.
ISBN 0-306-44272-8
1. Arenavirus diseases. 2. Arenaviruses. I. Salvato, Maria S.
II. Series.
[DNLM: 1. Arenaviridae. 2. Arenavirus Infections. QW 168 A6802]
QR201.A74A74 1993
616'.0194--dc20
DNLM/DLC
for Library of Congress 92-48280
CIP

ISBN 0-306-44272-8

A Division of Plenum Publishing Corporation
233 Spring Street, New York, N.Y. 10013

Printed in the United States of America

To my husband, C. David Pauza,
who has contributed a great deal of time and effort
and a great many pearls of wisdom
to the production of this volume

Contributors

Laura E. Alche, Laboratorio de Virologia, Departamento de Quimica Biologica, Facultad de Ciencias Exactas y Naturales, Ciudad Universitaria, Pabellón 2, Piso 4, 1428 Buenos Aires, Argentina

David D. Auperin, Special Pathogens Branch, Division of Viral and Rickettsial Diseases, Centers for Disease Control, Atlanta, Georgia 30333. *Present address:* Department of Molecular Genetics, Pfizer Central Research, Groton, Connecticut 06340

Michael J. Buchmeier, Division of Virology, Department of Neuropharmacology, The Scripps Research Institute, La Jolla, California 92037

John W. Burns, Palo Alto Veterans Administration Medical Center, Palo Alto, California 94304

James E. Childs, Department of Immunology and Infectious Diseases, The Johns Hopkins University School of Hygiene and Public Health, Baltimore, Maryland 21205. *Present address:* Viral and Rickettsial Zoonoses Branch, Centers for Disease Control and Prevention, Atlanta, Georgia 30333

J. Christopher S. Clegg, Public Health Laboratory Service, Centre for Applied Microbiology and Research, Division of Pathology, Porton Down, Salisbury, Wiltshire, SP4 0JG, England

Richard W. Compans, Department of Microbiology and Immunology, Emory University, Atlanta, Georgia 30322

Celia E. Coto, Laboratorio de Virologia, Departamento de Quimica Biologica, Facultad de Ciencias Exactas y Naturales, Ciudad Universitaria, Pabellón 2, Piso 4, 1428 Buenos Aires, Argentina

Elsa B. Damonte, Laboratorio de Virologia, Departamento de Quimica Biologica, Facultad de Ciencias Exactas y Naturales, Ciudad Universitaria, Pabellón 2, Piso 4, 1428 Buenos Aires, Argentina

Susan P. Fisher-Hoch, Special Pathogens Branch, Division of Viral and Rickettsial Diseases, Centers for Disease Control, Atlanta, Georgia

30333. *Present address:* Mycotic Diseases Branch, Division of Bacterial and Mycotic Diseases, Centers for Disease Control, Atlanta, Georgia 30333

Maria Teresa Franze-Fernandez, Centro de Virologia Animal (CONICET), Serrano 661, 1414 Buenos Aires, Argentina

Dominique Garcin, Département de Microbiologie, Université de Geneve, Faculté de Médecine, CH-1211 Geneva 4, Switzerland

Delsworth G. Harnish, Molecular Virology and Immunology Programme, Department of Pathology, McMaster University, Hamilton, Ontario, L8N 3Z5, Canada

Colin R. Howard, Department of Pathology and Infectious Diseases, Royal Veterinary College, London, NW1 0TU, England

Silvia Iapalucci, Centro de Virologia Animal (CONICET), Serrano 661, 1414 Buenos Aires, Argentina

Linda S. Klavinskis, Department of Neuropharmacology, Division of Virology, Research Institute of Scripps Clinic, La Jolla, California 92037. *Present address:* Department of Immunology, Guy's Hospital Medical School, London, SE1 9RT, England

Daniel Kolakofsky, Département de Microbiologie, Université de Geneve, Faculté de Médecine, CH-1211 Geneva 4, Switzerland

Nora Lopez, Centro de Virologia Animal (CONICET), Serrano 661, 1414 Buenos Aires, Argentina

Liliana A. Martinez Peralta, Catedra de Microbiologia, Parasitologia e Immunologia, Facultad de Medicina de la Universidad de Buenos Aires, 1121 Buenos Aires, Argentina

Michael B. A. Oldstone, Department of Neuropharmacology, Division of Virology, Research Institute of Scripps Clinic, La Jolla, California 92037

Clarence J. Peters, Disease Assessment Division, U. S. Army Medical Research Institute of Infectious Diseases, Fort Detrick, Frederick, Maryland 21701. *Present address:* Special Pathogens Branch, Centers for Disease Control and Prevention, Atlanta, Georgia 30333

Charles J. Pfau, Department of Biology, Rensselaer Polytechnic Institute, Troy, New York 12181

Stephen J. Polyak, Molecular Virology and Immunology Programme, Department of Pathology, McMaster University, Hamilton, Ontario, L8N 3Z5, Canada

William E. Rawls, Molecular Virology and Immunology Programme, Department of Pathology, McMaster University, Hamilton, Ontario, L8N 3Z5, Canada

Victor Romanowski, Instituto de Bioquimica y Biologia Molecular, Departamento de Ciencias Biológicas, Facultad de Ciencias Exactas, Universidad Nacional de La Plata, 1900 La Plata, Argentina

Carlos Rossi, Centro de Virologia Animal (CONICET), Serrano 661, 1414 Buenos Aires, Argentina

Maria S. Salvato, Department of Pathology and Laboratory Medicine, University of Wisconsin, Madison, Wisconsin 53706

Luis Scolaro, Laboratorio de Virologia, Departamento de Quimica Biologica, Facultad de Ciencias Exactas y Naturales, Ciudad Universitaria, Pabellón 2, Piso 4, 1428 Buenos Aires, Argentina

Marylou V. Solbrig, Lassa Fever Research Project, Centers for Disease Control, Atlanta, Georgia 30333. *Present address:* Department of Neurology, University of California, Irvine, Irvine, California 92717

A. Randrup Thomsen, Institute of Medical Microbiology and Institute of Experimental Immunology, University of Copenhagen, DK-2200 Copenhagen N, Denmark

Mercedes C. Weissenbacher, Catedra de Microbiologia, Parasitologia e Immunologia, Facultad de Medicina de la Universidad de Buenos Aires, 1121 Buenos Aires, Argentina

J. Lindsay Whitton, Department of Neuropharmacology, Division of Virology, Research Institute of Scripps Clinic, La Jolla, California 92037

Rolf M. Zinkernagel, Department of Pathology, Institute of Experimental Immunology, University of Zurich, Zurich, Switzerland

In Memoriam

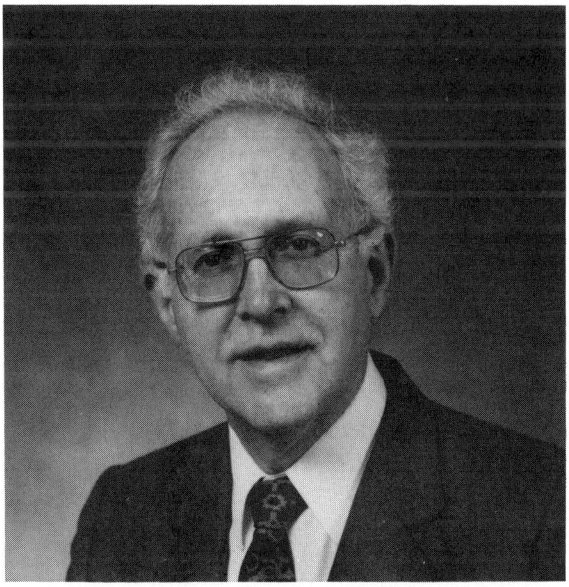

DR. WILLIAM E. RAWLS

Dr. William E. Rawls (June 7, 1933–October 8, 1990) was an exceptional friend, educator, and researcher. His death from liver cancer ended a long and productive career. Dr. Rawls received his M.D. from the University of Tennessee in 1958 and a Master of Science from the University of Minnesota (Mayo Foundation) in 1964. Following military duty as a Captain with the U.S. Army Medical Corps (1959–1961) and a year as an Assistant Staff member at the Mayo Clinic, Dr. Rawls devoted his exceptional skills to research as a Special NIH Fellow and a Research Career Development Awardee at the Baylor College of Medicine (1965–1973). From 1974 to 1987 he headed the Cancer Research Group in the Department of Pathology at McMaster University, and in this role was responsible for nurturing many young scientists. He was appointed Director of the Molecular Virology and Immunology Program at McMaster University in 1984 and maintained this position until his untimely death.

Dr. Rawls was an active member of several committees of the National Cancer Institute (Canada) and from 1986 to 1988 served as institute president. His scientific expertise was recognized in his contributions as a member of the editorial boards of *Intervirology, Journal of Clinical Microbiology, Infection and Immunity, Archives of Virology, Journal of Clinical Microbiology, Infection and Immunity, Archives of Virology, Journal of Immunology,* and *Viral Immunology,* as well as of the scientific review committees of the National Institutes of Health (Virus Cancer Program) and the Medical Research Council (Canada). Between 1963 and 1990, he published over 173 peer-reviewed articles and contributed an additional 69 invited papers, reviews, and book chapters. He was recognized for pivotal observations in congenital rubella, in which he demonstrated persistence of rubella virus in cultured postmortem tissues even though the tissues appeared normal (Rawls and Melnick, *J. Exp. Med.* **123**:795, 1966). In 1968 he proposed and supported a link between herpes virus 2 and cervical carcinoma (Rawls et al., *Science* **161**:1255, 1968). He remained active in this area of investigation until his death.

We are particularly indebted to Bill Rawls as one of the founders of modern arenavirus research. His substantial contributions in characterizing the immunological and virological features of arenaviruses were primarily based on studies of Pichinde, a member of the Tacaribe complex of arenaviruses.

Dr. Rawls's strengths were in research and education. He enjoyed supervision and guidance of graduate students (12 M.Sc. and 12 Ph.D.), postdoctoral fellows (26), and medical students. He was always available for support and discussion. His quiet, unassuming attitude promoted those things that are best in science.

Preface

The arenaviridae are distinguished by two salient features that capture the attention of virologists, pathologists, and immunologists. First, they have the ability to cause persistent as well as acute infections, and second, they display a peculiar "ambisense" gene arrangement that combines features of both the negative-strand and positive-strand RNA viruses. Ultimately, the arenaviruses adopt the replication strategy of the negative-strand viruses in that their genomic RNA is not infectious and transcriptional events precede translational events. Whether the ability to cause persistent infection presents a biological advantage that is somehow linked to the ambisense genome structure remains an intriguing question.

This volume presents the most current molecular, organismal, and epidemiological research on arenaviruses. In keeping with the history of this subject, leading researchers in the field describe the fascinating scenario of arenavirus infection; their investigations contribute greatly to our basic understanding of immunology, cell biology, biochemistry, and medicine. Although research on arenaviruses has been typified by biological observations that precede detailed molecular studies, structural studies are presented at the beginning of this volume to provide a framework for understanding the complex phenomena of arenavirus biology. Thus, the nineteen chapters of this book are distributed between the sections "Molecular Structure of the Arenaviruses" and "Immunopathology, Vaccines, and Epidemiology." Notably, some chapters, particularly those dealing with the identification of effective epitopes for subunit vaccines, contain the characteristics of both sections.

Our entré to the arenaviridae is the overview of ultrastructure and morphogenesis by Compans. Inasmuch as natural arenavirus infections are commonly acquired via the mucosa, Compans points out that the mechanism of arenavirus entry and budding from polarized epithelial cells remains an important and uncharted area of research. The well-known tendency of arenavirus particles to package ribosomes is described here, although the chapter by Romanowski suggests that ribosome-containing virions may be noninfectious. Processing and assembly of the envelope glycoproteins are described by Burns and Buchmeier, who also present excellent cryoelectron micrographs that reveal a spherically symmetrical (rather than pleiomorphic) morphology of arenaviruses.

Several chapters address mechanisms of arenavirus persistence. Howard proposes that variation of envelope antigens allows escape of the host's immune response and persistent infection. This idea is further developed in the chapters by Coto *et al.* and Martinez Peralta *et al.*, where neutralization-resistant variants of Junin and Tacaribe viruses are examined. Although escape from the humoral immune response is important for persistence of the Tacaribe complex group of arenaviruses, a cytotoxic T-lymphocyte (CTL) response more frequently determines acute or persistent infection for the Old World arenavirus, LCMV. In light of the landmark paper from Zinkernagel's group (Pircher *et al.*, "Viral escape by selection of cytotoxic T cell–resistant virus variants *in vivo*," *Nature* **346**:629, 1990), it is known that the CTL response selects for escape mutants in a manner similar to that for humoral immunity. Whereas some viral variants arise that can escape host immune surveillance, others acquire the ability to suppress immune response and thereby persist. The chapter by Salvato describes the CTL-"immunosuppressive" variants of LCMV, the chapter by Zinkernagel describes strains of LCMV that exhibit a generalized, CTL-mediated suppression of antibody responses, and the chapter by Pfau and Thomsen describes high-dose immune suppression (another means to this end).

New arenavirus variants have been isolated by several groups. Temperature-sensitive mutants of LCMV (Romanowski), Junin (Coto *et al.*), and Pichinde (Harnish *et al.*) display interesting defects in virus dissemination, in early transcription, or in protein synthesis; some of these mutations were mapped genetically to the large or small RNA segment. Virulent "macrophage-tropic" and attenuated "lympho/neurotropic" strains of Junin virus in guinea pigs are described in the chapter by Coto *et al.* Pfau and Thomsen discuss the Docile (viscerotropic) and Aggressive (neurotropic) isolates of LCMV (UBC strain), in which docility is associated with the tendency for high-titer replication in the viscera that leads to suppression of the CTL response and persistent infection. Although few of these interesting arenavirus variants have been subjected to sequence analysis, detailed comparisons of sequences for CTL^+ and CTL^- variants of LCMV are presented in the Salvato chapter.

Several groups analyzed the essential components for arenavirus gene expression and replication as a prelude to the anticipated development of reintroduction systems. The genomic sequences of LCMV and Tacaribe viruses were completed by the Salvato and Franze-Fernandez laboratories, respectively; genes and structures likely to control gene expression or replication have been identified. Most importantly, these investigators report a new metal-binding protein that may be crucial to the arenavirus life cycle. Significant progress has been made in the sequence analysis of Junin (Romanowski), Pichinde (Harnish *et al.*), and Lassa (Auperin and Clegg chapters); the phylogenetic relationships of arenavirus sequences are summarized by Clegg.

Because arenavirus transcription takes place in the cytoplasm, it has been particularly difficult to express arenavirus mRNA from DNA vectors that are transcribed only in the nucleus. Romanowski describes pitfalls and some successes in using SV40 to express Junin and LCMV NP protein. Transient gene expression was achieved with vaccinia as a cytoplasmic expression vector (see chapters by Auperin or Zinkernagel, or Klavinskis *et al.*). The involvement of host-derived molecules in arenavirus expression has been explored by the use of metabolic inhibitors (see Harnish *et al.*). Arenavirus transcription products have been analyzed carefully by the Franze-Fernandez group, who identified the 3′ termini of the four major arenavirus mRNAs. A well-defined *in vitro* transcription system, described in the Kolakofsky and Garcin chapter, constitutes a significant advance in our ability to study arenavirus gene expression. This laboratory is also pioneering the reintroduction of reporter genes flanked by arenavirus transcription control signals. Soon it may be routine to reintroduce genetically altered arenavirus genes for structure/function studies.

Although the first section of the book emphasizes the role of viral genes in determining the outcome of infection, the second section emphasizes the role of the host in mechanisms of pathogenesis. In a lively historical introduction, Pfau and Thomsen describe the beginnings of arenavirus research in the 1930s. They use the LCMV model to detail the roles of defective interfering particles and host genetics (both MHC and non-MHC genes) in determining the outcome of infection.

A practical application of arenavirus research is the study of vaccines in well-defined model systems. Mice are protected from lethal LCMV challenge either by peptide vaccination or by vaccinia vectors that express small epitopes (see chapters by Zinkernagel or by Klavinskis *et al.*). Guinea pigs are protected from lethal Lassa virus challenge by inoculation with vaccinia vectors that express either Lassa glycoprotein or Lassa nucleoprotein; primates, however, are protected only by the vector expressing glycoprotein (see chapters by Auperin and by Fisher-Hoch). Primates are protected from the Junin virus disease, hemorrhagic fever, by injection with attenuated strains of Junin or with the related strain Tacaribe (Martinez Peralta *et al.*). A strong case is made for the use of Tacaribe as a vaccine because it is attenuated, unlikely to revert to pathogenicity, and causes a brief infection.

Vaccine research using arenaviruses established important negative as well as positive results. Vaccination of guinea pigs with both the Lassa

NP and GP-expressing vector does not give double protection against virus challenge, but instead gives less protection than a single vaccination (see chapter by Auperin). Vaccination of mice with a GP-expressing vector can sometimes exacerbate virus-mediated pathogenesis (see chapter by Zinkernagel). Despite the setbacks, there is considerable optimism that effective subunit vaccines may eliminate some problems that arise when attenuated viral strains are used for vaccination.

Detailed clinical observations with special emphasis on human infection are presented in the chapter by Fisher-Hoch. We are made acutely aware of the gap between disease symptoms and our knowledge of the underlying pathogenic mechanisms. In view of the fact that the arenaviruses have been a model for viral effects on the CNS, a brief chapter by Solbrig suggests measurements of cerebrospinal fluid that could yield new information about the neuropathology of Lassa fever.

The human toll of arenavirus disease is strongly felt in the grain fields of Argentina and in the Lassa-endemic areas of West Africa. Worldwide epidemiological data have been surveyed by Childs and Peters in a sweeping account of fluctuations in reservoir populations as they relate to the maintenance and transmission of arenaviruses. A previously unrecognized form of arenavirus persistence, long-term virus presence without detectable viremia, has been documented for certain South American viruses. These observations deserve further study, particularly with regard to human infections. The emergence of new arenaviruses and identification of new host populations are also reported.

This volume captures our understanding of arenaviruses at one moment in time for a rapidly evolving field. Arenaviruses continue to serve as a model for persistent infection, and arenavirus research continues to elucidate basic mechanisms of immunological tolerance, viral immunosuppression, the nature of protective immune responses to vaccination, and viral effects on cell functions. Additionally, human arenavirus infection is an increasingly significant medical problem, and we can foresee the application of basic studies for detection, treatment, and prevention of these devastating viral diseases.

Maria Salvato

Madison, Wisconsin

Contents

I. Molecular Structure of the Arenaviruses

Chapter 1

Arenavirus Ultrastructure and Morphogenesis

Richard W. Compans

Chapter 2

Glycoproteins of the Arenaviruses

John W. Burns and Michael J. Buchmeier

Chapter 5

Genetic Variation in Junin Virus

Celia E. Coto, Elsa B. Damonte, Laura E. Alche, and Luis Scolaro

Chapter 6

The Unusual Mechanism of Arenavirus RNA Synthesis

Daniel Kolakofsky and Dominique Garcin

Chapter 7

Subgenomic RNAs of Tacaribe Virus

Maria Teresa Franze-Fernandez, Silvia Iapalucci, Nora Lopez, and Carlos Rossi

Chapter 9

Arenavirus Replication: Molecular Dissection of the Role of Viral Protein and RNA

Delsworth G. Harnish, Stephen J. Polyak, and William E. Rawls

Chapter 10

Molecular Phylogeny of the Arenaviruses and Guide to Published Sequence Data

J. Christopher S. Clegg

II. Immunopathology, Vaccines, and Epidemiology

Chapter 11

Lymphocytic Choriomeningitis Virus: History of the Gold Standard for Viral Immunobiology

Charles J. Pfau and A. Randrup Thomsen

Chapter 12

Influence of Host Genes on the Outcome of Murine Lymphocytic Choriomeningitis Virus Infection: A Model for Studying Genetic Control of Virus-Specific Immune Responses

A. Randrup Thomsen and Charles J. Pfau

Chapter 13

Molecular Anatomy of the Cytotoxic T-Lymphocyte Responses to Lymphocytic Choriomeningitis Virus

Linda S. Klavinskis, J. Lindsay Whitton, and Michael B. A. Oldstone

Chapter 14

Virus-Induced Acquired Immunosuppression by T-Cell-Mediated Immunopathology and Vaccine Strategies

Rolf M. Zinkernagel

Chapter 15

Construction and Evaluation of Recombinant Virus Vaccines for Lassa Fever

David D. Auperin

PART I

MOLECULAR STRUCTURE OF THE ARENAVIRUSES

CHAPTER 1

Arenavirus Ultrastructure and Morphogenesis

RICHARD W. COMPANS

I. INTRODUCTION

Although lymphocytic choriomeningitis (LCM) virus was identified as an infectious agent during the early decades of virology research (Armstrong and Lillie, 1934; Traub, 1935), no information on its structure or classification was obtained until many years had elapsed. The development of cell culture systems for virus growth and improved preparative procedures for electron microscopy enabled the initial identification of characteristic LCMV particles in thin sections of virus-infected cells (Dalton *et al.*, 1968). Soon thereafter, other viruses belonging to the Tacaribe complex were found to have a similar morphology (Murphy *et al.*, 1970), leading to the classification of these agents in a new virus family (Rowe *et al.*, 1970). Subsequently, some additional information has been obtained on the structure of the internal components of arenaviruses. This chapter reviews our current knowledge of the virion structure and the process of viral morphogenesis for this family of viruses.

II. VIRION MORPHOLOGY

The virions of the arenavirus family are lipid-enveloped, pleomorphic particles. When examined in thin sections, LCMV particles were

---

RICHARD W. COMPANS • Department of Microbiology and Immunology, Emory University, Atlanta, Georgia 30322.

The Arenaviridae, edited by Maria S. Salvato. Plenum Press, New York, 1993.

found to range from 50 to over 300 nm in size; the small to medium-sized particles were generally spherical, whereas larger particles were pleomorphic (Dalton *et al.*, 1968). A similar morphology was subsequently described for other members of the family (Murphy *et al.*, 1970). Virions possessed an electron-dense unit membrane with external surface projections. In the interior, one or more electron-dense granules, 20–25 nm in diameter, were observed in most particles, although occasional particles appeared to lack such structures. Incubation of sections prepared in a water-miscible embedding medium with ribonuclease resulted in the disappearance of the granules, indicating their content of RNA (Dalton *et al.*, 1968). The granules have subsequently been identified as host-cell-derived ribosomes (see below), and the characteristic granular internal structure was the basis for the name given to the virus family (arena=sandy, Rowe *et al.*, 1970). Some sectioned virions exhibit an apparently ordered, circular arrangement of ribosomes, which sometimes appeared to be connected by linear electron-dense structures (Murphy and Whitfield, 1975). It has been reported that under certain preparative conditions, LCM virus particles exhibit a homogeneous electron-dense core (Muller *et al.*, 1983), but the basis of these apparently different morphological features remains to be determined.

In negatively stained preparations, the surfaces of arenavirus particles were found to be covered with distinct spikes, which appeared to vary in spacing on the envelope (Murphy *et al.*, 1970). Examples of virus particles visualized by negative staining or thin-section electron microscopy are shown in Fig. 1. Some of the surface projections appear triangular when observed end-on (Fig. 1D); however, this is not uniformly observed. At high magnification, the surface spikes on Pichinde virions appear club-shaped (Murphy and Whitfield, 1975; Vezza *et al.*, 1977; Fig. 1B). The negatively stained virus particles are also generally spherical (Fig. 1C), although some pleomorphism is evident, particularly in the larger particles in virus preparations (Fig. 1D). No long filamentous virions of the type often seen with orthomyxoviruses and paramyxoviruses have been observed. Internal structural details have been difficult to resolve in intact virions; however, after disruption of the viral envelope, strand-like nucleoprotein complexes are observed (see below).

III. NUCLEOCAPSID STRUCTURE

Although the arrangement of viral internal components within intact arenavirus particles has not been determined, several studies have been carried out on the structure of nucleoprotein components isolated from disrupted virions. Elongated, strand-like structures, 9–15 nm in diameter were observed to be released from spontaneously disrupted Pichinde virions (Vezza *et al.*, 1977). The nucleocapsids of Tacaribe and Tamiami virions have been isolated by detergent disruption of virions and isopycnic centrifugation in CsCl, in which they band at a buoyant density of 1.31–1.36 g/cc (Palmer *et al.*, 1977; Gard *et al.*, 1977). When examined by staining with uranyl acetate (Palmer *et al.*, 1977), the nucleocapsids were found to be filamentous structures 5–10 nm in diameter. Most of the nucleocapsids appeared to be closed, circular structures (Fig. 2). Length distribution measurements revealed two predominant size classes of about 640 nm and 1300 nm length, presumably corresponding to the two size classes of viral genome RNAS. Two viral ribonucleoprotein size classes were also reported when LCM virus was disrupted with a detergent and analyzed in sucrose density gradients (Pedersen and Konigshofer, 1976). Host-cell-derived ribosomes were not found to be associated with these nucleocapsid structures, although they are present in virions (see below). By negative staining with sodium phosphotungstate (Gard *et al.*, 1977) the nucleocapsids were found to be 3–4 nm in diameter and had a beaded appearance indicating a series of globular subunits spaced at a periodicity of about 5 nm. The viral RNA in the nucleoprotein complexes is single-stranded and sensitive to digestion with ribonuclease (Gard *et al.*, 1977). The major polypeptide in the nucleoprotein is the 63-kDa polypeptide designated NP, which may represent the individual globular units; in addition, the L polypeptide (250 kDa), which is believed to possess transcriptase activity, has also been associated with nucleoprotein complexes (Buchmeier and Parekh, 1987). Evidence has been obtained as well for association of RNA polymerase enzyme activity with the nucleoprotein complexes (Leung *et al.*, 1977).

In preparations of Pichinde virus disrupted by osmotic shock and negatively stained with phosphotungstic acid, released nucleocapsids appeared as convoluted strands with a diameter of 12 nm (Young and Howard, 1983). Some examples of apparently helical structures were identified; these were 12–15 nm in diameter and were composed of 4–5 nm globular subunits. Uncoiling of these structures occurred after purification in a urograffin gradient revealing linear strands consisting of globular nucleosome-like structures, similar in appearance to the CsCl-purified nucleocapsids described above. Under conditions of high ionic strength, the nucleocapsids were found as more condensed, globular structures about 15 nm in diameter. Frequent appearance of circular forms of nucleocapsids was attributed to putative complementary sequences at the 5′- and 3′-ends of genomic RNA species (Young and Howard, 1983), which have subsequently been demonstrated by sequence analyses (Bishop and Auperin, 1987).

IV. INCORPORATION OF RIBOSOMES INTO VIRIONS

The characteristic electron-dense granules found in thin sections of arenavirus particles were subsequently shown to correspond to ribo-

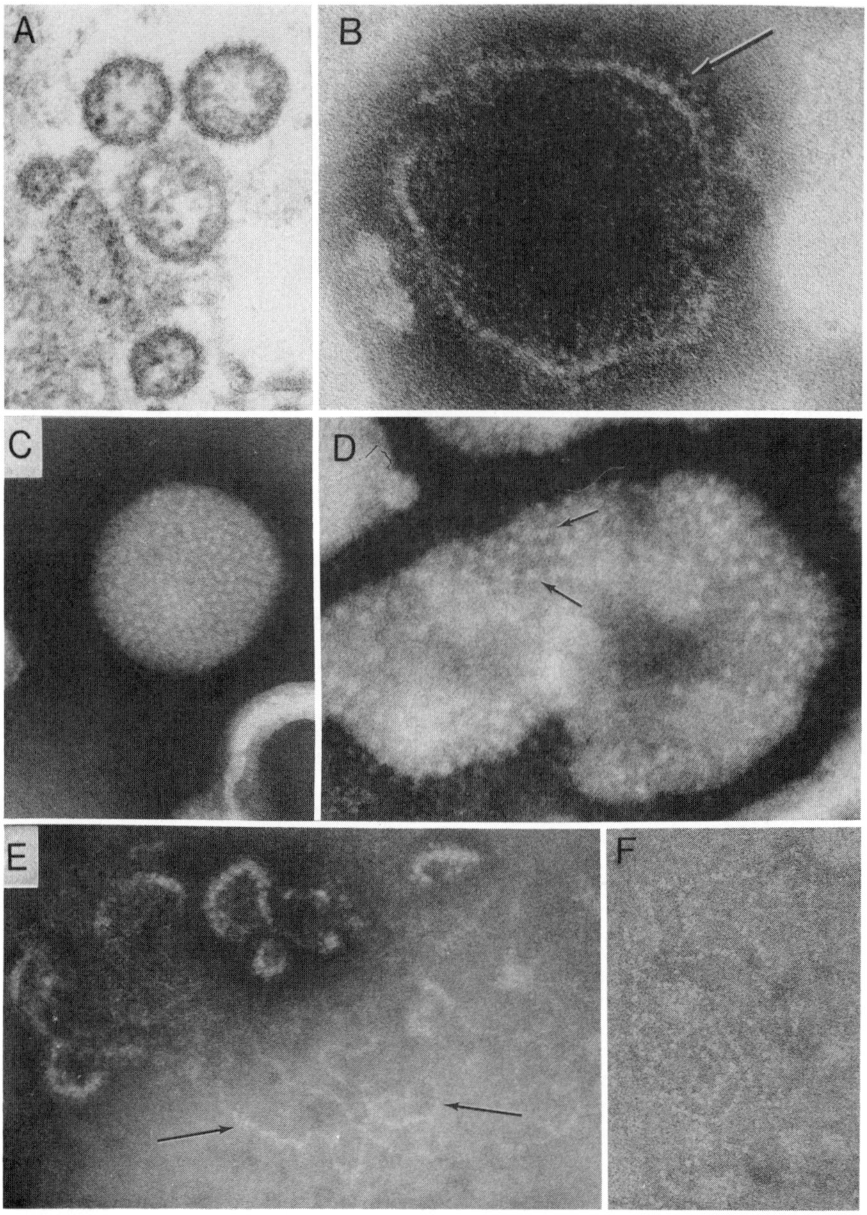

FIGURE 1 (A–F). Electron micrographs of arenavirus particles and subviral components. (A) Thin section of Pichinde virus particles released from infected BHK-21 cells. Many ribosomes are evident in the interior. (B) Single Pichinde virus particle penetrated by stain, revealing a distinct electron-lucent membrane. Individual surface spikes, 5 nm in length, can be resolved (arrow). (From Vezza *et al.*, 1977.) (C) Spherical Tacaribe virion covered with closely spaced surface spikes. (D) Higher magnification of a pleomorphic Tacaribe virion; some of the surface spikes (arrows) appear triangular. (E) Spontaneously disrupted Tacaribe virions, revealing fragments of the viral envelope and strand-like components believed to be the internal ribonucleoproteins (arrows). (F) Preparation of Tamiami ribonu-

somes, which are apparently indistinguishable from those found in normal uninfected cells. Ribosomal 60S and 40S subunits have been isolated from disrupted virions and were shown to have properties similar to those of cell ribosomes, including the presence of 28S and 18S ribosomal RNA species as demonstrated by sedimentation analyses, oligonucleotide fingerprinting, and the presence of methylated bases (Farber and Rawls, 1975; Pedersen and Konigshoffer, 1976; Vezza *et al.*, 1978). The ribosomes are capable of directing protein synthesis in vitro, when supplied with mRNA and the appropriate factors to promote translation (Chinault *et al.*, 1981).

The significance of ribosomes in arenavirus particles is not known. However, infectious virus particles have been prepared lacking detectable ribosomes (Vezza *et al.*, 1978), and studies employing cells with ts ribosomes (Leung and Rawls, 1977) showed that virus propagated in these cells grew essentially equally well at either high or low temperatures. It is therefore likely that ribosomes are not essential components of arenaviruses, and that they may be incorporated into virions during budding because of the lack of specificity in the assembly process.

V. FINE STRUCTURE AND TRANSMEMBRANE TOPOLOGY OF THE ENVELOPE GLYCOPROTEINS

In most arenaviruses, two surface glycoproteins designated GPl and GP2 have been identified, and are synthesized from a common precursor, whereas in Tacaribe, Tamiami, and Junin viruses, only a single glycoprotein similar in size to GPl has been resolved. Several additional glycoproteins have been reported in LCM virus by one group (Bruns *et al.*, 1983), but their identity remains uncertain. The amino acid sequences of the envelope glycoproteins of several arenaviruses have been deduced from the nucleotide sequences of cloned cDNA copies of viral RNAs (see Chapter 10). The sequence analyses reveal two long hydrophobic domains, which may be involved in anchorage of the glycoproteins to the viral membrane. A hydrophobic stretch of 36 amino acids is apparent near the N-terminus of the glycoprotein precursors, which is slightly longer than the signal-anchor domains involved in N-terminal anchorage of glycoproteins such as the influenza neuraminidase, and paramyxovirus hemagglutinin-neuraminidase glycoproteins (Nayak and Jabbar, 1989). A single arginine residue characteristically interrupts this hydrophobic domain in arenavirus glycoproteins, but it has been demonstrated that introduction of such a residue in the influenza NA

---

cleoproteins purified in a CsCl equilibrium gradient, showing beaded strand-like components. (C, D) Negative staining with potassium phosphotungstate. (C, E, and F are from Gard *et al.*, 1977.) (A) ×100,000; (B) ×400,000; (C) ×180,000; (D) ×340,000; (E) ×180,000; (F) ×220,000.

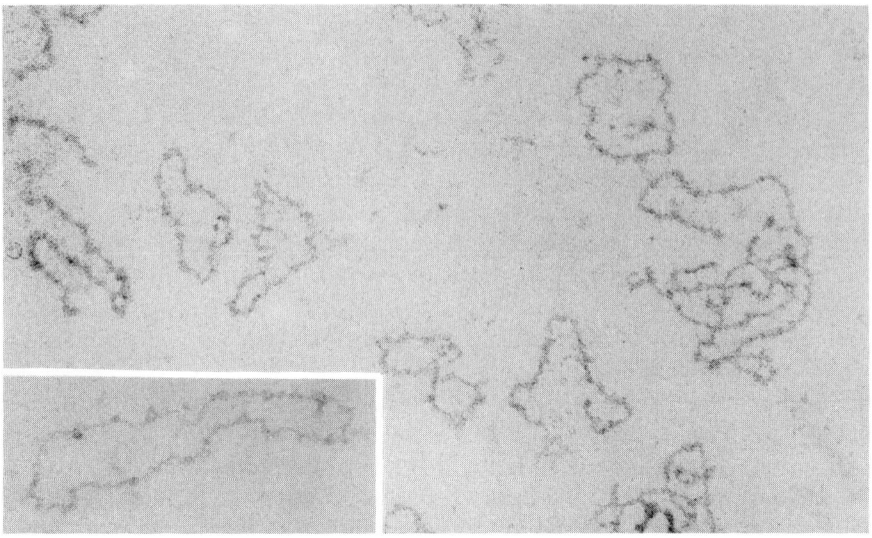

FIGURE 2. Tacaribe nucleocapsid structures from a CsCl gradient, stained with uranyl acetate, indicating a closed, circular structure. (Courtesy of Dr. Erskine Palmer; see Palmer et al., 1977.) Magnification: ×110,000.

signal-anchor domain had no effect on its function (Sivasubramanian and Nayak, 1987). Recent work by Burns *et al.* (1990) indicates that this N-terminal hydrophobic domain is absent in the mature virion form of the LCMV glycoprotein.

A second hydrophobic domain of 23 residues is present near the C-terminus of the glycoprotein precursor. Recent evidence from Burns and Buchmeier (1991) indicates that LCMV GP2 but not GP1 is an integral membrane protein, thereby implicating only the C-terminal hydrophobic domain in membrane anchorage. Thus only one of the two final glycoprotein cleavage products (GP2) is anchored to the membrane, whereas GP1 is associated to the virion by its noncovalent interactions with GP2. It has not been clearly established whether the two cleavage products remain associated as a single complex glycoprotein molecule, nor is it known whether the individual glycoprotein molecules are present as oligomeric structures in the virion. Some evidence suggesting the presence of glycoprotein oligomers has been obtained by analysis of LCMV glycoproteins using gel electrophoresis under nonreducing conditions (Wright *et al.,* 1989), in that monoclonal antibodies against the GPl (44 kDa) glycoprotein were found to react in Western blots not only with GPl monomers, but also with glycoprotein species of higher molecular weight found under these conditions. Evidence for oligomeric glycoprotein complexes has also been reported using a cross-linking procedure (Bruns and Lehmann-Grube, 1983) although these investigators observed several additional higher-molecular-weight glycoprotein species that have not been reported in studies from other laboratories.

Cross-linking studies by Burns and Buchmeier (1991) reveal homopolymeric forms of LCMV GP1 and GP2, but no mixed oligomers containing both GP1 and GP2.

The glycoproteins of arenaviruses were shown to be spike-like projections exposed on the external surface of the viral envelope by proteolytic digestion with pronase, bromelain, or chymotrypsin (Vezza *et al.*, 1977; Gard *et al.*, 1977; Buchmeier *et al.*, 1978). Such treatment yielded smooth-surfaced, spikeless particles, which were noninfectious and lacked the viral glycoproteins, whereas the NP protein was protected from digestion indicating its internal location.

Evidence has been obtained that the GP1 glycoprotein of LCMV is present in a more peripheral location in the viral envelope than is GP2. Only GP1 was found to be labeled by surface iodination procedures (Buchmeier *et al.*, 1981; Bruns *et al.*, 1983), and the GPl molecule also contains more highly processed oligosaccharides than GP2 (Buchmeier and Oldstone, 1979). Studies with monoclonal antibodies also have demonstrated that GP1 contains the major epitopes that interact with neutralizing antibodies (Buchmeier *et al.*, 1981).

The morphology of the glycoprotein appears to be club-shaped in some negatively stained images (see Fig. 1B), and in end-on views, a hollow structure has been reported (Murphy and Whitfield, 1975; Vezza *et al.*, 1977). In partially disrupted virions penetrated by negative stain, differences in electron density are sometimes apparent at positions in the envelope where surface spikes are localized, suggesting that the spike structure is embedded in, and traverses through the lipid bilayer (Fig. 1B).

VI. ASSEMBLY AND RELEASE

A. Maturation at the Cell Surface

The only stage in virus assembly that has been visualized using electron microscopy is the final process of budding at the plasma membrane (Dalton *et al.*, 1968; Murphy *et al.*, 1970). Presumably, viral nucleocapsids are assembled in the cytoplasm, since the nucleoprotein antigen can be identified throughout the cytoplasm using immunofluorescence (Zeller *et al.*, 1988). By analogy with other enveloped viruses (Stephens and Compans, 1988), it is likely that the viral glycoproteins are synthesized on membrane-bound polyribosomes in the rough endoplasmic reticulum and are transported through the Golgi complex, where simple, high-mannose oligosaccharides are processed into complex carbohydrate chains. The final movement of glycoproteins to the plasma membrane is presumably mediated by a vesicular transport process, as has been observed with other enveloped viruses.

Membrane changes seen at the sites of virus budding include an increase in density of both membrane lamellae in discrete areas large enough to form a viral envelope (Murphy *et al.*, 1970). Surface projections were observed on the exterior of emerging virus particles. In some cells observed at late stages of infection, extensive regions of the plasma membrane were involved in virus assembly. Virions are released after completion of budding by pinching off at the plasma membrane (Fig. 3).

In addition to demonstrating the assembly of virions, electron microscopic studies of arenavirus-infected Vero cells have revealed distinctive intracytoplasmic inclusions consisting of aggregations of electron-dense granules with the appearance of ribosomes (Murphy *et al.*, 1970; Murphy and Whitfield, 1975). The inclusions were variable in size and shape and appeared to become progressively more dense during the course of infection. Immunolabeling of infected cells indicated that these inclusions contain virus-specific antigens (Abelson *et al.*, 1969). The functional significance of such inclusions has not been determined.

No information has been obtained, as yet, on the molecular interactions between viral components that lead to the budding process at the plasma membrane. In contrast to most other enveloped viruses, arenaviruses do not appear to possess an internal protein that corresponds to a matrix (M) protein, which for other viruses is likely to play an important role in the organization of viral components during assembly. Thus, it is likely that the molecular interactions involved in virus assembly may be unusual. The following are some possible mechanisms that could be involved in the assembly process.

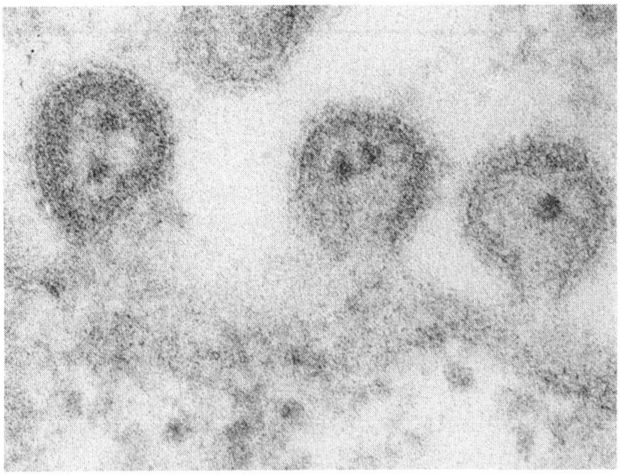

FIGURE 3. Tacaribe virions, containing one or two ribosomes, in the process of budding from an infected Vero cell. A diffuse layer of electron-dense material, which may represent the viral nucleoprotein, is localized beneath the envelope. (From Compans and Bishop, 1985.) Magnification: ×180,000.

B. Interaction between Nucleocapsids and Envelope Proteins

Since the only major virus-specified structural polypeptides that have been identified as components common to all the arenaviruses studied to date are the glycoproteins and the nucleoproteins, it seems most likely that assembly of virions may occur by direct interaction between the nucleoprotein subunits in the nucleocapsids and a cytoplasmic tail on the glycoproteins, which would be exposed on the internal surface of the plasma membrane. As shown in Fig. 3, thin sections of budding virions show a layer of electron-dense material that underlies the viral membranes and could correspond to an extended form of the nucleocapsid. Such a model for virus assembly would predict that the nucleocapsid is organized beneath the lipid bilayer in a regular arrangement, which has not, as yet, been demonstrated.

The NP proteins of LCM and Tacaribe virions have been found to a be heterogeneous with respect to their isoelectric points, which appears to be due to differences in extent of phosphorylation (Bruns *et al.*, 1986; Boersma and Compans, 1985). In the case of LCMV, evidence has been obtained that a subpopulation of NP molecules is associated with the envelope fraction of virions (Bruns *et al.*, 1986) and is exposed on surfaces of infected cells (Zeller *et al.*, 1988). Certain monoclonal antibodies directed against the NP protein were found to react with antigens expressed on the surfaces of infected cells as shown by immunofluorescence, and it was suggested that a relatively hydrophobic stretch of amino acids from residues 441–459 in the sequence of NP might serve as a membrane-spanning region (Zeller *et al.*, 1988). The precise transmembrane topology, as well as the possible functional significance of such membrane-associated NP molecules, remains to be determined. The surface expression of NP was reported to occur somewhat later in the viral replication cycle than the onset of virus release (Zeller *et al.*, 1988), and it remains to be determined whether the surface-associated NP molecules are involved in the assembly and release of budding virions.

C. Lateral Interactions between Viral Glycoproteins

An alternate possibility for organization of the budding virion could involve lateral interactions among the viral glycoproteins to form a domain on the plasma membrane, from which host cell member proteins are excluded. The nucleocapsids would then recognize the cytoplasmic domains of the viral glycoproteins, leading to the budding process. If a significant portion of the envelope protein were located on the cytoplasmic surface of the viral envelope, such a domain could play a role in assembly analogous to that of the matrix protein, in a manner similar to the E1 glycoprotein of coronaviruses (Sturman, 1977). However, neither the deduced amino acid sequences of the viral glycoproteins (Bishop and Auperin, 1987) nor analyses of protease-treated virions have yielded any suggestion that large protease-resistant cytoplasmic domains may exist (Vezza *et al.*, 1977; Gard *et al.*, 1977). In addition, no evidence has been obtained that viral glycoproteins are packed into viral envelopes in an organized manner, such as that observed, for example, with certain bunyaviruses (von Bonsdorff and Petterson, 1975). Such a regular arrangement of glycoproteins might be expected if lateral interactions among spikes played a major role in organizing the structure of the budding virion.

D. Involvement of Other Viral Proteins

The possibility may also exist that other virion proteins participate in the assembly process, and that they could play a role analogous to that of the M proteins of other enveloped viruses. For example, a polypeptide of 10–14 kDa has been reported in LCMV (Salvato and Shimomaye, 1989) and in Tacaribe (Iapalucci *et al.*, 1989), and a 12-kDa polypeptide has been reported in Pichinde virions (Ramos *et al.*, 1972). Nucleotide sequence analyses indicates that the LCM and Tacaribe gene products share a conserved zinc-binding motif, and immunological evidence has been obtained that the LCMV polypeptide represents a distinct viral gene product (Salvato *et al.*, 1992). Chemical cross-linking studies by this group showed an association of the 12 kDa protein with NP, but the role of this protein is still a mystery.

In the presence of tunicamycin, an inhibitor of N-linked glycosylation, the production of noninfectious Junin virus particles was observed at levels approximately 70% of normal amounts (Padula and de Martinez Segovia, 1984). Incorporation of radiolabeled sugar precursors into such particles was inhibited, and the amount of at least one of the viral glycoproteins (designated gp38) detected in the virus preparation was markedly diminished. It was not determined whether a nonglycosylated molecule corresponding to the polypeptide backbone of gp38 was present in the resulting particles. These results demonstrate that glycosylation of viral glycoproteins is not required for virus particle production, a result similar to that obtained for several other enveloped viruses (Gibson *et al.*, 1979; Basak and Compans, 1983). However, it remains to be established whether nonglycosylated forms of the glycoprotein are involved in virus assembly under these conditions.

Superinfection of Pichinde virus-infected cells with vesicular stomatitis virus (VSV) resulted in the production of pseudotypes that were not neutralized by antiserum to VSV, but were neutralized by antiserum to Pichinde virus (Sengupta and Rawls, 1979). Although these particles have not been characterized biochemically, by analogy to other VSV pseudotypes they presumably consist of typical VSV particles possessing the surface glycoproteins of Pichinde virus, rather than the VSV-G protein (McSharry *et al.*, 1971). These results suggest that the structural properties of arenavirus glycoproteins are compatible with their incorporation into the envelopes of VSV particles, even though VSV particles are assembled by a process involving the incorporation of an M protein into their envelopes, in apparent contrast to the arenaviruses.

The process of assembly for arenaviruses appears to be less precise than that of most other enveloped viruses. Such lack of precision in virus assembly is indicated by the observed pleomorphism of the virions and the incorporation of host components, such as ribosomes, into virus particles. The incorporation of other RNA species including mRNAs and viral complementary sense RNA species into arenaviruses has also been observed (Bishop, 1990). Nevertheless, the assembly process is sufficiently precise to produce particles containing only a limited number of virus-coded polypeptides as the major structural components of virions.

Some evidence for the incorporation of host-cell-derived antigens into the envelopes of LCM virions has also been obtained, in that the virus was reported to be neutralized by antiserum to host cells (Ofodile *et al.*, 1973). No evidence has been obtained concerning the nature of such host antigens. High-molecular-weight sulfated polysaccharides, presumably of host cell origin, have been detected in LCMV (Bruns *et al.*, 1983) and may represent a host antigen, as has been found for other enveloped viruses (Lee *et al.*, 1969).

VII. ASSEMBLY MECHANISMS AND VIRAL PATHOGENESIS

It is of interest to consider the possible interrelationships between the mechanism of virus release at the cellular level and the pathogenesis of virus infection. A common feature of arenaviruses is their propensity to establish persistent infections, both in cell culture systems and in their natural rodent hosts (Oldstone, 1987). The replication and assembly processes of arenaviruses are clearly compatible with the establishment of a persistent infection, in that synthesis of viral components occurs without any major alterations in biosynthesis of essential host cell macromolecules, and budding of progeny virions at cellular plasma membranes provides a mechanism for virus release without any cytopathic effect.

Initiation of the natural infection process by many viruses occurs at epithelial cell surfaces, such as those that line the respiratory and gastrointestinal tracts. Such body surfaces are lined by polarized epithelial cells, with biochemically distinct apical and basolateral membranes, and it has been demonstrated that entry as well as release of many viruses from such cells occurs exclusively at either the apical or basolateral membrane domains, depending on the virus type (Stephens and Compans, 1988). As yet, no information has been reported on the mode of entry or release of arenaviruses from such polarized epithelial cells, and it will be of interest to investigate this question in relation to the pathogenesis of virus infection. For example, in rodent hosts, the excretion of large amounts of virus in the saliva as well as in the urine might be consistent with selective release at apical membranes from appropriate cell types.

VIII. CONCLUDING REMARKS

Although substantial progress has been made in studies of the molecular biology of arenaviruses, many questions remain to be answered concerning the ultrastructure of the virion and the process of virus assembly. Determination of the transmembrane topology and quaternary structure of the envelope glycoproteins should be feasible using the same approaches that have been employed successfully with other membrane glycoproteins. Analysis of the precise arrangement of the nucleoprotein components within intact virions, as well as the sequence of events and precise molecular interaction involved in virus assembly, may, however, require novel approaches. Studies of the assembly process could benefit from the use of temperature-sensitive virus mutants, a number of which have been described (Vezza and Bishop, 1977).

ACKNOWLEDGMENTS. Research by the author was supported by Grants AI 12680 and CA 18611 from the National Institutes of Health.

REFERENCES

Abelson, H. T., Smith, G. H., Hoffman, H. A., and Rowe, W. P., 1969, Use of enzyme-labelled antibody for electron microscope localization of lymphocytic choriomeningitis virus antigens in infected cell cultures, *J. Natl. Cancer Inst.* **42**:497.

Armstrong, C., and Lillie, R. D., 1934, Experimental lymphocytic choriomeningitis of monkeys and mice produced by a virus encountered in studies of the 1933 St. Louis encephalitis epidemic, *Public Health Rep.* **49**:1019.

Basak, S., and Compans, R. W., 1983, Studies on the role of glycosylation in the functions and antigenic properties of influenza virus glycoproteins, *Virology* **128**:77.

Bishop, D. H. L., 1990, Arenaviridae and their replication, in: *Virology* (B. N. Fields and D. M. Knipe, eds.), pp. 1231–1243, Raven Press, New York.

Bishop, D. H. L., and Auperin, D. D., 1987, Arenavirus gene structure and organization, *Curr. Top. Microbiol. Immunol.* **133**:5.

Boersma, D. P., and Compans, R. W., 1985, Synthesis of Tacaribe virus polypeptides in an *in vitro* coupled transcription and translation system, *Virus Res.* **2**:271.

Bruns, M., and Lehmann-Grube F., 1983, Lymphocytic choriomeningitis virus. V. Proposed structural arrangement of proteins in the virion, *J. Gen. Virol.* **64**:2157–2167.

Bruns, M., Peralta, L. M., and Lehmann-Grube, F., 1983, Lymphocytic choriomeningitis virus. III. Structural proteins of the virion, *J. Gen. Virol.* **64**:599.
Bruns, M., Zeller, W., Rohdewohld, H., and Lehmann-Grube, F., 1986, Lymphocytic choriomeningitis virus. IX. Properties of the nucleocapsid, *Virology* **151**:77.
Buchmeier, M. J., and Oldstone, M. B. A., 1979, Protein structure of lymphocytic choriomeningitis virus: Evidence for a cell-associated precursor of the virion glycopeptides, *Virology* **99**:111.
Buchmeier, M. J., and Parekh, B. S., 1987, Protein structure and expression among arenaviruses, *Curr. Top. Microbiol. Immunol.* **133**:41.
Buchmeier, M. J., Elder, J. H., and Oldstone, M. B. A., 1978, Protein structure of lymphocytic choriomeningitis virus: identification of the virus structural and cell associated polypeptides, *Virology* **89**:133.
Buchmeier, M. J., Lewicki, H. A., Tomori, O., and Oldstone, M. B. A., 1981, Monoclonal antibodies to lymphocytic choriomeningitis and Pichinde viruses: Generation, characterization, and cross-reactivity with other arenaviruses, *Virology* **113**:73.
Burns, J. W., and Buchmeier, M. J., 1991, Protein-protein interactions in lymphocytic choriomeningitis virus, *Virology* **183**:620.
Burns, J. W., Salvato, M. S., and Buchmeier, M. J., 1990, Molecular architecture of lymphocytic choriomeningitis virus, *Proc. VIIIth Intl. Cong. Virol.*, Berlin, Abstr. W3–001.
Chinault, D. N., Thompson, H. A., and Gangemi, J. D., 1981, Polypeptide synthesis catalysed by components of Pichinde virus disrupted by detergent, *J. Gen. Virol.* **55**:213.
Compans, R. W., and Bishop, D. H. L., 1985, Biochemistry of arenaviruses, *Curr. Top. Microbiol. Immunol.* **114**:153.
Dalton, A. J., Rowe, W. P., Smith, G. H., Wilsnack, R. E., and Pugh, W. E., 1968, Morphological and cytochemical studies on lymphocytic choriomeningitis virus, *J. Virol.* **2**:1465.
Farber, F. E., and Rawls, W. E., 1975, Isolation of ribosome-like structures from Pichinde virus, *J. Gen. Virol.* **26**:21.
Gard, G. P., Vezza, A. C., Bishop, D. H. L., and Compans, R. W., 1977, Structural proteins of Tacaribe and Tamiami virions, *Virology* **83**:84.
Gibson, R., Schlesinger, S., and Kornfeld, S., 1979, The non-glycosylated glycoprotein of vesicular stomatitis virus is temperature-sensitive and undergoes intracellular aggregation at elevated temperatures, *J. Biol. Chem.* **254**:3600.
Iapalucci, S. Lopez, N., Rey, O., Zakin, M., Cohen, G. N., and Franze-Fernandez, M. T., 1989, The 5′ region of Tacaribe virus L RNA encodes a protein with a potential metal binding domain, *Virology* **173**:357.
Lee, L. T., Howe, C., Meyer, K., and Choi, H. U., 1969, Quantitative precipitin analysis of influenza virus host antigen and of sulfated polysaccharides of chicken embryonic allantoic fluid, *J. Immunol.* **102**:1144.
Leung, W-C., and Rawls, W. E., 1977, Virion-associated ribosomes are not required for the replication of Pichinde virus, *Virology* **81**:176–176.
Leung, WC., Ghosh, H. P., and Rawls, W. E., 1977, Strandedness of Pichinde virus RNA, *J. Virol.* **22**:235.
McSharry, J. J., Compans, R. W., and Choppin, P. W., 1971, Proteins of vesicular stomatitis virus and of phenotypically mixed VSV-SV5 virions, *J. Virol.* **8**:722.
Muller, G., Bruns, M., Martinez Peralta, L., and Lehmann-Grube, F., 1983, Lymphocytic choriomeningitis virus IV. Electron microscopic investigation of the virion, *Arch. Virol.* **75**:229.
Murphy, F. A., and Whitfield, S. G., 1975, Morphology and morphogenesis of arenaviruses, *Bull WHO* **52**:409.
Murphy, F. A., Webb, P. A., Johnson, K. M., Whitfield, S. G., and Chappell, W. A., 1970, Arenaviruses in Vero cells: Ultrastructural studies, *J. Virol.* **6**:507.
Nayak, D. P., and Jabbar, M. A., 1989, Structural domains and organizational conformation involved in the sorting and transport of influenza virus transmembrane proteins, *Ann. Rev. Microbiol.* **43**:465.

Ofodile, A., Padnos, M., Molomut, N., and Duffy, J. L., 1973, Morphological and biological characteristics of the M–P strain of lymphocytic choriomeningitis virus, *Infect. Immun.* **7**:309.
Oldstone, M. B. A., 1987, Arenaviruses: Biology and immunotherapy, *Curr. Top. Microbiol. Immunol.* **134**:1.
Padula, P. J., and de Martinez Segovia, Z., 1984, Replication of Junin virus in the presence of tunicamycin, *Intervirology* **22**:227.
Palmer, E. L., Obijeski, J. F., Webb, P. A., and Johnson, K. M., 1977, The circular, segmented nucleocapsid of an arenavirus–Tacaribe virus, *J. Gen. Virol.* **36**:541.
Pedersen, I. R., and Konigshofer, E. P., 1976, Characterization of ribonucleoproteins and ribosomes isolated from lymphocytic choriomeningitis virus, *J. Virol.* **20**:14.
Ramos, B. A., Courtney, R. J., and Rawls, W. E., 1972, Structural proteins of Pichinde virus, *J. Virol.* **10**:661.
Rowe, W. P. , Murphy, F. A., Bergold, G. H., Casals, J. Hotchin, J. Johnson, K. M. Lehmann-Grube, F. Mims, C. A., Traub, E., and Webb, P. A., 1970, Arenaviruses: Proposed name for a newly defined virus group, *J. Virol.* **5**:651.
Salvato, M. S., and Shimomaye, E. S., 1989, The completed sequence of lymphocytic choriomeningitis virus reveals a unique RNA structure and a gene for a zinc-finger protein, *Virology* **173**:1.
Salvato, M. S., Schweighofer, K. J., Burns, J. W., and Shimomaye, E. S., 1992, Biochemical and immunological evidence that the 11 kDa zinc-binding protein of lymphocytic choriomeningitis virus is a structural component of the virus, *Virus Res.* (in press).
Sengupta, S., and Rawls, W. E., 1979, Pseudotypes of vesicular stomatitis virus and pichinde virus, *J. Gen. Virol.* **42**:141.
Sivasubramanian, N., and Nayak, D. P., 1987, Mutational analysis of the signal-anchor domain of influenza virus neuraminidase, *Proc. Natl. Acad. Sci. USA* **84**:105.
Stephens, E. B., and Compans, R. W., 1988, Assembly of animal viruses at cellular membranes, *Annu. Rev. Microbiol.* **42**:489.
Sturman, L. S., 1977, Characterization of a coronavirus. I. Structural proteins: Effects of preparative conditions on the migration of protein in polyacrylamide gels, *Virology* **77**:637.
Traub, E., 1935, A filterable virus recovered from white mice, *Science* **81**:298.
Vezza, A. C., and Bishop, D. H. L., 1977, Recombination between temperature sensitive mutants of the arenavirus Pichinde, *J. Virol.* **24**:712.
Vezza, A. C., Gard, G. P., Compans, R. W., and Bishop, D. H. L., 1977, Structural components of the arenavirus Pichinde, *J. Virol.* **23**:776.
Vezza, A. C., Clewley, J. P., Gard, G. P., Abraham, N. Z., Compans, R. W., and Bishop, D. H. L., 1978, Virion RNA species of the arenaviruses Pichinde, Tacaribe, and Tamiami, *J. Virol.* **26**:485.
von Bonsdorff, C-H., and Patterson, R., 1975, Surface structure of Uukeniemi virus, *J. Virol.* **16**:1296.
Wright, K. E., Salvato, M. S., and Buchmeier, M. J., 1989, Neutralizing epitopes of lymphocytic choriomeningitis virus are conformational and require both glycosylation and disulfide bonds for expression, *Virology* **171**:417.
Young, P. R., and Howard, C. R., 1983, Fine structure analysis of Pichinde virus nucleocapsids, *J. Gen. Virol.* **64**:833.
Zeller, W., Bruns, M., and Lehmann-Grube, F., 1988, Lymphocytic choriomeningitis virus. X. Demonstration of nucleoprotein on the surface of infected cells, *Virology* **162**:90.

CHAPTER 2

Glycoproteins of the Arenaviruses

JOHN W. BURNS AND MICHAEL J. BUCHMEIER

I. INTRODUCTION

The Arenaviridae are characterized by spherical to pleomorphic particles, 50–300 nm in diameter, with club-shaped projections extending outward 5–10 nm from the virion surface. The viral genome contains two segments of single-stranded RNA transcribed in an ambisense manner. Each of the genome segments, designated L and S, encodes two primary translation products, which in most cases give rise to a total of five structural proteins. The prototype arenavirus, lymphocytic choriomeningitis virus (LCMV), contains a 200-kDa polymerase, L, a nucleocapsid protein, NP (63 kDa), two structural glycoproteins, GP-1 (44 kDa) and GP-2 (35 kDa), which are derived by cleavage of the glycoprotein precursor, GP-C (70–75 kDa), and an 11- to 14-kDa nonglycosylated protein, Z, of unknown function. Equivalent proteins have been reported for most of the other arenaviruses.

The arenaviruses are divided for taxonomic purposes into two categories, the Old World complex (*e.g.*, LCM, Lassa, Mozambique) and the New World, Tacaribe complex (*e.g.*, Tacaribe, Junin, Machupo) based on the geographic location in which they were isolated. Serologic cross-reactivity is greatest among members of a complex, and to a lesser extent between complexes (Howard and Simpson, 1980; Buchmeier *et al.*, 1982; Weber and Buchmeier, 1988).

JOHN W. BURNS • Palo Alto Veterans Administration Medical Center, Palo Alto, California 94304. MICHAEL J. BUCHMEIER • Division of Virology, Department of Neuropharmacology, The Scripps Research Institute, La Jolla, California 92037.

The Arenaviridae, edited by Maria S. Salvato. Plenum Press, New York, 1993.

Proteolytic digestion of purified PIC, TAC, or LCM virions resulted in "bald" or spikeless particles that had lost their structural glycoproteins, thus identifying the glycoproteins as the structural components of the spikes (Vezza *et al.*, 1977; Gard *et al.*, 1977; Buchmeier *et al.*, 1978). Further characterization of the arenavirus glycoproteins has established them as major targets of host immune responses. Serological analyses have demonstrated that neutralizing antibodies are elicited only against the glycoproteins (Buchmeier *et al.*, 1982; Bruns *et al.*, 1983a; Howard *et al.*, 1985; Parekh and Buchmeier, 1986; Ruo *et al.*, 1991). Evaluation of the cellular immune response has identified epitopes for murine cytotoxic T lymphocytes (CTL) on both structural glycoproteins (Whitton *et al.*, 1988a,b). Further, in conjunction with the various components of the immune response, the expression of the arenavirus glycoproteins (or the lack thereof) may play a major role in viral pathogenesis and persistence. These and other aspects of the glycoproteins of the arenaviruses will be the subject of this review.

II. GLYCOPROTEIN EXPRESSION AND PROCESSING

The arenavirus genome encodes only a single translation product that becomes glycosylated, the precursor glycoprotein, GP-C. GP-C is encoded in the message sense on the S RNA segment, and the nucleocapsid protein, NP, is encoded in the antimessage sense on S (Harnish *et al.*, 1983; Riviere *et al.*, 1985). Despite the message polarity of GP-C, because of the replication strategy of the ambisense genome, NP mRNA is transcribed and translated prior to GP-C gene expression. This regulatory process may play a major role in the establishment of arenavirus persistent infections (discussed below).

The LCMV GP-C open reading frame encodes 498 amino acids, including a 58-amino-acid leader sequence. The GP-C precursor has an apparent molecular weight of 70–80,000 kDa in its glycosylated form (see Table I). Analyses of the LAS, LCM, PIC, and TAC GP-C genes indicate they contain 8–16 potential N-linked glycosylation sites (Auperin *et al.*, 1984, 1986; Franze-Fernandez *et al.*, 1987; Southern *et al.*, 1987; Wright *et al.*, 1989). Wright *et al.* (1989) demonstrated that eight of nine potential N-linked glycosylation sites are utilized for LCMV. Harnish *et al.* (1981) estimated that up to 47% of the apparent molecular weight of PIC GP-C may be due to carbohydrate. Using radiolabeled sugars, it was demonstrated that both LCM and TAC GP-C initially contained only high-mannose residues, which were subsequently converted to complex carbohydrates (Buchmeier and Oldstone, 1979; Boersma *et al.*, 1982). Further, it was determined that GP-1 contained glucosamine, fucose, and galactose whereas GP-2 contained predominantly glucosamine and fucose. These findings were substantiated using

TABLE 1. Arenavirus Glycoproteins[a]

Virus	Nonstructural	Structural	Ref.
LCM	74–75 kDa	44, 35 kDa	Buchmeier and Oldstone, 1979
LCM	ND	130, 85, 60, 44, 35 kDa	Bruns *et al.*, 1983a
PIC	ND	63, 38 kDa	Vezza *et al.*, 1977
PIC	79 kDa	52, 36 kDa	Harnish *et al.*, 1981
JUN	ND	91, 72, 52, 38 kDa	De Martinez Segovia and De Mitri, 1977
JUN	ND	44, 34–39 kDa	Grau *et al.*, 1981
LAS	84 or 115 kDa	52, 39 kDa	Kiley *et al.*, 1981
LAS	72 kDa	45, 38 kDa	Clegg and Lloyd, 1983
MAC	ND	50, 41 kDa	Gangemi *et al.*, 1978
MOB	ND	48, 37 kDa	Gonzalez *et al.*, 1984a
MOP	ND	48, 35 kDa	Gonzalez *et al.*, 1984b
MOZ	ND	52, 39 kDa	Kiley *et al.*, 1981
TAC	70 kDa	42 kDa	Gard *et al.*, 1977 Saleh *et al.*, 1979
TAM	ND	44 kDa	Gard *et al.*, 1977

[a] ND = not described; kDa = kilodalton.

a series of drugs that inhibit sequential steps in the processing of N-linked sugars, conclusively establishing that GP-C utilizes the cellular secretory pathway for its processing (Wright *et al.*, 1990). In these experiments inhibition of glycosylation by tunicamycin resulted in blockage of glycoprotein processing and transport and failure to produce virions. Other inhibitors such as castantospermine, which allowed *en bloc* addition of the mannose-rich precursor chain, permitted processing transport and virion maturation. The Tacaribe glycoproteins reportedly also contain glucosamine, galactose, and mannose, as well as terminal sialic acid residues (Boersma *et al.*, 1982). Clearly, the carbohydrate content reflects that of the host cell. We are not aware of reports describing the sulfation or phosphorylation of arenavirus glycoproteins. A single report describes the palmitylation of the LCM structural glycoproteins gp60 and gpl30, reportedly a dimer of gp60 (Bruns and Lehmann-Grube, 1983); however, subsequently the entire LCM genome has been sequenced and there does not appear to be an open reading frame corresponding to gp60.

The kinetics of viral glycoprotein synthesis has been examined, at various multiplicities of infection (MOI), with LCM, PIC, and TAC. In cells acutely infected with LCM, NP expression was first observed at approximately 6 hr postinfection (p.i.) whereas GP-C expression was readily detectable at 24–48 hr p.i. at an MOI of 1.0 (Buchmeier *et al.*, 1978). Similar findings were reported for PIC with the exception that GP-C was detectable at 12 hours p.i. at high MOI infection (MOI 50), but not at lower MOIs (MOI 0.1; Harnish *et al.*, 1981). In TAC-infected cells NP was observed at 24–34 hr, increasing until 48 hr p.i. TAC GP-C was detected by 48 hr and increased until 60 hr p.i. (Saleh *et al.*, 1979).

Several studies have examined the cleavage of the arenavirus glycoprotein precursor. Pulse-chase studies with LCM and PIC have demonstrated that GP-C is cleaved approximately 75–90 min after synthesis, resulting in the appearance of the two structural glycoproteins, GP-1/G1 and GP-2/G2 (Harnish *et al.*, 1981; Wright *et al.*, 1990). It appears that cleavage of LCM GP-C is a two-step process (Fig. 1). Recent experiments in this laboratory have shown that GP-C of LCMV, TAC, and PIC all contain a long (58 amino acid) signal sequence, which is likely conserved among all the arenaviruses. This signal peptide is cleaved at a conserved signal peptidase site and removed prior to glycoprotein transport from the endoplasmic reticulum. We find no evidence that the long signal is incorporated into virions (Burns *et al.*, 1992, submitted for publication). This finding is especially interesting since a CTL epitope has been mapped to the cleaved signal sequence (amino acids 34–43; Klavinskis *et al.*, 1990) and thus constitutes the first reported CTL epitope contained within a signal sequence.

Cleavage of GP-C to form GP-1 and GP-2 occurs later in the secretory pathway, between the medial and the trans-Golgi network (Wright *et al.*, 1990). Using synthetic peptides, the second GP-C cleavage site has been localized to a stretch of nine amino acids that span the dibasic

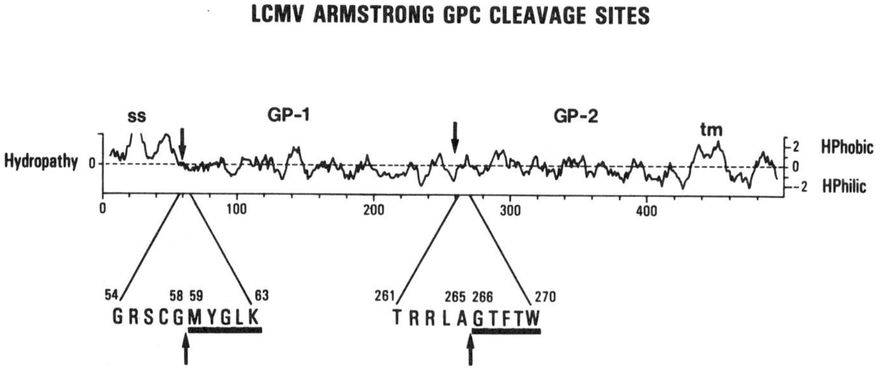

FIGURE 1. Structure of the proteolytic sites of the GP-C precursor glycoprotein. Topographical markers indicating signal sequence (ss), amino terminal GP-1 and carboxy terminal GP-2, and hydrophobic transmembrane-spanning domains (tm) are shown above, and structure of the cleavage sites of signal peptidase after amino acid 58 and a Golgi-associated protease after amino acid 265 are indicated by bold arrows. N-terminal sequences of GP-1 and GP-2 as confirmed by microsequencing are underlined.

residues -Arg-Arg- at amino acids 262–263 (Buchmeier *et al.*, 1987). Immune precipitation using antisera raised to synthetic peptides corresponding to LCM GP-C amino acids 59–79 and 378–391 allowed the mapping of GP-1 and GP-2 to the amino- and carboxy-terminal regions of GP-C (Buchmeier *et al.*, 1987) These findings have been substantiated by amino-terminal sequence analysis of GP-1 and GP-2, which established that the amino terminus of GP-1 was Met-59 and of GP-2 was Gly-266 (Burns *et al.*, unpublished data).

It is probable that the GP-C cleavage event is mediated by a Golgi-associated, furin-like protease acting at or following the dibasic residues, which are conserved among most arenaviruses (Fig. 2). Amino-terminal sequencing of the structural glycoproteins of PIC and TAC indicates that equivalent cleavage events occur in both viruses (Burns, Milligan, Burke, and Buchmeier, manuscript in preparation). Similar cleavage events are therefore likely in other members of the arenavirus family.

Cleavage of GP-C to yield GP-1 and GP-2 has been shown to require prior glycosylation (Wright *et al.*, 1989). In the presence of tunicamycin, an inhibitor of N-linked glycosylation, both LCM and PIC produced an unglycosylated form of GP-C but cleavage was not observed (Harnish *et al.*, 1981; Wright *et al.*, 1990). Further, GP-C cloned into a baculovirus vector and expressed in Spodoptera cells was not cleaved, presumably due to the lack of proper glycosylation and processing in the insect cells (Matsuura *et al.*, 1986).

```
                                         58↓                 265↓
Arm:        P H I I D E V I N I—L L A G R S C G M Y G L K—F T R R L A G T F T W T L S D

                                         58↓                 265↓
WE:         P H I I D E V I N I—L L A G R S C G M Y G L N—L T R R L S G T F T W T L S D

                                         58↓                 259↓
Lassa:      P H V I E E V M N I—L L C G R S C T T S L Y K—I S R R L L G T F T W T L S D

                                         58↓                 261↓
Tacaribe:   P I F L Q E A L N I—V L A G R S C S E E T F K—V G R T L K A F F S W S L T D

                     16↓             56↓ 59↓                 273↓
Pichinde:   P E V L Q E V F N V—I L S G R S C D S M M I D—V S R K L L G F F T W D L S D

Consensus:  P       E   N     L   G R S C                   R   L     F   W   L   D
```

FIGURE 2. Sequencing indicates a high degree of conservation of GP-C cleavage sites among several Old World and New World arenaviruses. Sequences are aligned at the signal peptide–GP-1 junction and the GP-1:GP-2 junction. Underlined sequences have been determined experimentally.

III. STRUCTURE AND ORGANIZATION OF THE GLYCOPROTEIN SPIKE

The arenavirus particle, visualized by electron microscopy, contains surface projections (spikes) that are reportedly 5–10 nm in length with a club-shaped appearance (Murphy *et al.*, 1970; Murphy and Whitfield, 1975; Young *et al.*, 1981) (Fig. 3). These spikes are closely spaced, but appear to be mobile in the viral membrane since osmotic swelling of the virion causes the spikes to become more widely spaced. When viewed end-on at higher magnification, the spikes appear to have a hollow central axis, suggesting a macromolecular organization of multiple polypeptide chains (Murphy and Whitfield, 1975; Young *et al.*, 1981). The arenavirus spike has been studied in detail and models have been proposed to explain its structure (Bruns and Lehmann-Grube, 1983; Young, 1987; Burns and Buchmeier, 1991; Burns *et al.*, 1993, submitted for publication).

The composition of the arenavirus spike has been established by treating purified PIC, TAC, and LCM virions with proteases. The resulting spikeless particles had lost their glycoproteins whereas the other structural proteins were unaffected by proteolysis (Vezza *et al.*, 1977; Gard *et al.*, 1977; Buchmeier *et al.*, 1978). Quantitation of the molar ratios of the structural proteins of PIC demonstrated that equal numbers of G1 and G2 molecules were present in the virion (Vezza *et al.*, 1977). Equal amounts of GP-1 and GP-2 were also found in LCM particles (Bruns and Lehmann-Grube, 1983b).

Surface iodination of LCM-infected cells or virions resulted in incorporation of the label into GP-1, whereas GP-2 was labeled poorly or not at all (Buchmeier and Oldstone, 1979; Bruns *et al.*, 1983a). These results suggested that GP-1 was more externally exposed than was GP-2, but this must be qualified since only exposed tyrosine residues would be expected to label under the conditions used. Direct support for this conclusion was provided by experiments in which LCM virions were extracted using Triton X-114, which has been used to distinguish peripheral from integral membrane proteins (Bordier, 1981). The results of these experiments demonstrated that GP-1 and GP-2 were peripheral and integral membrane glycoproteins, respectively (Burns and Buchmeier, 1991). The membrane-spanning domain of GP-2 has been established by amino-terminal sequencing of a polypeptide fragment protected by the membrane from proteolysis. The GP-2 membrane-spanning hydrophobic domain begins with amino acid 439 and extends for 18 residues; however, a longer fragment beginning with Gly_{430} is protected from proteolysis in the virion envelope (Burns *et al.*, 1993, submitted for publication). The C-terminal end of GP-2 is highly charged; 5 of the 12 C-terminal residues are Lys or Arg. It is likely that the cross-linking interactions we have observed between NP and GP-2

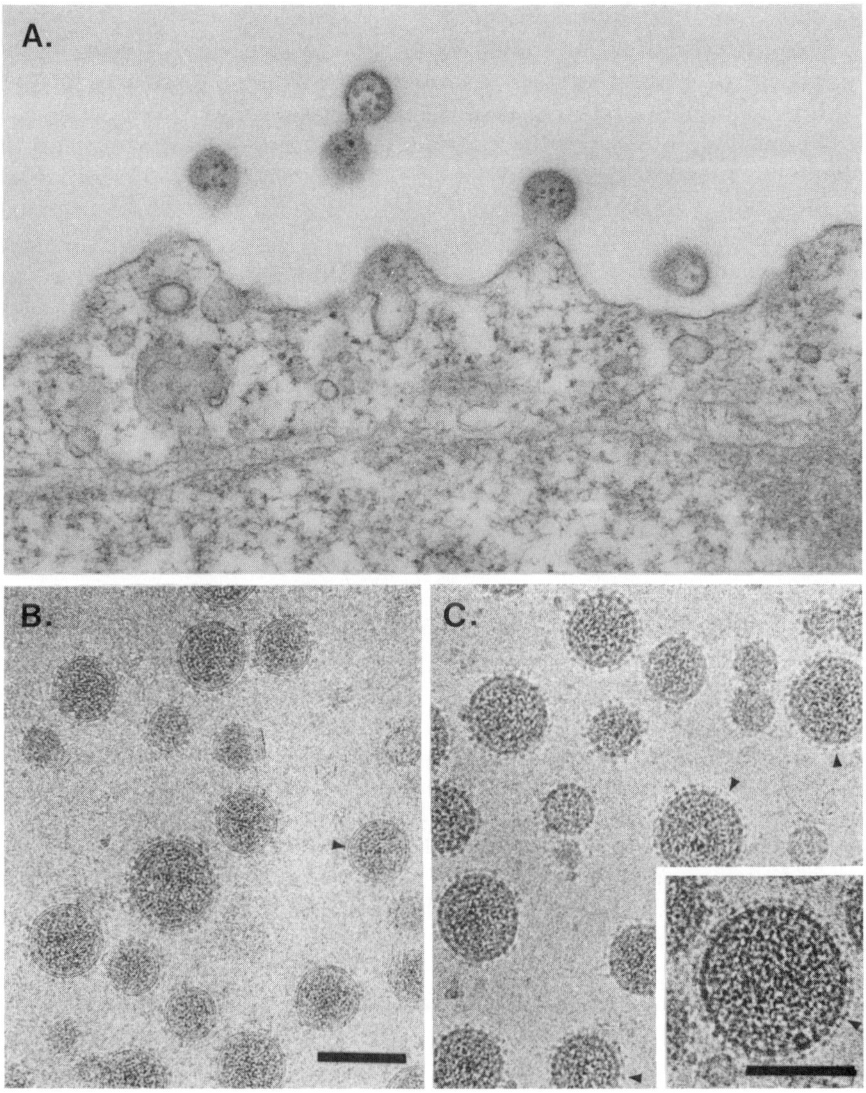

FIGURE 3. Electron microscopy of lymphocytic choriomeningitis virions. (A) Thin section showing virions budding from infected BHK-21 cells. Typical 110-nm virions containing numerous electron-dense 20-nm particles are evident. (B, C) Cryoelectron microscopy of unstained, purified LCM virions at 1.5 μ (B) and 3.0 μ (C) defocus levels. In B, the lipid bilayer is emphasized (arrow) and in the right panel the surface topography is more evident. Glycoprotein spikes are indicated by arrows. Magnification of B and C is $\times$167,100 and the inset is $\times$232,750. Bars are 1000 A.

involve these basic residues (Burns and Buchmeier, 1991). These findings have firmly established the integral membrane nature of GP-2 and its orientation in the viral envelope.

The macromolecular organization of the structural glycoprotein complexes of LCM has been examined using detergents and cross-linking reagents. Solubilization of LCM virions at 4C with SDS, CHAPS, or Triton X-100 in the absence of reducing agents caused the released of oligomeric structures consisting of GP-1 homopolymers in complexes as large as homotetramers (Wright *et al.*, 1989; Burns and Buchmeier, 1991). Further characterization using membrane-impermeable cross-linking reagents (DTSSP and Sulfo-DST) demonstrated that both GP-1 and GP-2 were assembled in separate homotetrameric complexes. The results varied when a membrane permeable cross-linker was used. Hetero-oligomeric complexes of GP-2 cross-linked with NP were observed following cross-linking with the membrane-permeable cross-linking reagent DMS (Burns and Buchmeier, 1991). Taken together these results demonstrated that GP-1 and GP-2 form separate homotetrameric complexes that are located outside the envelope. Within the viral envelope, GP-2 was cross-linked with NP, indicating that the cytoplasmic tail of GP-2 may interact with NP, a part of the ribonucleoprotein complex. The possibility that GP-2 and NP were associated within the virion had been proposed, and it was suggested this association may play a critical role in budding of the virion (Dubois-Dalcq *et al.*, 1984; Compans and Bishop, 1985).

The native conformation of both GP-1 and GP-2 has been investigated in the course of studies to establish the structure of the virion spike. Disulfide bonding is essential in maintaining the native conformation of GP-1 (Wright *et al.*, 1989; Burns and Buchmeier, 1991). Neutralizing anti-LCM monoclonal antibodies (MAbs) failed to react with reduced virions (Wright *et al.*, 1989). Likewise, polyclonal guinea pig and human anti-Lassa antisera failed to react with the reduced form of G1 in immunoblots (Clegg and Lloyd, 1983).

Computer prediction of the likely secondary of LAS GP-2 identified a domain likely to assume a coiled-coil conformation (Chambers *et al.*, 1990). Comparison of this region of LAS GP-2 with other arenaviruses for which sequence is available indicates they all share a similar heptad repeat region approximately 50 amino acids long (Fig. 4), suggesting that a coiled-coil "stalk" domain is common to the GP-2 molecules of all arenaviruses. Such a structure would be consistent with the club- or T-shaped spike seen in electron micrographs. Cumulatively, these data suggest that the 233-amino-acid GP-2: (1) forms a homotetrameric complex, (2) contains a coiled-coil domain that assumes an elongated conformation, (3) has a 164-amino-acid ectodomain and a 69-amino-acid endodomain, (4) is anchored in the membrane by a hydrophobic stretch of 20–25 hydrophobic amino acids near the amino terminus of the carboxy-terminal endodomain, and (5) has a charged cytoplasmic tail through

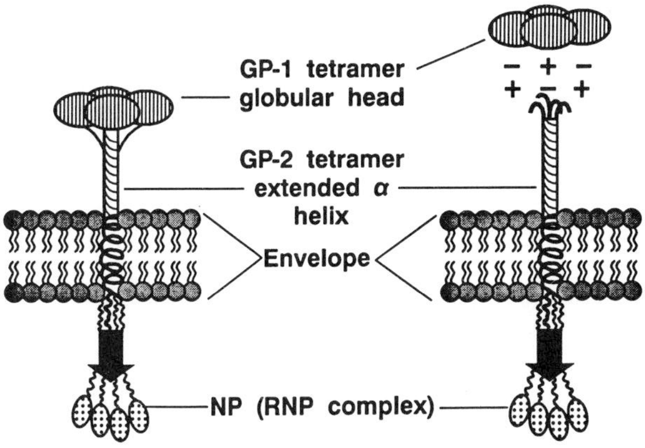

FIGURE 4. Proposed working model of the LCMV glycoprotein spike. The disulfide-linked GP-1 homotetramer forms the cross-member component of the spike that is associated via ionic interactions with the N-terminal portion of the Gp-2 transmembrane protein. The spike is anchored in the membrane by an 18-amino acid hydrophobic domain, and the C-terminal internal (cytoplasmic) tail may associate with the ribonucleoprotein complex within the virion.

which it can interact with NP, a component of the ribonucleoprotein complex.

Recently we began to address the nature of the interaction between GP-1 and GP-2. We have observed that GP-1 (G1) of LCM, Pichinde, and Tacaribe viruses is eluted from the virion by incubation in 1 M NaCl or LiCl (Burns *et al.*, 1993, submitted for publication). Preliminary cryoelectron microscopic analyses of these salt-stripped virions suggest that the club-shaped component of the spike has been lost. This observation suggests that GP-1 forms the club-chaped head of the glycoprotein spike and is associated via ionic interactions with the tetrameric GP-2 stalk.

Of further interest is the finding that a portion of the TAC "G" protein band is recovered in the same location in sucrose gradients that PIC G1 and LCM GP-1 are found. N-terminal sequencing of TAC "G" protein resulted in a mixture of two amino acids in each of five cycles of Edman degradation, which exactly correlated with the predicted amino termini of TAC GP-1 and GP-2. Thus it appears that the two structural glycoproteins of TAC comigrate in SDS polyacrylamide gel electrophoresis.

IV. HOST IMMUNE RESPONSES

Pathogenesis associated with acute LCM virus infection is characterized by a vigorous T-cell-mediated immune response against the virus. Indeed, it is this T-cell response directed against viral antigens which characterizes lethal acute LCM disease; if the response is suppressed by cyclophosphamide or gamma irradiation, little acute pathology is evident (Buchmeier *et al.*, 1980). The evidence for immunopathological disease with other members of the arenavirus family is not as well documented, but the host immune response is likely to play an important role in these diseases as well. As a result of these observations, the host immune response to arenavirus infection has been investigated extensively. Numerous studies have demonstrated that the viral nucleocapsid protein and glycoproteins serve as major targets of immune recognition at the cellular and humoral levels (see Klavinskis *et al.*, Chapter 13, this volume).

The cytotoxic T lymphocyte (CTL) response to LCM infection has been well characterized. Using reassortant viruses, CTL activity was mapped to the S RNA segment of LCMV (Oldstone *et al.*, 1985; Riviere *et al.*, 1986). Further characterization of LCM-specific CTLs led to the identification of CTL epitopes on both the nucleoprotein and glycoprotein components of the virion (Whitton *et al.*, 1988a,b, 1989). The recognition of individual epitopes was shown to be MHC haplotype specific. A detailed analysis of CTL recognition in C57BL/6 mice (H-2b) demonstrated the existence of two glycoprotein epitopes, one in GP-1 (now believed to be located in the signal sequence, see above) and one in GP-2 (Whitton *et al.*, 1988a). These two glycoprotein epitopes were engineered into recombinant vaccinia viruses and expressed as "minigenes." Tissue culture cells infected with either recombinant virus were recognized by cognate CTLs, but only one construction, representing GP-C amino acids 34–43, was able to induce protection from lethal virus challenge (Klavinskis *et al.*, 1990).

Specificity of the humoral immune response to arenavirus infections has been examined primarily by analysis of MAb reactivity. Hybridoma cell lines secreting MAbs have been generated to LCM, JUN, PIC, LAS, and MOP (Buchmeier *et al.*, 1981, 1982; Bruns and Lehmann-Grube, 1983; Howard *et al.*, 1985; Parekh and Buchmeier, 1986; Sanchez *et al.*, 1989; Ruo at al., 1991). The antigenic topography of the LCM glycoproteins was mapped using MAbs. It was determined there were two epitopes capable of eliciting neutralizing antibodies, both located on GP-1 (Parekh and Buchmeier, 1986). Competitive binding assays, using these neutralizing MAbs against hyperimmune sera, demonstrated that guinea pigs responded predominantly to one of these two neutralization epitopes whereas a battery of rat anti-LCMV-neutralizing MAbs recognized only the second epitope (Parekh and Buchmeier, 1986). Two neutralization epitopes were also identified on the TAC glycoprotein by competitive binding and neutralization kinetics assays. Analysis of the kinetics of neutralization of TAC by MAb provided evidence of multihit neutralization kinetics (Howard *et al.*, 1985; Howard, 1987).

To assess the role of antiviral antibody in acute LCM infection, passive protection models were investigated. Antiglycoprotein MAbs administered passively, either before or after intracerebral virus challenge, protected mice from subsequent CNS disease. In contrast, passive administration of anti-NP or anti-GP-2 MAbs had no effect on the outcome of infection (Wright and Buchmeier, 1991). Likewise, mouse pups that had received maternally derived anti-GP1 antibodies were protected from lethal challenge (Baldridge and Buchmeier, 1992). In all cases, mice that received protective levels of MAb prior to challenge had reduced CTL responses and less severe acute disease than their unprotected cohorts. It was proposed that the presence of antiglycoprotein antibodies limited the degree of T-cell-mediated immunopathology (Wright and Buchmeier, 1991).

The nature of a protective immune response in other arenavirus infections is less clear. Little is known concerning the CTL response to arenavirus infections other than LCM. In fatal cases of Lassa fever in humans, virus titer correlates well with prognosis; patients with high viremia have lower probability of survival than those with low levels (Johnson *et al.*, 1987; Fisher-Hoch and McCormick, 1987). Neutralizing antiserum to PIC or LAS was shown to either block or enhance virus infectivity in a concentration-dependent manner with the U937 monocyte cell line (Lewis *et al.*, 1988). Neutralizing antibodies can be detected in the acute phase of JUN and MAC infections (Webb *et al.*, 1969; Howard, 1987). The use of convalescent phase sera has led to mixed results in human cases of Lassa fever, but these sera have been used successfully in Argentine hemorrhagic fever (reviewed in Fischer-Hoch and McCormick, 1987) These antibodies may play a role in virus clearance, the establishment of persistent infection, or the selection of viral variants that are resistant to neutralization (Webb *et al.*, 1969; Howard, 1987; Alche and Coto, 1988).

Clearly, there is much to be learned concerning the host immune response during various arenavirus infections. Not all lessons learned with the LCM-infected mouse model can be extrapolated to other arenavirus infections, particularly those of humans. Future studies should provide a clearer picture. The development of successful vaccines and prophylactic therapies will depend heavily on these data.

V. ROLE OF THE GLYCOPROTEINS IN IMMUNOPATHOLOGY AND PERSISTENT INFECTIONS

Arenaviruses typically induce either acute or lifelong persistent infections. This dichotomy is seen both *in vivo* and *in vitro* (Welsh and Oldstone, 1977; Welsh and Buchmeier, 1979; Oldstone and Buchmeier, 1982; D'Aiutolo and Coto, 1986/87). Factors that determine the outcome of arenavirus infections include route of inoculation, immunological status of the host, age of the host, and virus strain (for review, see Buchmeier *et al.*, 1980; Buchmeier and Knobler, 1984; Southern *et al.*,

1986). Recently it has been demonstrated that the presence or absence of preexisting antiglycoprotein antibody may also play a role in the establishment of infection (Wright and Buchmeier, 1991; Baldridge and Buchmeier, 1992).

In the immunocompetent adult mouse, the outcome of acute intracerebral LCMV infection is either viral clearance with the development of protective immunity or fatal central nervous system (CNS) disease. The results of numerous studies strongly demonstrate the importance of CTLs in the course of acute LCM infection and the development of fatal CNS disease (reviewed in Allan *et al.*, 1987). Adult immunosuppressed mice do not respond to LCM infection, fail to develop fatal CNS disease, and become persistently infected. Adoptive transfer of splenocytes from immune mice into such persistently infected mice may result in the development of fatal CNS disease (Gilden *et al.*, 1972). In contrast, virus infection in neonatally infected carrier mice is usually cleared following adoptive transfer of immune T cells. Analysis at the cellular level in H-2b mice demonstrated that CTLs were responsible for clearance (Ahmed, 1988). Whitton and co-workers demonstrated that the vast majority of CTLs in H-2b mice are reactive toward the glycoproteins (1988a; see above). It is likely that these glycoprotein-reactive CTLs play a significant role in acute CNS disease. This conclusion is supported by the finding that the LCM glycoproteins, but not the nucleocapsid protein, is expressed on the surface of choroid, meningeal, and ependymal cells from several strains of LCM-infected mice (Oldstone and Buchmeier, 1982), targeting these cells for cytolysis by specific CTLs.

The establishment of persistent arenavirus infections has been carefully examined in animal and tissue culture systems. A major feature of the persistent infection is an apparent down-regulation of the expression of the viral glycoproteins (Welsh and Oldstone, 1977; Welsh and Buchmeier, 1979; Oldstone and Buchmeier, 1982). This decreased level of glycoprotein expression may allow persistently infected cells to escape immune surveillance (Oldstone and Buchmeier, 1982), although persistently infected animals mount specific antibody responses to both the viral nucleocapsid protein and the glycoproteins (Buchmeier and Oldstone, 1978; Oldstone *et al.*, 1983). The antibody produced is largely IgG-1 subclass and does not neutralize virus efficiently *in vitro*.

Defective interfering (DI) virus is rapidly generated during acute infection and has been implicated in the establishment of persistence with LCM and TAC (Welsh and Pfau, 1972; Welsh *et al.*, 1972; Popescu and Lehmann-Grube, 1977; Dutko and Pfau, 1978; Welsh and Buchmeier, 1979; D'Aiutolo and Coto, 1986/87). DI virus has been characterized at both the nucleic acid and structural protein levels. No direct correlation could be made between the appearance of subgenomic viral RNAs and an underrepresentation of genome segments in DI virions (Francis and Southern, 1988). In comparing DI with standard LCM virions, Welsh and Buchmeier found a slight difference in bouyant density but no difference in structural protein composition (1979). In contrast, Martinez Peralta and colleagues reported that DI LCM virions lacked the S RNA segment (23S RNA) and contained a novel 65-kDa glycoprotein instead of GP-1 and GP-2 (1981). These findings must be viewed with caution as their virus contained the 63-kDa nucleocapsid protein, which is also encoded on the 23S genomic segment. Recently it has been reported by this laboratory that the 65-kDa glycoprotein is actually a product of the S-genome segment and arises from the generation of a truncated form of the GP-C mRNA which, when translated, yields the 65-kDa glycoprotein (Bruns *et al.*, 1990). Further experimentation is needed to reconcile these data.

Several vaccinia virus recombinants, containing arenavirus glycoprotein gene segments, have recently been reported (Fisher-Hoch *et al.*, 1989; Morrison *et al.*, 1989; Klavinskis *et al.*, 1990). These potential vaccines reportedly generate antibody and/or CTL responses that protect vaccinees from subsequent live arenavirus challenge. In contrast, Oehen and co-workers (1991) report that a recombinant vaccinia virus of similar construction may in fact aggravate a subsequent arenavirus infection when used as a vaccine. The possible selection of resistant virus variants and differential responses with MHC haplotype variation must also be kept in mind when evaluating such potential vaccines (Alche and Coto, 1988; Klavinskis *et al.*, 1990; Pircher *et al.*, 1990). This is especially true when only a fragment of the glycoprotein gene will be included in the recombinant vector. The interactions between the immune system and the infected host are quite complex, but further studies of the immune responses to recombinant vaccines should provide a better foundation from which to design vaccine strategies.

VI. CONCLUSIONS

Emerging evidence suggests that all members of the arenavirus family contain two structural glycoproteins, derived from a single glycoprotein precursor. Sequence data from an increasing number of arenaviruses have consistently demonstrated the ambisense nature of the S-genome segment, encoding the glycoprotein precursor GP-C and nucleoprotein (NP) genes (Auperin *et al.*, 1984, 1986; Romanowski and Bishop, 1985; Franze-Fernandez *et al.*, 1987; Southern *et al.*, 1987; Salvato *et al.*, 1988; Auperin and McCormick, 1989; Wilson and Clegg, 1991). Comparison of deduced amino acid sequences predicted for the varying GP-C molecules indicates that several features of the GP-C structure are conserved among the arenaviruses. These include a long (ca. 58 amino acid) cleaved signal sequence at the amino-terminus, a second cleavage site near a paired basic amino acid site at the middle of the GP-C ORF, and a highly hydrophobic transmembrane spanning domain near the carboxy-terminus. Moreover, observations from this laboratory have shown that GP-1, the amino-terminal cleavage fragment, is a peripheral membrane protein, whereas GP-2 is an integral membrane protein, anchored in the viral envelope by the hydrophobic membrane-spanning domain and a cytoplasmic tail near its carboxy-terminus. Identification of a second structural glycoprotein in JUN and TAC (Grau *et al.*, 1981; Burns, Milligan, Burke, and Buchmeier, manuscript in preparation), both previously reported as containing only one structural glycoprotein, further supports the conclusion that arenavirus particles contain two structural glycoproteins.

The early electron microscopic analyses reported that the arenaviruses were morphologically indistinguishable particles containing similar club-shaped surface projections, 6–10 nm in length (Murphy *et al.*, 1970; Murphy and Whitfield, 1975; Young *et al.*, 1981). The surface projections were shown early on to be composed of the structural glycoproteins. We have recently provided evidence that the two LCM glycoproteins each form homo-oligomeric structures, presumably tetramers (Wright *et al.*, 1989; Burns and Buchmeier, 1991). Based on further characterization, we have proposed a schematic model of the LCM spike structure, which we believe to be composed of an oligomeric GP-2 stalk component that interacts via ionic interactions with an oligomeric head assembly composed of GP-1. We have extended some of our analyses to both PIC and TAC and have obtained consistent results. Therefore, it is our belief that the structure of this glycoprotein spike structure may be common to all members of the Arenaviridae, in agreement with the previous studies using electron microscopy. Assuming these conclusions prove accurate, attempts at X-ray crystallography of the spike and its individual glycoprotein components can begin in earnest. The determination of the three-dimensional structure of the glycoprotein spike will provide invaluable insight for future studies on the role of the glycoproteins in the biology of the virus and the host immune response.

Study of the murine-LCMV infection model has proven to be extremely productive in increasing our understanding of virus–host interactions. Recent observations concerning the *in vivo* generation of CTL-resistant variant virus (Pircher *et al.*, 1990), novel mechanisms of viral clearance in the CNS (Oldstone *et al.*, 1986), and the characterization of the mechanism of CTL killing (Welsh *et al.*, 1990) are but a few examples of the far-reaching utility of this model system in furthering our knowledge. As the viral glycoproteins have been repeatedly implicated as playing key roles in immune recognition, the initiation of infection, viral maturation, and persistence and as components of potential vaccine strategies, the continued study of these molecules will certainly provide us with a plethora of important observations.

ACKNOWLEDGMENTS. We wish to thank Dr. Ronald Milligan for the cryoelectron micrographs presented in Fig. 3, B and C. M. J. B. was supported by NIH Grant AI 16102 and Army Contract C6234. J. W. B.

was supported by NIH Training Grant GM07437. The views, opinions, and/or findings contained in this publication are those of the authors and should not be construed as an official Department of the Army position, policy, or decision unless so designated by other documentation.

REFERENCES

Ahmed, R., 1988, Curing of a congenitally acquired chronic virus infection:acquisition of T-cell competence by a previously "tolerant" host, in: *Immunobiology and Genesis of Persistent Virus Infections* (C. Lopez, ed.), pp. 159–167, American Society for Microbiology, Washington, D.C.

Alche, L. E., and Coto, C. E., 1988, Differentiation of Junin virus and antigenic variants isolated *in vivo* by kinetic neutralization assays, *J. Gen. Virol.* **69**:2123.

Allan, J. E., Dixon, J. E., and Doherty, P. C., 1987, Nature of the inflammatory process in the central nervous system of mice infected with lymphocytic choriomeningitis virus, *Curr. Top. Microbiol. Immunol.* **134**:131.

Auperin, D. D., and McCormick, J. B., 1989, Nucleotide sequence of the Lassa virus (Josiah strain) S genome RNA and amino acid sequence comparison of the N and GPC proteins to other arenaviruses, *Virology* **168**:421.

Auperin, D. D., Romanowski, V., Galinski, M., and Bishop, D. H. L., 1984, Sequencing studies of Pichinde arenavirus S RNA indicate a novel coding strategy, an ambisense viral S RNA, *J. Virol.* **52**:897.

Auperin, D. D., Sasso, D. R., and McCormick, J. B., 1986, Nucleotide sequence of the glycoprotein gene and intergenic region of the Lassa virus S genome RNA, *Virology* **154**:155.

Baldridge, J. R., and Buchmeier, M. J., 1992, Mechanisms of antibody-mediated protection against lymphocytic choriomeningitis virus infection: Mother-to-baby transfer of humoral protection. *J. Virol.* **66**:4252.

Boersma, D. P., Saleh, F., Nakamura, K., and Compans, R. W., 1982, Structure and glycosylation of Tacaribe viral glycoproteins, *Virology* **123**:452.

Bordier, C., 1981, Phase separation of integral membrane proteins in Triton X-114 solution, *J. Biol. Chem.* **256**:1604.

Bruns, M., and Lehmann-Grube, F., 1983, Lymphocytic choriomeningitis virus. V. proposed structural arrangement of proteins in the virion, *J. Gen. Virol.* **64**:2157.

Bruns, M., Cihak, J., Muller, G., and Lehmann-Grube, F., 1983a, Lymphocytic choriomeningitis virus. VI. isolation of a glycoprotein mediating neutralization, *Virology* **130**:247.

Bruns, M., Martinez Peralta, L., and Lehmann-Grube, F., 1983b, Lymphocytic choriomeningitis virus. III. structural proteins of the virion, *J. Gen. Virol.* **64**:599.

Bruns, M., Kratzberg, T., Zeller, W., and Lehmann-Grube, F., 1990, Mode of replication of lymphocytic choriomeningitis virus in persistently infected cultivated mouse L cells, *Virology* **177**:615.

Buchmeier, M. J., and Oldstone, M. B. A., 1978, Virus-induced immune complex disease: identification of specific viral antigens and antibodies deposited in complexes during chronic lymphocytic choriomeningitis virus infection, *J. Immunol.* **120**:1297.

Buchmeier, M. J., and Oldstone, M. B. A., 1979, Protein structure of lymphocytic choriomeningitis virus: evidence for a cell-associated precursor of the virion glycopeptides, *Virology* **99**:111.

Buchmeier, M. J., and Knobler, R. L., 1984, Experimental models for immune-mediated and immune-modulated diseases, in: *Neuroimmunology* (P. Behan and F. Spreafico, eds.), pp. 219–228, Raven Press, New York.

Buchmeier, M. J., Elder, J. H., and Oldstone, M. B. A., 1978, Protein structure of lymphocytic choriomeningitis virus: Identification of the virus structural and cell-associated polypeptides, *Virology* **89:**133.

Buchmeier, M. J., Welsh, R. M., Dutko, F. J., and Oldstone, M. B. A., 1980, The virology and immunobiology of lymphocytic choriomeningitis virus infection, *Adv. Immunol.* **30:**275.

Buchmeier, M. J., Lewicki, H. A., Tomori, O., and Oldstone, M. B. A.,1981, Monoclonal antibodies to lymphocytic choriomeningitis and Pichinde viruses: generation, characterization, and cross-reactivity with other arenaviruses, *Virology* **113:**73.

Buchmeier, M. J., Lewicki, H. A., Tomori, O., and Johnson, K. M., 1982, Monoclonal antibodies to lymphocytic choriomeningitis virus react pathogenic arenaviruses, *Nature* **5790:**486.

Buchmeier, M. J., Southern, P. J., Parekh, B. S., Wooddell, M. K., and Oldstone, M. B. A., 1987, Site-specific antibodies define a cleavage site conserved among arenavirus GP-C glycoproteins, *J. Virol.* **61:**982.

Burns, J. W., and Buchmeier, M. I., 1991, Protein–protein interactions in lymphocytic choriomeningitis virus, *Virology* **183:**620.

Burns, J. W., Milligan, R. A., Burke, T. A., and Buchmeier, M. J., 1993, Proteins of lymphocytic choriomeningitis virus: Structural studies of the glycoprotein spike (submitted for publication).

Chambers, P., Pringle, C. R., and Easton, A. J., 1990, Heptad repeat sequences are located adjacent to hydrophobic regions in several types of virus fusion glycoproteins, *J. Gen. Virol.* **71:**3075.

Clegg, J. C. S., and Lloyd, G., 1983, Structural and cell-associated proteins of Lassa virus, *J. Gen. Virol.* **64:**1127.

Compans, R. W., and Bishop, D. H. L., 1985, Biochemistry of arenaviruses, *Curr. Top. Microbiol. Immunol.* **114:**153.

D'Aiutolo, A. C., and Coto, C. E., 1986/87, Vero cells persistently infected with Tacaribe virus: Role of interfering particles in the establishment of infection, *Virus Res.* **6:**235.

De Martinez Segovia, Z. M., and De Mitri, M. I., 1977, Junin virus structural proteins, *J. Virol.* **21:**579.

Dubois-Dalcq, M., Holmes, K. V., and Rentier, B., 1984, Assembly of Arenaviridae, in: *Assembly of Enveloped RNA Viruses* (M. Dubois-Dalcq, K. V. Holmes, and B. Rentier, eds.), pp. 90–99, Springer-Verlag, New York.

Dutko, F. J., and Pfau, C. J., 1978, Arenavirus defective interfering particles mask the cell killing potential of standard virus, *J. Gen. Virol.* **38:**195.

Fisher-Hoch, S. P., and McCormick, J. B., 1987, Pathophysiology and treatment of Lassa fever, *Curr. Top. Microbiol. Immunol.* **134:**231.

Fisher-Hoch, S. P., McCormick, J. B., Auperin, D., Brown, B. G., Castor, M., Perez, G., Ruo, S., Conaty, A., Brammer, L., and Bauer, S., 1989, Protection of rhesus monkeys from fatal Lassa fever by vaccination with a recombinant vaccinia virus containing the Lassa virus glycoprotein gene, *Proc. Natl. Acad. Sci. USA* **86:**317.

Francis, S. J., and Southern, P. J., 1988, Deleted viral RNAs and lymphocytic choriomeningitis viruspersistence *in vitro, J. Gen. Virol.* **69:**1893.

Franze-Fernandez, M-T., Zetina, C., Iapalucci, S., Lucero, M. A., Bouissou, C., Lopez, R., Rey, O., Daheli, M., Cohen, G. N., and Zakin, M. M., 1987, Molecular structure and early events in the replication of Tacaribe arenavirus S RNA, *Virus Res.* **7:**309.

Gangemi, J. D., Rosato, R. R., Connell, E. V., Johnson, E. M., and Eddy, G. A., 1978, Structural polypeptides of Machupo virus, *J. Gen. Virol.* **41:**183.

Gard, G. P., Vezza, A. C., Bishop, D. H. L., and Compans, R. W., 1977, Structural proteins of Tacaribe and Tamiami virions, *Virology* **83:**84.

Gilden, D. H., Cole, G. A., and Nathanson, N., 1972, Immunopathogenesis of acute central nervous system disease produced by lymphocytic choriomeningitis virus. II. adoptive immunization of virus carriers, *J. Exp. Med.* **135:**874.

Gonzalez, J. P., McCormick, J. B., Georges, A. J., and Kiley, M. P., 1984a, Mobala virus:

Biological and physiochemical properties of a new arenavirus isolated in the Central African Republic, *Ann. Virol.(Institute Pasteur)* **135E**:145.
Gonzalez, J. P., Buchmeier, M. J., McCormick, J. B., Mitchell, S. W., Elliot, L. H., and Kiley, M. P., 1984b, Comparative analysis of Lassa and Lassa-like arenaviruses isolated from Africa, in: *Segmented Negative Strand Viruses* (R. W. Compans and D. H. L. Bishop, eds.), pp. 201–218, Academic Press, New York.
Grau, O., Franze-Fernandez, M. T., Romanowski, V., Rustici, S. M., and Rosas, M. F., 1981, Junin virus structure, in: *The Replication of Negative Stranded Viruses* (D. H. L. Bishop and R. W. Compans, eds.), pp. 11–14, Elsevier–North Holland, New York.
Harnish, D. G., Leung, W-C., and Rawls, W. E., 1981, Characterization of polypeptides immunoprecipitable from Pichinde virus-infected BHK-21 cells, *J. Virol.* **38**:840.
Harnish, D. G., Dimock, K., Bishop, D. H. L., and Rawls, W. E., 1983, Gene mapping in Pichinde virus: Assignment of viral polypeptides to genomic L and S RNAS, *J. Virol.* **46**:638.
Howard, C., 1987, Neutralization of arenaviruses by antibody, *Curr. Top. Microbiol. Immunol.* **134**:117.
Howard, C. R., and Simpson, D. I. H., 1980, The biology of the arenaviruses, *J. Gen. Virol.* **51**:1.
Howard, C. R., Lewicki, H., Allison, L., Salter, M., and Buchmeier, M. J., 1985, Properties and characterization of monoclonal antibodies to Tacaribe virus, *J. Gen. Virol.* **66**:1383.
Johnson, K. M., McCormick, J. B., Webb, P. A., Smith, E., Elliot, L. H., and King, I. J., 1987, Lassa fever in Sierra Leone: Clinical virology in hospitalized patients, *J. Infect. Dis.* **155**:456.
Kiley, M. P., Tomori, O., Regnery, R. L., and Johnson, K. M., 1981, Characterization of the arenaviruses Lassa and Mozambique, in: *The Replication of Negative Strand Viruses* (D. H. L. Bishop and R. W. Compans, eds.), pp. 1–9, Elsevier–North Holland, New York.
Klavinskis, L. S., Whitton, J. L., Joly, E., and Oldstone, M. B. A., 1990, Vaccination and protection from a lethal viral infection: Identification, incorporation, and use of a cytotoxic T lymphocyte glycoprotein epitope, *Virology* **178**:393.
Lewis, R. M., Cosgriff, T. M., Griffin, B. Y., Rhoderick, J., and Jahrling, P. B., 1988, Immune serum increases arenavirus replication in monocytes, *J. Gen. Virol.* **69**:1735.
Martinez Peralta, L., Bruns, M., and Lehmann-Grube, F., 1981, Biochemical composition of lymphocytic choriomeningitis virus interfering particles, *J. Gen. Virol.* **55**:475.
Matsuura, Y., Possee, R., and Bishop, D. H. L., 1986, Expression of the S-coded genes of lymphocytic choriomeningitis virus using a baculovirus vector, *J. Gen. Virol.* **67**:1515.
Morrison, H. G., Bauer, S. P., Lange, J. V., Esposito, J. J., McCormick, J. B., and Auperin, D. D., 1989, Protection of guinea pigs from lassa fever by vaccinia virus recombinants expressing the nucleoprotein orr the enveloped glycoproteins of Lassa virus, *Virology* **171**:179.
Murphy, F. A., and Whitfield, S. G., 1975, Morphology and morphogenesis of arenaviruses, *Bull. WHO* **52**:409.
Murphy, F. A., Webb, P. A., Johnson, K. M., Whitfield, S. G., and Chappell, W. A., 1970, Arenoviruses in Vero cells: Ultrastructural studies, *J. Virol.* **6**:507.
Oehen, S., Hengartner, H., and Zinkernagel, R. M., 1991, Vaccination for disease, *Science* **251**:195.
Oldstone, M. B. A., and Buchmeier, M. J., 1982, Restricted expression of viral glycoprotein in cells of persistently infected mice, *Nature* **300**:360.
Oldstone, M. B. A., Tishon, A., and Buchmeier, M. I., 1983, Virus-induced immune complex disease: genetic control of Clq binding complexes in the circulation of mice persistently infected with lymphocytic choriomeningitis virus, *J. Immunology* **130**:912.

Oldstone, M. B. A., Ahmed, R., Byrne, J., Buchmeier, M. J., Riviere, Y., and Southern, P., 1985, Virus and immune responses: Lymphocytic choriomeningitis virus as a prototype model of viral pathogenesis, *Br. Med. Bull.* **41**:70.

Oldstone, M. B. A., Blount, P., Southern, P. J., and Lambert, P. W., 1986, Cytoimmunotherapy for persistent virus infection reveals a unique clearance pattern from the central nervous system, *Nature* **321**:239.

Parekh, B. S., and Buchmeier, M. J., 1986, Proteins of lymphocytic choriomeningitis virus: Antigenic topography of the viral glycoproteins, *Virology* **153**:168.

Pircher, H., Moskophidis, D., Rohrer, U., Burki, K., Hengartner, H., and Zinkernagel, R. M., 1990, Viral escape by selection of cytotoxic T cell–resistant virus variants *in vivo*, *Nature* **346**:629.

Popescu, M., and Lehmann-Grube, F., 1977, Defective interfering particles in mice infected with lymphocytic choriomeningitis virus, *Virology* **77**:78.

Riviere, Y., Ahmed, R., Southern, P. J., Buchmeier, M. J., Dutko, F. J., and Oldstone, M. B. A., 1985, The S RNA segment of lymphocytic choriomeningitis virus codes for the nucleoprotein and glycoproteins 1 and 2, *J. Virol.* **53**:966.

Riviere, Y., Southern, P. J., Ahmed, R., and Oldstone, M. B. A., 1986, Biology of cloned cytotoxic T lymphocytes specific for lymphocytic choriomeningitis virus: recognition is restricted to gene products encoded by the viral S RNA segment, *J. Immunol.* **136**:304.

Romanowski, V., and Bishop, D. H. L., 1985, Conserved sequences and coding of two strains of lymphocytic choriomeningitis virus (WE and ARM) and Pichinde arenavirus, *Virus Res.* **2**:35.

Ruo, S. L., Mitchell, S. W., Kiley, M. P., Roumillat, L. F., Fisher-Hoch, S. P., and McCormick, J. B., 1991, Antigenic relatedness between arenaviruses defined at the epitope level by monoclonal antibodies, *J. Gen. Virol.* **72**:549.

Saleh, F., Gard, G. P., and Compans, R. W., 1979, Synthesis of Tacaribe viral proteins, *Virology* **93**:369.

Salvato, M., Shimomaye, E., Southern, P., and Oldstone, M. B. A., 1988, Virus–lymphocyte interactions IV: molecular characterization of LCMV Armstrong (CTL^+) small genomic segment and that of its variant, clone 13 (CTL^-), *Virology* **164**:517.

Sanchez, A., Pifat, D. Y., Kenyon, R. H., Peters, C. J., McCormick, J. B., and Kiley, M. P., 1989, Junin virus monoclonal antibodies: characterization and cross-reactivity with other arenaviruses, *J. Gen. Virol.* **70**:1125.

Southern, P. J., Buchmeier, M. J., Ahmed, R., Francis, S. J., Parekh, B., Riviere, Y., Singh, M. K., and Oldstone, M. B. A., 1986, Molecular pathogenesis of arenavirus infections, in: *Vaccines 86*, (F. Brown, R. Chanock, and R. Lerner, eds.), pp. 239–245, Cold Spring Harbor Lab, New York.

Southern, P. J., Singh, M. K., Riviere, Y., Jacoby, D. R., Buchmeier, M. J., and Oldstone, M. B. A., 1987, Molecular characterization of the genomic S RNA segment from lymphocytic choriomeningitis virus, *Virology* **157**:145.

Vezza, A. C., Gard, G. P., Compans, R. W., and Bishop, D. H. L., 1977, Structural components of the arenavirus Pichinde, *J. Virol.* **23**:776.

Webb, P. A., Johnson, K. M., and MacKenzie, R. B., 1969, The measurement of specific antibodies in Bolivian haemorrhagic fever by neutralization of virus plaques, *Proc. Soc. Exp. Biol. Med.* **130**:1013.

Weber, E. L., and Buchmeier, M. J., 1988, Fine mapping of a peptide sequence containing an antigenic site conserved among arenaviruses, *Virology* **164**:30.

Welsh, R. M., and Buchmeier, M. J., 1979, Protein analysis of defective interfering lymphocytic choriomeningitis virus and persistently infected cells, *Virology* **96**:503.

Welsh, R. M., and Oldstone, M. B. A., 1977, Inhibition of immunologic injury of cultured cells infected with lymphocytic choriomeningitis virus: Role of defective interfering virus in regulating viral antigenic expression, *J. Exp. Med.* **145**:1449.

Welsh, R. M., and Pfau, C. J., 1972, Determinants of lymphocytic choriomeningitis interference, *J. Gen. Virol.* **14**:177.

Welsh, R. M., O'Connell, C. M., and Pfau, C. J., 1972, Properties of defective lymphocytic choriomeningitis virus, *J. Gen. Virol.* **17**:355.

Welsh, R. M., Nishioka, W. K., Antia, R., and Dundon, P. L., 1990, Mechanism of killing by virus-induced cytotoxic T lymphocytes elicited *in vivo*, *J. Virol.* **64**:3726.

Whitton, J. L., Gebhard, J. R., Lewicki, H., Tishon, A., and Oldstone, M. B. A., 1988a, Molecular definition of a major cytotoxic T-lymphocyte epitope in a glycoprotein of lymphocytic choriomeningitis virus, *J. Virol.* **62**:687.

Whitton, J. L., Southern, P. J., and Oldstone, M. B. A., 1988b, Analyses of the cytotoxic T lymphocyte responses to glycoprotein and nucleoprotein components of lymphocytic choriomeningitis virus, *Virology* **162**:321.

Whitton, J. L., Tishon, A., Lewicki, H., Gebhard, J., Cook, T. Salvato, M., Joly, E., and Oldstone, M. B. A., 1989, Molecular analyses of a five-amino-acid cytotoxic T-lymphocyte (CTL) epitope: An immunodominant region which induces nonreciprocal CTL cross-reactivity, *J. Virol.* **63**:4303.

Wilson, S. M., and Clegg, J. C. S., 1991, Sequence analysis of the S RNA of the african arenavirus Mopeia: An unusual secondary structure feature in the intergenic region, *Virology* **180**:543.

Wright, K. E., and Buchmeier, M. J., 1991, Antiviral antibodies attenuate T-cell-mediated immunopathology following acute lymphocytic choriomeningitis virus infection, *J. Virol.* **65**:3001.

Wright, K. E., Salvato, M. S., and Buchmeier, M. J., 1989, Neutralizing epitopes of lymphocytic choriomeningitis virus are conformational and require both glycosylation and disulfide bonds for expression, *Virology* **171**:417.

Wright, K. E., Spiro, R. C., Burns, J. W., and Buchmeier, M. J., 1990, Post-translational processing of the glycoproteins of lymphocytic choriomeningitis virus, *Virology* **177**:175.

Young, P. R., 1987, Arenaviridae, in: *Animal Virus Structure* (M. Nermut and G. Steven, eds.), pp. 185–198, Elsevier, New York.

Young, P. R., Chanas, A. C., and Howard, C. R., 1981, Analysis of the structure and function of Pichinde virus polypeptides, in: *The Replication of Negative Strand Viruses* (D. H. L. Bishop and R. W. Compans, eds.), pp. 15–22, Elsevier–North Holland, New York.

CHAPTER 3

Antigenic Diversity among the Arenaviruses

COLIN R. HOWARD

I. INTRODUCTION

The close antigenic relationship between the arenaviruses has played a key role in the identification of these viruses as a distinct virus family. Two early observations of particular importance were the finding that Tacaribe virus from Trinidad was closely related to Junin virus, despite the wide geographical distance between these isolates (Murphy *et al.*, 1969) and the close relationship between the prototype LCM virus and Lassa virus (Casals, 1977). Although each member of the arenavirus family possesses some antigenic relationship to other members of the group, the degree of cross-reactivity largely depends on the assay system used and the specificity of the reagents. This leads to difficulties in classifying these viruses into genera, although in broad terms viruses within the Old World and New World groupings show varying degrees of cross-reaction. This is based largely on the presence of common epitopes on the major nucleocapsid protein, NP, although work with monoclonal antibodies has shown common antigens located on the smaller envelope glycoprotein, GP2.

The complement fixation (CF) test reveals the broadest relationships (Casals, 1975), and these are largely paralleled by the reactions obtained with the more sensitive immunofluorescence assay. In contrast, neutralization of infectivity is almost type-specific, with only lim-

COLIN R. HOWARD • Department of Pathology and Infectious Diseases, Royal Veterinary College, London, NW1 0TU, England.

The Arenaviridae, edited by Maria S. Salvato. Plenum Press, New York, 1993.

ited cross-reactions demonstrable. As with all serological tests, the potency of the antigen and antisera together with the species immunized to prepare hyperimmune reagents will determine the sensitivity of such methods. Qualitative determinations between different laboratories using separate reagents are thus difficult to compare, although considerable progress has been made with available murine monoclonal antibodies against many of these arenaviruses.

II. SEROLOGICAL RELATIONSHIPS AMONG ARENAVIRUSES

A. General Relationships among Arenaviruses as Defined by Use of Polyclonal Antisera

The 11 distinct arenaviruses described to date mostly belong to the so-called Tacaribe complex and are restricted to the New World. These form an antigenically related cluster of viruses distinct from those isolated in the Old World, such as Lassa and LCM viruses. These groupings are rather arbitrary, however, given the cross-reactions that may occur between Old and New World members and the extensive homologies in nucleotide sequences (Southern and Bishop, 1987).

A strong relationship between Tacaribe, Junin, and Machupo viruses can be readily demonstrated by CF, with a more distant cross-reactivity discernible between these viruses and other members of the Tacaribe complex; Pichinde and Tamiami appear not so closely related to each other or to other New World arenaviruses (Casals *et al.*, 1975). It should be mentioned, however, that immunofluorescence analysis of animal antisera against heterologous antigen substrates was an early indication of the relatedness of Tacaribe and LCM virus, even though only limited cross-reactions were observed by CF (Rowe *et al.*, 1970). Whereas LCM virus shows little cross-reactivity with members of the Tacaribe complex, Lassa virus exhibits a distant relationship with both this group and LCM virus (Casals, 1977).

All the available evidence suggests that the complement-fixing antigen is associated with the internal nucleoprotein; studies by Gshwender and Lehmann-Grube (1973) determined that this activity was associated with a soluble fraction obtained from LCM virus-infected cells. Repeated inoculations of animals resulted in antisera with a high titer by the CF test, but the antibodies failed to neutralize virus infectivity. Moreover, guinea pigs inoculated only with the soluble antigen were not protected against subsequent challenge with LCM virus. The CF antigen has been shown to be derived from the internal nucleoprotein by proteolytic cleavage (Buchmeier *et al.*, 1977; Mitri and Martinez-Segovia, 1980), an event that may correlate with the migration of nucleocapsid protein fragments to the nucleus (Young *et al.*, 1986). LCM and Lassa viruses show some relationship to each other by the CF test; these cross-reactions have now been more satisfactorily resolved by taking advantage of the monospecifity of monoclonal antibodies (see below).

The use of immunofluorescence (IF) tests reveals similar patterns of serological relatedness between arenaviruses, although the increased sensitivity shows clearer links, for example, between LCM and Lassa viruses (Wulff *et al.*, 1977). These relationships are summarized in Table 1. Much of the antigen load within infected cells represents nucleocapsid antigens rather than the type-specific envelope proteins. Serological cross-reactions by IF can be detected in patients infected with Machupo and Junin viruses. Antibody titers by IF against other members of Tacaribe complex are highest during early convalescence and decrease thereafter in direct proportion to the titer obtained with the homologous substrate (Peters *et al.*, 1973). Greatest cross-reactivity is seen with Junin antigens and at a level approximately the same as that seen using Machupo substrates for the assessment of antibody in patients convalescing from Argentinian hemorrhagic fever. Other cross-reactions against Tacaribe and Latino viral antigens have been demonstrated; perhaps surprisingly, positive results were also observed using an LCM virus-infected substrate. It is relevant to note that acetone-fixed substrates give predominantly cytoplasmic reactions. Antibodies to the envelope proteins are best detected by examining for immunofluorescence at the plasma membrane of unfixed, infected cells mixed in suspension with antisera. The viral envelope glycoproteins expressed at the cell surface bear type-specific antigens, and using polyclonal antisera, few cross-reactions can be demonstrated.

B. Fine Analysis Using Monoclonal Antibodies

The availability of monoclonal antibodies has permitted a greater analysis of the fine specificity differences between arenavirus antigens. For example, Buchmeier *et al.* (1980) characterized 46 monoclonal reagents raised against the Armstrong CA 1371 strain of LCM virus. Cross-reactivity was detected with five antibodies by indirect immuno-

TABLE I. Cross-Reactions with Monoclonals to Tacaribe N Protein

Monoclonal antibodies	Junin	Machupo	Pichinde	Amapari
2.16.2	+	−	+	+
2.48.3	+	+	+	+
2.52.2	+	−	−	+
2.74.3	+	−	−	+
2.100.3	+	+	−	+

fluorescence using cells infected with Lassa or Mopeia viruses, and a sixth reacted with the Mopeia virus only. Interestingly, this reagent specific for LCM and Mopeia viruses alone was shown to recognize the external GP2 glycoprotein of LCM virus, whereas the remaining antibodies that recognized both Lassa- and Mopeia-infected cell substrates were directed against epitopes on the nucleocapsid of the LCM virion. Monoclonal antibodies to the GP2 glycoprotein of the WE strain of LCM virus also may recognize Mopeia but not Lassa viral antigens. An additional four antibodies specific for the larger GP1 glycoprotein of LCM virus did not cross-react with either of the African arenaviruses (Buchmeier, 1984). These results suggest a degree of conservation among epitopes on the GP2 molecule, and the finding of cross-reactivity against Pichinde virus also suggests that this conservation is not restricted to the Old World arenaviruses.

Direct confirmation of the immunofluorescence findings has been obtained by immunoprecipitation of viral proteins from infected cells and analysis of the immune complexes by SDS–polyacrylamide gel electrophoresis (Buchmeier *et al.*, 1981; Kiley *et al.*, 1981; Buchmeier, 1984). A problem often encountered in these studies, however, is the use of *Staphylococcus aureus* protein A for the collection of antigen–antibody complexes, as often this results in nonspecific binding to nucleocapsid protein (Kiley *et al.*, 1981; Howard *et al.*, 1985).

Monoclonal antibodies have also been prepared against several members of the Tacaribe complex of New World arenaviruses, the vast majority of which are specific for epitopes on the internal nucleocapsid antigens. Antibodies to Pichinde virus frequently cross-react with other members of the complex, and in one instance cross-reactions have been recorded against Lassa viral antigens, suggesting that at least some epitopes situated on the internal nucleocapsid may be shared by both New and Old World arenaviruses (Buchmeier *et al.*, 1981). In the latter study, the extent of cross-reactivity differed between antibodies examined, suggesting that each recognized a unique determinant. Similar findings have also been reported with a collection of monoclonal antibodies prepared against Tacaribe virus (Allison *et al.*, 1984; Howard *et al.*, 1985). Among those reagents specific for the nucleocapsid of Pichinde virus, a particularly close relationship could be discerned between Pichinde, Parana, and Tamiami viruses, with a more distant relationship being evident using Amapari and Junin substrates. These findings reflect in broad terms those relationships identified by immunofluorescence using polyclonal sera or monoclonal antibodies to Tacaribe and Junin viruses and confirm the particularly close relationship between these two viruses and Amapari and Machupo viruses. Pichinde and Parana viruses appear to be more distantly related (Allison *et al.*, 1984; Howard *et al.*, 1985).

Monoclonal antibodies clearly recognize distinct epitopes on the nucleocapsid protein, although the exact locations have yet to be defined. Young *et al.*, (1986) showed that monoclonal antibodies to the Pichinde virus nucleoprotein differentiated between NP-related species in terms of their localization within the infected cell; this suggests that at least some epitopes are located on the 28,000 mol. wt. fragment that is detected in nuclear inclusion bodies (Fig. 1).

A particularly fascinating close antigenic relationship exists between the envelope glycoproteins of Tacaribe and Junin viruses. For example, both guinea pigs and marmoset monkeys inoculated with Tacaribe virus are protected against challenge with the normally lethal Junin virus (Weissenbacher *et al.*, 1975/1976, 1982). Yet these viruses were originally isolated from regions some 3000 miles apart and from totally different natural host species. The recent availability of monoclonal antibodies to the envelope glycoproteins of these viruses together with progress in genome sequencing of both viruses should do much to reveal the immunological basis of this cross-protection and the nature of the antigens involved. For example, Sanchez *et al.* (1989) have prepared seven monoclonal antibodies against the Junin virus GP2 protein, all of which neutralize infectivity *in vitro*. However, their protective capacity

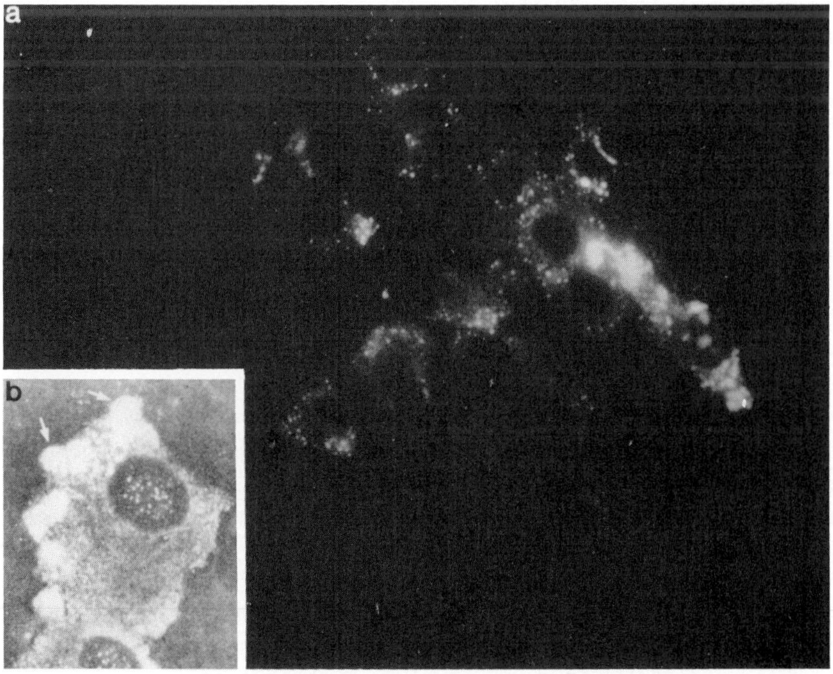

FIGURE 1. Immunofluorescence pattern seen with a nucleocapsid-specific antibody and fixed Vero cells infected with Pichinde virus. The granular cytoplasmic fluorescence is typical of nucleocapsid-specific antibody reactions with infected cells. This antibody (PV1-1-3) also detects nuclear inclusion bodies containing a proteolytic cleavage fragment of the nucleocapsid protein. (From Young *et al.*, 1986.)

in vivo was found to more closely approximate the extent of antibody-mediated cell lysis in the presence of complement. Six of these antibodies also neutralize Tacaribe virus, but this cross-reactivity is lost for Tacaribe variants grown in the presence of neutralizing antibodies to Tacaribe virus (M. W. Salter and C. R. Howard, unpublished observations).

Monoclonal antibodies are increasingly used to differentiate virus strains by use of reagents selective for epitopes that go unrecognized when polyclonal antisera are used. Buchmeier (1984) has summarized these patterns of reactivity obtained with a panel of monoclonal antibodies against laboratory strains of LCM virus and also Lassa and Mopeia viruses. Reagents directed against the smaller GP2 envelope glycoprotein were found to cross-react by immunofluorescence with all substrates examined, whereas antibodies directed against the larger GP1 glycoprotein were either strain-specific or reacted with only a subset of the different strains examined. More recent work has shown the presence of at least four antigenic sites on LCM virus GP2 (Parekh and Buchmeier, 1986). Two of these sites induced cross-reacting, but non-neutralizing antibodies whereas the remaining two were involved in strain-specific and common antibody neutralization, respectively. The observation that certain of these antibodies for GP2 are broadly cross-reactive suggested the conservation of epitopes on surface envelope structures among Old World and New World arenaviruses. This has since been confirmed by the definition of these epitopes using synthetic peptides (Weber and Buchmeier, 1988).

A comparison of Lassa with Mopeia and Mobala viruses from Africa has also been undertaken using monoclonal antibodies (Gonzalez *et al.*, 1983; 1984). Again, various degrees of cross-reactivity were observed using reagents specific for the GP2 external glycoprotein. Some distinction may be made with Mobala virus from the Central African Republic, however, as it was found that several cross-reactive monoclonal antibodies originally prepared against LCM virus failed to recognize Mobala-infected substrates.

Clegg and Lloyd (1984) analyzed an extensive range of different Lassa and Mopeia isolates and found antigenic determinants common to strains of both viruses. These are located on the internal nucleocapsid and one of the two envelope glycoproteins. Interestingly, sera raised against Mopeia virus showed a pattern of reactivity restricted to the nucleocapsid of this virus, with no reactivity against Lassa virus.

Although the introduction of monospecific antibodies using hybridoma techniques has begun to unravel the antigenic complexity among the arenaviruses, the preparation of more reagents is required to make these comparisons meaningful. This is hampered somewhat by the difficulty in preparing monoclonal antibodies specific for the envelope virion glycoproteins, where the greatest antigenic diversity may be expected.

III. PROPERTIES OF VIRUS-NEUTRALIZING ANTIBODIES

The neutralization test is very specific for all members of the arenavirus family, with only limited examples of cross-reactivity, notably using high-titer animal antisera raised against Junin, Tacaribe, and Machupo viruses. However, the ease of quantifying neutralizing antibodies differs with each virus, and the difficulty in assessing neutralizing antibodies for some arenaviruses limits their usefulness. For example, neutralizing antibodies to Lassa virus can only be detected *in vitro* with great difficulty (Buckley and Casals, 1970). Apparently, neutralizing antibodies to Lassa virus appear only late in convalescence (Peters, 1984). No cross-reactions have been observed between Junin and Machupo viruses in plaque-reduction tests using human convalescent sera despite the close relationship that can be demonstrated. With LCM and Lassa sera, both viruses are readily distinguishable from one another by this technique (Henderson and Downs, 1965; Johnson *et al.*, 1973).

The sensitivity of the neutralization test for LCM virus can be increased by incorporating either complement or anti-gamma-globulin into the test system (Oldstone, 1975; Lehmann-Grube and Ambrassat, 1977). Buchmeier (1984) reported neutralization of LCM virus by monoclonal antibodies in the absence of complement. One of the three antibodies studied neutralized both the Armstrong and WE strains, demonstrating a sharing of epitopes involved in the relevant antigenic site between closely related viruses. Neutralizing antibodies to LCM virus are directed against the major GP1 envelope glycoprotein (Buchmeier and Oldstone, 1979; Bruns *et al.*, 1983). Competitive inhibition assays using a panel of neutralizing and nonneutralizing monoclonal antibodies have shown that the LCM GP1 and GP2 proteins contain a minimum of four and one antigenic domains, respectively (Parekh and Buchmeier, 1986). At least two of the GP1 antigens appear involved in virus neutralization, as neutralizing antibodies in guinea pig antisera react predominantly with these two sites on GP1. The antigenic domain on GP2 is conserved among many arenaviruses from both the Old and New World and has been analyzed in depth (Weber and Buchmeier, 1988, and Fig. 2).

An analysis of arenavirus neutralization by monoclonal antibodies in the absence of complement has been reported using Tacaribe virus (Allison *et al.*, 1984; Howard *et al.*, 1985). Five monoclonal antibodies were prepared which neutralized infectivity at high dilution. In none of the reactions was the one-way cross-neutralization observed using polyclonal mouse antisera as reported by Henderson and Downs (1965); in the latter instance anti-Junin sera failed to neutralize Tacaribe virus whereas anti-Tacaribe immune ascites neutralized both Junin and the homologous virus. None of the monoclonal antibodies with neutralizing activity for Tacaribe virus significantly reduced the infectivity of Junin or any other arenavirus, despite the fact that these antibodies are directed to a conserved region of the GP2 protein and recognize heterolo-

Reactivity of LCM-specific monoclonal antibodies
with conserved region peptide analogues

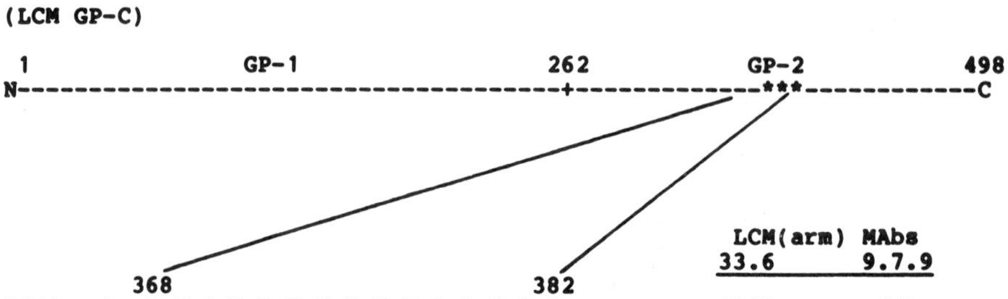

		LCM(arm) MAbs 33.6	LCM(arm) MAbs 9.7.9
LCM(arm)	P Y C N Y S K F W Y L E H A K	2560	640
Lassa	- - - - - - - - Y - - - N - T T	2560	160
Pichinde	- - - - - T - - - - I N D T I	5120	2560
Tacaribe	- - - - - T R - - - V N - T L	5120	320

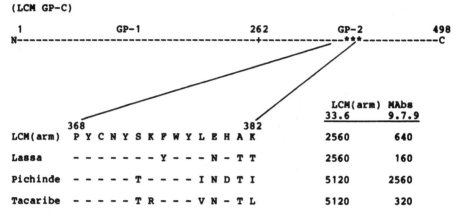

FIGURE 2. Amino acid sequence of the conserved antigenic domain of GP2 and titers of cross-reactive antibodies as measured by ELISA using synthetic peptide analogues.

gous synthetic peptide analogues by ELISA. This suggests that conformation plays an important role in antibody recognition and virus neutralization in this site. Such subtle differences are also suggested from studies with neutralizing antibodies to the WE and Armstrong strains of LCM virus; the WE strain appears to have a single, immunodominant neutralizing antigenic determinant on GP1, but the Armstrong strain has a second, topologically related site on the same molecule (Parekh and Buchmeier, 1986; Wright *et al.*, 1989).

Kinetic neutralization experiments showed that antibodies differed in their ability to destroy infectivity during a 20-min incubation with virus (Allison *et al.*, 1984). At high antibody dilution, neutralization occurred after a longer incubation period, suggesting that several molecules of antibody are required for neutralization; whereas a linear relationship between antibody concentration and incubation time would suggest a stoichiometric relationship between neutralizing antibody and virion (Della-Porta and Westaway, 1978). The remaining antibodies gave significant nonneutralized fractions after an initial linear response (Howard *et al.*, 1985). Given the monoclonal nature of the antibodies used in these studies, it is unlikely that the persistent fraction is due to steric hindrance from nonneutralizing antibodies. Incomplete neutralization may be the result of an equilibrium between free and bound antibody molecules at a point prior to complete neutralization (Volk *et al.*, 1980). However, it has been shown for other enveloped viruses that a lower avidity of neutralizing antibody does not necessarily result in an increase in the size of the nonneutralized fraction (Iorio and Bratt, 1984). This latter study did show, however, that a mixture of monoclonal antibodies directed against four distinct epitopes on the virus surface showed neutralization kinetics similar to those obtained with polyclonal animal antisera. A similar situation may be the case for Tacaribe virus in that monoclonal antibodies directed against other, hitherto unidentified sites may singly or in combination with other epitopes constitute important domains for eliciting neutralizing antibodies.

It is well recognized that high-titer antisera against either Pichinde or Lassa viruses possess minimal neutralizing capacity when examined by the mouse neutralization test (Lehmann-Grube *et al.*, 1979) or by tissue culture tests (Buckley and Casals, 1970; Sengupta and Rawls, 1979). The use of an alternative and simple method for detecting antibodies to Pichinde virus has been described whereby the incorporation of antiserum into the overlay of an infected cell culture results in plaque-size reduction (Chanas *et al.*, 1980). Such antisera may be deficient in antibodies directed against an essential virus surface component with the result that the covering of critical areas required for virus neutralization does not occur. Whether this may be the case in convalescent sera to Lassa virus is unclear, but the studies of Chanas *et al.* showed clearly that aggregation and extensive sensitization of Pichinde virus particles does occur after reaction with homologous antiserum. Despite the presence of antibodies directed against all the major structural proteins of the virus, absorption and penetration of virus antibody complexes were not prevented. The cell type may also be important; Lewis et al (1988) have shown that antibody-mediated enhancement of infection may occur with both Pichinde and Lassa viruses mixed with monkey antisera and inoculated onto cultured human monocytes.

IV. PROPERTIES OF ANTIGENIC SITES

A. Definition of a Conserved Envelope Antigenic Site

The use of synthetic peptides and antisera has allowed the definition of antigenic sites on the envelope glycoprotein molecules of LCM and other arenaviruses. Considerable homology is seen among arenaviruses in the carboxyl half of the glycoprotein (GPC) gene (Southern and Bishop, 1987; Clegg, Chapter 10, this volume). Buchmeier and colleagues have studied in detail a group-specific antigen conserved on all arenaviruses so far characterized. The broadly cross-reactive antibody 33.6 (Buchmeier *et al.*, 1980) recognized five amino acids (374–378) of the LCM glycoprotein GP2 in a region that is broadly conserved between Old and New World arenaviruses (Fig. 2). Antibody 9–7.9 also was specific for this region but showed fewer cross-reactions, recognizing only Mopeia but not Lassa virus. This is apparently due to the change at residue 373 in the Lassa protein from phenylalanine to tyrosine (Weber and Buchmeier, 1988).

Anti-LCM antibodies 9–7.9 and 33.6 also bind to the Tacaribe-specific peptide analogue of this region; furthermore, the series of neutralizing monoclonal antibodies raised against Tacaribe virus (Howard *et al.*, 1985) reacted with this peptide as well as the heterologous sequences from this region representing Lassa, Pichinde, and LCM viruses (M. W. Salter and C. R. Howard, unpublished results). It is worth noting that antibodies directed to this region on GP2 do not generally neutralize infectivity; these Tacaribe virus–specific antibodies appear to be an exception. Until recently, there has been no demonstration of the presence of a GP2 molecule on Tacaribe virions, although the S RNA strand sequence predicts the expression of both GP1 and GP2 glycoproteins.

B. Analysis of Antigenic Variants

Some insight into the antigenic variation of arenaviruses may be gained by the selection of virus variants grown in the presence of neutralizing monoclonal antibody. Howard *et al.* (1985) have reported that Tacaribe virus variants may be selected in the presence of antibody at a frequency greater than 10^{-4}. Reaction of the variants with heterologous monoclonal antibodies manifesting incomplete neutralization suggests that point mutations in or around the epitope recognized by the selecting antibody may alter the conformation of nonoverlapping epitopes recognized by the heterologous antibodies. These variants have been analyzed by the production of double-stranded cDNA by reverse transcription and polymerase chain reaction (PCR) using oligonucleotide primers flanking nucleotides 1140–1185 (encoding the epitope recognized by the selecting antibody, 2.25.4). Sequence data show that this region remains conserved among variants but significant changes are induced upstream between nucleotides 900 and 966 that would be expected to alter the conformation of GP2 in its major antigenic site (Allison and Salter, 1991). There appears to be a parallel situation in cricetid rodents experimentally infected with Junin virus. Alche and Coto (1988) found that virus isolated from the blood and brains of persistently infected animals was not readily neutralized *in vitro* by immune sera that possessed neutralizing antibodies against the parental virus. The role of antibody in the selection of variants from a phenotypically mixed virus stock may be a significant factor in the persistence of these viruses.

REFERENCES

Alche, L. E., and Coto, C. E., 1988, Differentiation of Junin virus and antigenic variants isolated *in vivo* by kinetic neutralization assays, *J. Gen. Virol.* **69**:2123.

Allison, L. M., and Howard, C. R., 1984, Neutralization of arenaviruses: reaction of Tacaribe virus and variants with monoclonal antibodies, in: *Segmented Negative Strand*

Viruses (R. W. Compans and D. H. L. Bishop, eds.), pp. 209–216, Academic Press, Orlando, FL.

Allison, L. M., Salter, M. W., Kiguwa, S., and Howard, C. R., 1991, Analysis of the glycoprotein gene of Tacaribe virus and neutralization-resistant variants, *J. Gen. Virol.* **72**:2025.

Bruns, M., and Lehmann-Grube, F., 1983, Lymphocytic choriomeningitis virus. VI. Isolation of a glycoprotein mediating neutralization, *Virology* **130**:247.

Buchmeier, M. J., 1984, Antigenic and structural studies on the glycoproteins of lymphocytic choriomeningitis virus, in: *Segmented Negative Strand Viruses* (D. H. L. Bishop and R. W. Compans, eds.), pp. 193–200, Academic Press, New York.

Buchmeier, M. J., and Oldstone, M. B. A., 1979, Protein structure of lymphocytic choriomeningitis virus: evidence for a cell-associated precursor of the virion glycopeptides, *Virology* **99**:111.

Buchmeier, M. J., and Rawls, W. E., 1977, Antigens of Pichinde virus. I. Relationship of soluble antigens derived from BHK-21 cells to the structural components of the virion, *J. Virol.* **22**:175.

Buchmeier, M. J., and Johnson, K. M., 1980, Monoclonal antibodies to lymphocytic choriomeningitis virus react with pathogenic arenaviruses, *Nature (London)* **288**:486.

Buchmeier, M. J., Lewicki, H. A., Tomori, O., and Oldstone, M. B. A., 1981, Monoclonal antibodies to lymphocytic choriomeningitis virus and Pichinde viruses: generation, characterization and cross-reactivity with other arenaviruses, *Virology* **113**:73.

Buckley, S. M., and Casals, J., 1970, Lassa fever, a new virus disease of man from West Africa. III. Isolation and characterization of the virus, *Am. J. Trop. Med. Hyg.* **19**:680.

Casals, J., 1975, Arenaviruses, *Yale J. Biol. Med.* **48**:115.

Casals, J., 1977, Serological reactions with arenaviruses, *Medicina (Buenos Aires)* **37**(Suppl. 3):59.

Casals, J., Buckley, S. M., and Cedeno, R., 1975, Antigenic properties of the arenaviruses, *Bull. WHO* **52**:421.

Chanas, A. C., Mann, G., Young, P., Ellis, D. S., Stamford, S., and Howard, C. R., 1980, Evaluation of plaque size reduction as a method for the detection of Pichinde virus antibody, *Arch. Virol.* **65**:157.

Clegg, J. C. S., and Lloyd, G., 1984, The African arenaviruses Lassa and Mopeia: Biological and immunochemical comparisons, in: *Segmented Negative Strand Viruses* (R. W. Compans and D. H. L. Bishop, eds.), pp. 341–347, Academic Press, Orlando, FL.

Della-Porta, A. J., and Westaway, E. G., 1978, A multi-hit model for the neutralization of animal viruses, *J. Gen. Virol.* **38**:1.

Gonzalez, J. P., McCormick, J. B., Saluzzo, J. F., Herve, J. P., Georges, A. J., and Johnson, K. M., 1983, An arenavirus isolated from wild-caught rodents (Praomys species) in the Central African Republic, *Intervirology* **19**:105.

Gonzalez, J. P., Buchmeier, M. J., McCormick, J. B., Mitchell, S. W., Elliott, L. H., and Kiley, M. P., 1984, Comparative analysis of Lassa and Lassa-like arenavirus isolates from Africa, in: *Segmented Negative Strand Viruses* (R. W. Compans and D. H. L. Bishop, eds.), pp. 210–208, Academic Press, New York.

Gschwender, H. H., and Lehmann-Grube, F., 1973, Antigenic properties of the LCM virus: virion and complement-fixing antigen, in: *Lymphocytic Choriomeningitis Virus and Other Arenaviruses* (F. Lehmann-Grube, ed.), pp. 25–35, Springer Verlag, Berlin.

Henderson, J. R., and Downs, W. G., 1965, Junin and Tacaribe plaque reduction in rhesus monkey kidney cell monolayers, *Am. J. Trop. Med. Hyg.* **14**:796.

Howard, C. R., Lewicki, H., Allison, L., Salter, M., and Buchmeier, M. J., 1985, Properties and characterization of monoclonal antibodies to Tacaribe virus, *J. Gen. Virol.* **66**:1383.

Iorio, R. M., and Bratt, M., 1984, Neutralization of Newcastle disease virus by monoclonal antibodies to haemagglutinin neuraminidase glycoprotein: requirement for antibodies to four sites for complete neutralization, *J. Virol.* **51**:445.

Johnson, K. M., Webb, P. A., and Justines, G., 1973, Biology of Tacaribe-complex viruses, in: *Lymphocytic Choriomeningitis Virus and Other Arenaviruses* (F. Lehmann-Grube, ed.), pp. 241–258, Springer/Verlag, Vienna.

Kiley, M. P., Tomori, O., Regnery, R. L., and Johnson, K. M., 1981, Characterization of the arenaviruses Lassa and Mozambique, in: *The Replication of Negative Strand Viruses* (D. H. L. Bishop and R. W. Compans, eds.), pp. 1–9, Elsevier/North Holland, Amsterdam.

Lehmann-Grube, F., and Ambrassat, J, 1977, A new method to detect lymphocytic choriomeningitis virus-specific antibody in human sera, *J. Gen. Virol.* **37**:85.

Lehmann-Grube, F., Kallay, M., Ibscher, B., and Schwartz, R., 1979, Serologic diagnosis of human infections with lymphocytic choriomeningitis virus: comparative evaluation of seven methods, *J. Med. Virol.* **4**:125.

Lewis, R. M., Cosgriff, T. M., Griffin, B. Y., Rhoderick, J., and Jahrling, P. B., 1988, Immune serum increases arenavirus replication in monocytes, *J. Gen. Virol.* **69**:1735.

Mitri, M. I. de, and Martinez-Segovia, Z. M., 1980, Biological activities of Junin virus proteins. II. Complement fixing polypeptides associated with the soluble antigen and purified virus particles, *Intervirology* **14**:84.

Murphy, F. A., Webb, P. A., Johnson, K. M., and Whitfield, S. G., 1969, Morphological comparison of Machupo and lymphocytic choriomeningitis virus: basis for a new taxonomic group, *J. Virol.* **4**:535.

Oldstone, M. B. A., 1975, Virus neutralization and virus-induced immune complex disease: Virus–antibody union resulting in immunoprotection or immunologic injury—Two different sides of the same coin, *Prog. Med. Virol.* **19**:84.

Parekh, B., and Buchmeier, M. J., 1986, Proteins of lymphocytic choriomeningitis virus: antigenic topography of the viral glycoproteins, *Virology* **153**:168.

Peters, C. J., 1984, Arenaviruses, in: *Textbook of Human Virology* (R. B. Belshe, ed.), pp. 513–545, PSG, Littleton, MA.

Peters, C. J., Webb, P. A., and Johnson, K. M., 1973, Measurement of antibodies to Machupo virus by the indirect immunofluorescent technique, *Proc. Soc. Exp. Biol. Med.* **142**:526.

Rowe, W. P., Pugh, W. E., Webb, P. A., and Peters, C. J., 1970, Serological relationship of the Tacaribe complex of viruses to lymphocytic choriomeningitis virus, *J. Virol.* **5**:289.

Sanchez, A., Pifat, D. Y., Kenyon, R. H., Peters, C. J., McCormick, J. B., and Kiley, M. P. 1989, Junin virus monoclonal antibodies: characterization and cross-reactivity with other arenaviruses, *J. Gen. Virol.* **70**:1125.

Sengupta, S., and Rawls, W. E., 1979, Pseudotypes of vesicular stomatitis virus and Pichinde virus, *J. Gen. Virol.* **42**:141.

Southern, P., and Bishop, D. H. L., 1987, Arenavirus gene structure and organization, *Curr. Top. Microbiol. Immunol.* **133**:5.

Volk, W. A., Snyder, R. M., Benjamin, D. C., and Wagner, R. R., 1980, Monoclonal antibodies to the glycoprotein of vesicular stomatitis virus: comparative neutralizing activity, *J. Virol.* **42**:220.

Weber, E. B., and Buchmeier, M. J.,1988, Fine mapping of a peptide sequence containing an antigenic site conserved among arenaviruses, *Virology* **164**:30.

Weissenbacher, M. C., Coto, C. E., and Calello, M. A., 1975/1976, Cross-protection between Tacaribe complex viruses. Protection of neutralizing antibodies against Junin virus (Argentine haemorrhagic fever) in guinea pigs infected with Tacaribe virus, *Intervirology* **6**:42.

Weissenbacher, M. C., Coto, C. E., Calello, M. A., Rondinone, S. N., Damonte, E. B., and Frigerio, M. J., 1982, Cross-protection in non-human primates against Argentine haemorrhagic fever, *Infection and Immunity* **35**:425.

Wright, K. E., Salvato, M. S., and Buchmeier, M. J., 1989, Neutralizing epitopes of lymphocytic choriomeningitis virus are conformational and require both glycosylation and disulfide bonds for expression, *Virology* **171**:417.

Wulff, H., McIntosh, B. M., Hammer, D. B., and Johnson, K. M., 1977, Isolation of an arenavirus closely related to Lassa virus from Mastomys natalensis in south-east Africa, *Bull. WHO* **52**:609.

Young, P. R., Chanas, A. C., Lee, S. R., and Howard, C. R., 1986, Localization of an arenavirus protein in the nuclei of infected cells, *J. Gen. Virol.* **68**:2281.

Young, P. R., Chanas, A. C., and Howard, C. R., 1987, Localization of an arenavirus protein in the nuclei of infected cells, *J. Gen. Virol.* **68**:246.

CHAPTER 4

Genetic Organization of Junin Virus, the Etiological Agent of Argentine Hemorrhagic Fever

VICTOR ROMANOWSKI

I. INTRODUCTION

Most of the data on the genome organization of Junin virus presented in this chapter come from our laboratory and include results that are at the stage of manuscript preparation. Molecular cloning and sequencing studies on this human pathogenic arenavirus were focused on the S RNA, which codes for the major structural proteins.

In the first part of this chapter the biochemical structure of Junin virus is reviewed after a short historical background. In some instances, where information specific for Junin virus is ambiguous, references are made to data published for other arenaviruses.

II. HISTORICAL CONSIDERATIONS

The symptoms and clinical forms of Argentine hemorrhagic fever (AHF) were first described by Aribalzaga (1955) after the observations of the 1953 and 1954 epidemic outbreaks in Bragado, Buenos Aires. The viral etiology of the disease was established in 1958 by two groups led by Parodi and Pirosky, respectively, who isolated the virus from the blood

VICTOR ROMANOWSKI • Instituto de Bioquimica y Biologia Molecular, Departamento de Ciencias Biológicas, Facultad de Ciencias Exactas, Universidad Nacional de La Plata, 1900 La Plata, Argentina.

The Arenaviridae, edited by Maria S. Salvato. Plenum Press, New York, 1993.

of patients and organs obtained from necropsies at the city of Junin Regional Hospital (Parodi *et al.*, 1958; Pirosky *et al.*, 1959). The etiological agent was named Junin virus after the geographical site of its initial isolation.

At that time the epidemic outbreaks were restricted to an area of 16,000 km^2. Since then, AHF has spread to the provinces of Cordoba, Santa Fe, and La Pampa, increasing the endemic area to five times its initial size by 1975 (Maiztegui and Sabattini, 1977). In 1986 the progressive extension of the endemic area was reported to cover more than 120,000 km^2 of the richest farming land in Argentina with more than 2 million inhabitants (Maiztegui *et al.*, 1986). Consequently, the disease has had considerable impact on the welfare and economy of Argentina and has stimulated research on AHF and Junin virus to solve this sanitary problem.

As is obvious, early research on AHF was focused on reproducing the disease in animal models (Boxaca *et al.*, 1961; 1965) and establishing serological relationships with other viruses (Mettler *et al.*, 1963).

The propagation of Junin virus in tissue culture (Mettler *et al.*, 1961; Martinez Segovia *et al.*, 1967; 1974; Murphy *et al.*, 1970; Lascano and Berria, 1970) allowed a better morphological and serological characterization of the virus.

In 1965 Junin virus was found to be related to Tacaribe virus and Machupo virus, the etiological agent of Bolivian hemorrhagic fever, by serological means (Johnson *et al.*, 1973). These viruses and, subsequently, others that were isolated on the American continent, were grouped into what became known as the Tacaribe complex. All members of the Tacaribe complex were found to share the morphological features of LCM and Machupo viruses (Murphy *et al.*, 1970). The relatedness of these viruses was further evidenced by the cross reactivity between antigens of LCM and antisera to viruses of the Tacaribe complex (Rowe *et al.*, 1970a), leading to the establishment of a new taxonomic group known as Arenaviridae (Rowe *et al.*, 1970b; Pfau *et al.*, 1974).

Considerable effort has been devoted to study of the pathobiology of Junin virus both in human patients and animal models, as reviewed by Weissenbacher and Damonte (1983). However, the biochemical characterization of Junin virus and the molecular biology of the infection have received less attention in the same period, as is apparent from the list of published results.

III. STRUCTURAL COMPONENTS OF JUNIN VIRUS

A. Virus Growth and Purification

The biochemical studies on Junin virus were initially precluded by the low titers obtained in tissue cultures (Martinez Segovia *et al.*, 1967).

In 1974 Martinez Segovia *et al.* described a convenient procedure for the production of Junin virus in BHK-21 cell cultures. Daily changes of the supernatant culture medium resulted in maximum virus yields at days 4, 5, and 6 postinfection quantitated as 10^6 to 10^7 LD_{50}/ml by intracerebral inoculation in suckling mice. Several methods have been used to concentrate the viral particles, including an aqueous biphasic polymer system (Martinez Segovia and Diaz, 1968), precipitation with protamine sulfate (Coto and Parodi, 1968) or ammonium sulfate (Help *et al.*, 1970), and ultracentrifugation (Anon *et al.*, 1976). The latter method has been the most popular owing to its simplicity and the good recovery of virus infectivity. Junin virus purification described by Anon *et al.* (1976) included two successive sucrose gradient steps and yielded virus particles sufficiently pure for RNA composition analysis. However, this type of gradient did not eliminate the bovine serum proteins that came from the tissue culture medium. A more efficient purification procedure using a potassium tartrate–glycerol step gradient and a cesium chloride continuous gradient produced virions completely devoid of foreign proteins as judged by Coomassie blue staining or *in vitro* iodination (Grau *et al.*, 1981; Romanowski, 1981). These data suggest that alternative purification procedures should always include a high-salt step to strip off the contaminating serum proteins.

Buoyant densities of virus particles from different sources were reported to range from 1.12 to 1.17 g/cm^3 in sucrose gradients (Coto *et al.*, 1972; Johnson *et al.*, 1973; Anon *et al.*, 1976). The reason for this variation could be more related to the different amounts of contaminating proteins than to the pleomorphic nature of the viruses. Purified Junin virus particles float at a density of 1.20 g/cm^3 in CsCl (Romanowski, 1981).

B. Proteins

The structural polypeptides of Junin virus have been examined in preparations of virus particles metabolically labeled with ^{3}H or ^{14}C amino acids. In these studies the MC2 strain of Junin virus was grown in BHK-21 cell monolayers and recovered from the supernatant culture medium. Martinez Segovia and De Mitri (1977) reported six structural proteins with estimated molecular weights of 91, 71, 64, 52, 38, and 25 kDa. The 91-kDa, 71-kDa, 52-kDa, and 38-kDa polypeptides were also labeled with [^{14}C]glucosamine.

On the other hand, Grau *et al.* (1981) showed a more simple protein pattern for Junin virus particles (MC2 and XJCl3 strains). Three polypeptides with molecular weights of 60, 44, and 35 kDa were labeled with [^{35}S]methionine or [^{3}H]leucine. When [^{3}H]glucosamine was used as radioactive precursor, the 35-kDa polypeptide was intensely labeled and smaller amounts of [^{3}H]glucosamine were found in the 44-kDa protein. In this study the SDS–PAGE pattern of labeled proteins remained unchanged throughout the purification procedure, rendering coincident patterns of bands in the stained and labeled gels at the last purification step.

The apparent discrepancy on the protein composition of Junin virus published by two laboratories is not very big if one considers that both report two major polypeptides: a 64-kDa or 60-kDa sharp and most prominent band and a 38-kDa or 35-kDa broad glycoprotein band. The 52-kDa glycoprotein in Martinez Segovia and De Mitri's paper (1977) could well be the 44-kDa polypeptide reported by Grau *et al.* (1981), since different polyacrylamide gel electrophoresis systems were employed in both studies. The 91-kDa glycoprotein is thought to be an aggregation artifact in the gel system used by Martinez Segovia and De Mitri (1977). They also report that the proportion of the 25-kDa polypeptide is variable and, sometimes, is not detected. A polypeptide of a similar size has been shown in some preparations of other arenaviruses (Clegg and Lloyd, 1983; Young *et al.*, 1987) and might be a proteolytic cleavage product of the 60-kDa protein. And, at last, the 72-kDa glycosylated polypeptide could reflect the incorporation of the uncleaved glycoprotein precursor GPC (Rustici, 1984; De Mitri and Martinez Segovia, 1985) into virions. Buchmeier *et al.* (1987) have occasionally observed traces of GPC in purified LCM virions.

After all, several laboratories have produced data for other arenaviruses that were also dissimilar to some extent (Gard *et al.*, 1977; Gangemi *et al.*, 1978; Ramos *et al.*,, 1972; Vezza *et al.*, 1977; Young *et al.*, 1981; Kiley *et al.*, 1981; Clegg and Lloyd, 1983).

C. Nucleocapsids

Electron microscopic studies have not shown an electron-dense, highly structured nucleocapsid in the arenavirus particles. However, when virions disrupted with nonionic detergents, such as Nonidet P-40, were centrifuged to equilibrium in CsCl gradients, a ribonucleoprotein structure banded at densities of 1.31–1.37 g/cm^3. Examination of these structures derived from Tacaribe virus under the electron microscope revealed closed circular structures which appeared as strings of beads (Palmer *et al.*, 1977). The length of the circles fell into two predominant size classes, 640 nm and 1300 nm, which most likely correspond to both genomic RNAs known to be roughly 3.5 kb (S RNA) and 7 kb (L RNA) in length (Franze-Fernandez *et al.*, 1987; Iapalucci *et al.*, 1989 a,b). Vezza *et al.* (1977) also reported circular filamentous structures 10–15 nm in diameter for Pichinde virus ribonucleoproteins. More recently, Young and Howard (1983) were able to reveal a more detailed structure consisting of globular 4- to 5-nm subunits that appeared to fold into helical fiber structures 12–15 nm in diameter.

Martinez-Segovia and De Mitri (1977) reported that Junin virions disrupted with 2% Triton X-100 and banded in a CsCl gradient contained RNA associated with the major 64-kDa structural protein. When analyzed by ultracentrifugation in sucrose gradients, Junin virus treated with 0.12% Nonidet P-40 was shown to dissociate into a soluble protein fraction, containing only the glycoproteins, and a fast sedimenting (>170 S) nucleoprotein component containing RNA and only the 60-kDa major structural polypeptide (Grau *et al.*, 1981; Romanowski, 1981). This 60-kDa polypeptide will be called N (for nucleocapsid protein) throughout the rest of this text.

The N protein as well as its proteolytic cleavage products occurring in infected cells have been shown to react in complement fixation (CF) tests (De Mitri and Martinez Segovia, 1980). This test was earlier used to examine serological interrelationships between members of the Arenaviridae family (Rowe *et al.*, 1970a; Casals *et al.*, 1975). Viruses belonging to the New World arenavirus group or Tacaribe complex exhibit more extensive cross-reactivities than those shown between members of this group and the Old World arenaviruses LCM and Lassa.

Using immunoprecipitation of radiolabeled virus proteins with homologous and heterologous antisera, it was demonstrated that the N protein was responsible for the antigenic cross-reactivity among the Tacaribe complex arenaviruses (Buchmeier and Oldstone, 1978). Moreover, among the monoclonal antibodies raised against Pichinde virus, several were shown to react with one or more heterologous viruses and one of them cross-reacted broadly with five New World and two Old World arenaviruses (Buchmeier *et al.*, 1981). Although, the epitopes responsible for the serological cross-reactivity among N proteins of different arenaviruses have not been identified, cloning and sequencing of Pichinde and LCM S RNA showed a high degree of conservation in several regions of the N protein amino acid sequence (Romanowski and Bishop, 1985).

D. Viral Envelope

Electron microscopic studies show round or oval particles of Junin virus surrounded by a lipid bilayer with 10-nm surface projections or spikes (Lascano and Berria, 1974). The viral envelope is acquired during virus budding at the cellular plasma membrane (Murphy *et al.*, 1970; Murphy and Whitfield, 1975; Lascano and Berria, 1974). This aspect of Junin virus morphogenesis is supported by comparative data on the phospholipid composition of viruses and host cells (Rosas *et al.*, 1988; Rosas, 1984).

The solubilization of the viral envelope using nonionic detergents removes the 35- to 38-kDa major glycoprotein (Grau *et al.*, 1981; Cresta *et al.*, 1980). This protein is exposed on the surface of the viral envelope as shown by *in vitro* iodination of intact highly purified virions (Romanowski, 1981). Removal of glycoproteins by protease treatment of Pichinde virus coincided with the disappearance of spike structures on the virus envelope (Vezza *et al.*, 1977). By analogy, it is deduced that the 35- to 38-kDa glycoprotein builds the spikes on the surface of Junin virions.

Pulse-chase experiments of Junin virus–infected BHK-21 cells have demonstrated that a 34-kDa glycoprotein is formed from a 57-kDa precursor. No other glucosamine-labeled polypeptides have been observed during the processing of the 57-kDa protein (GPC?) (Rustici, 1984). However, in most of the gel electrophoretic analyses of mature virus proteins the major glycoprotein band appears often as a doublet peaking at 34 kDa and 39 kDa (Grau *et al.*, 1981). This observation has been *a priori* attributed to a difference in the glycosylation of a single polypeptide, but has not been confirmed by peptide mapping. In relation to this possibility, it is worth mentioning than only one structural glycoprotein has been reported for Tacaribe virus, a very close serological relative of Junin virus (Gard *et al.*, 1977; Boersma *et al.*, 1982). More recent studies have indicated that the broad glycoprotein band found in Tacaribe virions actually contains two different polypeptide species, one of which is partially lost during some purification procedures using high-salt solutions (Burns and Buchmeier, personal communication, and see their Chapter 2, this volume).

The antigenic potential of Junin virus major glycoprotein has been explored by Cresta *et al.* (1980), who have shown that it elicits neutralizing antibodies. In this respect, this glycoprotein resembles the behavior of LCM virus GP-1 and Lassa virus G1. The inoculation of the purified protein protected guinea pigs from a challenge with 10 LD_{50} of a pathogenic Junin virus strain. Therefore, it is reasonable to expect that the cloned glycoprotein gene, or a fragment of it, expressed in a suitable host–vector system could be used as a candidate vaccine.

Regarding the 44-kDa minor glycoprotein of Junin virus, no data are available on its localization in the viral envelope or its relationship to the GPC.

E. Viral RNA

Indirect evidence arguing for an RNA genome in Junin virus was published in 1969 (Martinez Segovia and Graziola, 1969; Coto and Vombergar, 1969). Only after a large-scale virus growth was optimized (Martinez Segovia *et al.*, 1974; Anon *et al.*, 1976) was biochemical analysis of the genome possible.

The genome of Junin virus consists of two single-stranded RNA species described in the original report by Anon *et al.* (1976) as 33S and 25S. Their molecular weights were estimated to be 2.4×10^6 and 1.3×10^6, respectively, from the electrophoretic mobilities in 2.7% polyacrylamide gels. The apparent molecular weights of the two genomic

RNA species, known now as L (large) and S (small), estimated from their mobility in denaturing agarose gels using CH_3HgOH, are very similar to the ones previously described (Romanowski *et al.*, unpublished results).

In addition to the genomic RNA, 28S, 18S, and 4–6S RNAs of cellular origin were found to account for 50% or even more of the [^{3}H]uridine incorporated into the viruses. These RNAs come from the ribosomes and tRNAs incorporated into the virions in quite variable proportions in different virus preparations (Rawls and Leung, 1979).

Electron micrographs normally show that variable numbers of ribosomal granules are packaged in virions that range in diameter from 50 to 300 nm (Murphy and Whitfield, 1975). Using cells with a ts mutation that affected the 60S ribosomal subunit, Leung and Rawls (1977) demonstrated that Pichinde virus–associated ribosomes were not required for virus replication. In these experiments, the virus was first grown in the ts cells and its progeny, containing ts ribosomes, was subsequently replicated successfully in wild-type cells at either the permissive or the nonpermissive temperature.

On the other hand, Mannweiler and Lehmann-Grube (1973) have observed small ribosome-less LCM virus particles and suggested that these were the truly infective units, whereas the larger particles containing the electron-dense granules were not infectious. A large proportion of similar 60-nm particles were observed in tissues from AHF patients in addition to the more classical larger particles with ribosomes (Maiztegui *et al.*, 1975). Consistent with these data, filtration of Junin virus through membranes with different pore-size distributions suggested that infectivity was associated with 40 to 70 nm particles (Coto *et al.*, 1972).

In addition to the variability in the amount of ribosomal RNAs between different virus populations (preparations) and among virions within the same population, a variability in the L RNA to S RNA ratio can be noted in the literature reports (see Table I). In all the arenaviruses studied, S RNA molecules are present in excess relative to the L RNA.

TABLE I. Molar Ratios of L and S RNAs in Purified Virions[a]

Virus	Radioactivity ratio (L-RNA/S-RNA)	Molar ratio (S-RNA/L-RNA)	Ref.
Junin	1.3	1.5	Anon *et al.* (1976)
LCM	0.7–0.4	2.9–5.0	Pedersen (1973)
LCM	0.55	3.6	Romanowski and Bishop (1983)
Pichindé	1.0	2.0	Carter *et al.* (1973)
Pichindé	1.1	1.8	Farber and Rawls (1975)
Pichindé	<1.0	>2.0	Dutko *et al.* (1976)
Pichindé	<1.0	>2.0	Vezza *et al.* (1978)

[a] The incorporation of radioactivity precursors ([^{32}P]-orthophosphate or [^{3}H]-uridine) into the different RNA species of purified virions was used to calculate the relative molar amounts of L and S RNA. To this end, the molecular size ratio of L to S RNA was considered to be roughly 2 and the proportion of uridine in the RNAs was assumed to be 25%.

The nonequimolar proportions of L and S RNAs could be explained through the formation of virions containing multiple copies of viral nucleocapsids, or different amounts of virions that have packaged S or L nucleocapsid only. A not very accurate packaging mechanism for the arenavirus particles would not be surprising since different amounts of ribosomes are included in the virions budding from the infected cells. In fact, the formation of arenaviruses that are genetically diploid for S RNA has been demonstrated in the progeny of cells coinfected with complementing ts mutants of two LCM virus strains (Romanowski and Bishop, 1983).

The formation of virus particles was also studied from a kinetic perspective using a double isotope label technique (Lopez *et al.*, 1986). In this approach, Junin virus-infected BHK-21 cells were first labeled with [^{14}C]uridine until a constant specific activity was reached in the RNA of the virions recovered from the supernatant medium. Only then the medium was changed to one containing both [^{14}C]uridine (with a specific activity equal to the one used in the previous step) and [^{3}H]-uridine. The variation of the ^{3}H/^{14}C ratio in the virus RNA collected at different time intervals after the addition of [^{3}H]uridine was used to calculate the generation time of a virus particle. This time was estimated to be 5 hr for Junin virus and includes the replication, assembly, and release of a virion to the medium at the stationary phase of virus growth.

The same authors also demonstrated that actinomycin D affects neither arenavirus RNA synthesis nor virus yields in tissue culture, when the times of exposure to the drug are shorter than the time when cell-toxic effects become evident. These results indicate that continuous host transcription is not required for replication of the arenaviruses.

IV. JUNIN VIRUS S RNA

A. Molecular Cloning and Sequence Analysis of Junin Virus S RNA

As most of the biochemical data were derived from the MC2 strain of Junin virus, it was also chosen for the first cDNA cloning and sequencing studies. RNA extracted from plaque-cloned virus grown in BHK-21 cells was used for cDNA synthesis in several rounds. Based on the published 3′ end sequence of different arenaviruses (Auperin *et al.*, 1982; i.e., GCGUGUCACCUAGGAUCCG), a complementary synthetic oligodeoxyribonucleotide was used to prime cDNA synthesis in a protocol similar to the one described by Gubler and Hoffman (1983). Double-stranded cDNA was inserted into pUC plasmid vectors and cloned in *Escherichia coli* DH5α. Initial screening of the cDNA library using a short copy cDNA probe (Bishop *et al.*, 1982) yielded clones that belonged to both L and S viral RNAs as shown by Northern blot analysis.

Analogous results have also been observed during the cloning of LCM-WE S RNA (Romanowski and Bishop, 1985). The recovery of L cDNA clones using an S RNA primer was presumably due to the similarity of the L and S 3′ end sequences.

Since full-length cDNA clones were not obtained, a primer complementary to the distal end of the first set of S RNA clones was used for a second round of cDNA synthesis and cloning (see Fig. 1). A third round of cDNA synthesis and cloning was necessary to reach the 5′ end of Junin virus S RNA.

The DNA sequences of partially overlapping clones were determined according to the method of Sanger *et al.* (1977). Some regions of these clones were also sequenced using the chemical modification method of Maxam and Gilbert (1980). Nucleotide sequences in the overlapping regions of the inserts were identical (Fig. 1). The 5′ and 3′ terminal nucleotide sequences were carefully confirmed in additional experiments. The 5′ end of the S RNA was confirmed by primer extension using an oligonucleotide complementary to a sequence near this

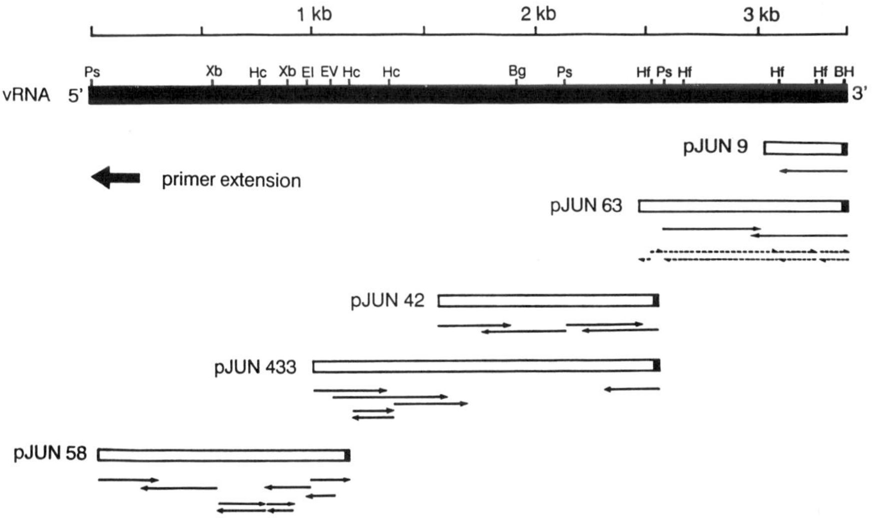

FIGURE 1. Cloning and sequencing strategy of Junin virus S RNA. The sequence of S RNA nucleotide is represented in the viral sense with a size scale in kilobases above. Some of the overlapping clones spanning the entire S RNA are shown as rectangles. Phosphorylated synthetic oligonucleotides that were used to prime cDNA synthesis are indicated as black boxes at the beginning of each clone. Subcloning into M13 mpl8 and mpl9, using the restriction sites that are indicated, was performed in order to generate the nucleotide sequence data by the method of Sanger *et al.* (1977). The sequences were obtained using a modified T7 DNA polymerase (Sequenase ™, USB) as shown by the continuous arrows. A region of the sequence (nucleotides 2472–3400) determined by the chemical cleavage method of Maxam and Gilbert (1980) is indicated by dashed arrows representing the restriction fragments involved. The thick arrow indicates the 5′ end sequence confirmed by primer extension using a synthetic oligonucleotide complementary to Junin virus S RNA nucleotides 143–194 (Ghiringhelli *et al.*, 1991).

end. The 3′ end sequence of the viral RNA was deduced from primer extension products generated using intracellular viral-complementary full-length S RNA as template and was found to be coincident with the cloned cDNA sequence derived from viral genomic RNA. The 19 nucleotides at the 3′ terminus were identical to the direct RNA sequence data derived for other arenaviruses by Auperin *et al.* (1982). This analysis showed that the complete S RNA sequence comprised 3400 nucleotides with a base ratio of 26.6% A, 21.0% G, 23.8% C, and 28.6% U (Fig. 2) (Ghiringhelli *et al.*, 1991).

The analysis of the nucleotide sequence indicates the existence of two nonoverlapping long open reading frames of opposite polarities (Fig. 3). They were deduced to code for the N protein and GPC protein, by analogy to the other arenaviruses and confirmed by Western blot analyses of recombinant products using bacterial and eukaryotic expression systems (Rivera-Pomar *et al.*, 1990, 1991; Romanowski *et al.*, unpublished results). One open reading frame, in the viral-complementary

GPC (genome polarity,v)

N (antigenome polarity, vc)

```
   1 UGCAGUAAGG GGAUCCUAGG CGAUUUUGGU AACGCUAUAA GUUGUUACUG CUUUCUAUUU GGACAACAUC AAACCAUCUA
  81 UUGUACAAUG GGGCAAUUCA UCAGCUUCAU GCAAGAAAUA CCUACCUUUU UGCAGGAAGC UCUGAAUAUU GCUCUUGUUG
 161 CAGUCAGUCU CAUUGCCAUC AUUAAGGGUG UAGUAAACCU GUACAAAAGU GGUUGUUCAA UUCUUGAUCU AGCAGGAAGA
 241 UCCUGCCCGA GAGCUUUUAA AAUCGGACUG CACACAGAGG UUCCAGACUG UGUCCUUCUU CAAUGGUGGG UCUCUUUUUC
 321 CAACAAUCCA CAUGACCUGC CUCUGUUGUG UACCUUAAAC AAGAGCCAUC UUUACAUUAA GGGGGGCAAU GCUUCAUUCA
 401 AGAUCAGCUU UGAUGACAUC GCAGUGUUGU UACCAGAAUA UGACGUUAUA AUUCAGCAUC CGGCAGAUAU GAGCUGGUGU
 481 UCUAAAAGUG AUGAUCAAAU UAGGCUGUCU CAGUGGUUCA UGAAUGCUGU GGGGCAUGAU UGGUAUCUAG ACCCACCAUU
 561 UCUGUGUAGG AACCGUACAA AGACAGAAGG CUUCAUCUUU CAAGUCAAUA CCUCCAAGAC UGGUAUCAAU GAAAACUAUG
 641 CCAAGAAGUU UAAGACUGGU AUGCACCAUU UAUAUAGAGA AUACCCCGAC UCUUGCUUGG AUGGCAAACU GUGUUUGAUG
 721 AAGGCACAAC CCACCAGUUG GCCUCUCCAA UGUCCACUUG ACCAUGUCAA CACAUUACAU UUCCUCACAA GAGGCAAGAA
 801 CAUUCAGCUU CCAAGGAGGU CUUUAAAAGC AUUCUUUUCC UGGUCUCUGA CAGACUCAUC CGGCAAGGAC ACCCCUGGAG
 881 GCUAUUGUCU AGAAGAGUGG AUGCUCGUUG CAGCCAAAAU GAAGUGUUUU GGCAAUACUG CUGUAGCAAA AUGCAAUCUG
 961 AAUCAUGACU CUGAAUUCUG UGACAUGCUG AGGCUUUUUG AUUACAACAA AAAUGCUAUC AAAACCUUAA AUGAUGAAAC
1041 UAAGAAACAA GUAAAUCUGA UGGGACAGAC AAUCAAUGCG CUGAUAUCUG ACAAUUUAUU GAUGAAAAAC AAAAUUAGGG
1121 AAUUGAUGAG UGUCCCUUAC UGCAAUUACA CAAAAUUUUG GUAUGUCAAC CACACACUUU CAGGACAACA CUCAUUACCA
1201 AGGUGCUGGU UAAUAAAAAA CAACAGCUAU UUGAACAUUU CUGACUUCCG UAAUGACUGG AUACUAGAAA GUGACUUCUU
1281 AAUUUCUGAA AUGCUAAGCA AAGAGUAUUC GGACAGGCAG GGCAAAACUC CCUUGACUUU AGUUGACAUC UGUUUUUGGA
1361 GCACAGUAUU CUUCACAGCG UCCCUCUUCC UUCACUUGGU GGGCAUACCC ACCCAUAGGC ACAUCAGAGG CGAGGCAUGC
1441 CCUCUGCCCC ACAGGCUAAA UAGCUUGGGU GGUUGCAGAU GUGGUAAGUA CCCCAAUCUA AAGAAACCAA CAGUUUGGCG
1521 CAGAGGACAC UAAGACCUCC CGAAGGUCCC CACCAGCCCG GGCAUUGCCC GGGCUGGUGU GGCCCCCCAG UCCGCGGCCU
1601 GGCCGCGGAC UGGGGAGGCA CUGCUUACAG UGCAUAGGCU GCCUUCGGGA GGAACAGCAA GCUCGGUGGU AAUAGAGGUG
1681 UAAGUUCUUC UUCAUAGAGC UUCCCAUCCA ACACUGACUG AAACAUUAUG CAGUCUAGCA GAGCACAGUG UGGCUCACUG
1761 GAGGCCAACU UAAAGGGAGC AUCCUUAUCU CUCUUUUUCU UGCUGACAAC CACUCCAUUG UGAUGUUUGC AUAGGUGGCC
1841 AAAUUUCUCC CAGACCUGUU GGUCGAACUG CCUGGCUUGU UCUGAUGUAA GCCUAACAUC AACCAGCUUA AGAUCUCUUC
1921 UUCCAUGGAG GUCAAACAAC UUCCUGAUGU CAUCGGACCC UUGAGUGGUC ACAACCAUGU CCGGAGGCAG CAAGCCAAUC
2001 ACGUAACUAA GAACUCCUGG CAUUGCAUCU UCUAUGUCUU UCAUUAAGAU GCCGUGAGAG UGUCUGCUAC CAUUUUUAAA
2081 CCCUUUCUCA UCAUGUGGUU UUCUGAAGCA GUGAAUAUAC UUGCUACCUG CAGGCUGGAA CAACGCCAUC UCAACAGGGU
2161 CAGUAGCUGG UCCUUCAAUG UCGAGCCAAA GGGUAUUGGU GGGGUCGAGU UUCCCCACUG CCUCUCUGAU GACAGCUUCU
2241 UGUAUCUCUG UCAAGUUAGC CAAUCUCAAA UUCUGACCGU UCUUUUCCGG UUGUCUAGGU CCAGCAACUG GUUUCCUUGU
2321 CAGAUCAAUA CUUGUGUUGU CCCAUGACCU GCCUAUGAUU UGUGAUCUGG AACCAAUAUA AGGCCAACCA UCGCCAGAAA
2401 GGCAAAGUUU GUACAGAAGG UUUUCAUAAG GGUUUCUAUU GCCUGGUUUC UCAUCAAUAA ACAUGCCUUC UCUUCGUUUA
2481 ACCUGAAUGG UUGAUUUUAU GAGGGAAGAA AAGUUAUCUG GGGUGACUCU GAUUGUCUCC AACAUAUUUC CAUCAUCAAG
2561 AAUGGAUGCA CCAGCCUUUA CUGCAGCUGA AAGACUAAAG UUGUAGCCAG AAAUGUUGAU GGAGCUUUCA UCCUUAGUCA
2641 CAAUCUGGAG GCAGUCAUGU UCCUGAGUCA AUCUGUCAAG GUCACUCAAG UUUGGAUACU UCACAGUGUA UAGAAGCCCA
2721 AGAGAGGUUA AAGCCUGUAU GACACUGUUC AUUGUCUCAC CUCCUUGAAC AGUCAUGCAU GCAAUUGUCA AUGCAGGAAC
2801 AGAACCAAAC UGAUUGUUAA CUUUUGAAGG AUCUUUAACA UCCCAUACCC UCACCACACC AUUUCCCCCA GUUCCUUGCU
2881 GUUGAAAUCC CAGUGUUCUC AAUAUCUCUG AUCUCUUGGC CAGUUGUGAC UGAGACAAGU UACCCAUGUA AACCCCUUGA
2961 GAGCCUGUCU CUGCUCUUCU AAACUUGGUU UUUAAAUUCC CAAGGCCAGA CGCCAACUCC AUCCGCUCAA CCCUCCCCAG
3041 AUCUCCCGCC UUGAAAACCG UGUUUCGUUG AACACUCCUC AUGGACAUGA GUCUGUCAAC CUCUUUAUUC AGGUCCCUCA
3121 ACUUAUUGAG GUCUUCUUCC CCCCUUUUAG UCUUUCUGAG UGCCCGCUGC ACCUGUGCCA CUUGGUUGAA GUCAAUGCUG
3201 UCAGCAAUUA GCUUGGCAUC CUUCAGAACA UCCGACUUGA CAGUCUGAGU AAAUUGACUC AAACCUCUCC UUAAGGACUG
3281 AGUCCAUCUA AAGCUUGGAA CCUCUUUGGA GUGUGCCAUG CCAGAAGAUC UGGUGGUUUU GAUCUGAGAA AAAAUUGCUC
3361 AGUGAAAGUG UUAGACACUA UGCCUAGGAU CCACUGUGCG
```

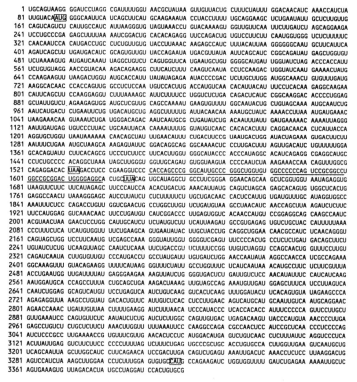

FIGURE 2. Nucleotide sequence of Junin virus S RNA. The 3400-nucleotide-long sequence of the genome or viral S RNA of Junin virus is presented and the orientation of the open reading frames corresponding to the GPC and N genes is indicated by arrows on the sides. The translational initiation and termination codons of GPC and the corresponding anticodons of N (encoded in the viral-complementary sequence) are boxed. The inverted repeats at the intergenic region are underlined.

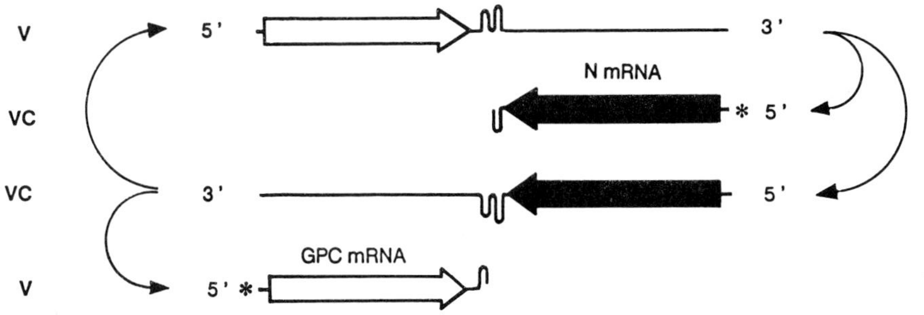

FIGURE 3. Schematic of the gene arrangement in Junin virus S RNA. The "ambisense" S RNA of Junin virus is shown divided into a negative (−) and positive (+) region encoding the N gene in the (−) or viral complementary (VC) and the GPC gene in the (+) or viral (genomic) polarity RNA (V). The open reading frames corresponding to the N and GPC genes are shown as filled and open rectangles, respectively, with arrowheads indicating the direction of translation. The subgenomic mRNAs (correlating well with size and polarity detected in cytoplasmic RNA of infected cells) are also shown, the asterisks indicating nontemplated nucleotide sequences at their 5′ ends (Rivera-Pomar, 1991). The thin curved arrows indicate explicitly the template–transcript relationships among the different RNA species.

(vc) RNA sequence, extends from nucleotides 82–84 (AUG) to nucleotides 1774–1776 (UAA) and codes for the nucleocapsid protein, N (Fig. 4, Ghiringhelli *et al.*, 1989). The other open reading frame, encoding the GPC (glycoprotein precursor) in the viral genome-sense (v) RNA sequence starts at positions 88–90 (AUG; CAU vc sequence residues 3310–3312) and stops at positions 1531–1533 (UAA, UUA vc sequence residues 1867–1869, Fig. 7).

B. The Nucleocapsid Protein Gene (N)

The first AUG in the vc S RNA at positions 82–84 is probably the true initiation codon for the N protein. Examination of the surrounding nucleotide sequences indicates that this AUG is placed in a favorable context (CTGGCAUGG) according to the consensus nucleotides identified by Kozak (1978, 1984) as the most frequent flanking sequence for eukaryotic translation initiation codons (i.e., $CC^{G}_{A}CC$ AUG G). The 1692-nucleotides-long open reading frame specifies a polypeptide of 564 amino acids with a molecular weight of 63,033 daltons, in good agreement with the estimates based on electrophoretic mobility (Martinez Segovia and De Mitri, 1977; Grau *et al.*, 1981). Assuming, for simplicity, the charge of amino acids at neutral pH for K or R as +1, for H as +0.5, and for D and E as −1, the N protein has a net charge of +14.5. Although most of the positively charged amino acid residues are scattered throughout the protein sequence, some clustering of Ks and/or Rs may be noted. A strongly positive KKKR center is located at amino acid

1	MAHSKEVPSF	RWTQSLRRGL	SQFTQTVKSD	VLKDAKLIAD	SIDFNQVAQV
51	QRALRKTKRG	EEDLNKLRDL	NKEVDRLMSM	RSVQRNTVFK	AGDLGRVERM
101	ELASGLGNLK	TKFRRAETGS	QGVYMGNLSQ	SQLAKRSEIL	RTLGFQQQGT
151	GGNGVVRVWD	VKDPSKLNNQ	FGSVPALTIA	CMTVQGGETM	NSVIQALTSL
201	GLLYTVKYPN	LSDLDRLTQE	HDCLQIVTKD	ESSINISGYN	FSLSAAVKAG
251	ASILDDGNML	ETIRVTPDNF	SSLIKSTIQV	KRREGMFIDE	KPGNRNPYEN
301	LLYKLCLSGD	GWPYIGSRSQ	IIGRSWDNTS	IDLTRKPVAG	PRQPEKNGQN
351	LRLANLTEIQ	EAVIREAVGK	LDPTNTLWLD	IEGPATDPVE	MALFQPAGSK
401	YIHCFRKPHD	EKGFKNGSRH	SHGILMKDIE	DAMPGVLSYV	IGLLPPDMVV
451	TTQGSDDIRK	LFDLHGRRDL	KLVDVRLTSE	QARQFDQQVW	EKFGHLCKHH
501	NGVVVSKKKR	DKDAPFKLAS	SEPHCALLDC	IMFQSVLDGK	LYEEELTPLL
551	PPSLLFLPKA	AYAL*			

FIGURE 4. Amino acid sequence of Junin virus N protein. Translation product of the N ORF from Fig. 2. The basic amino acid residues K and R are boxed and clusters of two, three, and four such residues are shaded.

residues 507–510, a triple basic KRR sequence at amino acid positions 281–283, and nine groups of two basic K and/or R residues.

The relatively high frequency of positively charged amino acid residues in the N protein sequence conforms well to the idea of interaction with the negatively charged RNA in the viral nucleocapsids. However, no homology to the eukaryotic RNP consensus sequence was found (Dreyfuss *et al.*, 1988). It has also been suggested that N could play a role in the regulation of the transcription/replication process during arenavirus infection (Ghiringhelli *et al.*, 1991; Iapalucci *et al.*, 1991). On the other hand, there is some evidence that N-derived polypeptides can enter the cell nucleus (Young *et al.*, 1987). The essence of a nuclear targeting sequence appears to be a short run of basic amino acids such as those found in Junin virus N residues 507–510 or 55–59 (Hunt, 1989). However, this sequence, while probably necessary, is not sufficient to ensure nuclear entry. At the present time, the nuclear targeting of N requires further investigation.

The alignment of the amino acid sequences of the N gene products of Junin and Tacaribe viruses reveals a sequence identity as high as 76%. If the same comparison is made with other New World and Old World arenaviruses the sequence homologies fall within the range of 53–47%

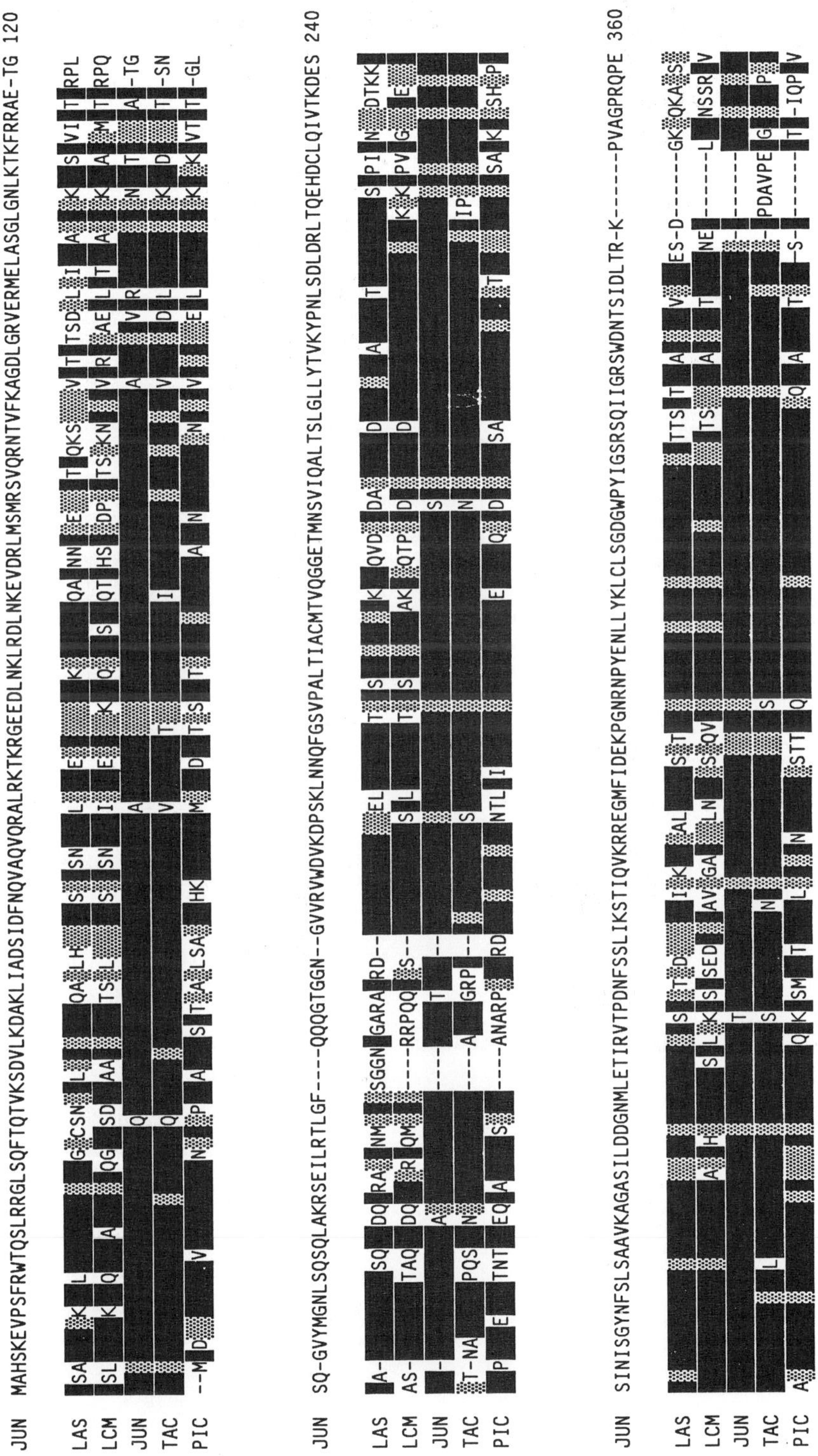

JUN MAHSKEVPSFRWTQSLRRGLSQFTQTVKSDVLKDAKLIADSIDFNQVAQVQRALRKTKRGEEDLNKLRDLNKEVDRLMSMRSVQRNTVFKAGDLGRVERMELASGLGNLKTKFRRAE-TG 120
LAS
LCM
JUN
TAC
PIC
JUN SQ-GVYMGNLSQSQLAKRSEILRTLGF----QQQGTGGN--GVVRVWDVKDPSKLNNQFGSVPALTIACMTVQGGETMNSVIQALTSLGLLYTVKYPNLSDLDRLTQEHDCLQIVTKDES 240
LAS
LCM
JUN
TAC
PIC
JUN SINISGYNFSLSAAVKAGASILDDGNMLETIRVTPDNFSSLIKSTIQVKRREGMFIDEKPGNRNPYENLLYKLCLSGDGWPYIGSRSQIIGRSWDNTSIDLTR-K------PVAGPRQPE 360
LAS
LCM
JUN
TAC
PIC

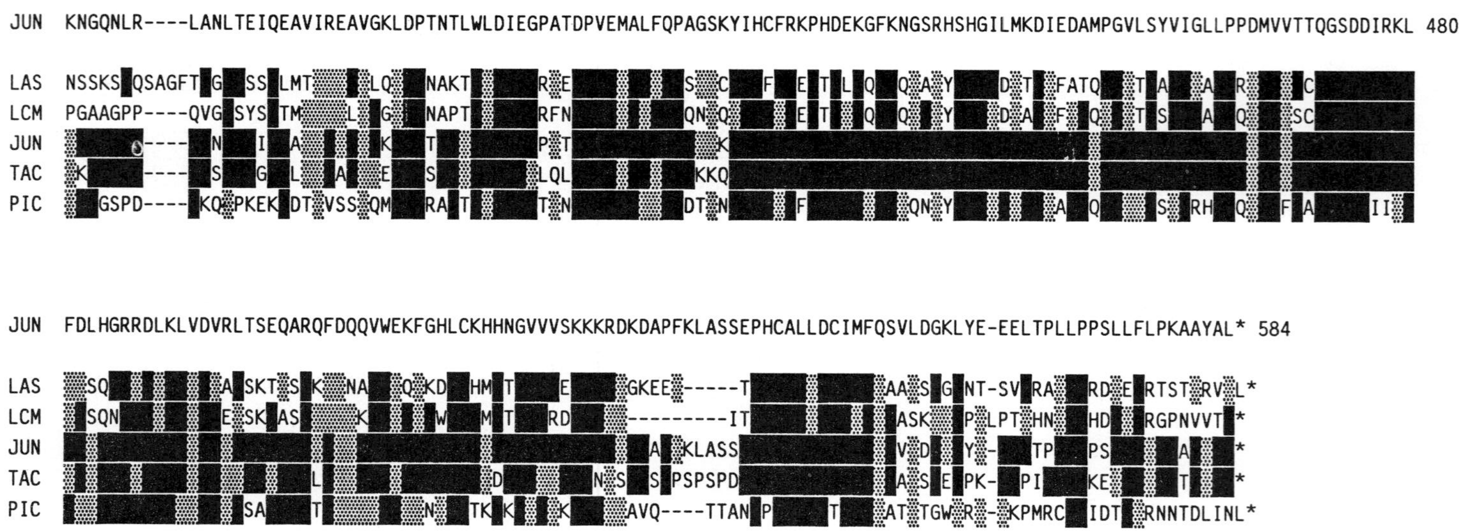

FIGURE 5. Comparison of amino acid sequence of the N proteins. Sequences of the Old World arenaviruses Lassa (LAS; Clegg and Oram, 1985) and LCM-WE (LCM; Romanowski and Bishop, 1985) and the New World arenaviruses Tacaribe (TAC; Franze-Fernandez *et al.*, 1987) and Pichinde (PIC; Auperin *et al.*, 1984) with that of Junin N polypeptide (JUN; Ghiringhelli *et al.*, 1989). The amino acid residues identical to those of JUN N protein are indicated by black boxes; those that represent conservative changes according to Schwartz and Dayhoff (1979) appear as shaded areas. Gaps were incorporated in the sequences to obtain a best-fit alignment. The complete amino acid sequence of JUN N protein, which was used as reference, is printed above the comparison of the five N proteins (Ghiringhelli *et al.*, 1991).

TABLE II. Amino Acid Sequence Homology of Arenavirus Structural Polypeptides

		N				G1 (GPC)				G2 (GPC)				
	TAC	PIC	LCM	LAS	TAC	PIC	LCM	LAS	TAC	PIC	LCM	LAS	
JUN	76	54	49	47	50	20	21	21	82	57	53	57	A
TAC		54	48	48		24	16	17		51	51	52	A
PIC			49	49			20	26			52	54	A
LCM				60				27				69	A
JUN	87	70	66	65	69	43	42	44	92	72	70	68	B
TAC		69	64	64		41	38	36		70	69	67	B
PIC			65	66			41	47			68	69	B
LCM				73				44				83	B

(A) The figures represent the percentages of amino acid identities obtained by the comparison between the different arenavirus proteins. Gaps were introduced in the amino acid sequences of pairs of homologous proteins to produce a best-fit alignment, but conservative changes were not considered for the calculation of homology percentages (the names of the different arenaviruses are abbreviated as in Fig. 5). N: nucleocapsid protein. The amino acid sequence of the glycoprotein precursor (GPC) was formally divided into the amino-terminal half G1 and the carboxy-terminal half G2. The cleavage site demonstrated for LCM virus GPC and the homologous sequences in the other arenaviruses were used to formally define the G1 and G2 regions (see Fig. 8).
(B) Sequence homology (%) including conservative changes according to Schwartz and Dayhoff (1979).

(Fig. 5 and Table II). These homology figures do not take into account substitutions representing amino acid residues with similar biochemical characteristics (e.g., R-K, D-E, V-I, etc.). In addition, the conservation of many cysteine (C), glycine (G), and proline (P) residues suggests that the three-dimensional structures of the arenavirus N proteins are more conserved than indicated by simple comparison of their primary structures. The amino acid sequence similarities between the N proteins are in agreement with the monoclonal antibodies cross-reactivities reported by Howard *et al.* (1985), Sanchez et al. (1989), and Buchmeier *et al.* (1981). In particular, the outstanding degree of cross-reactivity between Junin and Tacaribe viruses is highlighted by the percentage of sequence homology in Table II.

C. Expression of the Nucleocapsid Protein Gene in Transfected Cells

When the cloned genes for the nucleocapsid proteins of Junin and LCM viruses were inserted into an SV40-derived expression vector and transfected into BHK-21 and CV-1 cells, the transient expression yielded a polypeptide biochemically and immunologically identical to the N protein synthesized during viral infection (Rivera-Pomar *et al.*, 1991). The N polypeptide was detected by indirect immunofluorescence in 5–20% of the cells, so that the rest of the cells could be used as internal negative controls. The level of expression of Junin and LCM virus nucleocapsid protein in these transfected cells was found to be roughly similar to that observed in infected BHK-21 and CV-1 cell lines. Therefore, one would expect the N protein synthesized in the transfected cell to behave in a manner similar to that observed in virus-infected cells (Weissenbacher and Damonte, 1983).

A cytoplasmic localization was determined for the N polypeptide expressed in cells transfected with Junin as well as LCM virus N genes. There are some reports of the nucleocapsid protein found in the nucleus of Pichinde virus–infected cells (Young *et al.*, 1987) and in the membrane of L cells infected with LCM virus (Zeller *et al.*, 1986, 1988). However, Rivera-Pomar *et al.*, (1990, 1991) could not demonstrate these localizations either for the transfected cells or for the infected ones. The patterns of immunofluorescence are variable and match those found in cells infected with these arenaviruses, i.e., varying from diffuse cytoplasmic staining to granules either distributed throughout the cytoplasm or concentrated in the perinuclear region. In particular, a speckled pattern of immunofluorescence has been observed coincident with previous reports on arenavirus-infected cells (Weissenbacher and Damonte, 1983).

These studies suggest that N protein is associated with basophilic granules. Such basophilic particles have previously been described in infected cells as part of the cytopathic effect (Bruno-Lobo *et al.*, 1968; Weissenbacher and Damonte, 1983). The above mentioned association is possibly related to the biochemical properties of the polypeptides, exhibiting clusters of basic amino acids (Romanowski and Bishop, 1985; Ghiringhelli *et al.*, 1989 and 1991). These polypeptides might interact with cytoplasmic RNA to form ribonucleoprotein complexes.

The association of N protein with cytoplasmic granules that undergo rearrangements at different stages of the cell cycle suggests a behavior similar to that of a newly described group of ribonucleoprotein particles. These large virus-like structures named "vaults" have been found in the cytoplasm of higher and lower eukaryotes and reported to associate with cytoskeletal elements (Kedersha and Rome, 1990). The regular appearance of extensive granulation in the cytoplasm of N-expressing cells suggests a possible link between the synthesis of the N protein and the development of the cytopathic effect (CPE) during arenavirus infection.

The use of SV40-derived expression vectors bearing the *E. coli* neomycin resistance gene permitted the isolation of BHK-21 and CV-1 cell lines stably expressing N. CV1 and Vero cells that expressed LCM or Junin virus N protein showed a marked CPE, similar to that observed in infected cells. The CPE led to cell lysis after few days, thus preventing the isolation of stably transformed clones. On the contrary, it was possible to isolate G418-resistant clones from transfected BHK-21 cultures.

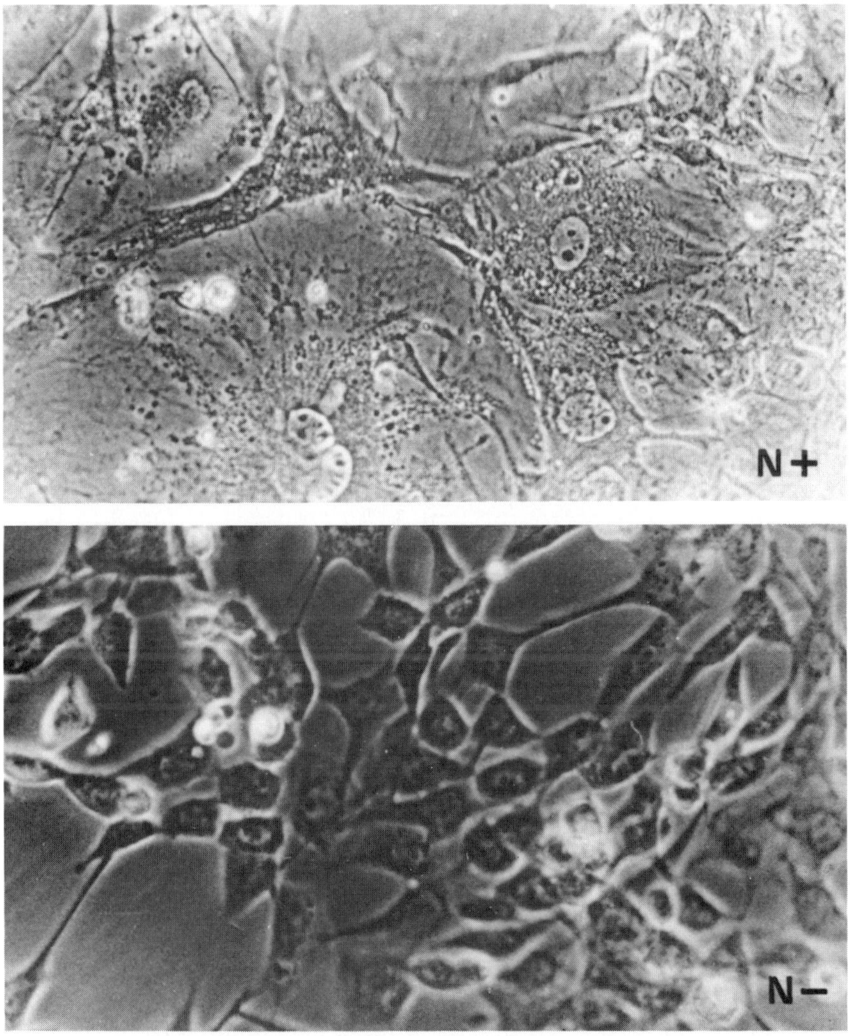

FIGURE 6. Transfected BHK-21 cells stably expressing Junin virus N protein. Cell monolayers were transfected with either the SV40-derived vector pKG4 or the recombinant expression plasmid pKG4-NJUN containing the N ORF of Junin virus in the appropriate orientation. Stable transfectant clones were selected using the neomycin analog G418 and propagated in selective medium. The BHK-21 cells transfected with pKG4-NJUN yielded clones where all or none of the cells expressed N, depending on the mode of integration of the plasmids. Shown are G418-resistant BHK cells that do not express N (N−) and those that do (N+). The phase-contrast micrographs (200× magnification) were taken without prior fixation of the cell monolayers to prevent artifactual distortion of the cell morphology. Examination of fixed cells using indirect immunofluorescence detected N in 100% of the cells of the G418-resistant clones that exhibited altered cell morphology, such as shown in (N+), whereas the other G418-resistant clones that did not express N (such as the one in N−), derived from pKG4 or pKG4.N JUN, had no signs of CPE (Rivera-Pomar *et al.*, 1991; Rivera-Pomar, 1991).

BHK-21 clones expressing the N gene showed cytoplasmic granulation, vacuolization, polykaryosis, and altered cell shape (Fig. 6; Rivera-Pomar *et al.*, 1990, 1991). Further studies will improve our understanding of the mechanisms leading to these dramatic changes in cell morphology.

D. The Glycoprotein Precursor Gene (GPC)

The first AUG codon of the viral-sense S RNA of Junin virus is found at 88–90 nucleotides from the 5′ end. Its flanking nucleotides (i.e., GUACAAUGG) match the consensus purine residue at position −3 and the G at +4, observed in all arenavirus genes, and also contain one of the C residues identified in the upstream region of Kozak's most frequent flanking sequence to eukaryotic translation initiation codons (Kozak, 1984). A second AUG is found within the next seven codons in frame with the first one, but lacking the G at position +4 and the A or G at position −3. Therefore, the first AUG triplet is deduced to be the initiating codon for a 481-amino-acids long polypeptide.

The calculated molecular weight of the GPC gene product (i.e., 55,126 daltons) is smaller than the molecular weight determined by gel electrophoresis for the intracellular glycoprotein precursor (De Mitri and Martinez Segovia, 1985; Rustici, 1984). The cotranslational glycosylation renders an obvious explanation to this experimental observation. Eight potential N-linked glycosylation sites are found in the GPC sequence (Fig. 7). Whether all are actually used for attachment of carbohydrate side chains is not known. However, one should note that four of the potential glycosylation sites are conserved among all the arenavirus GPC genes analyzed to date (Fig. 8), suggesting their functional relevance. When all the available amino acid sequences of arenavirus GPC genes are compared, it is found that the region proximal to the carboxyl-terminus contains a high proportion of homologous stretches, while the amino-terminal sequences diverge considerably (see Fig. 8, Table II). However, if Junin virus GPC gene product is compared to the homologous protein of Tacaribe virus, it is noticed that the first 40 amino acids from the amino-terminus are conserved with just five amino acid substitutions, only two of which could be considered nonconservative changes (i.e., Ala for Cys, Thr for Ile).

Recently, a live Junin virus vaccine, named Candid 1, was developed to control AHF. The GPC gene of this highly attenuated virus was cloned and sequenced, and its ORF was compared to the homologous regions in the S RNA of the wild-type MC2 strain. The nucleotide changes concentrate in the NH_2-proximal (20 nucleotide changes) and the COOH-proximal (19 nucleotide substitutions) regions. However, major changes in the amino acid sequence occur only in the amino-terminal region of GPC as a result of several insertions and deletions in the nucleotide sequence affecting codons 43–80. After proteolytic cleav-

```
1    MGQFISFMQE IPTFLQEALN IALVAVSLIA IIKGVVNLYK SGCSILDLAG
51   RSCPRAFKIG LHTEVPDCVL LQWWVSFSNN PHDLPLLCTL NKSHLYIKGG
101  NASFKISFDD IAVLLPEYDV IIQHPADMSW CSKSDDQIRL SQWFMNAVGH
151  DWYLDPPFLC RNRTKTEGFI FQVNTSKTGI NENYAKKFKT GMHHLYREYP
201  DSCLDGKLCL MKAQPTSWPL QCPLDHVNTL HFLTRGKNIQ LPRRSLKAFF
251  SWSLTDSSGK DTPGGYCLEE WMLVAAKMKC FGNTAVAKCN LNHDSEFCDM
301  LRLFDYNKNA IKTLNDETKK QVNLMGQTIN ALISDNLLMK NKIRELMSVP
351  YCNYTKFWYV NHTLSGQHSL PRCWLIKNNS YLNISDFRND WILESDFLIS
401  EMLSKEYSDR QGKTPLTLVD ICFWSTVFFT ASLFLHLVGI PTHRHIRGEA
451  CPLPHRLNSL GGCRCGKYPN LKKPTVWRRG H*
```

FIGURE 7. Amino acid sequence of Junin virus GPC gene product. The putative N-glycosylation sites are boxed; those that are also conserved among the rest of the arenavirus GPC proteins are indicated by the shaded boxes. The double basic amino acid sequence identified by Buchmeier *et al.* (1987) as the putative proteolytic cleavage site is underlined.

age of GPC, these alterations appear in the G1 polypeptide, which is thought to be located on the surface of the virion in association with the more internal G2 protein. Although these findings are strongly suggestive of the genetic changes that lead to an attenuated phenotype, they must be complemented with studies on the direct predecessors of Candid 1, which exhibit gradually more virulent phenotypes, in order to locate precisely the mutations responsible for attenuation.

E. Proteolytic Cleavage Site for the Processing of GPC

Most of the arenaviruses have been shown to exhibit a similar glycoprotein-processing pattern, differing only in the apparent molecular weights of the GPC precursor and the processed viral glycoproteins G1 (or GP1) and G2 (or GP2) (Buchmeier and Oldstone, 1978; Compans and Bishop, 1985; Saleh *et al.*, 1979; Grau *et al.*, 1981; Harnish *et al.*, 1981; Lukashevich *et al.*, 1985; De Mitri and Martinez Segovia, 1985; Clegg and Lloyd, 1983). However, only one structural glycoprotein (G) has been reported for Tacaribe and Tamiami virions (Gard *et al.*, 1977; Boersma *et al.*, 1982). This apparent discrepancy is due to the selective loss of GP1 during virus purification (see Chapter 2, this volume).

In LCM virus a mannose-rich precursor, GPC (75 kDa) is posttrans-

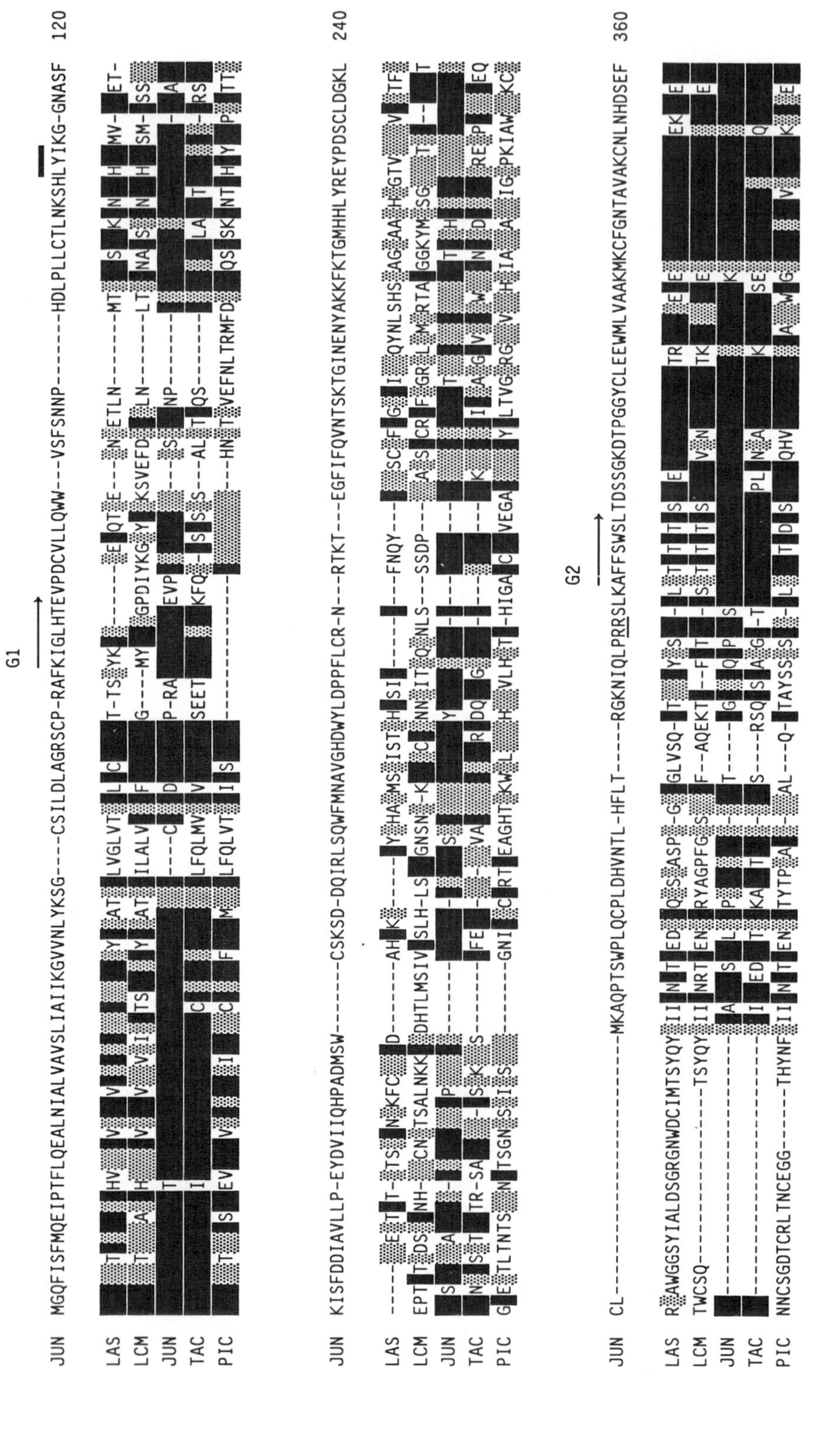
G1
G2
JUN MGQFISFMQEIPTFLQEALNIALVAVSLIAIIKGVVNLYKSG----CSILDLAGRSCP-RAFKIGLHTEVPDCVLLQWW---VSFSNNP------HDLPLLCTLNKSHLYIKG-GNASF 120
LAS
LCM
JUN
TAC
PIC
JUN KISFDDIAVLLP-EYDVIIQHPADMSW------CSKSD-DQIRLSQWFMNAVGHDWYLDPPFLCR-N---RTKT---EGFIFQVNTSKTGINENYAKKFKTGMHHLYREYPDSCLDGKL 240
LAS
LCM
JUN
TAC
PIC
JUN CL-----------------MKAQPTSWPLQCPLDHVNTL-HFLT-----RGKNIQLPRRSLKAFFSWSLTDSSGKDTPGGYCLEEWMLVAAKMKCFGNTAVAKCNLNHDSEF 360
LAS
LCM
JUN
TAC
PIC

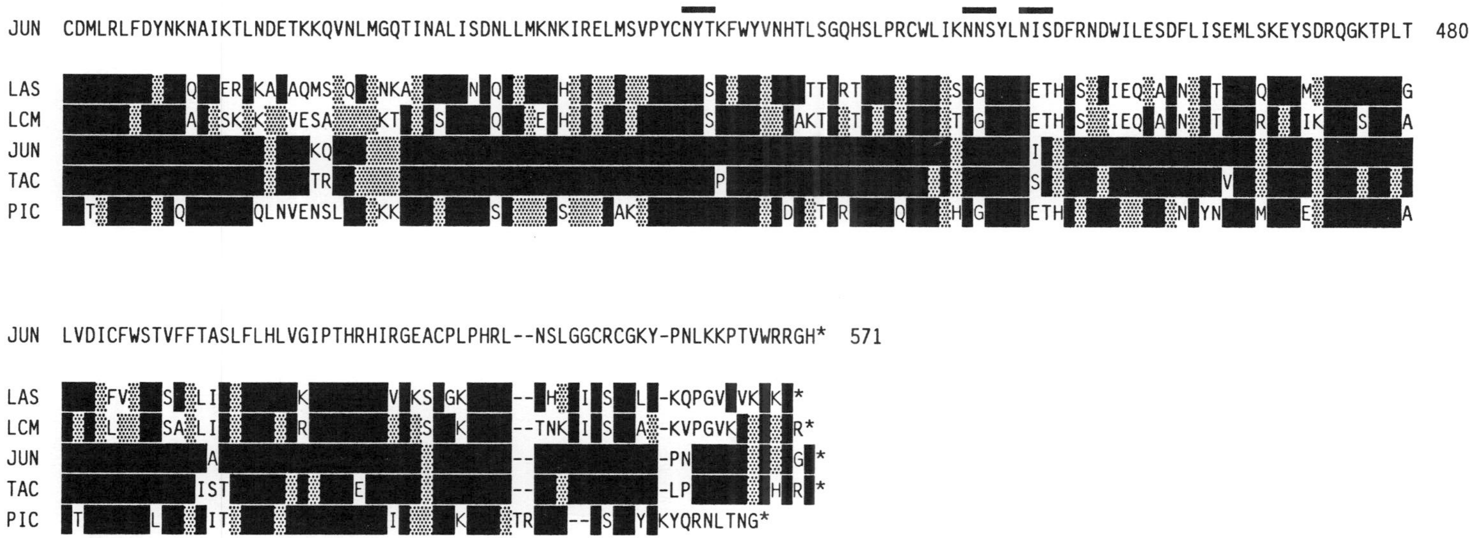

FIGURE 8. Comparison of arenavirus GPC proteins. The predicted amino acid sequences of the GPC proteins of LAS (Auperin *et al.*, 1986), LCM (Romanowski *et al.*, 1985), TAC (Franze-Fernandez *et al.*, 1987), and PIC (Auperin *et al.*, 1984) are compared to that of JUN GPC. The identities and homologies of amino acid residues are indicated as in Fig. 5, and the reference JUN GPC sequence is printed above the aligned sequences. In addition, the conserved N-glycosylation sites (NX^{S}_{T}) are indicated by thick lines above the JUN GPC sequence. The approximate position of the proteolytic cleavage site is underlined (the amino acid sequence to the left of this site is referred to as G1 and that to the right, as G2). The arrows start at the amino-termini of G1 and G2 of LCM sequenced by Burns *et al.* (1990).

lationally modified by proteolytic cleavage and oligosaccharide processing to yield the glycoproteins GP1 (44 kDa) and GP2 (35 kDa). Buchmeier *et al.* (1987) have mapped the cleavage site to be a paired basic amino acid group Arg–Arg corresponding to amino acids 262–263. This Arg–Arg pair is followed by a predominantly hydrophobic stretch of some 23 amino acids. These two features are highly conserved among the LCM (strains WE and ARM), Lassa, Pichinde, and Junin virus GPC proteins (Fig. 8). In Pichinde virus GPC the Arg–Arg pair is replaced by Arg-Lys, of similar characteristics.

The hypothesis, confirmed for LCM–ARM GPC, is that a protease of cellular origin cleaves the GPC at or near this Arg–Lys or Arg–Arg sequence generating two structural glycoproteins designated GP-1 (for LCM) or G1 (for Pichinde and Lassa) and GP-2 (for LCM) or G2 (for Pichinde and Lassa). Immunoprecipitation with antibodies raised against synthetic peptides (Buchmeier *et al.*, 1987) and limited microsequencing of G1 and G2 tryptic peptides (Auperin *et al.*, 1984) demonstrated that GP-1 or G1 comprises the amino-terminal half of GPC. It is also known that this protein is the antigen that elicits the synthesis of antiviral neutralizing antibodies. On the other hand, GP2 or G2 corresponds to the most conserved carboxy-terminal half of GPC. The amino terminus of GP2/G2, which would be liberated by cleavage at the Arg–Arg or Arg–Lys site, contains the above mentioned consensus hydrophobic sequence.

In this scenario, Junin virus major 35-kDa glycoprotein, which elicits a neutralizing antibody response, should be homologous to LCM, Lassa, and Pichinde virus G1 protein, whereas the 44-kDa protein could be the counterpart of G2. At this point, the functional or biological properties as well as the unambiguous sequence data should prevail over the electrophoretic mobility to designate the Junin virus glycoproteins as G1 and G2.

A particular consideration should be devoted to the comparison of Junin and Tacaribe virus glycoprotein genes. A best fit alignment of the GPC genes shows that 66% of the amino acids are conserved in identity and position. When the G1- and G2-homologous regions are analyzed separately, the percentages of amino acid sequence identity are 50% and 82%, respectively. These figures are by far higher than any of the others presented in Table II.

The putative proteolytic cleavage site in Tacaribe virus GPC does not contain a doublet of basic, amino acid residues. This pair of residues is replaced by one basic amino acid (Arg) followed by a Thr and the consensus 23-amino-acids-long hydrophobic sequence. An alignment of this hydrophobic region with that of Junin virus places the above-mentioned Arg_{258}-Thr_{259} pair in register with an Arg_{244}-Ser_{245} of Junin virus GPC. The fact that Tacaribe virus GPC is proteolytically processed, together with the existence of other dibasic amino acid sequences in nonconserved regions of arenavirus GPC and N proteins, which are not known to be cleaved intracellularly, suggests that the double basic amino acid cluster is not an essential feature of the proteolytic target. Most likely, the protease involved in the processing of GPC requires only one basic amino acid in a sequence context that makes the particular peptide bond accessible. The sequence requirements might include all or part of the predominantly hydrophobic 23-amino-acid-long sequence conserved in all the arenaviruses, including Tacaribe (Figure 9).

F. Noncoding Sequences in S RNA 5′ and 3′ Ends

The N and GPC open reading frames do not overlap and account for 3135 of the total 3400 nucleotides of Junin virus S RNA. The noncoding regions upstream from the N and GPC translation initiation codons comprise 81 and 87 nucleotides, respectively. The lengths of the same regions in different arenavirus S RNAs vary, but are generally shorter than the ones described here for Junin virus.

The first 19 nucleotides at the 3′ end of the viral S RNA are known

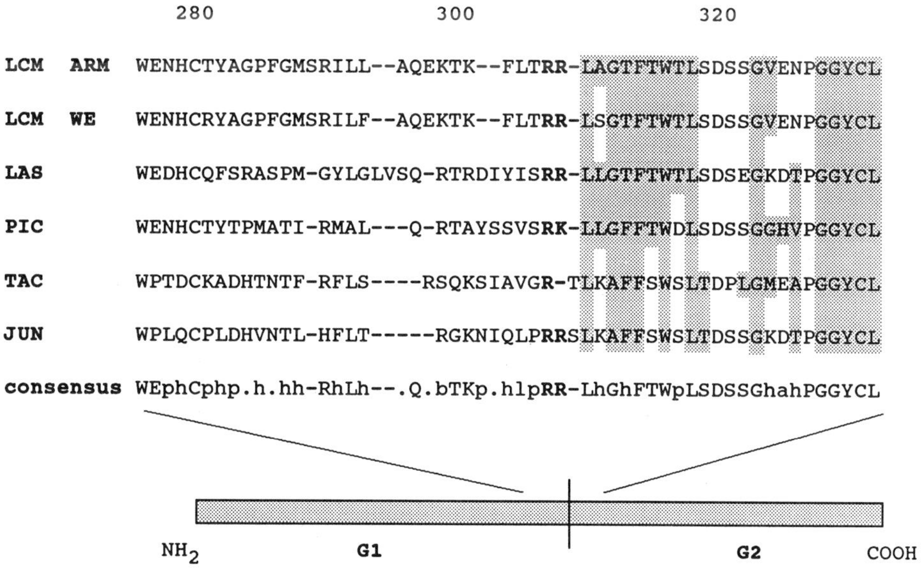

FIGURE 9. Arenavirus GPC sequence region surrounding the cleavage site (RR). The polypeptides G1 (amino-terminal half) and G2 (carboxy-terminal half) are predicted to flank the cleavage site (Buchmeier *et al.*, 1987). The figures above the sequences indicate the amino acid residue numbers starting at the first met codon of the GPC ORF and including the gaps introduced to obtain the best-fit alignment. A consensus amino acid sequence was derived from the comparison of the aligned arenavirus GPC sequences, grouping the amino acid residues, according to Taylor (1987): hydrophobic (h), hydrophobic–aliphatic (l), charged–acidic (a), charged–basic (b), polar (p); no observation is indicated by dots (.) and gaps are shown using dashes.

to be conserved in the Arenaviridae family. This information was obtained by direct RNA sequencing of Pichinde (prototype and Munchique strain), LCM (Armstrong strain), and Tacaribe virus genomic RNA (Auperin *et al.*, 1982), and by Maxam and Gilbert sequencing of cDNA clones derived from *in vitro* polyadenylated LCM (WE strain) virus S RNA (Romanowski and Bishop, 1985). The characteristic 19-nucleotide-long 3′ terminal sequence was also found to be conserved in Junin virus by sequencing the primer extension products complementary to the 5′ end of the antigenomic copy of the full-length S RNA (Rivera-Pomar, 1991).

In general, the sequence homology for the 3′ termini of arenavirus S RNA does not extend beyond the 19 nucleotides consensus sequence. In this context, the homology between Junin and Tacaribe stands out, extending through nucleotide 53 with only six mismatches. Lassa and LCM viruses are the only other pair of arenaviruses that exhibit a sequence homology beyond the first 19 nucleotides, reaching nucleotide 34 with three mismatches (Clegg and Oram, 1985; Romanowski and Bishop, 1985).

In contrast with the conserved 3′ terminus, the 5′-terminal-most sequence of Junin virus S RNA is slightly different from the sequence conserved in Pichinde, LCM, Lassa, and Tacaribe virus (Fig. 10) (Auperin *et al.*, 1984, 1986; Romanowski *et al.*, 1985; Raju *et al.*, 1990). Nevertheless, both ends can be aligned to produced a partially double-stranded secondary structure with a stabilization free energy of ca. −58.0 kcal/mole calculated according to Tinoco *et al.* (1973). The fact that other arenavirus S RNA ends can be also aligned to form a homologous secondary structure is most likely related to the observation of circular forms in preparations of RNAs and nucleocapsids of Pichinde and Tacaribe virions (Vezza *et al.*, 1978; Palmer *et al.*, 1977). Recently, the actual existence of similar panhandle structures has been demonstrated in cross-linking experiments of influenza and La Crosse virus RNAs and nucleocapsids (Hsu *et al.*, 1987; Raju and Kolakofsky, 1989).

On the other hand, the unexpected difference found in a region of the S RNA that was presumed to be more conserved might indicate that the viral RNA polymerase does not recognize an invariant nucleotide sequence at the 3′ end when viral or viral-complementary RNA is used as template. In turn, the sequence specificity could be less stringent, or the conserved panhandle secondary structure could be recognized as the RNA polymerase docking signal.

G. Intergenic Region

As mentioned in the previous sections, the coding sequences and part of the noncoding upstream regions exhibit varying degrees of nucleotide sequence homology among the different arenaviruses. How-

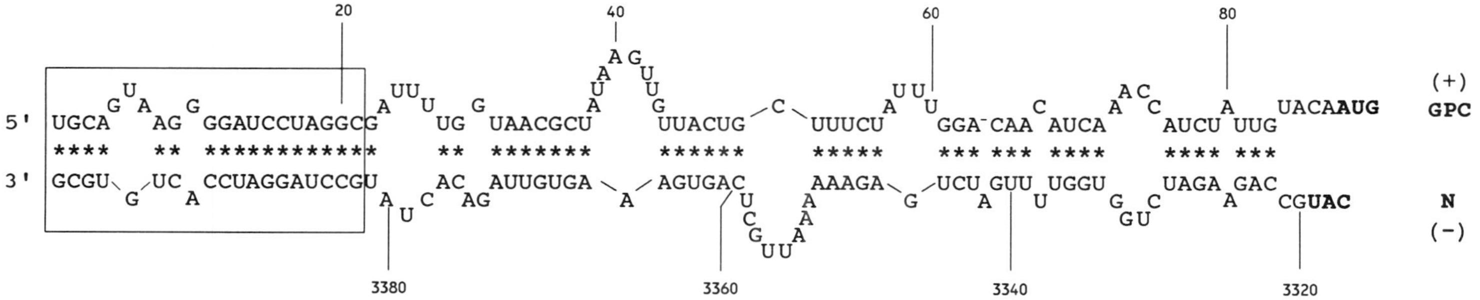

FIGURE 10. Predicted panhandle structure in Junin virus S RNA. The noncoding sequences at the 5′ and 3′ ends were aligned. The first 19 nucleotides at the 3′ end base-pair with the complementary sequence at the 5′ end forming a secondary structure that is conserved among the arenaviruses (boxed, $\Delta G° = -37.8$ kcal/mole). Complementary nucleotides, indicated by asterisks, hold the 5′ and 3′ ends of this RNA molecule with a stabilization free energy of -58.0 kcal/mole (Tinoco *et al.*, 1973). G–U pairs are also indicated.

ever, no such sequence conservation can be inferred from the comparison of the noncoding region downstream from the translation termination signals of the N and GPC genes. Conversely, this intergenic region can be arranged in a relatively stable hairpin-loop structure. It is this type of secondary structure that has been conserved throughout the evolution of the arenaviruses, pointing to its functional significance (Auperin *et al.*, 1984, 1986; Romanowski and Bishop, 1985; Franze-Fernandez *et al.*, 1987).

Junin virus exhibits two consecutive very stable hairpin structures ($\Delta G° = -57.2$ and -39.0 kcal/mole) separated by a hinge of two nucleotides instead of a single hairpin loop reported for most of the other arenaviruses (Ghiringhelli *et al.*, 1987, 1991) (Fig. 11). However, more recently a corrected version of Tacaribe virus S RNA and that of the African arenavirus Mopeia were shown to contain also two potential hairpin-loop structures in the intergenic region (Iapalucci *et al.*, 1991; Wilson and Clegg, 1991).

Northern blot analyses indicate that Junin virus–infected cells contain genomic and antigenomic copies of S RNA as well as subgenomic mRNAs (Fig. 3). Nevertheless, when cells are infected in the presence of protein synthesis inhibitors, only the subgenomic N mRNA is transcribed from the S RNA template (Franze-Fernandez *et al.*, 1987; Rivera-Pomar, 1991). This observation and the mRNA size indicate that an antiterminator function is required for the viral RNA polymerase to synthesize a full-length copy of the S RNA, and the antiterminator function is probably supplied by the N protein binding to and destabilizing the intergenic hairpin structures. Although other viral or cellular proteins with high turnover rate could be alternative antiterminators, Junin

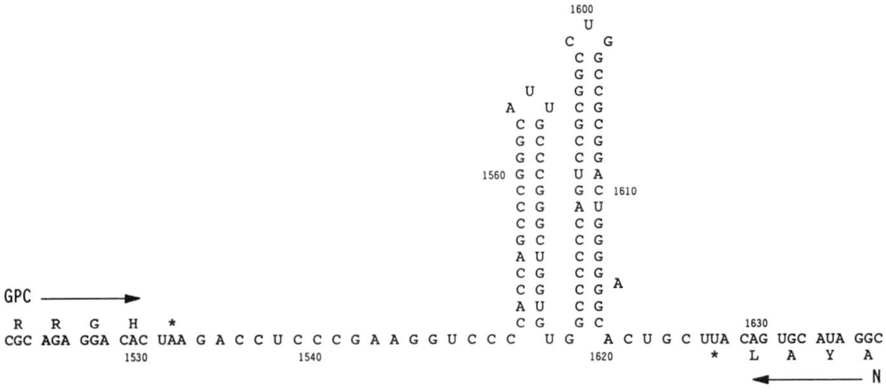

FIGURE 11. Intergenic region of Junin virus S RNA. A portion of the S RNA (nucleotides 1519–1639) comprising the ends of the GPC and N genes (sense and anti-sense strand, respectively) and the noncoding intergenic region is shown. The directions of translation are indicated by arrows, and the stop codon for the GPC protein and the stop anticodon for the N gene are marked with asterisks. Two potential hairpin-loop structures are stabilized by 13 and 17 base pairs ($\Delta G° = -39.0$ and -54.2 kcal/mole).

virus N protein contains a basic amino-acid-rich sequence (RALRKRGE) similar to those found in several RNA-binding proteins such as N antiterminators of phage lambda, P21 and P22, some ribosomal proteins, the HIV Tat, and others (Lazinski *et al.*, 1989).

The proposed mechanism of transcription termination consists in the formation of a hairpin loop at the 3′ end of N mRNA, when the intergenic region is copied, leading to the release of RNA polymerase. The binding of N to the RNA would partially melt the RNA structure, allowing the polymerase to continue the transcription process. In this perspective, RNase protection experiments indicated that the 3′ termini of the subgenomic mRNAs are complementary to the first hairpin-loop-forming sequence encountered by the RNA polymerase during transcription of the intergenic region. This information correlates with the proposed model, and the role of N is currently being evaluated using mammalian expression systems (Rivera-Pomar, 1991).

V. CONCLUDING REMARKS

Figure 12 shows a scheme of the organization of Junin virus S RNA based on the information presented in this chapter. Our knowledge of the molecular biology of Junin virus is still fragmentary, but cDNA cloning and sequencing of the genomic RNA have given some clarifying

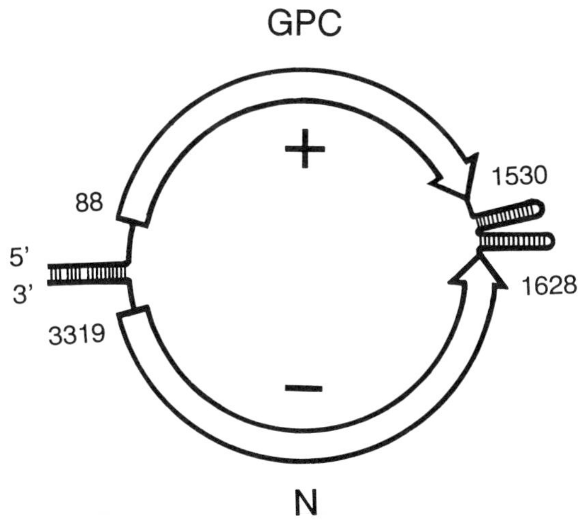

FIGURE 12. Molecular organization of Junin virus S RNA. The schematic summarizes the conclusions and predictions of the nucleotide sequence data. The nucleotide numbering starts at the 5′ end of the ambisense S RNA. The S RNA molecule is shown to form a circle closed by a panhandle structure (5′ and 3′ terminal complementary regions are not drawn to scale). The positive-sense GPC gene and the negative-sense N gene regions of this ambisense RNA are indicated (coding regions are shown as arrows). The double hairpin structure in the intergenic region is not drawn to scale.

insights, raised new questions, and yielded some solutions to diagnostic problems.

The insertion of the N and GPC open reading frames in prokaryotic and eukaryotic expression vectors provides safe-to-handle antigens for immunoassays to be used in confirmatory AHF diagnosis or epidemiological screening studies. On the other hand, making use of the sequence data, a simplified nucleic acid amplification assay has been developed for the early and rapid detection of Junin virus RNA in clinical samples, a critical issue considering that the only effective therapy that reduces AHF mortality rate from 30% to 1% consists of treatment with immune plasma in the first few days after the onset of symptoms that are not always clear (Romanowski *et al.*, 1991).

ACKNOWLEDGMENTS. The author acknowledges the enthusiasm and dedication with which P. D. Ghiringhelli undertook the experimental work in the molecular cloning and sequencing of Junin virus RNA. Other present and past members of the laboratory at the University of La Plata: R. V. Rivera-Pomar, M. E. Lozano, N. I. Baro, M. F. Rosas, and E. M. Manzella—also contributed, with varying inputs, to the cDNA cloning of the genome of Junin virus. The author would also like to kindly acknowledge the encouragement and continuous support of Dr. O. Grau. Finally, S. A. Moya deserves special thanks for her careful work at the word processor.

The project was financially supported by Grants PID 3916301/85, PID 3053700/88 from CONICET, 349–0275/86 from SECYT, 1348/85 from CIC BA (Argentina), and DAMD 17–89-Z-9024 from USAMRDC (USA). The author is a career investigator of CONICET and Adjunct Professor of Biological Chemistry at the University of La Plata.

REFERENCES

Anón, M. C., Grau, O., Martínez Segovia, Z. M., and Franze-Fernández, M. T., 1976, RNA composition of Junin virus, *J. Virol.* **18**:833.

Aribalzaga, R. A., 1955, Una nueva enfermedad epidémica a germen desconocido: Hipertermia nefrotóxica, leucopénica y enantemática, *Día Médico* **27**:1204.

Auperin, D. D., Compans, R. W., and Bishop, D. H. L., 1982, Nucleotide sequence conservation at the 3′ termini of the virion RNA species of New World and Old World arenaviruses, *Virology* **121**:200.

Auperin, D. D., Romanowski, V., Galinski, M. S., and Bishop, D. H. L., 1984, Sequencing studies of Pichinde arenavirus S RNA indicate a novel coding strategy, an ambisense viral S RNA, *J. Virol.* **52**:987.

Auperin, D. D., Sasso, D. R., and McCormick, J. B., 1986, Nucleotide sequence of the glycoprotein gene and intergenic region of the Lassa virus S genome RNA, *Virology* **154**:155.

Bishop, D. H. L., Gould, K. G., Akashi, H., and Clerx-Van Haaster, C. M., 1982, The complete sequence and coding content of snowshoe hare bunyavirus small (S) viral RNA species, *Nucleic Acids Res.* **10**:3703.

Boersma, D. P., Saleh, F., Nakamura, K., and Compans, R. W., 1982, Structure and glycosylation of Tacaribe viral glycoproteins, *Virology* **123**:452.

Boxaca, M. C., Parodi, A. S., Rugiero, H., and Blay, R., 1961, Fiebre hemorrágica experimental en el cobayo (virus Junin), *Rev. Asoc. Argent. Biol.* **37**:170.

Boxaca, M. C., Guerrero, L. B. de, Parodi, A. S., Rugiero, H. R., and Gonzalez Cappa, S., 1965, Viremia en enfermos de fiebre hemorragica argentina, *Rev. Asoc. Med. Argent.* **79**:230.

Bruno-Lobo, G. G., Bruno-Lobo, M., Johnson, K. M., Webbs, P. A., and De Paola, V., 1968, Pathogenesis of Junin virus infection in the infant hamster, *An. Microbiol. (Rio de Janeiro)* **15**:11.

Buchmeier, M. J., and Oldstone, M. B. A., 1978, Identity of the viral protein responsible for serological cross-reactivity among the Tacaribe complex arenaviruses, in: *Negative Strand Viruses and the Host Cell* (B. J. W Mahy and R. D. Barry, eds.), pp. 91–97, Academic Press, New York.

Buchmeier, M. J., Lewicki, H. A., Tomori, O., and Oldstone, M. B. A., 1981, Monoclonal antibodies to Lymphocytic choriomeningitis and Pichinde viruses: generation, characterization and cross-reactivity with other arenaviruses, *Virology* **113**:73.

Buchmeier, M. J., Southern, P. J., Parekh, B. S., Wooddell, M. K., and Oldstone, M. B. A., 1987, Site-specific antibodies define a cleavage site conserved among arenaviruses GP-C glycoproteins, *J. Virol.* **61**:982.

Burns, J., Salvato, M., and Buchmeier, M. J., 1990, Molecular architecture of lymphocytic choriomeningitis virus. VIII International Congress of Virology (Berlin), abstract W3-001, p. 27.

Carter, M. F., Biswal, N., and Rawls, W. E., 1973, Characterization of nucleic acid of Pichindé virus, *J. Virol.* **11**:61.

Casals, J., Buckley, S. M., and Cedeno, R., 1975, Antigenic properties of the arenaviruses, *Bull WHO* **52**:421.

Clegg, J. C. S., and Lloyd, G., 1983, Structural and cell-associated proteins of Lassa virus, *J. Gen. Virol.* **64**:1127.

Clegg, J. C. S., and Oram, J. D., 1985, Molecular cloning of Lassa-virus RNA: Nucleotide sequence and expression of the nucleocapsid protein gene, *Virology* **144**:363.

Compans, R. W., and Bishop, D. H. L., 1985, Biochemistry of arenaviruses, *Curr. Top. Microbiol. Immunol.* **114**:153.

Coto, C. E., and Parodi, A. S., 1968, Purificacion del virus Junin (FHA) a partir de cerebro de raton infectado, *Rev. Soc. Argent. Biol.* **44**:77.

Coto, C. E., and Vombergar, M. D., 1969, The effect of 5-iododeoxyuridine and actinomycin D on the multiplication of Junin virus, *Arch. Ges. Virusforsch.* **27**:307.

Coto, C. E., Help, G. I., and Tkaczevski, L. Z., 1972, Biological properties of Junin virus purified from infected mouse brain, *Medicina (Buenos Aires)* **32**:281.

Cresta, B., Padula, P., and Martinez Segovia, Z. M., 1980, Biological properties of Junin virus proteins. I. Identification of the immunogenic glycoprotein, *Intervirology* **13**:284.

De Mitri, M. I., and Martinez Segovia, Z., 1980, Biological activities of Junin virus proteins. II. Complement fixing polypeptides associated with the soluble antigen and purified virus particles, *Intervirology* **14**:84.

De Mitri, M. I., and Martinez Segovia, Z. M., 1985, Polypeptide synthesis in Junin virus-infected BHK-21 cells, *Acta Virologica* **29**:97.

Dreyfuss, G., Swanson, M. S., and Pinol-Roma, S., 1988, Heterogeneous nuclear ribonucleoprotein particles and the pathway of mRNA formation, *Trends Biochem. Sci.* **13**:86.

Dutko, F. J., Wright, E. A., and Pfau, C. J., 1976, The RNAs of defective interfering Pichinde virus, *J. Gen. Virol.* **31**:417.

Farber, F. E., and Rawls, W. E., 1975, Isolation of ribosome-like structures from Pichinde virus, *J. Gen. Virol.* **26**:21.

Franze-Fernandez, M. T., Zetina, C., Iapalucci, S., Lucero, M. A., Boissou, C., Lopez, R.,

Rey, O., Daheli M., Cohen, G., and Zalein, M., 1987, Molecular structure and early events in the replication of Tacaribe arenavirus S RNA, *Virus Res.* **7**:309.

Gangemi, J. D., Rosato, R. R., Connell, E. V., Johnson, E. M., and Eddy, G. A., 1978, Structural polypeptides of Machupo virus, *J. Gen. Virol.* **41**:183.

Gard, G. P., Vezza, A. C., Bishop, D. H. L., and Compans, R. W., 1977, Structural proteins of Tacaribe and Tamiami virions, *Virology* **83**:85.

Ghiringhelli, P. D., Rivera, R., Baro, N., Grau, O., and Romanowski, V., 1987, Molecular cloning of Junin virus RNA. VII International Congress of Virology (Edmonton, Canada), abstract OP3.10, p. 253.

Ghiringhelli, P. D., Rivera-Pomar, R. V., Baro, N. I., Rosas, M. F., Grau, O., and Romanowski, V., 1989, Nucleocapsid protein gene of Junin arenavirus (cDNA) sequence, *Nucleic Acids Res.* **17**:8001.

Ghiringhelli, P. D., Rivera-Pomar, R. V., Lozano, M., Grau, O., and Romanowski, V., 1991, Molecular organization of Junin virus S RNA: Complete nucleotide sequence, relationship with the other members of Arenaviridae and unusual secondary structures, *J. Gen. Virol.* **72**:2129.

Grau, O., Franze-Fernández, M. T., Romanowski, V., Rustici, S. M., and Rosas, M. F., 1981, Junin virus structure, in: *The Replication of Negative Strand Viruses* (D. H. L. Bishop and R. W. Compans, eds.), pp. 11–14, Elsevier, North Holland.

Gubler, U., and Hoffman, B. J., 1983, A simple and very efficient method for generating cDNA libraries, *Gene 25:*263.

Harnish D., Dimock, K., Leung, W. C., and Rawls, W., 1981, Immunoprecipitable polypeptides in Pichinde virus-infected BHK-21 cells, in: *The Replication of Negative Strand Viruses* (D. H. L. Bishop and R. W. Compans, eds.), pp. 23–29, Elsevier, New York.

Help, G. I., Coto, C. E., and Tkaczevski, L. Z., 1970, Evaluacion de un metodo de purificacion del virus Junin, *Medicina (Buenos Aires)* **30**:15.

Howard, C. R., Lewicki, H., Allison, L., Slater, M., and Buchmeier, M. J., 1985, Properties and characterization of monoclonal antibodies to Tacaribe virus, *J. Gen. Virol.* **66**:1385.

Hsu, M., Parvin, J. D., Gupta, S., Krystal, M., and Palese, P., 1987, Genomic RNAs of influenza viruses are held in a circular conformation in virions and in infected cells by a terminal panhandle, *Proc. Natl. Acad. Sci. USA* **84**:8140.

Hunt, T., 1989, Cytoplasmic anchoring proteins and the control of nuclear localization, *Cell* **59**:949.

Iapalucci, S., López, R., Rey, O., López, N., Franze-Fernández, M. T., Cohen, G. N., Lucero, M., Ochoa, A., and Zakin, M. M., 1989a, Tacaribe virus L gene encodes a protein of 2210 amino acid residues, *Virology* **170**:40.

Iapalucci, S., Lopez, N., Rey, O., Zakin, M. M., Cohen, G. N., and Franze-Fernandez, M. T., 1989b, The 5′ region of Tacaribe virus L RNA encodes a protein with a potential metal binding domain, *Virology* **173**:357.

Iapalucci, S., Lopez, N., and Franze-Fernandez, M. T., 1991, The 3′ end of the Tacaribe arenavirus subgenomic RNAs, *Virology* **182**:269.

Johnson, K. M., Webb, P. A., and Justines, G., 1973, Biology of Tacaribe-complex viruses, in: *Lymphocytic Choriomeningitis Virus and Other Arenaviruses* (F. Lehmann-Grube, ed.), pp. 241–258, Springer, Berlin.

Kedersha, N. L., and Rome, L. H., 1990, Vaults: Large cytoplasmic RNPs that associate with cytoskeletal elements, *Mol. Biol. Rep.* **14**:121–122.

Kiley, M. P., Tomori, O., Regnery, R. L., and Johnson, K. M., 1981, Characterization of the arenaviruses Lassa and Mozambique, in: *Replication of Negative Strand Viruses* (D. H. L. Bishop and R. W. Compans, eds.), pp. 11–14. Elsevier–North Holland, New York.

Kozak, M., 1978, How do eukaryotic ribosomes select initiation regions in messenger RNA? *Cell* **15**:1109.

Kozak, M., 1984, Compilation and analysis of sequences upstream from the translational start site in eukaryotic mRNAs, *Nucleic Acids Res.* **12**:857–872.
Lascano, E. F., and Berria, M. I., 1970, Microscopia electronica de cultivos primarios de fibroblastos de raton inoculados con virus Junin, *Medicina (Buenos Aires)* **29**:487.
Lascano, E. F., and Berria, M. I., 1974, Ultrastructure of Junin virus in mouse whole brain and mouse brain tissue culture, *J. Virol.* **14**:965.
Lazinski, D., Grzadzielska, E., and Das, A., 1989, Sequence-specific recognition of RNA hairpins by bacteriophage antiterminators requires a conserved arginine-rich motif, *Cell* **59**:207.
Leung, W. C., and Rawls, W. E., 1977, Virion-associated ribosomes are not required for the replication of Pichinde virus, *Virology* **81**:174.
Lopez, R. A., Grau, O., and Franze-Fernandez, M. T., 1986, Effect of actinomycin D on arenavirus growth and estimation of the generation time for a virus particle, *Virus Res.* **5**:213.
Lukashevich, T. S., Lemeshko, N. N., Stelmakh, T. A., Golubev, V. P., and E. P. Stcheslenok, 1985, Some biochemical properties of Lassa virus RNA and polypeptides, *Med. Microbiol. Immunol.* **175**:73.
Maiztegui, J. I., and Sabattini, M. S., 1977, Extensión progresiva del área endémica de fiebre hemorrágica argentina, *Medicina (Buenos Aires)* **37**(Suppl. 3):162.
Maiztegui, J. I., Laguens, R. P., Cossio, P. M., Casanova, M. B., de la Vega, M. T., Ritacco, V., Segal, A., Fernández, N. J., and Arana, R. M., 1975, Ultrastructural and immunohistochemical studies in five cases of Argentine hemorrhagic fever, *J. Infect. Dis.* **132**:35.
Maiztegui, J. I., Feuillade, M., and Briggiler, A., 1986, Progressive extension of the endemic area and changing incidence of AHF, *Med. Microbiol. Immunol.* **175**:149.
Mannweiler, K., and Lehmann-Grube, F., 1973, Electron microscopy of LCM virus-infected L cells, in: *Lymphocytic Choriomeningitis Virus and Other Arenaviruses* (F. Lehmann-Grube, ed.), pp. 37–48, Springer, Berlin.
Martinez Segovia, Z. M., and De Mitri, M. I., 1977, Junin virus structural proteins, *J. Virol.* **21**:579.
Martinez Segovia, Z. M., and Diaz, A., 1968, Purification of Junin virus by an aqueous biphasic polymer system, *Appl. Microbiol.* **15**:1602.
Martinez Segovia, Z. M., and Grazioli, F., 1969, The nucleic acid of Junin virus, *Acta Virol.* **13**:264.
Martinez Segovia, Z. M., Holstein, B. A., and Grazioli, F., 1967, Multiplicacion del virus Junin en cultivo de tejidos, *Cie. Investi.* **23**:35.
Martinez Segovia, Z. M., De Mitri, M. I., and Berdesky, S., 1974, Nutritional requirements for Junin virus production in BHK cultured cells, *Acta Physiol. Lat. Am.* **24**:656.
Maxam, A., and Gilbert, M., 1980, Sequencing end-labeled DNA with base specific chemical cleavages, in: *Methods in Enzymology* (L. Grossman and K., Moldave, eds.), Vol. 65, pp. 499–560, Academic Press, New York.
Mettler, N. E., Buckley, S. M., and Casals, J., 1961, Propagation of Junin virus, the etiological agent of Argentine hemorrhagic fever, in HeLa cells, *Proc. Soc. Exp. Biol. Med.* **107**:684.
Mettler, N. E., Casals, J., and Shope, R. E., 1963, Study of antigenic relationship between Junin virus, the etiological agent of Argentinian hemorrhagic fever, and other arthropod-borne viruses, *Am. J. Trop. Med. Hyg.* **12**:647.
Murphy, F. A., and Whitfield, S. G., 1975, Morphology and morphogenesis of arenaviruses, *Bull. WHO* **52**:409.
Murphy, F. A., Webb, P. A., Johnson, K. M., Whitfield, S. G., and Chappel, W. A., 1970, Arenaviruses in Vero cells: Ultrastructural studies, *J. Virol.* **6**:507.
Palmer, E. L., Obijeski, J. F., Webb, P. A., and Johnson, K. M., 1977, The circular segmented nucleocapsid of an arenavirus: Tacaribe virus, *J. Gen. Virol.* **36**:541.
Parodi, A. S., Greenway, D. J., Rugiero, H. R., Rivero, E., Frigerio, M. J., Mettler, N. E.,

Garzon, F., Boxaca, M., Guerrero, L. B. de, and Nota, N. R., 1958, Sobre la etiologia del brote epidemico en Junin, *Dia Med.* **30**:2300.
Pederson, I. R., 1973, Different classes of ribonucleic acids isolated from lymphocytic chorimeningitis virus, *J. Virol.* **11**:416.
Pfau, C. J., Bergold, G. H., Casals, J., Johnson, K. M., Murphy, F. A., Pedersen, I. R., Rawls, W. E., Rowe, W. P., Webb, P. A., and Weissenbacher, M. C., 1974, Arenaviruses, *Intervirology* **4**:207.
Pirosky, I., Zuccarini, J., Molinelli, E. A., Di Pietro, A., Martini, P., Ferreyra, B., Gutman Frugone, L. F., and Vazquez, T., 1959, Virosis hemorrágica del noroeste bonaerense. Endemoepidémica, febril, enantemática y leucopénica. I. La primera inoculación experimental al hombre, *Orient. Med.* **8**:144.
Raju, R., and Kolakofsky, D., 1989, The ends of La Crosse virus genome and antigenome RNAs within nucleocapsids are base paired, *J. Virol.* **63**:122.
Raju, R., Raju, L., Hacker, D., Garcin, D., Compans, R., and Kolakofsky, D., 1990, Non templated bases at the 5′ ends of Tacaribe virus mRNAs, *Virology* **174**:53.
Ramos, B. A., Courtney, R. J., and Rawls, W. E., 1972, The structural proteins of Pichinde virus, *J. Virol.* **10**:661.
Rawls, W. E., and Leung, W-C., 1979, Arenaviruses, in: *Comprehensive Virology* (H. Frankel-Conrat and R. Wagner, eds.), Vol. 14, pp. 157–192, Plenum Press, New York.
Rivera-Pomar, R. V., 1991, Análisis molecular del gen de la proteina de la nucleocápside del virus Junin, Doctoral thesis, Facultad de Ciencias Exactas, Universidad Nacional de La Plata, Argentina.
Rivera-Pomar, R. V., Manzelia, E. M., Ghiringhelli, P. D., Grau, O., and Romanowski, V., 1990, Expression of the arenavirus nucleocapsid protein: its possible role in the development of cytopathic effect. **VIII**th International Congress of Virology, Berlin, Abstract W3–005.
Rivera-Pomar, R. V., Manzella, E. M., Ghiringhelli, P. D., Grau, O., and Romanowski, V., 1991, Patterns of transient expression of the arenavirus nucleocapsid protein gene in transfected cells, *Microscop. Electron. Biol. Cel.* **15**:41.
Romanowski, V., 1981, Estructura bioquimica del virus Junin, Doctoral thesis, Facultad de Ciencias Exacta Universidad Nacional de La Plata, Argentina.
Romanowski, V., and Bishop, D. H. L., 1983, The formation of arenaviruses that are genetically diploid, *Virology* **126**:87.
Romanowski, V., and Bishop, D. H. L., 1985, Conserved sequences and coding of two strains of lymphocytic choriomeningitis virus (WE and ARM) and Pichinde arenavirus, *Virus Research* **2**:35.
Romanowski, V., Matsuura, Y., and Bishop, D. H. L., 1985, Complete sequence of the S RNA of lymphocytic choriomeningitis virus (WE strain) compared to that of Pichinde arenavirus, *Virus Research* **3**:101.
Romanowski, V., Lozano, M. E., Ghiringhelli, P. D., and Grau, O., 1991, A simple nucleic acid amplification assay for the early and rapid detection of Junin virus in clinical specimens, International Conference on Negative Strand Viruses, Charleston, SC.
Rosas, M. F., 1984, Bioquimica de virus: estructura de la envoltura del virus Junín, Doctoral thesis, Facultad de Ciencias Exactas, Universidad Nacional de La Plata, Argentina.
Rosas, M. F., Romanowski, V., and Grau, O., 1988, The phospholipid composition of arenaviruses, *Anal. Asoc. Quim. Argent.* **76**:269.
Rowe, W. P., Pugh, W. F., Webb, P. A., and Peters, C. J., 1970a, Serological relationships of the Tacaribe complex of viruses to lymphocytic choriomeningitis virus, *J. Virol.* **5**:289.
Rowe, W. P, Murphy, F. A., Bergold, G. H., Casals, J., Hotchin, J., Johnson, K. M., Lehmann-Grube, F., Mans, C. A., Traub, E., and Webb, P. A., 1970b, Arenoviruses: Proposed name for a newly defined virus group, *J. Virol.* **5**:651.
Rustici, S. M., 1984, Desarrollo *in vitro* del virus Junin, proteinas intracelulares, Doctoral thesis, Facultad de Ciencias Exactas, Universidad Nacional de La Plata, Argentina.

Saleh, F., Gard, G. P., and Compans, R. W., 1979, Synthesis of Tacaribe viral proteins, *Virology* **93**:369.
Sanchez, A., Pifat, D. Y., Kenyon, R. H., Peters, C. J., McCormick, J. B., and Kiley, M. P., 1989, Junin virus monoclonal antibodies: Characterization and cross-reactivity with other arenaviruses, *J. Gen. Virol.* **70**:1125.
Sanger, F., Nicklen, S., and Coulson, A. R., 1977, DNA sequencing with chain-terminating inhibitors, *Proc. Natl. Acad. Sci. USA* **74**:5463.
Schwartz, R. M., and Dayhoff, M. O., 1979, in: *Atlas of Protein Sequence and Structure* (M. O. Dayhoff, ed.), pp. 353–358, National Biomedical Research Foundation, Washington, D.C.
Taylor, W. R., 1987, Protein structure prediction, in: *Nucleic Acid and Protein Sequence Analysis* (M. J. Bishop and C. J. Rawlings, eds.), pp. 285–322, IRL Press, Oxford.
Tinoco, I., Borer, P. N., Dengler, B., Levine, M. D., Uhlenbeck, O. C., Crothers, D. M., and Gralla, J., 1973, Improved estimation of secondary structure in ribonucleic acids, *Nature (New Biol.)* **246**:40.
Vezza, A. C., Gard, G. P., Compans, R. W., and Bishop, D. H. L., 1977, Structural components of the arenavirus Pichinde, *J. Virol.* **23**:776.
Vezza, A. C., Clewley, J. P., Gard, G. P., Abraham, N. Z., Compans, R. W., and Bishop, D. H. L., 1978, Virion RNA species of the arenaviruses Pichinde, Tacaribe and Tamiami, *J. Virol.* **26**:485.
Weissenbacher, M. C., and Damonte, E. B., 1983, Fiebre hemorrágica argentina, in: *Adel. Microbiol. Enf. Infecc.* (C. Coto, J. Esparza, and R. A. de Torres, eds.), Vol. 2, pp. 119–171, Buenos Aires–Caracas.
Wilson, S. M., and Clegg, J. C. S., 1991, Sequence analysis of the S RNA of the African arenavirus Mopeia: An unusual secondary structure feature in the intergenic region, *Virology* **180**:543.
Young, P. R., and Howard. C., 1983, Fine structure analysis of Pichinde virus nucleocapsids, *J. Gen. Virol.* **64**:833.
Young, P. R., Chanas, A. C., and Howard, C. R., 1981, Analysis of the structure and function of Pichinde virus polypeptides, in: *Negative Strand Virus* (D. H. L. Bishop and R. W. Compans, eds.), pp. 11–14, Elsevier–North Holland, New York.
Young, P. R., Chanas, A. C., Lee, S. R., and Howard, C. R., 1987, Localization of an arenavirus protein in the nuclei of infected cells, *J. Gen. Virol.* **68**:2465.
Zeller, W., Bruns, M., and Lehmann-Grube, F., 1986, Viral nucleoprotein can be demonstrated on the surface of lymphocytic choriomeningitis virus-infected cells, *Med. Microb. Immunol.* **175**:89.
Zeller, W., Bruns, M., and Lehmann-Grube, F., 1988, Lymphocytic choriomeningitis virus. X. Demonstration of the nucleoprotein on the surface of infected cells, *Virology* **162**:90.

CHAPTER 5

Genetic Variation in Junin Virus

CELIA E. COTO, ELSA B. DAMONTE, LAURA E. ALCHE, AND LUIS SCOLARO

I. INTRODUCTION

Junin virus (JV) is a member of Arenaviridae able to cause a severe disease in humans known as Argentine hemorrhagic fever (AHF). The first isolation of this agent was performed in 1958 by Parodi *et al.* from a human case. Epidemics of AHF have occurred each year in the central area of the country, with incidence peaks coincident with harvest time of maize crops (Maiztegui, 1975).

The viral genome consists of two single-stranded RNA segments: a large species (L, 33S MW of 2.4×10^6 daltons) and a small species (S, 25S MW of 1.3×10^6 daltons) (Añón *et al.*, 1976). Recently, the sequence for the S RNA of JV has been determined and comparative analysis has shown 50–63% homology with the other four arenaviruses sequenced so far, Pichinde, lymphocytic choriomeningitis virus (LCMV), Lassa, and Tacaribe viruses (Ghiringhelli *et al.*, 1989; Clegg *et al.*, 1990; Ghiringhelli *et al.*, 1991).

The virion contains a major nucleocapsid-associated protein (NP, MW approximately 60–64 kDa) and two glycoproteins (G1 or GP52, MW 44–52 kDa and G2 or GP38, MW 36–38 kDa) (Grau *et al.*, 1981).

CELIA E. COTO, ELSA B. DAMONTE, LAURA E. ALCHE, AND LUIS SCOLARO • Laboratorio de Virologia, Departamento de Quimica Biologica, Facultad de Ciencias Exactas y Naturales, Ciudad Universitaria, Pabellón 2, Piso 4, 1428 Buenos Aires, Argentina.

The Arenaviridae, edited by Maria S. Salvato. Plenum Press, New York, 1993.

The presence of G1 is controversial since it has not been regularly detected in all JV preparations. On the basis of radioactive glucosamine incorporation in JV-infected cells, it was concluded that GP38 is the most heavily glycosylated protein (Damonte *et al.*, 1985) and, by external iodination, it seems to be the only glycoprotein exposed on the viral surface (Mersich *et al.*, 1988). Furthermore, GP38 is responsible for the induction of neutralizing antibodies to infectious virus (Cresta *et al.*, 1980). Also, a glycosylated precursor of the glycoproteins (GPC, MW 63–70 kDa) has been identified in virus-infected cells (De Mitri and Martinez Segovia, 1985; Damonte *et al.*, 1985). As detected by specific lectin binding, the oligosaccharide side chains of the precursor GPC and the mature protein GP38 have been shown to contain mannose, *N*-acetylglucosamine, and galactose residues (Mersich *et al.*, 1988), indicating the presence of complex-type oligosaccharides.

In addition to the three major structural proteins, other minor polypeptides have been observed in JV-infected cells: a high-molecular-weight protein (L, MW 200 kDa) and a variety of small proteins (MW 18–25 kDa) (De Mitri and Martinez Segovia, 1985; Damonte *et al.*, 1985). Although there is no direct evidence for the function of L protein, it was proposed to be the viral replicase–transcriptase whereas the small proteins seem to derive from NP (De Mitri and Martinez Segovia, 1980).

As a salient feature of their biological properties, arenaviruses are characterized by their ability to originate persistent infections in cell cultures and in rodent hosts. For JV, cell carrier cultures are readily established *in vitro*. The main properties of these persistently infected cells are a cyclical pattern of infectious virus release to the supernatant, the presence of interfering particles, and a complete cell resistance to homotypic virus superinfection (Damonte and Coto, 1979; Coto *et al.*, 1981; Damonte *et al.*, 1983; Candurra and Damonte, 1985; Weber *et al.*, 1985). *In vivo*, experimentally JV-infected *Calomys musculinus* (field mouse), its main reservoir in nature, develops a chronic infection characterized by the simultaneous presence of circulating anti-JV antibodies and high viral titers in brain and blood (Laguens *et al.*, 1982; Alche and Coto, 1986, 1988).

The mechanisms involved in JV persistence are still an intriguing unresolved problem. As will be discussed below, increasing evidence has accumulated to demonstrate an association of chronic infection with the appearance of virus variants. This observation suggests that, as it occurs for several groups of RNA viruses, genetic variation is also a frequent phenomenon among arenaviruses. Another line of evidence of JV variability has been provided by the isolation from nature of a wide spectrum of viral strains differing in virulence and antigenic characteristics (Candurra *et al.*, 1989; Kenyon *et al.*, 1986).

This chapter is concerned with *in vitro* and *in vivo* studies performed on JV variability. To this end several approaches have been used, such as analysis of a number of natural strains of JV, analysis of evolving virus populations during persistent infections, and isolation and use of chemically induced mutants to analyze virus functions, particularly those related to virulence and pathogenicity.

II. VARIABILITY OF JUNIN VIRUS STRAINS

Since 1958, JV has been repeatedly isolated from severe and mild human cases as well as from *C. musculinus.* Studies performed on JV strains have initially focused on their pathogenic properties. Several animal models have been developed with the aim of reproducing the typical disease of human AHF and obtaining a marker system to differentiate pathogenic and attenuated strains. The virulence of JV strains for humans and adult guinea pigs seems to be very closely related and, therefore, this animal model has been the most widely used to characterize the virulence pattern of JV strains. Alternative attenuation markers have been proposed, such as 14-day-old mice, primates, and 2-day-old rats (Contigiani and Sabattini, 1977; Avila *et al.*, 1981a; Weissenbacher *et al.*, 1985).

A broad spectrum of virulence among JV strains of human and rodent origin has been demonstrated analyzing their behavior in the above-mentioned animal models (Avila *et al.*, 1981b; Kenyon *et al.*, 1986; Candurra *et al.*, 1989). Low, intermediate, and high degrees of virulence were exhibited by the different isolates studied so far (Weissenbacher *et al.*, 1987), denoting that attenuated and virulent strains coexist in nature.

According to their lethality for guinea pigs, two highly pathogenic isolates [the prototype XJ strain of human origin (Parodi *et al.*, 1958) and the AN 9446 strain isolated from a rodent in 1967], the rodent-derived MC2 strain of intermediate pathogenicity (Berria *et al.*, 1967) and two strongly attenuated strains [XJC13 strain derived from XJ by serial passage in the laboratory (Guerrero *et al.*, 1969) and the naturally attenuated IV4454 isolated from a mild human case in 1970 (Contigiani and Sabattini, 1977)] were comparatively studied in their antigenic, biological, and biochemical properties.

Antigenic relatedness among these strains was determined by neutralization assays with hyperimmune polyclonal antisera raised in rabbits. Representative data obtained for each immunoserum in cross-neutralization reactions against homologous and heterologous viruses are shown in Table I. Independently of their virulence, antigenic heterogeneity among JV strains has been demonstrated by these studies, since polyclonal antisera could distinguish them (Candurra *et al.*, 1989). Clearly, the serum raised against one JV strain discriminates between the homologous strain and the other ones since the ratio of titer against homologous virus versus titer against heterologous virus is greater than 10. Similar results were obtained with sera raised against the other JV

TABLE I. Virulence and Cytopathogenicity of Laboratory and Field Strains of Junin Virus

	Virus			
	IV4454	XJ	XJC13	MC2
Pathogenicity for guinea pigs (% mortality)	30	100	20	80
Neutralizing titer against XJC13 antiserum	407	584	9058	901
Cytopathology				
% viable cells[a]	18	22	78	90
Plaque morphology	Lytic	Lytic	Turbid	Very turbid
% inhibition of cell protein synthesis[b]	65	35	25	0
Maximum yield in Vero cells (PFU/ml)	9.10^6	5.10^6	4.10^6	8.10^5

[a] Monolayers were infected at a multiplicity of infection (moi) of 0.02 PFU/cell. After 8 days, cells were counted and viability was determined by the exclusion of trypan blue dye.
[b] Cell protein synthesis was determined by the amount of [^{3}H]-leu incorporated in infected cultures and expressed as percentage of a mock infected control.

isolates (data not shown), allowing the quantitative expression of the antigenic relationships among JV strains through a dendrogram based on taxonomic distance coefficients (Candurra *et al.*, 1989).

Although a definition for serotype has not been established in Arenaviridae, according to the values of homologous and heterologous neutralizing titers it seems that all JV strains tested would have to be grouped into a single serotype but showing antigenic diversity. Similarly, several strains of Lassa virus were distinguished by polyclonal antisera in a plaque-reduction neutralization test (Jahrling and Peters, 1986; Peters *et al.*, 1987). Variation among JV strains was also expressed in their cytolytic potential *in vitro* measured by plaque morphology, cell viability after infection, and ability to induce inhibition of host cell protein synthesis in infected cells (Candurra *et al.*, 1990). As summarized in Table I, the cytopathogenicity of JV strains *in vitro* for Vero cells was independent of their *in vivo* virulence and was very variable among the tested strains. The attenuated IV4454 was highly cytopathic for Vero cells, inducing a strong inhibition of cell protein synthesis and reducing cell viability more than 80% (Table I). On the other hand, the pathogenic MC2 strain produced very slight alterations on Vero cell monolayers, whereas XJ and XJCl3 exhibited intermediate cytolytic potential. In many virus systems, variability in antigenic properties and virulence pattern as observed for JV strains is associated with alterations at the level of the external surface proteins. This seems to be also valid for JV, since the patterns of the peptides obtained by partial proteolysis of GP38 allow the differentiation of JV strains, showing that there is a variation in the surface glycoprotein that could be correlated with the evolution of virus properties (Candurra *et al.*, 1990).

Although these studies were intended to relate the degree of pathogenicity of JV strains with a biological or biochemical characteristic, no efficient *in vitro* markers of JV virulence have been found. However, clear differences have been detected in many properties among JV strains isolated several years apart, supporting the finding that considerable variability has occurred in JV natural populations. Further studies on genome sequencing of JV strains will provide information to elucidate the molecular basis of this phenomenon.

III. EXPERIMENTAL INDUCTION OF CONDITIONALLY LETHAL MUTANTS

Difficulties have been encountered in obtaining conditionally lethal mutants of arenaviruses. Ts mutants have been derived by chemical mutagenesis from only three members of the family, Pichinde (Vezza and Bishop, 1977; Shivaprakash *et al.*, 1988), Junin (Ceriatti *et al.*, 1983), and LCMV (Romanowski and Bishop, 1983). In all cases mutant isolation rates were very low, with values ranging from 0.1 to 1.8% for Pichinde virus (Vezza and Bishop, 1977), Junin virus (Ceriatti *et al.*, 1983), and LCMV strain WE (Romanowski and Bishop, 1983). Only the Armstrong strain of LCMV had a slightly higher mutation frequency of 7.8% (Romanowski and Bishop, 1983). The reason for this difference in the effectiveness of mutant induction between two strains of LCMV is not known. The detection of diploid viruses with respect to their S RNA species in LCMV populations may account for this behavior (Romanowski and Bishop, 1983). This finding is in agreement with previous observations on the variable relative proportions of S and L RNA species in purified viral preparations and the pleomorphic characteristics of the virions. Furthermore, the possible genetic stability due to diploidy has been postulated as responsible for the inefficient recovery of arenavirus mutants (Bishop, 1988).

After mutagenesis with 100 μg/ml of 5-fluoruracil, three ts mutants, designated C132, C140, and C167, were isolated from the attenuated strain XJCl3 of JV and partially characterized (Ceriatti *et al.*, 1986; Scolaro *et al.*, 1987). The efficiency of plaquing at the restrictive temperature (39°C) relative to that at the permissive temperature (an estimate of the reversion frequency) varied from 10^{-3} to 10^{-4}. Compared to ts mutants isolated from several animal viruses, JV mutants exhibited relatively high plaque-forming ability at the nonpermissive temperature and a certain degree of leakiness when the time of incubation at the restrictive temperature was extended to more than 48 hr, making it difficult to do advanced genetic studies with them. This was particularly true for C167, which displayed a relatively high ability to replicate at 39°C (Scolaro *et al.*, 1987).

Despite the above-mentioned difficulties, the analysis of temperature shift experiments and the pattern of labeled proteins immunoprecipitated from infected cell extracts indicate that C132 and C140 expressed their defects at early stages in the course of infection (Ceriatti *et al.*, 1986). In fact, highly reduced levels of viral proteins at the restrictive temperature have been shown for C140, whereas the infection with C132 was associated with undetectable synthesis of viral polypeptides. According to these phenotypic patterns of proteins synthesized at the nonpermissive temperature, C132 and C140 could be respectively ascribed to the groups 2 and 4 defined for Pichinde virus ts mutants. An altered transcriptase activity has been the postulated defect by Shivaprakash *et al.* (1988) for these two mutant groups. This is in agreement with our proposal that the affected function in C132 and C140 mutants is early transcription (Ceriatti *et al.*, 1986). To support this hypothesis, shift experiments have shown that the critical period during the replicative cycle of the mutants seemed to fall between 2 and 4 hr postinfection. Thus, for arenaviruses, it seems that the isolation of transcription-altered mutants is very common, in accordance with other negative-strand viruses.

Another altered phenotype in JV mutants with respect to wild-type XJC13 strain was their reduced pathogenicity for 2-day-old mice. Correlation between thermosensitivity and attenuation was observed only for C132, whereas C140 and C167 revertants in their ts phenotypes still maintained their reduced virulence (Ceriatti *et al.*, 1986). The attenuated phenotype exhibited by these mutants was a striking property, since newborn mouse has been found highly susceptible to any known strain of JV (Weissenbacher *et al.*, 1987). Consequently, these mutants represent valuable tools for the study of various aspects of JV pathogenesis.

Experiments performed with C167 have shown that the reduced virulence of this variant was independent of the mouse strain (Table II)

TABLE II. Pathogenicity of the Host-Range Mutant C167 and Its Parental Strain XJC13

Virus	XJC13			C167		
Mice strain	OF1	C3H/HeJ	BALB/cJ	OF1	C3H/HeJ	BALB/cJ
PFU/LD50	6.6	2.4	1.1	>1800	>400	>1050
PFU/g of brain[a]	2.10^6	2.10^7	2.10^7	$<5.10^1$	$<5.10^1$	$<5.10^1$
Spreading[b]	Yes	Yes	Yes	No	No	No

[a] Recovered infectivity from mice brains 3 days after infection with a dose of 100 PFU by i.c. route.
[b] Detection of infectivity in liver, kidneys, thymus, and spleen of mice after i.c. inoculation with 100 PFU.

and route of inoculation (intracerebral or intraperitoneal). Furthermore, it was associated with a highly delayed and restricted virus multiplication at the site of inoculation and a defective viral spread to other tissues (Table II). High titers of virus ranging from 10^4 to 10^8 PFU/g organ were detected in brain, kidney, spleen, liver, and thymus after intracerebral inoculation with XJCl3. By contrast, in mice infected with C167, virus was not recovered from either of these organs, except for brain at later times (Scolaro *et al.*, 1989). Likewise, no viral replication was detected in peritoneal cells obtained from neonatal mice intraperitoneally infected with C167 and maintained *in vitro*, although 10^4 PFU/ml could be recovered from macrophages of XJCl3-infected mice (Scolaro *et al.*, 1989). Even though no signs of illness were observed in C167-infected surviving animals, mice developed a humoral response to the virus and were protected against a challenge with XJCl3 before the appearance of neutralizing antibodies, probably due to an interference mechanism (Scolaro *et al.*, 1991).

Analysis of multiplication kinetics of C167 and XJCl3 in different cell cultures indicated that the above-mentioned inability of C167 to replicate in mouse tissues was not due to a general growth defect, but to a specific host-range restriction in murine cells. Patterns of growth curves were similar for both viruses in Vero cells, whereas C167 was highly restricted for replication in primary cultures of suckling mouse brain and mouse embryo fibroblasts (Scolaro *et al.*, 1989). To define the defective stage in C167 infection of murine cells, different steps of the replicative cycle in the permissive and restrictive cell systems, Vero cells, and murine embryo fibroblasts (MEF), respectively, were comparatively analyzed. As summarized in Table III, no viral specific proteins detected by radioimmune precipitation and gel electrophoresis were present in murine cells, indicating that the host-range restriction occurs previously to protein synthesis. Accordingly, adsorption and penetra-

TABLE III. Multiplication of C167 and XJC13 in Vero Cells and MEF

Virus	Cells	Yield[a] (PFU/ml)	Protein synthesis[b]	Adsorption[c] PFU	Adsorption[c] cpm	Penetration (IC)[d]
XJC13	Vero	3.10^6	Yes	1500	2683	$7,3.10^2$
	MEF	4.10^5	Yes	3200	3037	$1,0.10^3$
C167	Vero	2.10^6	Yes	2000	2821	$1,1.10^3$
	MEF	<5	No	100	967	<5

[a] Maximum yield obtained in culture supernatants after infection at a moi of 0.1 PFU/cell and incubation at 37°C.
[b] Viral protein synthesis was detected by the appearance of [^{35}S]met–labeled NP.
[c] Adsorption was assayed as the cell-associated infectivity after 60 min of incubation of the inoculum at 4°C or cell-associated radioactivity after 30 min of incubation of radioiodinated virus at the same temperature.
[d] Penetration was scored as the number of infectious centers (IC) nonneutralized by anti-XJC13 hyperimmune serum.

tion assays have shown that while XJCl3 adsorbed efficiently to murine cells, the mouse-attenuated C167 bound very poorly, leading to a reduced entry of the virus in the cell (Table III).

The molecular basis of the attenuated phenotype of C167 has been attributed to an alteration in the external surface glycoprotein GP38: the mutant and the parental viruses showed different GP38 peptide mapping after limited proteolysis (Scolaro *et al.*, 1990). Thus, amino acid changes in GP38 are responsible for an altered interaction between the virus and the cell receptor in the initial stage of the replicative cycle, and the blockade in viral adsorption determines murine cell resistance to C167 infection and the emergent lack of virulence for suckling mice.

IV. SPONTANEOUS MUTANTS GENERATED DURING PERSISTENT INFECTION

A. *In Vitro*

Although arenaviruses replicate in a wide variety of mammalian cell types, the cytolytic potential of their genome is expressed in only a limited number of host cell systems. By contrast, persistently infected cultures may be readily established using most of the cell lines suitable for virus growth (Howard, 1986). Arenavirus *in vitro* persistence is often associated with the production of interfering virus particles that replicate in a cyclical pattern in parallel with infectious virus (Weber *et al.*, 1983).

Appearance of viral mutants was also reported in cells persistently infected with LCMV (Lehmann-Grube *et al.*, 1969; Hotchin *et al.*, 1971; Jacobson *et al.*, 1979; Bruns *et al.*, 1990), Junin (Damonte and Coto, 1979; Coto *et al.*, 1981), and Tacaribe viruses (Coto *et al.*, 1981; Damonte *et al.*, 1981).

A clear association between the generation of ts mutants and the establishment of persistent infection in Vero cells with JV has been observed (Damonte and Coto, 1979; Coto *et al.*, 1979). Virus released from these cells could not be distinguished from wt parental virus by plaque morphology but showed a reduced plaquing efficiency at 39°C with respect to 33°C and a diminished replication rate, since virus yields at 39°C were 3 log lower than those obtained at 33°C. This variant virus did not interfere with wt virus growth at 37°C and was able to establish persistent infection in normal Vero cells after a period of marked cytolysis (Damonte and Coto, 1979).

Similar results were obtained on analysis of three Vero cell lines persistently infected with Tacaribe virus (Damonte *et al.*, 1981), except that Tacaribe ts mutants isolated from these cells developed a lytic infection in normal Vero cells characterized by a complete destruction of the monolayer without regrowth of the cultures.

Further analysis of eight Vero cell sublines persistently infected with Tacaribe virus allowed us to conclude that virus released from these cultures required the coinfection with interfering particles to initiate a persistent infection (D'Aiutolo and Coto, 1986). Besides the occurrence of ts mutants, it was also possible to isolate a slow-growth variant from these persistently infected cells (D'Aiutolo and Coto, 1986). Weber *et al.* (1985) established a persistent infection of human MRC-5 cells with JV, and the virus population shed from these carrier cultures also contained ts and plaque-morphology mutants.

In conclusion, although the occurrence of viral variants during *in vitro* persistent infections indicates a high tendency of virus variability, it does not provide clues to understanding virus-host cell interaction.

B. *In Vivo*

Perpetuation of arenaviruses in nature depends on their ability to establish a long-term persistent infection in rodents, with the exception of Tacaribe virus, which was isolated from bats and is not associated with other animal species.

Attempts to understand the mechanism responsible for viral persistence in their natural hosts have been confined to LCMV and Junin viruses by means of developing appropriate laboratory models: wild mice of the *Mus musculus* strain are LCMV carriers in nature (Traub, 1935, 1938), whereas JV is perpetuated in the cricetid *Calomys musculinus*, its main reservoir, which contaminates the environment by elimination of virus through saliva and urine (Sabattini *et al.*, 1977; Sabattini and Contigiani, 1982).

C. Junin Virus

To elucidate virus–host interactions in nature, an experimental infection with the attenuated strain XJC13 of JV was established in *C. musculinus* from an outbred colony. When neonatal cricetids were injected intraperitoneally with 4000 LD_{50} of XJC13, approximately 70% of infected animals died between 14 and 28 days postinfection after developing a clinical neurological disease. This acute stage of infection was also characterized by the detection of high viral titers in brain, an increased frequency of viremia, and the onset of a humoral immune response from day 15 postinfection onward (Fig. 1). Since mononuclear cells were permissive for JV multiplication *in vitro* (Alche *et al.*, 1985) and a marked lymphopenia was observed in 40% of infected cricetids during this period (Steyerthal *et al.*, 1990), it is tempting to suggest a possible association between JV and circulating lymphocytes and monocytes.

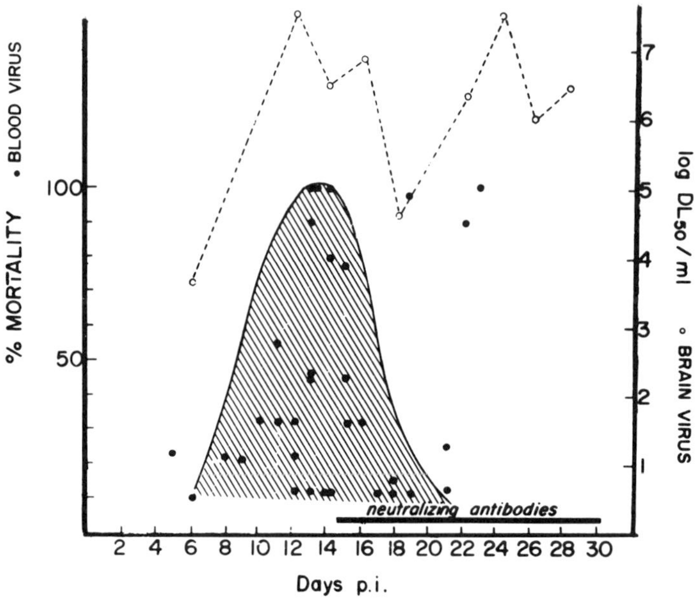

FIGURE 1. Detection of viral infectivity in brain and blood of *Calomys musculinus* inoculated with XJCl3. (O---O---O) Brain viral isolates were obtained from infected cricetids at different days post-infection and titered by LD_{50} in newborn mice injected by the intracerebral route. (●) Mortality percentage of mice inoculated intracerebrally with blood viral isolates obtained from infected *Calomys musculinus* at different days post-infection.

Among the survivors, some recovered completely and others presented persisting neurological symptoms related, in some cases, to growth perturbation (Lampuri *et al.*, 1982). One of the main features observed in chronically infected cricetids was the simultaneous presence of circulating neutralizing antibodies and high virus titers in brain (10^6 LD_{50}/g) up to 350 days postinfection (Laguens *et al.*, 1982; Alché and Coto, 1988). In addition, viremia was sporadically detected until 60 days postinfection (Alché and Coto, 1986, 1988).

The inefficiency of circulating neutralizing antibodies in clearing the virus during persistence could be explained by the emergence of antigenic variants that would escape the immune surveillance, a phenomenon already described for other RNA viruses. This assumption was supported by the results obtained performing neutralization assays with viral isolates from brain of persistently infected cricetids tested with sera obtained from the same animals (autologous sera). As shown in Table IV, neutralizing antibody titers were eightfold lower against homologous persistent viruses than those obtained against virus employed to initiate infection, indicating that, during chronic infection, circulating antibodies were poorly reactive against virus born in brain despite strongly recognized parental virus antigens. Similarly, the estimation of affinity constant values (K) by kinetic neutralization assays corrobo-

TABLE IV. Comparative Neutralizing Antibody Titers in Sera of Chronically Infected *Calomys musculinus* against Parental and Homologous Virus

	Antibody titer in serum[a]			
Virus	#114	#115	#119	#129
XJC13	2560[b]	2560	2560	1280
Homologous virus	<320	<320	640	<160

[a] Four cricetids were inoculated with 4.10^3 LD_{50} of XJC13 by intranasal (*C. musculinus* 114 and 115) and intraperitoneal route (*C. musculinus* 119 and 129). Animals were sacrificed at 80 days post inoculation, virus was isolated from brain homogenates, and sera were processed. Neutralizing antibody titers were determined against parental XJC13 virus and each homologous viral isolate.
[b] Numbers represent reciprocal of the highest dilutions of sera that neutralized 80% of plaque counts of each virus assayed.

rated a differential reactivity of autologous serum as well as anti-XJC13 hyperimmune serum between parental virus and antigenic variants isolated from brain at 19 days post-infection (Alché and Coto, 1988).

The appearance of serological variants as a regular event in JV persistence in *C. musculinus* was further confirmed when blood viral isolates from acute and chronically infected animals were assayed against an anti-XJCl3 hyperimmune serum prepared in guinea pigs. The dose–response curves plotted in Fig. 2 show that all viral isolates (#192, #385, #117, and #215) obtained from 8 to 61 days postinfection were resistant to neutralization by anti-XJCl3 antibodies. On the other hand, an antiserum raised against the viral mutant 215 did not show a significant differential reactivity among homologous virus, parental virus, and the variant 192 obtained at 19 and 35 days postinfection (Fig. 3). Consequently, these viral variants could be distinguished from virus employed to initiate infection only by means of an antiserum prepared against the initial virus.

The role of circulating neutralizing antibodies in the selection of mutant viral populations still remains obscure. There may be a determining factor in the emergence of viral variants *in vivo*, although their inability to abolish infection has been previously demonstrated (Coulombié and Coto, 1985). The selective pressure displayed by these antibodies *in vivo* would be questioned by the fact that the earliest mutant viral isolation from blood occurred at day 8 postinfection before the onset of humoral immune response. However, it was possible to obtain antigenic variants of JV *in vitro* under a strong pressure of polyclonal antibodies, suggesting that the parental viral stock would contain a phenotypically mixed population even after plaque purification (data not shown). In this hypothetical case, JV clones would constitute a quasispecies similar to those described for different RNA viruses exhibiting high mutation rates (Domingo *et al.*, 1985).

It is noteworthy that antigenic variants of JV could be discriminated by polyclonal antisera and, hence, would be of epidemiological signifi-

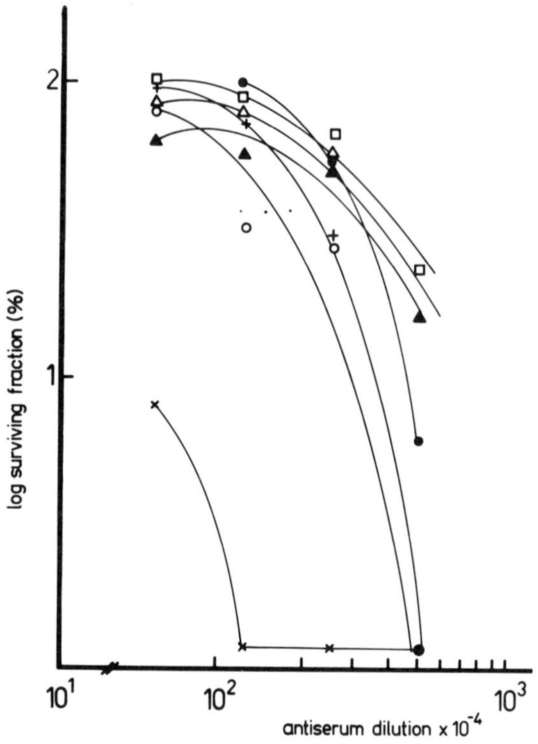

FIGURE 2. Dose response curves of Junin virus and blood viral variants against an anti-XJC13 serum. Neutralization assays were done with a hyperimmune anti-XJCI3 serum against (x) parental JV and viral variants: (▲) 385 (8), (△) 215 (61), (●) 215 (47), (□) 192 (19), (○) 192 (35), and (+) 117 (23) obtained from blood of infected *Calomys musculinus*, inoculated intraperitoneally with 4000 LD_{50} of XJC13 strain of JV. Numbers in parentheses represent the post-infection day of bleeding.

cance, since selection with monoclonal antibodies frequently produces virus with no survival advantage in nature (Webster and Laver, 1980).

Studies on the biological properties of viral mutants showed that the antigenic alteration was not associated with a ts phenotype, nor was it associated with altered pathogenicity for neonatal cricetids, even though they showed a higher virulence than parental JV (Alché and Coto, 1988). Considering that JV antigenic variants were detected early during infection, it seemed important to determine whether peritoneal macrophages were involved as a source of virus variants. Previous studies have demonstrated that macrophages from newborn cricetids were permissive for viral multiplication, whereas in adult animals they would act as a barrier against viral infection by means of an extrinsic macrophage-mediated antiviral activity (Coulombié *et al.*, 1986). Peritoneal nonstimulated exudate cells were obtained from infected *C. musculinus* at 26 days postinfection and cultivated *in vitro* for 10 days. Three of six established macrophage cultures were viral producers, releasing 10^3–10^5

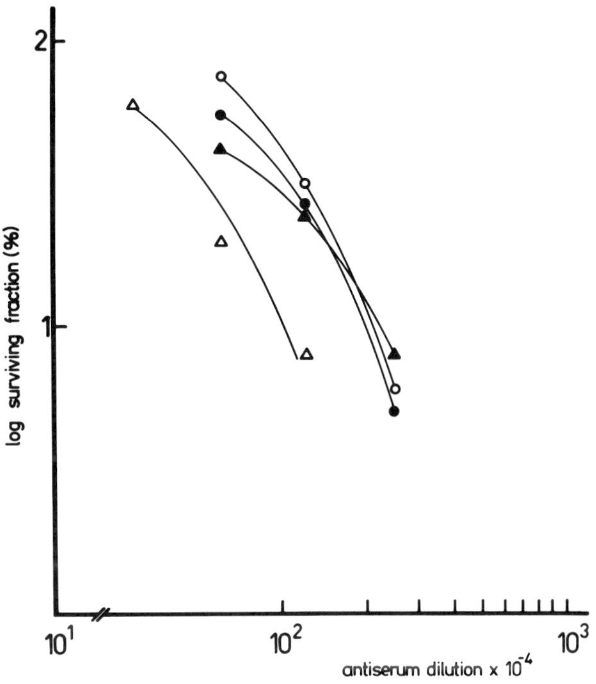

FIGURE 3. Dose response curves of Junin virus and blood viral variants against an anti 215 (61) serum. Neutralization assays were done with a hyperimmune serum anti-215 (61) viral variant of JV, against (△) 215 (61), (▲) XJC13, (○) 192 (19), and (●) 192 (35) viruses.

PFU/ml to the culture fluids. Supernatants containing virus were tested in neutralization assays against anti-XJCl3 hyperimmune serum. Neutralization titers against macrophage released virus were 8–16 times lower than those against parental virus (unpublished data). These results clearly indicate that viruses yielded by macrophages from infected cricetids were antigenic variants. The selection of these variants only occurred *in vivo* since macrophages obtained from uninfected *C. musculinus* at 20 days of age, and then infected *in vitro* with XJCl3, released into the supernatant a viral population indistinguishable by neutralization assay from parental virus.

In consequence, all available data indicate that the avoidance of immune elimination of the infecting virus allows JV antigenic variants to appear, to replicate in peritoneal macrophages, and, later, probably associated with peripheral blood cells, to reach the brain, the major target organ of JV persistent infection. Findings thus far obtained point out that viral variability is inherent not only to JV, but also to LCMV (Ahmed *et al.*, 1984; Ahmed and Oldstone, 1988). In both cases, the detection of antigenic and genetic variants by means of different experimental approaches allows us to conclude that they participate actively as regulators of virus–host interactions during arenavirus persistence.

V. CONCLUDING REMARKS

Genetic variation has been corroborated as a frequent event in arenavirus infections, particularly associated with the progression of persistent infections. By experimental induction or by spontaneous appearance during *in vitro* and *in vivo* infections, several kinds of arenavirus mutants have been isolated in the laboratory. These mutants exhibited altered phenotypes in their antigenic properties, tissue tropism, and/or pathogenicity. The biological significance of point mutations, genome reassortments, and selection of viral variants during arenavirus replication in the natural host is still a rewarding area of study. Further experimental work is required, especially with the human pathogenic Lassa and Junin viruses, to elucidate the role of these mechanisms during virus passage between the rodent reservoir and the human host and their effects on strain virulence.

REFERENCES

Ahmed, R., and Oldstone, M. B. A., 1988, Organ-specific selection of viral variants during chronic infection, *J. Exp. Med.* **167**:1719.

Ahmed, R. A., Salmi, A., Butler, L. D., Chiller, J. M., and Oldstone, M. B. A., 1984, Selection of genetic variants of lymphocytic choriomeningitis virus in spleens of persistently infected mice, *J. Exp. Med.* **60**:521.

Alché, L. E., and Coto, C. E., 1986, Antigenic variants of Junin virus isolated from *Calomys musculinus, Arch. Virol.* **90**:343.

Alché, L. E., and Coto, C. E., 1988, Differentiation of Junin virus and antigenic variants isolated *in vivo* by kinetic neutralization assays, *J. Gen. Virol.* **69**:2123.

Alché, L. E., Coulombie, F. C., and Coto, C. E., 1985, Aislamiento de virus Junin a partir de sangre y linfo-monocitos perifericos de *Calomys musculinus* infectados, *Rev. Arg. Microbiol.* **17**:177.

Añon, M. C., Grau, 0., Martinez Segovia, Z., and Franze-Fernandez, M. T., 1976, RNA composition of Junin virus, *J. Virol.* **18**:833.

Avila, M. M., Galassi, N. V., and Weissenbacher, M. C., 1981a Argentine hemorrhagic fever: A biologic marker, *Intervirology* **15**:97.

Avila, M. M., Laguens, R. M., Laguens, R. P., and Weissenbacher, M. C., 1981b, Selectividad tisular e indicadores de virulencia de 3 cepas de virus Junin, *Medicina (Buenos Aires)* **41**:157.

Berria, M. I., Gutman Frugone, L. F., Girda, R., and Barrera Oro, J. G., 1967, Estudios immunologicos con virus Junin. I. Formación de anticuerpos en cobayós inoculados con virus vivo, *Medicina (Buenos Aires)* **27**:93.

Bishop, D. H. L., 1988, Lymphocytic choriomeningitis virus ambisense coding: a strategy for persistent infection? in: *Immunobiology and Pathogenesis of Persistent Virus Infections* (C. Lopez, ed.), pp. 79–90, American Society for Microbiology, Washington, D.C.

Bruns, M., Kratzberg, T., Zeller, W., and Lehmann-Grube, F., 1990, Mode of replication of lymphocytic choriomeningitis virus in persistently infected cultivated mouse L cells, *Virology* **177**:615.

Candurra, N. A., and Damonte, E. B., 1985, Influence of cellular functions on the evolution of persistent infections with Junin virus, *Arch. Virol.* **86**:275.

Candurra, N. A., Damonte, E. B., and Coto, C. E., 1989, Antigenic relationships between attenuated and pathogenic strains of Junin virus, *J. Med. Virol.* **27**:145.

Candurra, N. A., Scolaro, L. A., Mersich, S. E., Damonte, E. B., and Coto, C. E., 1990, A comparison of Junin virus strains; growth characteristics, cytopathogenicity and viral polypeptides, *Res. Virol.* **141**:505.

Ceriatti, F. S., Damonte, E. B., and Coto, C. E., 1983, Induccion de mutantes termosensibles del virus Junin por efecto del 5-fluoruracilo, *Rev. Arg. Microbiol.* **15**:105.

Ceriatti, F. S., Damonte, E. B., Mersich, S. E., and Coto, C. E., 1986, Partial characterization of two temperature sensitive mutants of Junin virus, *Microbiologica* **9**:343.

Clegg, J. C. S., Wilson, S. M., and Oram, J. D., 1990, Nucleotide sequence of the S RNA of Lassa virus (Nigerian strain) and comparative analysis of arenavirus gene products, *Virus Res.* **18**:151.

Contigiani, M. S., and Sabattini, M. S., 1977, Virulencia diferencial de cepas de virus Junin por marcadores biológicos en ratones y en cobayos, *Medicina (Buenas Aires)* **37** (Suppl. 3):244.

Coto, C. E., León, M. E., Martinez Peralta, L., Help, G. I., and Laguens, R. P., 1979, Induction of infectious virus and viral surface antigens in Vero cells persistently infected with Junin virus, in: *Humoral Immunity and Neurological Diseases* (D. Karcher, A. Lowenthal, and A. D. Strosber, eds.), pp. 405–415, Plenum Press, New York.

Coto C. E., Vidal, M. del C., D'Aiutolo, A. C., and Damonte, E. B., 1981, Selection of spontaneous ts mutants of Junin and Tacaribe viruses in persistent infections, in: *The Replication of Negative Strand Viruses* (D. H. L. Bishop and R. W. Compans, eds.), pp. 59–64, Elsevier–North Holland, New York.

Coulombié, F. C., and Coto, C. E., 1985, Efecto de la ciclofosfamida en el desarrollo de la infección de *Calomys musculinus* con virus Junin, *Medicina (Buenos Aires)* **45**:487.

Coulombié, F. C ., Alché, L. E., Lampuri, J. S., and Coto, C. E., 1986, Role of *Calomys musculinus* peritoneal macrophages in age-related resistance to Junin virus infection, *J. Med. Virol.* **18**:289.

Cresta B., Padula P., and Martinez Segovia, Z. M., 1980, Biological properties of Junin virus proteins. 1. Identification of the immunogenic glycoproteins, *Intervirology* **13**:284.

D'Aiutolo, A. C., and Coto, C. E., 1986, Vero cells persistently infected with Tacaribe virus: Role of interfering particles in the establishment of the infection, *Virus Res.* **6**:235.

Damonte, E. B., and Coto, C. E., 1979, Temperature sensitivity of the arenavirus Junin isolated from persistently infected Vero cells, *Intervirology* **11**:282.

Damonte, E. B., D'Aiutolo, A. C., and Coto, C. E., 1981, Persistent infection of Vero cells with Tacaribe virus, *J. Gen. Virol.* **56**:41.

Damonte, E. B., Mersich, S. E., and Coto, C. E., 1983, Response of cells persistently infected with arenaviruses to superinfection with homotypic and heterotypic viruses, *Virology* **129**:474.

Damonte, E. B., Mersich, S. E., Candurra, N. A., and Coto, C. E., 1985, Junin and Tacaribe viruses: neutralizing and immunoprecipitating cross-reactivity, *Med. Microbiol. Immunol.* **175**:85.

De Mitri, M. I., and Martinez Segovia, Z. M., 1980, Biological activities of Junin virus proteins. II. Complement fixing polypeptides associated with the soluble antigen and purified virus particles, *Intervirology* **14**:84.

De Mitri, M. I., and Martinez Segovia, Z. M., 1985, Polypeptide synthesis in Junin virus infected BHK 21 cells, *Acta Virol.* **29**:97.

Domingo, E., Martinez-Salas, E., Sobrino, F., de la Torre, J. C., Portela, A., Ortin, J., Lopez-Galindez, C., Peres-Brena, P., Villanueva, N., Najera, R., VandePol, S., Steinhauer, D.,

DePolo, N., and Holland, J., 1985, The quasi-species (extremely heterogeneous) nature of viral RNA genome populations: Biological relevance, a review, *Gene* **40**:1.

Ghiringhelli, P. D., Rivera-Pomar, R. V., Baro, N. I., Rosas, M. F., Grau, 0., and Romanowski, V., 1989, Nucleocapsid protein gene of Junin arenavirus (cDNA sequence), *Nucleic Acids Res.* **17**:8001.

Ghiringhelli, P. D., Rivera-Pomar, R. V., Lozamo, M. E., Grau, O., and Romanowski, V., 1991, Molecular organization of Junin virus S RNA: Complete nucleotide sequence relationship with other members of the arenaviridae and unusual secondary structures, *J. Gen. Virol.* **72**:2129.

Grau, D., Franze-Fernandez, M. T., Romanowski, V., Rustici, S. M., and Rosas, M. F., 1981, Junin virus structure, in: *The Replication of Negative Strand Viruses* (D. H. L. Bishop and R. W. Compans, eds.), pp. 11–14, Elsevier–North Holland, New York.

Guerrero, L. B., Weissenbacher, M. C., and Parodi, A. S., 1969, Immunización contra la FHA con una cepa atenuada del virus Junin. I. Estudio de una cepa modificada del virus Junin. Immunización de cobayós, *Medicina (Buenos Aires)* **29**:1.

Hotchin, J., Kinch, W., and Benson, L., 1971, Lytic and turbid plaque-type mutants of lymphocytic choriomeningitis virus as a cause of neurological disease or persistent infection, *Infect. Immun.* **4**:281.

Howard, C. R., 1986, Arenavirus replication, in: *Perspectives in Medical Virology* (A. J. Zuckerman, ed.), pp. 145–168, Elsevier, Amsterdam.

Jacobson, S., Dutko, F. J., and Pfau, C. J., 1979, Determinants of spontaneous recovery and persistence in MDCK cells infected with lymphocytic choriomeningitis virus, *J. Gen. Virol.* **44**:113.

Jahrling, P. B., and Peters, C. J., 1986, Serology and virulence diversity among Old World arenaviruses and the relevance to vaccine development, *Med. Microbiol. Immunol.* **175**:165.

Kenyon, R. H., McKee, K. T., Maiztegui, J. I., Green, D. E., and Peters, C. J., 1986, Heterogeneity of Junin virus strains, *Med. Microbiol. Immunol.* **175**:169.

Laguens, R. M., Lampuri, J. S., Coto, C. E., and Laguens, R. P., 1982, Patologia de la infección persistente del *Calomys musculinus* con una cepa atenuada de virus Junin, *Medicina (Buenos Aires)* **42**:273.

Lampuri, J. S., Vidal, M. del C., and Coto, C. E., 1982, Respuesta del *Calomys musculinus* a la infeccion experimental con virus Junin, *Medicina (Buenos Aires)* **42**:61.

Lehmann-Grube, F., Slenczka, W., and Tees, R., 1969, A persistent and unapparent infection of L cells with the virus of lymphocytic choriomeningitis, *J. Gen. Virol.* **5**:63.

Maiztegui, J. I., 1975, Clinical and epidemiological patterns of Argentine hemorrhagic fever, *Bull. WHO* **52**:567.

Mersich, S. E., Castilla, V., and Damonte, E. B., 1988, Lectin affinity of Junin virus glycoproteins, *Ann. Inst. Pasteur/Virol.* **139**:277.

Parodi, A. S., Greenway, D. J., Rugiero, H. R., Rivero, E., Frigerio, M. J., Mettler, N. E., Garzon, F., Boxaca, M., Guerrero, L. B., and Nota, N. R., 1958, Sobre la etiologia del brote epidemico de Junin, *Dia Medico* **30**:2300.

Peters, C. J., Jahrling, P. B., Liu, C. T. Kenyon, R. H., McKee, K. T., and Barrera Oro, J. G., 1987, Experimental studies of arenaviral hemorrhagic fevers, *Curr. Top. Microbiol. Immunol.* **134**:5.

Romanowski, V., and Bishop, D. H. L., 1983, The formation of arenaviruses that are genetically diploid, *Virology* **126**:87.

Sabattini, M. S., and Contigiani, M. S., 1982, Ecological and biological factors influencing the maintenance of arenavirus in nature, with special reference to the agent of argentine hemorrhagic fever, in: *International Symposium on Tropical Arboviruses and Hemorrhagic Fevers* (F. deP. Pinheiro, ed.), pp. 251, Academia Brasilera de Ciencias, Rio de Janeiro.

Sabattini, M. S., González del Rio, L., Diaz, B., and Vega, V. R., 1977, Infección natural y experimental de roedores con virus Junin, *Medicina (Buenos Aires)* **37**:149.

Scolaro, L. A., Mersich, S. E., and Damonte, E. B., 1987, Atenuación en la virulencia para ratón lactante de una mutante de virus Junin, *Rev. Arg. Microbiol.* **19**:9.

Scolaro, L. A., Mersich, S. E., and Damonte, E. B., 1989, Reduced virulence of a Junin virus mutant is associated with restricted multiplication in murine cells, *Virus Res.* **13**:283.

Scolaro, L. A., Mersich, S. E., and Damonte, E. B., 1990, A mouse attenuated mutant of Junin virus with an altered envelope glycoprotein, *Arch. Virol.* **111**:257.

Scolaro, L. A., Mersich, S. E., and Damonte, E. B., 1991, Experimental infection of suckling mice with a host range mutant of Junin virus, *J. Med. Virol.* **34**:237.

Shivaprakash, M., Harnish, D., and Rawls, W., 1988, Characterization of temperature-sensitive mutants of Pichinde virus, *J. Virol.* **62**:4037.

Steyerthal, N. L., Lampuri, J. S., and Coto, C. E., 1990, Variación de parámetros sanguineos en *Calomys musculinus* infectados con virus Junin, cepa XJC13, *Medicina (Buenos Aires)* **50**:335.

Traub, E., 1935, A filtrable virus recovered from white mice, *Science* **81**:298.

Traub, E., 1938, Factors influencing the persistence of choriomeningitis virus in blood of mice after clinical recovery, *J. Exp. Med.* **68**:229.

Vezza, A. C., and Bishop, D. H. L., 1977, Recombination between temperature-sensitive mutants of the arenavirus Pichinde, *J. Virol.* **24**:712.

Weber, C., Martinez Peralta, L., and Lehmann-Grube, F., 1983, Persistent infection of cultivated cells with lymphocytic choriomeningitis virus: regulation of virus replication, *Arch. Virol.* **77**:271.

Weber, E. L., Guerrero, M. B., and Boxaca, M. C., 1985, MRC-5 cells: A model for Junin virus persistent infection, *J. Gen. Virol.* **66**:1179.

Webster, R. G., and Laver, W. G., 1980, Determination of the number of nonoverlapping antigenic areas on Hong Kong (H3N2) influenza virus hemagglutinin with monoclonal antibodies and the selection of variants with potential epidemiological significance, *Virology* **104**:139.

Weissenbacher, M. C., Avila, M. M., Calello, M. A. Frigerio, M. J., and Guerrero, L. B., 1985, The marmoset *Callithrix jacchus* as an attenuation marker for arenaviruses, *Comun. Biol.* **3**:375.

Weissenbacher, M. C., Laguens, R. P., and Coto, C. E., 1987, Argentine hemorrhagic fever, *Curr. Top. Microbiol. Immunol.* **134**:79.

CHAPTER 6

The Unusual Mechanism of Arenavirus RNA Synthesis

DANIEL KOLAKOFSKY AND DOMINIQUE GARCIN

I. ARENAVIRUSES IN RELATION TO OTHER RNA VIRUSES

Animal viruses that contain single-stranded RNA genomes are broadly divided into two groups;

1. Those whose genomes are of positive polarity and which function directly as mRNAs. The capsids of these viruses are icosahedral, and in general these genomes contain 5′ cap groups and 3′ poly(A) tails, reflecting their roles as mRNA. These genomes as naked RNA are infectious.
2. Those whose genomes are the complements of mRNA, or of negative polarity. These are also distinguished from (+) RNA viruses in that
   a. The capsids of these viruses have helical symmetry and are packaged into virions together with the viral polymerase. This complex, the nucleocapsid (NC), allows mRNA synthesis to begin immediately upon infection, without *de novo* protein synthesis.
   b. In general, (−) RNA genomes do not contain the hallmarks of mRNA (5′ caps and 3′ poly(A) tails).
   c. These genomes as naked RNA are not infectious. The minimum unit of infectivity here is the viral NC, including its polymerase; otherwise there is no way to begin viral gene expression.

By most of these criteria, arenaviruses are (−) RNA viruses even

DANIEL KOLAKOFSKY AND DOMINIQUE GARCIN • Département de Microbiologie, Université de Geneve, Faculté de Médecine, CH-1211 Geneva 4, Switzerland.

The Arenaviridae, edited by Maria S. Salvato. Plenum Press, New York, 1993.

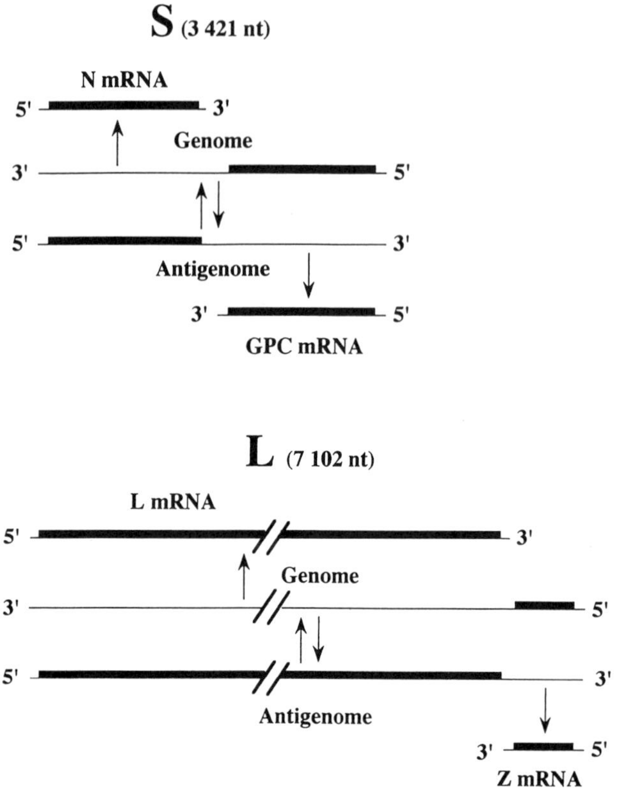

FIGURE 1. Genome organization of arenaviruses. The two genome segments, S (small) and L (large), of Tacaribe arenavirus are shown schematically as horizontal lines. Those regions which can be directly translated into proteins are shown as black rectangles. The arrows indicate the flow of RNA synthesis. (The sequences of these segments and their expression were determined by Franze-Fernandez et al., 1987; Iapalucci *et al.*, 1989a,b.).

though this term is not entirely appropriate. As shown in Fig. 1, the 3′ portions of the two arenavirus genome segments are (−) strands, whereas the 5′ portions can directly code for proteins and are (+) strands. The term "ambisense" has therefore been coined to describe this coding strategy (Auperin *et al.*, 1984). However, the 5′ ends of arenavirus genomes and antigenomes are not in fact directly translated into proteins. The genomes and antigenomes are found only as NC complexes tightly bound to N protein, and they do not contain the 5′ cap groups normally required for translation (Leung *et al.*, 1977; Raju *et al.*, 1990; Garcin and Kolakofsky, 1990). Rather, the coding sequences at the 5′ portions of these genomes are expressed from mRNAs transcribed from the 3′ portions of the antigenome segments, in the same way as mRNAs are transcribed from the 3′ portion of the genome templates (Bishop and Auperin, 1987). Thus, like bona fide (−) RNA viruses, arenavirus genomes and antigenomes never participate directly in protein synthesis.

II. THE 5′ ENDS OF ARENAVIRUS mRNAs

A central feature of (−) RNA viruses is that the same template is used for mRNA synthesis or transcription and genome replication. Similar to the other segmented viruses such as influenza and bunyaviruses, only a single mRNA is transcribed from each arenavirus genome (or antigenome) segment. In contrast to the nonsegmented (−) RNA viruses, these viral polymerases never reinitiate RNA synthesis on the same template.

For influenza and bunyaviruses (Lamb and Choppin, 1983; Kolakofsky and Hacker, 1991), the 5′ end of the genomes was found to be pppA at position 1, i.e., opposite the U residue at the precise 3′ end of the template. Genome synthesis would then begin with ATP at the end of the template as expected. Both influenza and bunyavirus mRNAs, however, were found to contain 10–18 extra bases at their 5′ ends, which were also heterogeneous in sequence and capped. According to the numbering system used, the mRNA ends are at positions −10 to −18 with respect to the genome terminus. The manner in which these 5′ ends are formed was investigated in detail by Krug and co-workers for influenza virus (Krug, 1981, 1983). The influenza polymerase also contains a methylated cap-dependent endonuclease activity, which cleaves host mRNAs 11–19 nt from their 5′ ends. These capped fragments are then used as primers for mRNA synthesis, with only the last base of the primer aligned with the 3′ U of the template. The term "capsnatching" is commonly used to describe this mechanism. Bunyaviruses were then found to initiate mRNA synthesis by capsnatching, in a manner that appears to be mechanistically identical to that of influenza viruses, even though bunyavirus mRNA is made in the cytoplasm whereas influenza mRNA is made in the nucleus (Patterson *et al.*, 1984; Kolakofsky and Hacker, 1991).

The manner in which arenavirus RNA synthesis is initiated was the last to be investigated, because arenaviruses grow more poorly than the other viruses. One of the problems has been to obtain mRNAs separated from genomes and antigenomes in sufficient amounts for characterization of their 5′ ends. This was finally achieved by using a CsCl density gradient centrifugation protocol that was successful for other (−) RNA viruses. When cytoplasmic extracts of infected cells are centrifuged in these gradients, the pellet fraction contains the mRNA, whereas genomes and antigenomes are found in the middle of the gradient, as judged by Northern blotting (Raju *et al.*, 1990). Arenavirus genomes and antigenomes are presumably found predominantly (or exclusively) as assembled NCs that are resistant to the high salt and pressure of these gradients and band at a density of 1.31 g/ml. Intracellular mRNAs, on the other hand, although they probably exist as mRNPs, are found as free RNAs in the pellet fraction because the mRNPs disassemble during centrifugation.

When the 5′ ends of the N protein mRNA in the CsCl pellet RNA were determined by primer extension, they were found at positions −1 to −5, with positions −3 and −4 predominating (see Fig. 2). These 5′ ends were apparently capped, as the mRNA:cDNA hybrids of the primer extension reactions could be specifically immunoselected with anticap antibodies. The 5′ ends of the N mRNA were then cloned via an anchored PCR protocol, and 29 were sequenced. The sequences confirmed the position range of the 5′ ends and showed that the extra bases were also heterogenous in sequence (Garcin and Kolakofsky, 1992).

The available data then suggest that arenaviruses also capsnatch to initiate mRNA synthesis. The major difference between arenaviruses and the others is the length of the extra bases at the 5′ end (1–5 vs. 10–18 for influenza and bunyaviruses). However, this can easily be explained if the arenavirus endonuclease generates shorter primers from host mRNAs than the others. The only other difference we noted was the complete lack of heterogeneity at position +1 of arenavirus mRNAs, whereas the others have some (but limited) heterogeneity at this position as well. Thus, although there are some differences, the evidence nevertheless still favors a form of capsnatching. However, the reader would be wise to suspend final judgment on how mRNAs are initiated until the initiation of genome synthesis is discussed.

III. THE 5′ ENDS OF GENOMES AND ANTIGENOMES

The 5′ end of the Tacaribe S genome was first determined by primer extension, followed by sequencing the extended primer by the chemical method, and this completed the sequence of the S segment (Raju *et al.*, 1990). Unexpectedly, there appeared to be one extra G residue at the 5′ end, i.e., relative to the presumed 3′ end of its template (the antigenome). Actually, the 3′ end of the antigenome has never been directly determined (only the 3′ end of the genome strand has been determined); it has

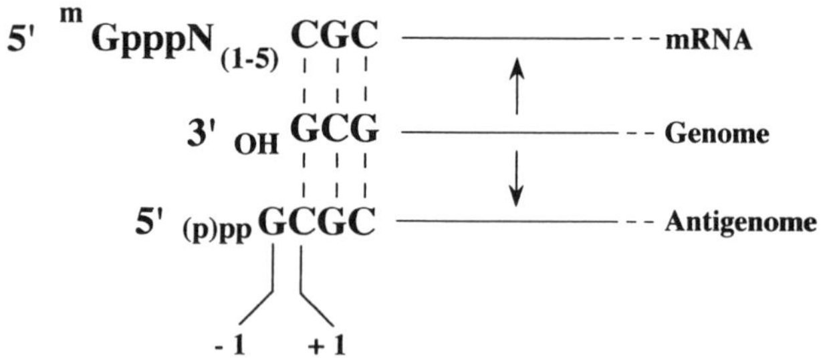

FIGURE 2. Alignment of the 3′ end of the genome and the 5′ ends of the RNAs transcribed from this template. The 3′ end of the genome is shown in the middle, and the relative positions and nature of the 5′ ends of the mRNA and antigenome are shown on either side. Positions 1 and −1 are indicated below.

simply been an article of faith that both 3′ ends are identical, as it is here that RNA synthesis begins. However, when the 5′ end of the antigenome was determined by primer extension, it too had one extra base, i.e., one base beyond that which could be templated at the 3′ end of its genome template (Fig. 2).

Considering that there was no precedent for this unusual situation, we wondered whether our results could be an artifact of the method used, i.e., whether the reverse transcriptase (RTase) was adding one nontemplated base to the end of the extended primer, like Q-beta replicase or T7 polymerase (Weber and Weissmann, 1970; Milligan *et al.*, 1987). Even though our RTase could correctly determine the ends of control RNAs in parallel experiments, it remained possible that an extra base was added only when certain sequences were encountered. If there was in fact an extra base at the 5′ end of genomes and antigenomes, it would exist as a 5′ overhang when these chains are annealed to form a complete ds-RNA (Fig. 2).

Alternatively, as the 5′ and 3′ ends of each of these chains are highly complementary, the 5′ overhang would also be present on the ds-panhandle structures formed by the intramolecular annealing of each chain. As the proposed overhanging base would be a G residue, it would be sensitive to digestion with RNase Tl, but not with RNase A. When this test was carried out with either ds-RNA form outlined above, predigestion with RNase A had no effect on the extended primer, which stopped at position −1, whereas when the ds-RNA was predigested with RNase Tl, the primer now extended only to position +1.

Thus, although unprecedented, there does appear to be one extra G residue at the 5′ ends of genome and antigenome chains, and this result has also been confirmed by cloning and sequencing the 5′ end of antigenomes (Garcin and Kolakofsky, 1992). Further experiments have also shown that these ends cannot be immunoselected with anticap antibodies, but can be capped *in vitro* with the vaccinia virus capping enzyme. Thus, the 5′ ends of arenavirus genomes and antigenomes contain an apparently nontemplated (p)ppG.

IV. CHARACTERIZATION OF AN *IN VITRO* SYSTEM

To learn more about how arenavirus RNA synthesis is initiated, we established an *in vitro* system for Tacaribe (TAC) virus (Garcin and Kolakofsky, 1992). To examine both genome replication and transcription if possible, we used cytoplasmic extracts of infected cells rather than purified virions. Moreover, an *in vitro* system in which lymphocytic choriomeningitis virus (LCMV) RNA was synthesized from intracellular nucleocapsids (NCs) had previously been reported (Fuller-Pace and Southern, 1989). Our *in vitro* system was relatively active and appeared to reflect viral RNA synthesis *in vivo* in large part. Both N and

GPC mRNAs, which terminate at the earliest of the sites used *in vivo*, are made and found predominantly in an unencapsidated form (CsCl pellet fraction) (Iapalucci *et al.*, 1991; Garcin and Kolakofsky, 1992). Genome-sense RNAs that have read through the intergenic region are also made, and these are found predominantly in the encapsidated region of the gradient. Moreover, similar to *in vitro* systems based on influenza virus and bunyavirus, TAC RNA synthesis *in vitro* is also strongly dependent on the addition of oligonucleotides to the reaction. For influenza virus and bunyavirus, whose templates begin at 3′ UCG and 3′ UCA, respectively, the dinucleotide ApG works best because it is complementary to the precise 3′ ends of these templates (see Fig. 3). The 3′ end of arenavirus templates is 3′ GCG, so other oligonucleotides were tried. Of the six dinucleotides (GpC, GpA, GpU, CpG, CpU, and UpC) and two trinucleotides (ApApC and ApCpC) we tested, only GpC, CpG, and ApApC stimulated the reaction, and in the order given. However, it is important to determine whether these oligonucleotides stimulated the reaction by acting as primers. If so, one might expect the 5′ ends of the resulting transcripts to be at different positions. This in fact turned out to be the case; ApApC stimulation led to chains whose 5′ ends were at position −2, GpC at −1, and CpG at +1 (Fig. 3). GpC stimulation was by far the strongest (ca. 70∼-fold) and led to transcripts whose 5′ ends were similar to genomes and antigenomes.

Encapsidated transcripts that had read through the intergenic region

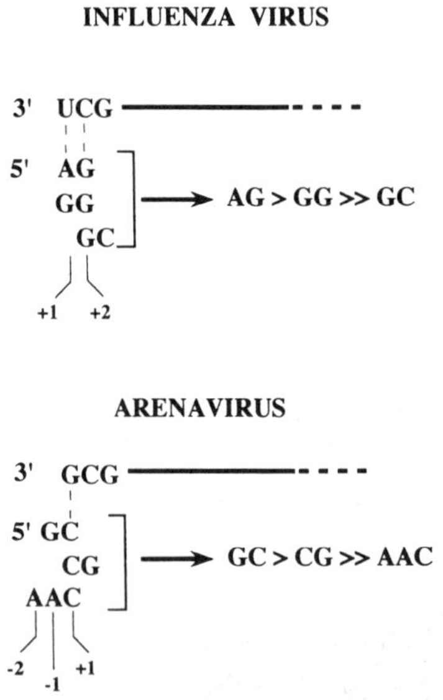

FIGURE 3. Oligonucleotide priming of influenza virus and arenavirus RNA synthesis *in vitro*. The di- and trinucleotides that prime RNA synthesis are shown below the templates and aligned according to where they start. The positions of the 5′ ends of the resulting transcripts, and the order of activity of the oligonucleotides, are also shown.

(i.e., genomes), however, represented only about 10% of the products; 90% of the product pelleted through the CsCl gradient and their 3′ ends mapped within the intergenic region, like those of *in vivo* mRNAs. The synthesis of transcripts that ended within the intergenic region was strongly stimulated by GpC, but genome synthesis was stimulated only slightly. It is possible that genome synthesis in our system is limited by the availability of N protein required for encapsidation. This would account for why the majority of transcripts whose 5′ ends are at position −1 are found in an unassembled form and terminate like mRNAs, whereas the opposite is true *in vivo*. These transcripts have been initiated by artificially high concentrations of GpC *in vitro*, but in the absence of sufficient N protein to carry out NC assembly, the majority would terminate as mRNAs.

V. AN UNUSUAL MODEL FOR THE INITIATION OF ARENAVIRUS GENOME REPLICATION

With this present work, *in vitro* RNA synthesis with all segmented negative-strand RNA viruses is highly dependent on oligonucleotide priming, and the major product is mRNA-like in length. For influenza and bunyaviruses, the requirement for ApG is thought to simply be a way of bypassing the natural route of chain initiation, i.e., capsnatching (mRNA) or initiation with pppA (genome replication), which operates poorly *in vitro*. Most RNA polymerases will initiate with dinucleotides representing the precise start sequence instead of NTPs (Downey *et al.*, 1971). For influenza virus, only 3 of the 16 possible NpNs were found to be active, in the order APG > GpG ≫ GpC (McGeoch and Kitron, 1975; Plotch and Krug, 1977). There is evidence that ApG primes influenza RNA synthesis at the precise 3′ end of the template (3′ UCG . . .) (Plotch and Krug, 1978). We assume that GpG acts similarly to ApG, except that the first base pair would be a G:U rather than an A:U pair, and that GpC acts by pairing with positions +2 and +3 of the template (Fig. 3), consistent with their order of activity. The 3′ end of arenavirus templates is similar, but is 3′ G̲CCG . . . . If the initiation of arenavirus and influenza virus RNA synthesis were mechanistically identical [as appears to be the case for influenza and bunyaviruses (Kolakofsy and Hacker, 1991)], one would have expected CpG to be more active than GpC, and to have initiated chains at positions +1 and +2, respectively. However, whereas CpG initiates at +1 as expected, GpC initiates chains at position −1 rather than +2 and is more active than CpG. We assume that ApApC is the least active because it can only form one base pair with the 3′ end of the template (Fig. 3) and cannot first anneal internally like GpC.

One possible explanation for these results is shown in Fig. 4b. The 5′ end of arenavirus genomes and antigenomes appears to be the nontemplated (p)ppG. We assume that this viral polymerase, like all others

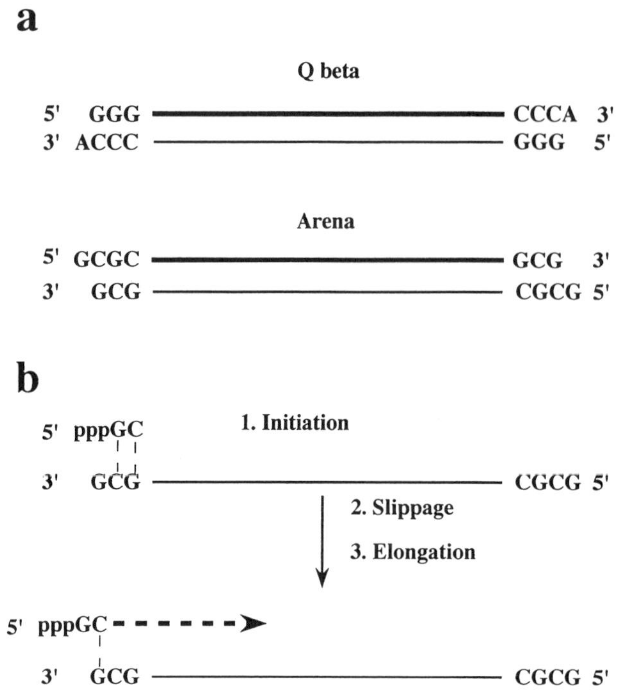

FIGURE 4. A model for the initiation of arenavirus genome replication. (a) Alignment of Q-beta and arenavirus genomes and antigenomes; only the terminal three or four bases are shown. (b) Model of how the overhanging 5′ end of arenavirus genomes is created, as discussed in the text.

known to date, cannot initiate with a pyrimidine triphosphate (Banerjee, 1980) and so initiates with GTP at position +2, by analogy with Q-beta RNA polymerase (Weber and Weissmann, 1970) (Fig. 4a). However, when the first phosphodiester bond is formed, the pppGpC must slip backward on the template, realigning to create a chain whose 5′ end is position −1, before the polymerase can continue downstream. There is evidence that paramyxovirus mRNA editing occurs via similar controlled slippage events, but during elongation rather than initiation (Vidal *et al.*, 1990), and other RNA polymerases are known to slip during initiation at certain promoters (reviewed in Jacques and Kolakofsky, 1991). In this scheme, the extra 5′ G is not nontemplated, but another example of pseudotemplated synthesis (Jacques and Kolakofsky, 1991). GpC is presumably the most potent stimulator because it follows the proposed natural route. It first aligns on the template at positions +2 and +3 to which it is complementary, before slipping backward by two bases such that only the 3′ C is aligned with the 3′ G of the template (Fig. 4b): hence the order of activity (GpC > CpG > ApApC) as well as the positions of the 5′ ends in the oligostimulated reactions.

Although the above explanation is not unreasonable, the rationale for such a mechanism seems obscure, and there is the added problem of how the overhanging ends are maintained. This latter problem is not unlike that of the telomeres of chromosomes. Because DNA synthesis begins on a RNA primer that is then removed, the 3′ ends of chromosome would get progressively shorter with continued replication if some mechanism did not exist to correct this: hence the continual growth of the 3′ ends due to a telomerase (Greider and Blackburn, 1989). A closer example is that of bacteriophage Q-beta, where the opposite but remarkably symmetrical situation holds (Weber and Weissmann, 1970) (See Fig. 4a). During replication, a nontemplated A residue is added to the 3′ ends, presumably because this polymerase cannot simply run off the end of its template, and adds the A in the act of terminating synthesis. Again, if some counteracting mechanism did not exist, the ends of the genomes would grow. The mechanism for Q-beta is to initiate chains not at the precise 3′ end of the template with UTP, but at the penultimate C with GTP, presumably because viral polymerases can only initiate with ATP or GTP. With this example in mind, it is possible that arenavirus polymerases can only terminate chains by removing the last base. The curious mechanism we propose for how arenaviruses create the extra 5′ G would then be part of the mechanism to prevent the natural ends of the genome from getting shorter with continued replication.

REFERENCES

Auperin, D. D., Romanowski, V., Galinski, M., and Bishop, D. H. L., 1984, Sequencing studies of Pichinde arenavirus S RNA indicate a novel coding strategy, an ambisense viral S RNA, *J. Virol.* **52**:897.

Banerjee, A. K., 1980, 5′ Terminal cap structure in eucaryotic messenger ribonucleic acids, *Microbiol. Rev.* **44**:175.

Bishop, D. H. L., and Auperin, D. D., 1987, Arenavirus gene structure and organization, *Curr. Top. Microbiol. Immunol.* **133**:5.

Downey, K. M., Jurmark, B. S., and So, A. G., 1971, Determination of nucleotide sequences at promoter regions by the use of dinucleotides, *Biochemistry* **10**:4970.

Franze-Fernandez, M. T., Zetina, C., Iapalucci, S., Lucero, M. A. , Bouissou, C., Lopez, R., Rey, 0., Daheli, M., Cohen, G. N., and Zakin, M. M., 1987, Molecular structure and early events in the replication of Tacaribe arenavirus S RNA, *Virus Res.* **7**:309.

Fuller-Pace, F. V., and Southern. P. J., 1989, Detection of virus-specific RNA-dependent RNA polymerase activity in extracts from cells infected with lymphocytic choriomeningitis virus: *In vitro* synthesis in full-length viral RNA species, *J. Virol.* **63**:1938.

Garcin, D., and Kolakofsky, D., 1990, A novel mechanism for the initiation of Tacaribe arenavirus genome replication, *J. Virol.* **64**:6196.

Garcin, D., and Kolakofsky, D., 1992, Tacaribe arenavirus RNA synthesis *in vitro* is primer dependent, and suggests an unusual model for the initiation of genome replication, *J. Virol.* **66**:1370.

Greider, C. W., and Blackburn, E. H., 1989, A telomeric sequence in the RNA of *Tetrahymena* telomerase required for telomere repeat synthesis, *Nature* **337**:331.

Iapalucci, S., Lopez, R., Rey, 0., Lopez, N., Franze-Fernandez, M. T., Cohen, G. N., Lucero, M., Ochoa, A., and Zakin, M. M., 1989a, Tacaribe virus L gene encodes a protein of 2210 amino acid residues, *Virology* **170**:40.

Iapalucci, S., Lopez, N., Rey, 0., Zakin, M. M., Cohen, G. N., and Franze-Fernandez, M. T., 1989b, The 5′ region of Tacaribe virus L RNA encodes a protein with a potential metal binding domain, *Virology* **173**:357.

Iapalucci, S., Lopez., N., and Franze-Ferndndez, M. T., 1991, The 3′ end termini of the Tacaribe arenavirus subgenomic RNAS, *Virology 182:269.*

Jacques, J. P., and Kolakofsky, D., 1991, Pseudo-templated transcription in prokaryotic and eukaryotic organisms, *Genes Dev.* **5**:707.

Kolakofsky, D., and Hacker, D., 1991, Bunyavirus RNA synthesis: Genome transcription and replication, *Curr. Top. Microbiol. Immunol.* **169**:143.

Krug, R. M., 1981, Priming of influenza viral RNA transcription by capped heterologous RNAS, *Curr. Top. Microbiol. Immunol.* **23**:125.

Krug, R. M., 1983, Transcription and replication of influenza viruses, in: *Genetics of Influenza Viruses* (P. Palese and D. W. Kingsbury, eds.), pp. 70–98, Springer-Verlag, Vienna.

Lamb, R. A., and Choppin, P. W., 1983, The gene structure and replication of influenza virus, *Annu. Rev. Biochem.* **52**:467.

Leung, W. C., Gosh, H. P., and Rawls, W. E., 1977, Strandedness of Pichinde virus RNA. *J. Virol.* **22**:235.

McGeoch, D., and Kitron, N., 1975, Influenza virion RNA-dependent RNA polymerase: stimulation by guanosine and related compounds, *J. Virol.* **4**:686–695.

Milligan, J. F., Groebe, D. R., Witherell, G. W., and Uhlenbeck, O. C. .1987, Oligoribonucleotide synthesis using T7 RNA polymerase and synthetic DNA templates. *Nucleic Acids Res.* **11**:8783.

Patterson, J. L., Holloway, B., and Kolakofsky, D., 1984, La Crosse virions contain a primer-stimulated RNA polymerase and a methylated cap-dependent endonuclease, *J. Virol.* **52**:215.

Plotch, S. J., and Krug, R. M., 1977, Influenza virion transcriptase: synthesis *in vitro* of large, polyadenylic acid-containing complementary RNA, *J. Virol.* **1**:24.

Plotch, S. J., and Krug, R. M., 1978, Segments of influenza virus complementary RNA synthesized *in vitro. J. Virol.* **2**:579.

Raju, R., Raju, L., Hacker, D., Garcin, D., Compans, R., and Kolakofsky, D., 1990, Non-templated bases at the 5′ ends of Tacaribe virus mRNAs, *Virology* **174**:53.

Vidal, S., Curran, J., and Kolakofsky, D., 1990, A stuttering model for paramyxovirus P MRNA editing, *EMBO J.* **2**:2017.

Weber, H., and Weissmann, C., 1970, The 3′-termini of bacteriophage Q-beta plus and minus strands, *J. Mol. Biol.* **51**:215.

CHAPTER 7

Subgenomic RNAs of Tacaribe Virus

MARIA TERESA FRANZE-FERNANDEZ, SILVIA IAPALUCCI, NORA LOPEZ, AND CARLOS ROSSI

I. INTRODUCTION

Tacaribe virus (TV) is the prototype of the Tacaribe group of serologically defined viruses geographically distributed in the Americas. Two of these viruses are human pathogens responsible for hemorrhagic fever, i.e., Junin virus in Argentina and Machupo virus in Bolivia. A particularly close antigenic relationship exists between TV and Junin virus manifested by their cross-reactivity at the level of antibody and by the ability of TV to protect animals from subsequent lethal challenges with Junin virus. TV, however, appears to be nonpathogenic for humans (reviewed by Howard, 1986).

Like all arenaviruses, TV is an enveloped virus with genetic information encoded in two segments of single-stranded RNA: a large segment, designated L, and a small segment, designated S. Since TV RNAs were completely sequenced (Franze-Fernandez *et al.*, 1987; Iapalucci *et al.*, 1989a,b, 1991; Raju *et al.*, 1990), it is now known that S RNA comprises 3421 nt and L RNA 7102 nt. In this chapter we will describe the genomic organization of TV, the identification of viral RNAs in

---

MARIA TERESA FRANZE-FERNANDEZ, SILVIA IAPALUCCI, NORA LOPEZ, AND CARLOS ROSSI • Centro de Virologia Animal (CONICET), Serrano 661, 1414 Buenos Aires, Argentina.

The Arenaviridae, edited by Maria S. Salvato. Plenum Press, New York, 1993.

infected cells, and the characterization of the 3′ termini of the subgenomic RNAs.

II. GENOMIC ORGANIZATION OF TACARIBE VIRUS

A. Proteins Encoded by the S RNA

Analysis of the nucleotide sequence of TV S RNA demonstrated two open reading frames (ORF) of significant size, one in the sense of the virus genome at the 5′ end and another in the virus-complementary or antigenome sense at the 3′ end of the S RNA. The two ORF do not overlap being separated by a noncoding intergenic region (Fig. 1). Comparison of the predicted primary structure of the two proteins with the predicted sequence of the corresponding proteins of other arenaviruses showed that the sequence at the 5′ region encoded the viral glycoprotein precursor (GPC), and that at the 3′ region the viral nucleoprotein (N) (Franze-Fernandez *et al.*, 1987). Thus, the TV S RNA, like the S RNA of a number of arenaviruses (reviewed by Bishop and Auperin, 1987), encodes the two major structural proteins, GPC and N, in an ambisense coding arrangement.

Consistent with the close antigenic relationship between TV and Junin virus, the N proteins of these two viruses share 77% identitiy, whereas the identity with N proteins of other arenaviruses ranges from 46% to 52% (Franze-Fernandez *et al.*, 1987; Ghiringhelli *et al.*, 1989). Comparison of the predicted sequence of TV GPC with the published primary sequences of other arenavirus GPCs showed an identity ranging from 33% to 41%. The distribution of the identical residues showed maximum homology in the C-terminal region and a clear minimum in the N-terminal moiety. This presumably reflects exposure of the less conserved region at the surfaces of the virus and of the infected cells (Franze-Fernandez *et al.*, 1987; Ghiringhelli *et al.*, 1989; Clegg *et al.*,

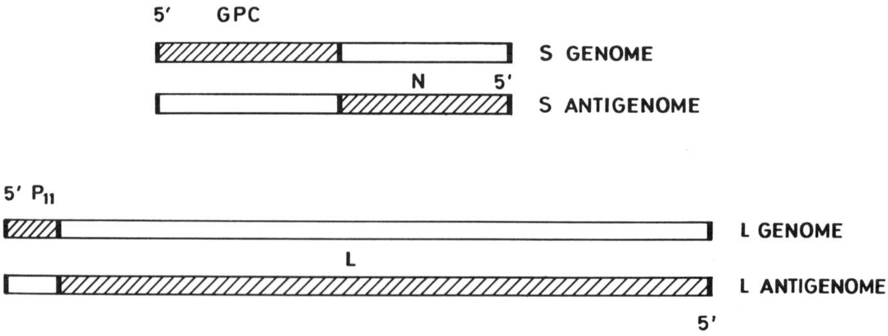

FIGURE 1. Schematic representation of the genomic organization of TV S and L-RNA species. The hatched lines indicate the sequences that can potentially be translated into proteins. The noncoding sequences are indicated in black.

1990). It is striking that the conserved, putative site of GPC cleavage, Arg-Arg or Arg-Lys (Buchmeier *et al.*, 1987), corresponds to an Arg-Thr sequence in TV GPC (Franze-Fernandez *et al.*, 1987).

B. Proteins Encoded by the L RNA

The L RNA segment, which accounts for about 70% of the coding potential of the virus genome, has been cloned and sequenced (Iapalucci *et al.*, 1989a,b). Analysis of the structure revealed a genetic organization similar to that of the S RNA (Fig. 1). An ORF located at the 3′ region is complementary to the L genome RNA and encodes a protein, the L protein, of 2210 amino acids. The other ORF at the 5′ region has the same sense as the genome and codes for a 94-amino-acid-long polypeptide, which we have called P11. The ambisense coding nature of the L RNA segment has been reported for another arenavirus, i.e., the lymphocytic choriomeningitis virus (LCMV) (Salvato and Shimomaye, 1989).

TV L protein (Iapalucci *et al.*, 1989a) is similar in size to both the unsegmented–negative strand viruses and the segmented–negative strand Bunyamwera virus L proteins (Poch *et al.*, 1990; Elliot, 1989), but bears little sequence homology with these putative RNA polymerases. If TV L protein sequence is aligned with that of LCMV L protein (Salvato *et al.*, 1989) the overall amino acid identity is 40%. The more conserved region lies in the central part of the molecules, spanning positions 680–8130 (56% overall homology) and 1000–1600 (62% overall homology). In these regions, the runs of consecutive invariant amino acids are up to 20 residues in length. It is noteworthy that when the L proteins of unsegmented–negative strand viruses are aligned, the profile of conservation obtained (Tordo *et al.*, 1988) is similar to that of TV and LCMV L proteins.

Since arenavirus L protein should operate in transcription and replication of the virus RNA, it might exhibit motifs with invariant or highly similar residues conserved in RNA-dependent polymerases. One of these motifs is an amino acid sequence consisting of an Asp-Asp sequence flanked by hydrophobic residues (Kamer and Argos, 1984). There is a sequence of this sort (residues 356–368) in TV L protein, but in the corresponding region of the predicted LCMV L protein the sequence motif is not conserved. Another motif is the pentapeptide Gln-Gly-Asp-Asn-Gln, invariably present in the L protein of unsegmented-negative strand viruses (Poch *et al.*, 1990). This motif is not found in TV L protein.

The polymerase protein PBI of influenza viruses A, B, and C exhibits a conserved 15-residue sequence motif, the central part of which is the tetrapeptide Ser-Ser-Asp-Asp (Yamashita *et al.*, 1989). In both, TV and LCMV L proteins the Ser-Ser-Asp-Asp motif is found within a con-

TABLE I. Homology between a Predicted Sequence in TV and LCMV L Proteins and a Conserved Sequence in the PB1 Protein of Influenza Viruses

	Amino acid residues											
TV L[a]	1325–1335	Y	T	S	S	D	D	Q	V	T	L	I
LCMV[b]	1317–1327	Y	T	S	S	D	D	Q	I	T	L	F
Influenza A virus PB1[c]	441–450	L	Q	S	S	D	D	–	F	A	L	I
Influenza B virus PB1[c]	440–449	L	Q	S	S	D	D	–	F	A	L	F
Influenza C virus PB1[c]	442–451	L	Q	S	S	D	D	–	F	V	L	F

[a] Iapalucci *et al.*, 1989a.
[b] Salvato *et al.*, 1989.
[c] Taken from Yamashita *et al.*, 1989.

served sequence located in the highly homologous region in the central part of the molecule (Table I).

The viral-sense sequence at the 5′ region of TV L RNA encodes a protein (P11) with an estimated molecular weight of 10,838 (Iapalucci *et al.*, 1989b). The TV P11 protein corresponds to the Z protein described by Salvato and Shimomaye (1989) in different strains of LCMV. Comparison of the deduced amino acid sequences of the P11 and Z proteins revealed only a 32% identity with no identical stretches of more than five amino acids. However, the P11 and Z proteins share some structural properties, the most striking being the distribution of the Cys residues in a conserved structure of the type Cys aa_2-Cys aa_9-Cys aa_5-Cys aa_2-Cys aa_{10}-Cys aa_2-Cys. This sequence bears a great similarity to the zinc finger protein sequences initially described by Miller *et al.* (1985). Another feature that is conserved in the P11 and Z proteins is a hydrophilic domain in the aminoterminal region. Within this region there is a cluster of four Ser residues in TV P11 (position 11–14). This tetrapeptide matches the sequences Thr-Asn-Ser-Thr and Ser-Gly-Thr-Ser in the Z protein of LCMV Armstrong and Traub strains, respectively (positions 11–14 and 12–15) (Salvato and Shimomaye, 1989). The Ser and Thr residues in the conserved sequences might be presumptive phosphorylation sites.

Salvato and Shimomaye (1989) have described the presence of ORFs other than those described above in the 5′ region of the L RNA. One of these, named X1, is in the viral-sense sequence and might correspond to an ORF in TV L RNA comprising nt 447–635 (Iapalucci *et al.*, 1989a,b). The amino acid sequences of the predicted proteins, however, exhibit no homology.

C. The 3′ and 5′ Termini of the S and the L RNA Segments

The 3′-end nucleotide sequence of both TV S and L genomic RNAs, directly determined by Auperin *et al.* (1982), show a 19-nucleotide sequence conserved in the arenaviruses, the sequence at the 3′ end of the S RNA species being identical to the corresponding L RNA sequence except for single base substitutions at positions 6 and 8. The sequence at the 5′ terminus of both the TV S and L genomes was determined by primer extension of the RNAs (Iapalucci *et al.*, 1989b; Raju *et al.*, 1990); the 5′-end 19-nt sequence is identical in both RNA segments (Fig. 2). As in other negative strand viruses, the 5′ and 3′ ends of arenavirus RNAs contain inverted complementary sequences in such an orientation that the ends can anneal to form a panhandle structure. In fact, structures resembling circularized RNA have been observed by electron microscopy of TV RNA and nucleocapsid preparations (Vezza *et al.*, 1978; Young and Howard, 1983). The predicted panhandle structures for TV S and L genome and antigenome are shown in Fig. 2. In the S genome, the ends are complementary for 20 nt, with mismatches at positions 6 and 13. These mismatches are preserved in the S antigenome. The complementarity of the 5′ and 3′ ends of the L genome extends for 20 nt with no mismatches, whereas in the L antigenome a mismatch appears in position 13. This arises from a UG pair in the genome, changing to an AC pair in the antigenome. If, as thought in other systems, the mismatched bases are important elements for RNA recognition by proteins (Wickens

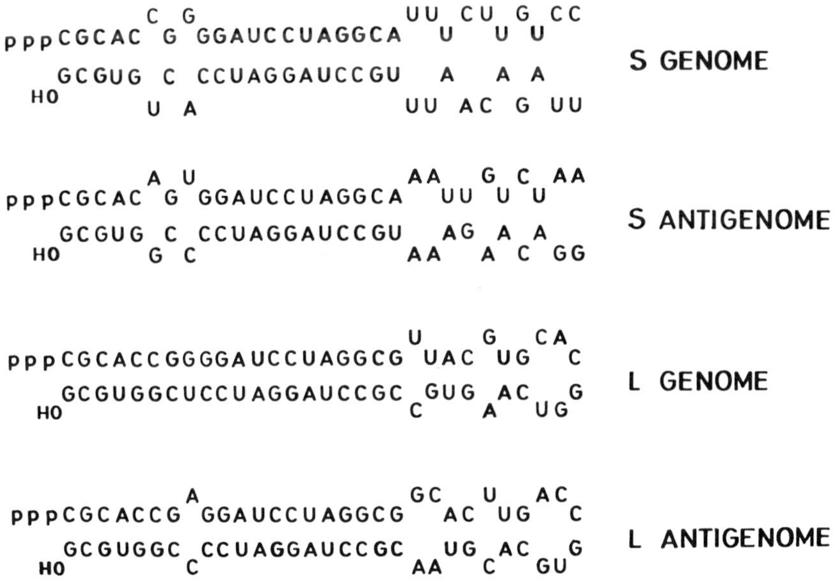

FIGURE 2. Predicted panhandle structures at the ends of TV S- and L-genomes and antigenomes. The estimated free energy of each hybrid complex (Tinoco *et al.*, 1973) is −43 kcal each for the S genome and the S antigenome and −55 kcal and −53 kcal for the L genome and the L antigenome, respectively. Base-pairing possibilities beyond the conserved 19 nt sequences have no major contribution to the stabilization of the ds RNA structures. The nontemplated extra G at the 5′ end of the S genome and the S antigenome (Garcin and Kolakofsky, 1990) was omitted.

and Dahlberg, 1987), the different structures at the predicted panhandles might play a role in the control of the transcription and replication of the virus RNAs.

D. Secondary Structures at the Intergenic Regions

The ambisense ORFs in both the S and the L RNAs are separated by noncoding intergenic regions, comprising 111 nt and 82 nt, respectively. As in all arenavirus S RNAs sequenced to date, the TV intergenic region presents a strong secondary structure that, in this virus, can be arranged into two hairpins (Iapalucci *et al.*, 1991) (Fig. 3) instead of the single hairpin structure found in Pichinde virus, LCMV, and Lassa virus (reviewed by Bishop and Auperin, 1987). Recently, a similar double-hairpin structure has been proposed for the S RNA intergenic region of Junin and Mopeia viruses (Ghiringhelli *et al.*, 1991; Wilson and Clegg, 1991). The similarity between the sequences of the two hairpins in Mopeia virus lends support to the suggestion that they were formed by a duplication event (Wilson and Clegg, 1991). In TV, however, the sequences of the two hairpins differ. The significance, if any, of a single- or a double-hairpin structure in the virus biology is a matter of speculation. At least, for the closely related TV and Junin viruses, these structures do not seem to be related to pathogenicity, since the sequences of the two potential hairpins in Junin virus S-intergenic region are very similar to those of TV.

The strong secondary structure of the intergenic region is conserved in the TV L segment. However, at variance with that occurring in the S RNA, the intergenic sequence of the L RNA can be arranged in a single hairpin configuration (Fig. 3). There is no sequence homology between the L RNA hairpin and the hairpins in the S RNA; the similarity lies instead in their high GC content. Interestingly, the predicted hairpin loop in the L RNA is stabilized by an estimated ΔG of -102 kcal; this value is similar to that obtained by adding the ΔGs corresponding to the hairpins of the S RNA ($\Delta G = -44$ kcal and -51 kcal for the hairpins closest to the GPC and the N stop codons, respectively).

III. TACARIBE VIRUS SUBGENOMIC RNAs

A. Characterization of the Subgenomic RNAs

Although both the S and the L genome and antigenome segments contain protein coding sequences at their 5′ region, they do not appear to be translated (Leung *et al.*, 1977). Instead, at least in Pichinde virus and LCMV, the GPC and the N proteins are expressed from subgenomic mRNAs of different polarities (Auperin *et al.*, 1984; Southern *et al.*,

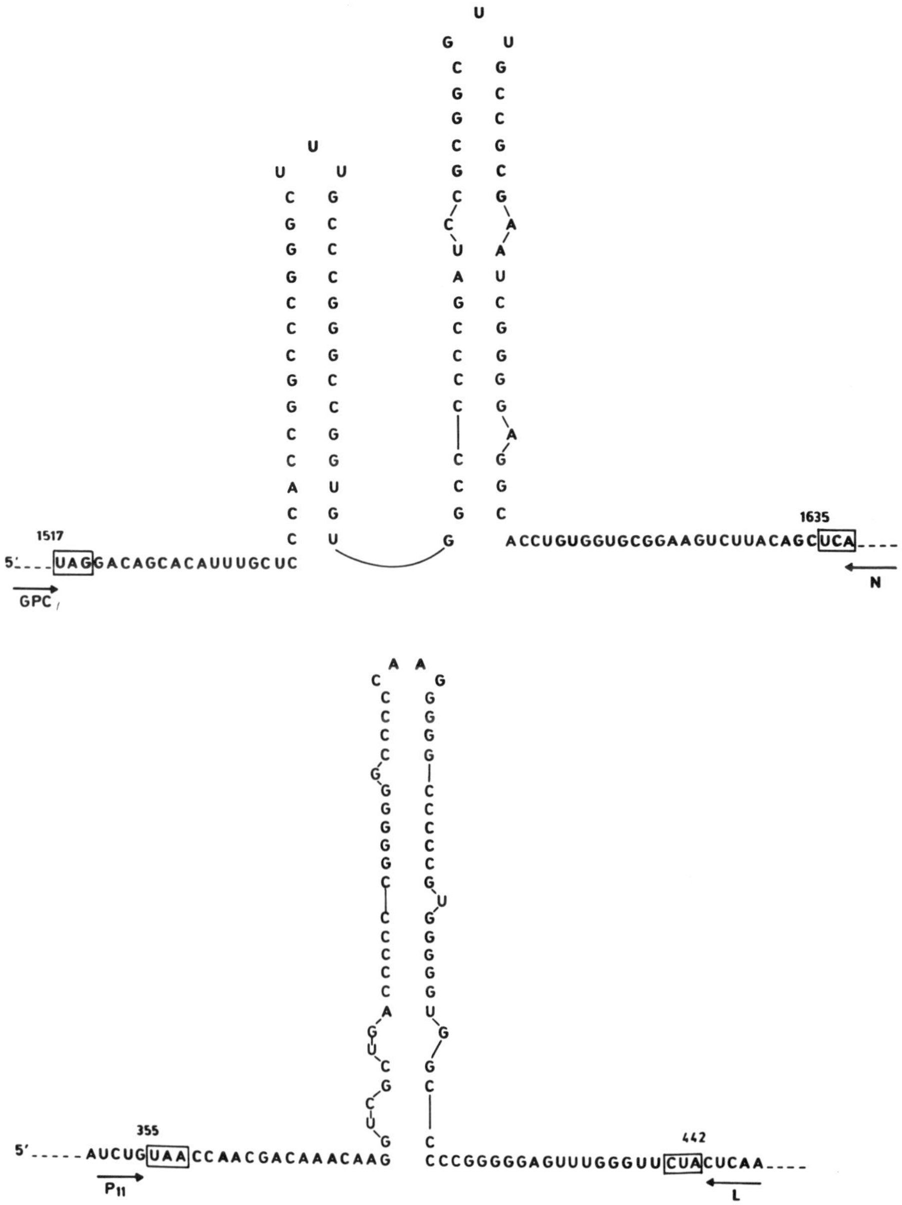

FIGURE 3. Intergenic region of TV S and L RNAs. Nucleotide sequence and predicted secondary structure of the S RNA (top) and the L RNA (bottom) intergenic regions. Nucleotides are numbered from the 5′ to the 3′ end of the viral-sense RNA. The predicted hairpins in the S RNA are stabilized by 13 and 16 base pairs ($\Delta G = -44$ kcal and -51 kcal) and the single hairpin in the L RNA by 20 base pairs ($\Delta G = -102$ kcal) (Tinoco *et al.*, 1973). Arrows indicate the orientations of the coding sequences for the GPC and N genes in the S RNA and for the P11 and L genes in the L RNA. The boxed nucleotide sequences represent the stop codons; those of the N and L genes should be read in the viral-complementary sense sequence.

1987). The translatability of subgenomic RNA over genomic RNA may lie in differences in secondary structure; for instance, the ORF may be obscured by interaction with the 3′ half of the molecule. On the other hand, the subgenomic RNA may be translatable because of 5′ end sequence modifications.

Northern blot analysis of TV-infected cell extracts using strand-specific probes for the S RNAs revealed, in addition to the S genome and antigenome, two subgenomic RNAS: one in the viral-sense GPC sequence and the other in the viral-complementary sense N sequence (Franze-Fernandez *et al.*, 1987). These last RNA species should represent the TV GPC and N mRNAs (Fig. 5).

The ambisense genetic organization of the L RNA segment (Iapalucci *et al.*, 1989a,b) suggested the existence of mRNAs of different polarities for the P11 and the L proteins. The presence of subgenomic L RNAs in TV-infected cells was investigated by Northern blot analysis using strand-specific probes covering different regions of the L RNA (Fig. 4). The viral-complementary DNA probe, and in much less proportion the viral-sense probe, prepared from the region encoding the P11 gene annealed to a small RNA species of about 400 nt. Neither RNA of this size nor other subgenomic RNA bands were detected when probes corresponding to the L gene were used in the hybridization experiments. Considering that the stop codon of the P11 ORF is located at nt 355–357

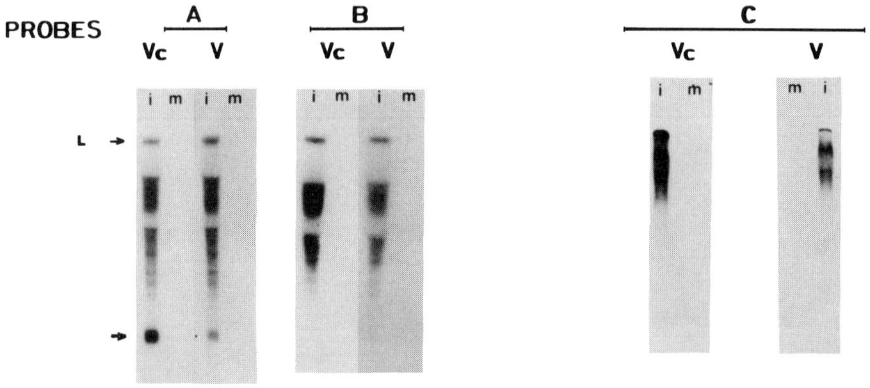

FIGURE 4. Identification of TV L-RNAs in infected Vero cells. Infection of the cells and northern blot analysis were performed as previously described (Iapalucci *et al.*, 1989b). Probe A represents nucleotides 173–322 (numbering from the 5′ end of the genome) and corresponds to the P11 gene. Probes B and C span nucleotides 618–961 and 4772–7072, respectively, and derive from the L gene. The probes are named according to their sense; thus, the viral-complementary sense (Vc) probe detected viral-sense RNA, and the viral-sense (V) probe detected complementary-sense RNA; i and m indicate RNA from infected and mock-infected cells, respectively. The arrow at the bottom of the leftmost panel denotes the approximately 400-base subgenomic RNA hybridizing only to probe A. The cross-hybridizing RNAs below the L RNA might represent defective RNAs generated during infection or degradation products of the L RNA.

from the 5′ end of the L genome (Iapalucci *et al.*, 1989b), the viral-sense, approximately 400 nt RNA should represent the P11 mRNA. The failure to detect a subgenomic RNA corresponding to the L gene might be explained by the fact that the putative L mRNA, with an estimated length of at least 6663 nt (Iapalucci *et al.*, 1989a), was not separated from the L antigenome (7102 nt) by analysis on gels, as a consequence of their small size difference in the high-molecular-weight range. It will be seen below that the 3′ end terminus of the putative L mRNA was defined by mapping with nuclease S1.

The intracellular TV S and L RNA species are schematically represented in Fig. 5, together with a proposed mechanism for transcription and replication. The ambisense coding arrangement of the arenaviruses determines that, at variance with other negative-strand viruses, both the genomes and the antigenomes serve as templates for mRNA transcription and for RNA replication.

In the negative-strand viruses both genomes and antigenomes are found encapsidated, and in some of the viruses the association between the viral nucleocapsid protein and the RNA is so tight that nucleocapsids (NCs) resist disaggregation on CsCl gradients. This is also the case for the virion genome of the arenavirus Tamiami (Gard *et al.*, 1977). Recently, Raju *et al.* (1990) detected the separation of the encapsidated TV S genome and antigenome from the unencapsidated GPC and NP mRNAs by centrifugation of the virus-infected cell extracts on CsCl density gradients. In the experiment depicted in Fig. 6, we have used a similar protocol for investigating the resolution of both the S and the L genomes and antigenomes from the putative mRNAs. To this end, the cytoplasmic extracts from TV-infected cells were fractionated on CsCl density gradients under conditions in which the encapsidated RNAs should band and the unencapsidated RNAs should sediment in the pellet. Then the RNAs from the Nc fraction and the pellet fraction of the

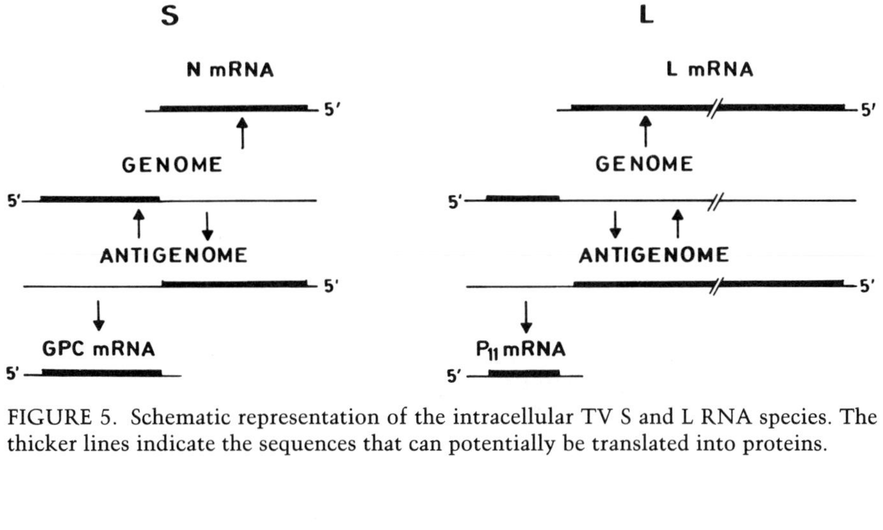

FIGURE 5. Schematic representation of the intracellular TV S and L RNA species. The thicker lines indicate the sequences that can potentially be translated into proteins.

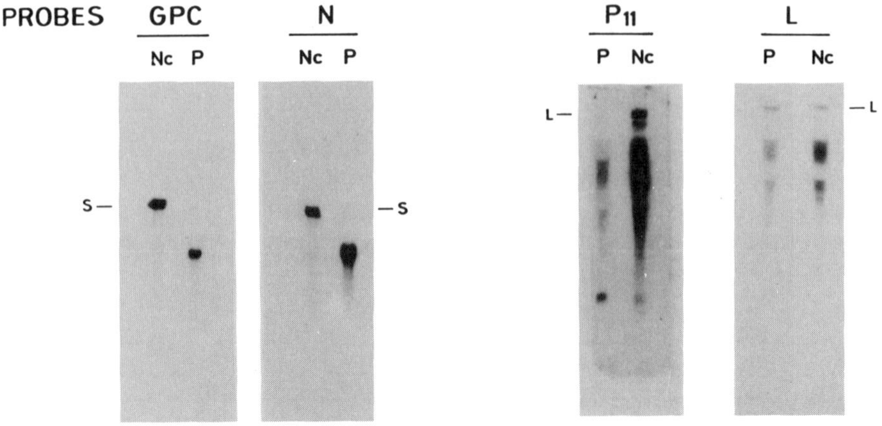

FIGURE 6. Separation of nucleocapsids from mRNAs by CsCl density gradient centrifugation. CsCl gradient fractionation of cytoplasmic extracts from TV-infected cells was performed as previously described (Iapalucci *et al.*, 1991). CsCl gradient pellet RNA (P) or nucleocapside RNA (Nc) were resolved on agarose gels and analyzed by Northern blot as indicated (Iapalucci *et al.*, 1991). The GPC probe hybridizes with the GPC mRNA and the S genome RNA. The N probe detects the N mRNA and the S antigenome. The P11 probe hybridizes with the P11 RNA and with the L genome. The L probe detects the L antigenome and the L mRNA. The derivation of the probes has been described previously (Iapalucci *et al.*, 1991).

gradient were analyzed by Northern blot using specific probes for recognition of the different RNAs. When the blot was hybridized with probes for the S RNA, bands representing the S genome and antigenome RNAs were found only in the Nc fraction, the subgenomic RNAs representing the N and the GPC mRNAs being detected only in the pellet fraction (left panel).

Analysis of the blot with the P11 probe, which anneals to the P11 mRNA and the L genome, led to the finding that the P11 RNA sedimented mostly in the pellet fraction and to a lesser extent in the Nc fraction (right panel). Since the experiment allowed the neat resolution of the S segment mRNAs, it seems unlikely that unencapsidated P11 RNA contaminated the Nc fraction. The observation that a P11 RNA species is packaged into virions (N. Lopez, unpublished results) supports the notion of an encapsidated P11 RNA as well. The subgenomic RNA encoding Z is also found encapsidated in virions of LCMV (Salvato and Shimomaye, 1989). The P11 probe also detected the L genome RNA sedimenting in the Nc fraction. The L probe led to the detection of bands with the mobility of the L antigenome in both the pellet RNA and the Nc RNA lanes. These bands might represent the L mRNA and the antigenome RNA, respectively. It should be noticed that under conditions in which the S RNAs looked intact, genome- and antigenome-sense L RNAs migrating faster than the full-length L segment were detected.

Whether they represent DI RNAs generated during the infection or degradation products of the L RNAs remains to be determined.

B. The 3′ Termini of the Subgenomic RNAs

In all arenavirus RNAs sequenced to date, the intergenic region conserves a strong secondary structure, regardless of their sequence diversity (Bishop and Auperin, 1987; Iapalucci *et al.*, 1989b, 1991; Salvato and Shimomaye, 1989; Wilson and Clegg, 1991; Ghiringhelli *et al.*, 1991). On the other hand, in closely related arenaviruses such as the African pair Mopeia and Lassa viruses (Wilson and Clegg, 1991) and the South American TV and Junin viruses (Iapalucci *et al.*, 1991; Ghiringhelli *et al.*, 1991), homology in the intergenic sequence is restricted to the predicted hairpins, but not to the flanking sequences. These observations suggest that the hairpin structures might somehow be involved in the generation of the subgenomic RNAs. In fact, the roles played by secondary structures at the 3′ end of the transcripts on the termination of transcription by prokaryotic and eukaryotic RNA polymerases are well known (Von Hippel *et al.*, 1984; Platt, 1986). The characterization of the termination sites of the TV subgenomic RNAs would, therefore, provide an insight into the signals involved in arenavirus transcription termination.

1. The 3′ Ends of the mRNAs Encoded by the L Segment

The 3′ ends of TV putative mRNAs were determined by mapping with nuclease Sl. Experiments were performed with the unencapsidated RNAs sedimenting in the pellet fraction of the CsCl density gradient (see Fig. 6). For the analysis of the 3′ end of the putative P11 mRNA, the RNA was hybridized with a probe consisting of a double-stranded DNA fragment labeled only at the 3′ end of the antigenome-sense strand (Fig. 7A). The hybrids were digested with nuclease S1 and the resistant products were analyzed in a sequencing gel; the sequence of the probe by the chemical method was run simultaneously to determine the size of the protected fragments (Fig. 7B). As shown in lane P, two bands migrating faster than the probe appeared. The major band, with the higher mobility, mapped in the vicinity of positions 410–412; the minor band migrated some 3–4 nt slower than the former band. Two similarly migrating, very faint bands, in addition to the expected band of the size of the original probe, were detected when the CsCl gradient Nc RNA was used for protection. The two faint bands should represent protection by the minor amount of P11 RNA sedimenting in the Nc fraction (see Fig. 6).

The L probe, used to map the 3′ terminus of the putative L mRNA, consisted of a 462-nt restriction fragment that was labeled at the 3′ end

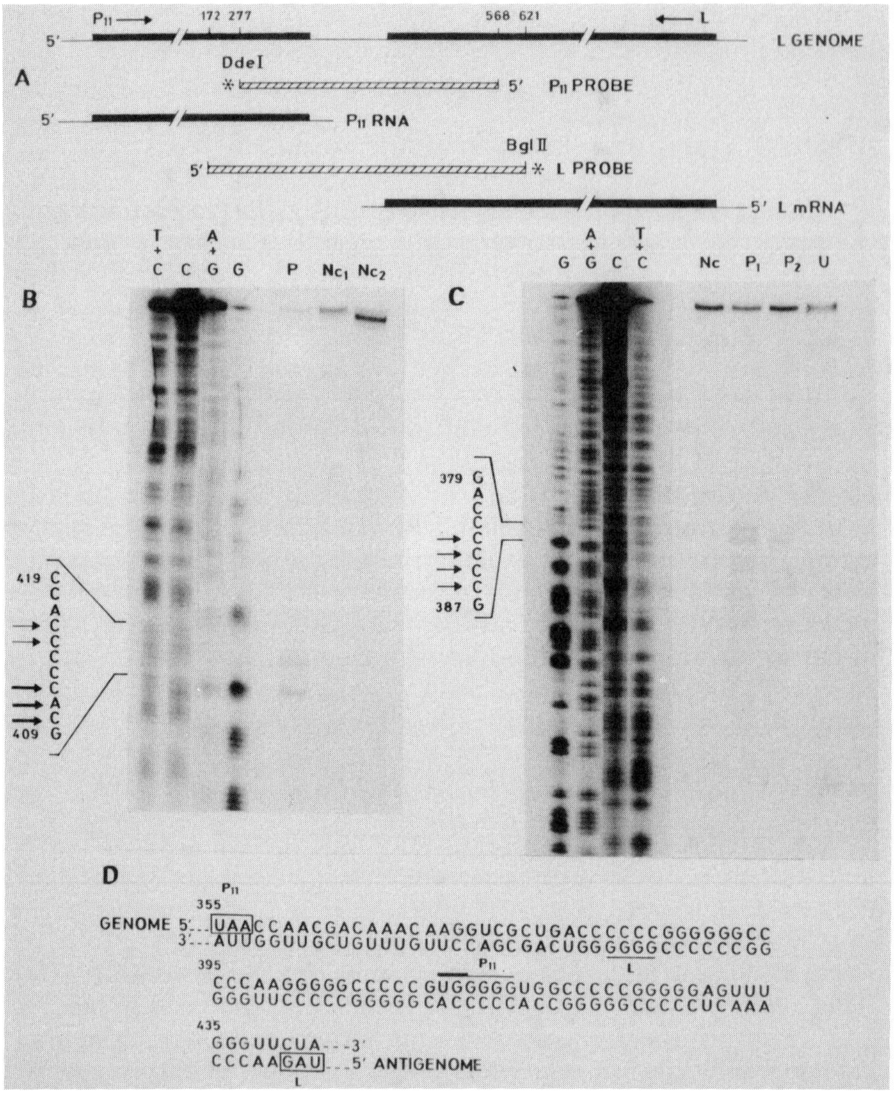

FIGURE 7. Analysis of the 3′ end of the mRNAs encoded by the L segment. (A) Organization of TV L genome RNA; the thicker lines indicate the coding sequences of the P11 and the L genes and the arrows show the orientations of the coding sequences. The numbers indicate the positions of the restriction sites used and the generation of the cDNA fragments for S1 mapping. The probes are represented as the cDNA strand that was labeled at the 3′ end (*). (B) Pellet RNA (P) or Nc RNA (Nc_1 and Nc_2) from a CsCl density gradient fractionation was hybridized with the P11 probe and digested with 450 units (P and Nc_1) or 600 units (Nc_2) of S1 nuclease. The protected fragments were analyzed in a sequencing gel along with the sequence of the probe by the chemical method. (C) Polyacrylamide gel electrophoresis analysis of the protected L probe after Sl mapping of the following RNAS: CsCl pellet RNA (P_1 and P_2) Nc RNA (Nc), and RNA from uninfected cells (U). The amounts of nuclease Sl used for mapping were 250 units (Nc; P_1; U) and 400 units (P_2). The leftmost lanes show the sequences of the probe by the chemical method. In (B) and (C) the numbers represent the positions from the 5′ end of the L genome and the arrows indicate

of the genome-sense cDNA strand (Fig. 7A). Protection of the L probe with unencapsidated RNA led to the appearance of two stronger bands and a smeared region down in the gel after digestion with the lower amount of the nuclease Sl (Fig. 7C, lane P_1). After digestion with the higher amount of the enzyme (lane P_2), a single band appeared the migration of which coincided with the reading of a stretch of cytidines in the sequence of the probe, which represent positions 386–383 relative to the 5′ end of the L genome. No band of this size was detected when the probe was protected with RNA from the Nc fraction of the CsCl gradient or with RNA from uninfected cells. A noticeable result of this experiment is the presence of an unencapsidated–antigenome-sense L RNA, the 3′ end of which mapped within the intergenic sequence (see Fig. 7D). This RNA should represent the L mRNA and is the first indication of a subgenomic mRNA for this gene ever seen among the arenaviruses.

As indicated in Fig. 7D, the sites where the 3′ end of the P11 and L mRNAS mapped are localized at positions 410–412 and 386–383, respectively. As a result, the noncoding region at the 3′ end of the transcripts overlapped by 24–34 nt in complementary sequences.

2. The 3′ Ends of the mRNAs Encoded by the S Segment

For the determination of the 3′ ends of the TV N and GPC mRNAs, protocols similar to those described for the L segment mRNAs were applied. The probes consisted of a double-stranded DNA fragment labeled at the 3′ end of either the genome-sense strand (NP probe) or the antigenome-sense strand (GP probe) (Fig. 3A).

When the N probe was protected with the unencapsidated mRNAs (i.e., CsCl pellet), a band migrating faster than the original probe was detected. The migration of this band paralleled the reading of a stretch of cytidines in the sequence of the probe representing positions 1573–1571 relative to the 5′ end of the S genome.

The experiment depicted in Fig. 8C was intended to map the 3′ end of the GPC mRNA. To this end, the antigenome-sense G probe was annealed with RNA from the pellet fraction of a CsCl gradient and the hybrids were digested with two different amounts of S1 nuclease. The nuclease-resistant products were then electrophoresed in a sequencing gel, using sequencing reactions from M13 mpl9 recombinant DNA as markers. Two bands were resolved, corresponding to 290–291 and 292–293 baselong Sl resistant fragments. This indicated that the GPC mRNA

---

the sites where the 3′ end of the mRNAs mapped. Procedures are detailed in Iapalucci *et al.* (1991). In (D) the sequence of both the L genome and antigenome intergenic region are represented. The numbers indicate the positions relative to the 5′ end of the genome. The stop codons are boxed and the mapped termination sites for the mRNAs are either overlined (P_{11}) or underlined (L).

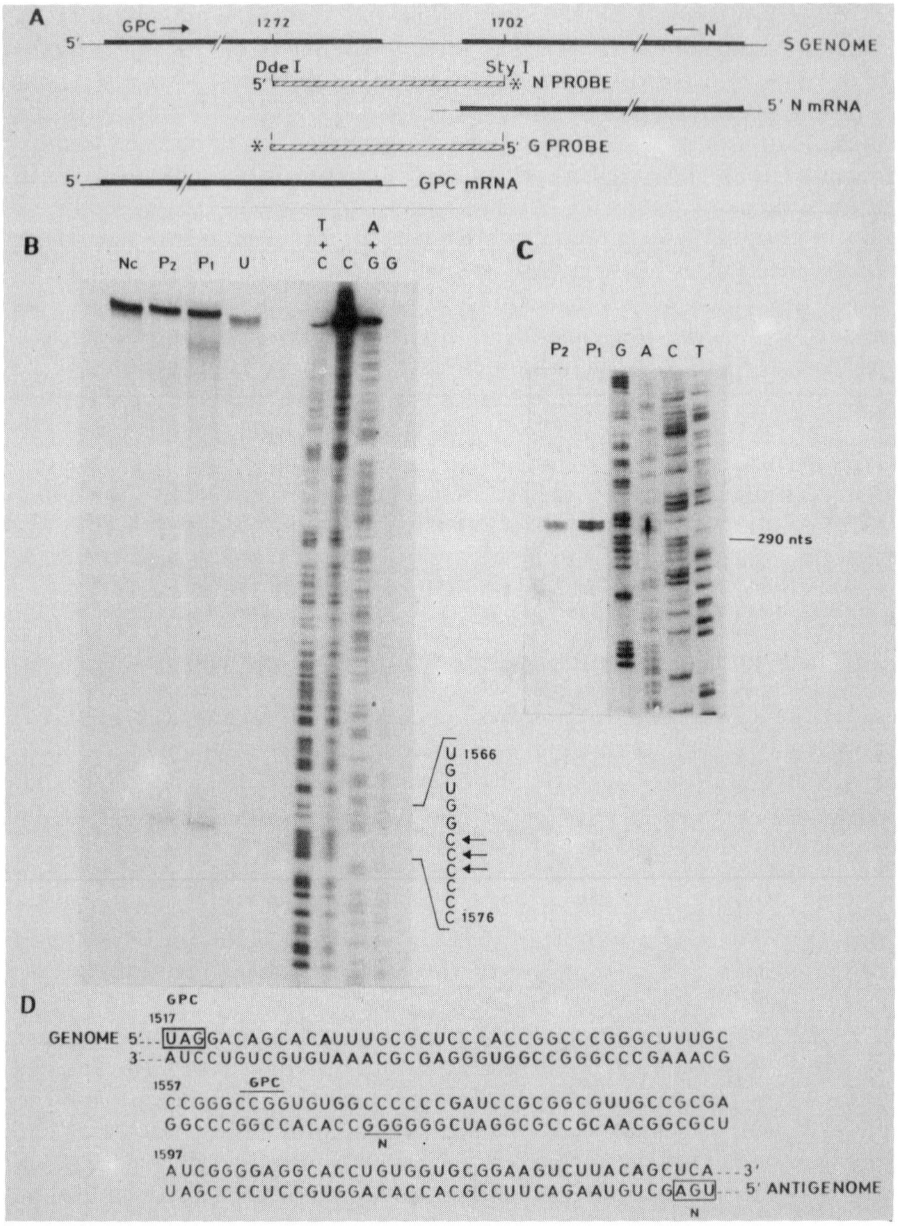

FIGURE 8. Analysis of the 3′ termini of the mRNAs encoded on the S segment. (A) Organization of TV S genome RNA; the thicker lines indicate the coding sequences of the GPC and the N genes and the arrows show the orientations of the coding sequences. The N and GP probes were used for mapping the N and GPC mRNAs, respectively. The star indicates the labeled end of the probe. (B) RNA from either the pellet fraction (P_1; P_2) or the Nc fraction (Nc) of a CsCl density gradient or RNA from uninfected cells (U) was mapped with nuclease S1 using the N probe as indicated by Iapalucci et al (1991). Digestions were performed with 200 units. (P_1) or 300 units (P_2; Nc; U) of S1 nuclease. The rightmost lanes show the sequence of the probe by the chemical method; the numbers indicate the posi-

mapped at positions 1562–1565 from the 5′ end of the S genome. As shown in Fig. 8D, the termination sites of both the GPC and the N mRNAs mapped within the S-intergenic sequence. At variance with the P11 and L mRNAs, the 3′ ends of the mRNAs encoded by the S segment do not overlap.

Since the intergenic regions where the subgenomic RNAs mapped have a strong secondary structure, we investigated the possibility of artifacts arising in the mapping experiments. To this end, we first synthesized run-off transcripts *in vitro* using vectors containing TV cDNA inserts downstream of the T7 promoter, the transcripts ending at known sites within the intergenic regions. Then, the 3′ ends of the *in vitro* transcripts were mapped applying the protocols used for mapping the 3′ end of the natural TV subgenomic RNAS. The results showed an excellent agreement between the actual 3′ end of the *in vitro* RNAs and the termination sites as mapped with nuclease S1 (Iapalucci *et al.*, 1991), thus validating the use of this technique for our specific problem.

3. Secondary Structure at the 3′ End of the Transcripts

In accordance with the sequence diversity in arenavirus RNA intergenic regions, no special sequences that might function as termination signals in the vicinity of the 3′ end of the transcripts were evident. For instance, the guanosine-rich 3′-end sequences of the transcripts are inserted in regions with a high GC content. The template sequence CAAAC located about 13–18 nt downstream of the P11 and the L mRNA 3′-end sites has no counterpart either in the S RNA intergenic sequence of TV or in that of other arenaviruses (Bishop and Auperin, 1987; Wilson and Clegg, 1991; Clegg *et al.*, 1990; Ghiringhelli *et al.*, 1991). The 3′ ends show, however, a common feature: a region of almost perfect dyad symmetry that can theoretically take the form of a self-complementary hairpin loop (Fig. 9). Although the sequences within the hairpins exhibit no homology, common properties of the proposed structures are their high GC content and their predicted stable hairpin configuration ($\Delta G \geq -25$ kcal).

Secondary structure in the nascent RNA seems to have a determining influence on the fate of the transcription complex. For instance, accumulated data in rho-independent terminators suggest that formation of a hairpin in the transcript causes the prokaryotic RNA polymer-

---

tions relative to the 5′ end of the genome and the arrows show the sites were the N mRNA mapped. (C) The G probe was hybridized with CsCl gradient pellet RNA and the hybrids were digested with 400 units (P_1) or 500 units (P_2) of S1 nuclease. To estimate the size of the protected fragment four sequence reactions (lanes T, C, A, G) of an S-specific M13 DNA were electrophoresed simultaneously. The sequences of both the S genome and antigenome intergenic region are represented in (D). The numbers indicate the positions relative to the 5′ end of the genome. The stop codons are boxed. The mapped termination sites for the GPC and N mRNAs are indicated.

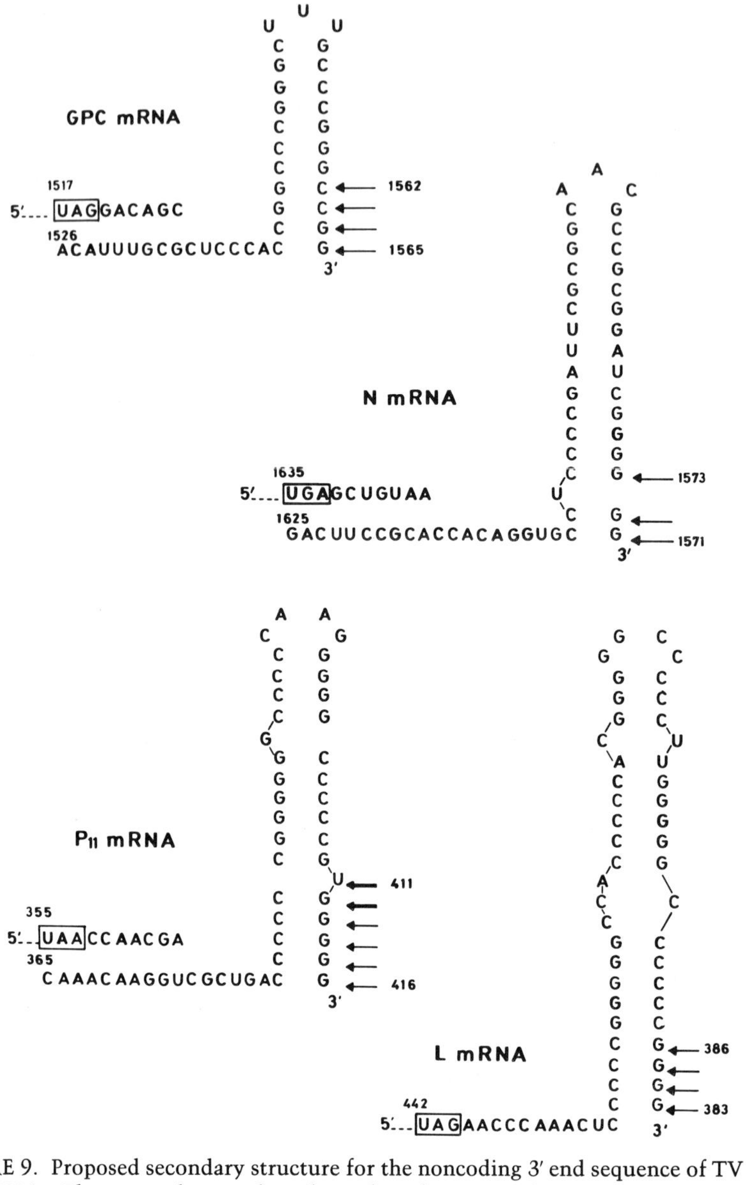

FIGURE 9. Proposed secondary structure for the noncoding 3′ end sequence of TV putative mRNAs. The stop codons are boxed. Nucleotides are numbered considering the 5′ end of the genome RNA to be 1. The arrows indicate the stop sites as mapped with nuclease S1. The proposed hairpins are stabilized by estimated, ΔG values of −35 to −55, −43 to −58, −33 to −40, and −25 to −37 kcal for the P11 RNA, the L mRNA, the NP mRNA, and the GPC mRNA, respectively (Tinoco *et al.*, 1973). The range of estimated Δ G values for each hairpin reflects the different termination sites considered.

ase to pause, and that the stretch of uridines downstream of the hairpin facilitates dissociation from the template (Platt, 1986). A similar mechanism appears to operate, at least in certain virus systems, in the termination of transcription by the eukaryotic RNA polymerase II (Bengal and Aloni, 1989). By analogy with these systems, it is tempting to speculate that the conserved hairpin structures at the 3′ end of the TV transcripts might at least be part of the transcription termination signal for the viral polymerase. On these bases, a mechanism for transcription termination and antitermination in TV is proposed (Fig. 10). When the polymerase complex, in the transcription mode, is copying the intergenic sequence, formation of the intramolecular hairpin structure in the nascent RNA might cause the polymerase to stall, leading to disruption of the transcription complex and termination of RNA synthesis (Fig. 10A). According to this proposal, any factor that prevents the folding at the 3′ end of the transcript will function as an antiterminator, leading to a switch from mRNA transcription to replication. Since TV genomes and antigenomes are found encapsidated, it is possible that in the replicative mode, the N protein initiates encapsidation at the conserved sequence at the 5′ end of the nascent RNA chains; then, subsequent addition of N molecules to the growing chain would lead to disruption of the hairpin structure when the intergenic region is reached (Fig. 10B). In this way,

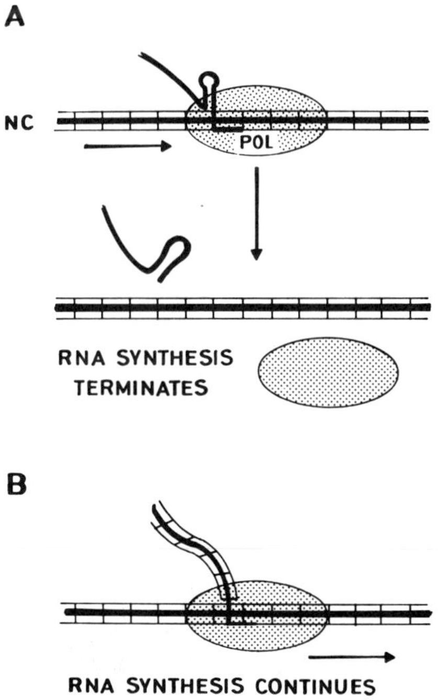

FIGURE 10. Proposed scheme for a mechanism of transcription termination/antitermination in TV. Nc = nucleocapsid; POL = polymerase complex.

the N protein would play the role of antiterminator and no special sequences for recognition of the hairpin structure would be needed. The observation that the inhibition of protein synthesis blocks S RNA replication while allowing primary transcription of the N mRNA (Franze-Fernandez *et al.*, 1987) supports the notion of a viral protein required for switching from transcription to replication. In this context, it should be mentioned that although transcription of the viral mRNAs should be coupled to translation, the position of the ribosome at the stop codon of the mRNAs would not disturb the formation of the hairpin (see Fig. 9).

Although the suggestion outlined in Fig. 10 on how transcription termination/antitermination would operate in TV is in accordance with the scarce data available so far, other explanations can also be envisaged. For instance, the polymerase might read through the intergenic sequences in both the transcriptive and the replicative modes, and the 3′ ends of the readthrough transcripts, unless encapsidated, would be degraded by a 3′ exonucleolytic processing mechanism that pauses at the stable hairpin structure. Evidence on whether termination, rather than processing, generates the 3′ end of the subgenomic RNAs will be obtained when reliable *in vitro* systems become available.

IV. SUMMARY

The genome structure of Tacaribe virus (TV) was completed. It was found that the S RNA, like the S RNA of a number of arenaviruses, encodes the major structural proteins; i.e. the glycoprotein precursor (GPC) and the nucleoprotein (N) is an ambisense coding arrangement. Analysis of the structure of TV L RNA revealed a genetic organization similar to that of the S RNA. The ORF that codes for the L protein is located at the 3′ region and is complementary to the L genome. The other ORF at the 5′ region has the same sense as the genome and codes for a polypeptide with a zinc finger-like sequence that we have called P11. The nucleotide sequence of the intergenic region in both the S and the L RNAs leads to the prediction of strong secondary structures.

Four unencapsidated subgenomic viral RNA species were detected in TV-infected cells in addition to the encapsidated S and L genomes and antigenomes. Both the sizes and orientations of the subgenomic RNAs correspond to those predicted for the mRNAs of the four viral proteins. The 3′ ends of the putative mRNAs mapped within the intergenic region in each RNA segment. The 3′ end sequences of the four subgenomic RNAs show a common feature: a region of almost perfect dyad symmetry that can theoretically take the form of a self-complementary hairpin loop. Although the sequences within the hairpins exhibit no homology, common properties of the proposed structures are their high GC content and their predicted stable hairpin configuration ($\Delta G \geq -25$ kcal).

REFERENCES

Auperin, D. D., Compans, R. W., and Bishop, D. H. L., 1982, Nucleotide sequence conservation at the 3′ termini of the virion RNA species of New World and Old World arenaviruses, *Virology* **121**:200.

Auperin, D. D., Romanowski, V., Galinski, M., and Bishop, D. H. L., 1984, Sequencing studies of Pichinde arenavirus S RNA indicate a novel coding strategy, an ambisense viral S RNA, *J. Virol.* **52**:897.

Bengal, E., and Aloni, Y., 1989, A block of transcription elongation by RNA polymerase II at synthetic sites *in vitro, J. Biol. Chem.* **264**:9791.

Bishop, D. H. L., and Auperin, D. D., 1987, Arenavirus gene structure and organization, in: *Aqenaviruses—Current Topics in Microbiology and Immunology* (M. B. A. Oldstone, ed.), Vol. 133, pp. 5–17, Springer-Verlag, Berlin.

Buchmeier, M. J., Southern, P. J., Parekh, B. S., Wooddell, M. K., and Oldstone, M. B. A., 1987, Site-specific antibodies define a cleavage site conserved among arenavirus GPC glycoproteins, *J. Virol.* **61**:982.

Clegg, J. C. S., Wilson, S. M., and Oram, J., 1990, Nucleotide sequence of the S RNA of Lassa virus (Nigerian strain) and comparative analysis of arenavirus gene products, *Virus Res.* **18**:151.

Elliott, R. M., 1989, Nucleotide sequence analysis of the large (L) genomic RNA segment of Bunyamwera virus, the prototype of the family Bunyaviridae, *Virology* **173**:426.

Franze-Fernandez, M. T., Zetina, C., Iapalucci, S., Lucero, M. A. Bouissou, C., Lopez, R., Rey, O., Daheli, M., Cohen, G. N., and Zakin, M. M., 1987, Molecular structure and early events in the replication of Tacaribe arenavirus S RNA, *Virus Res* **7**:309.

Garcin, D., and Kolakofsky, D., 1990, A novel mechanism for the initiation of Tacaribe arenavirus genome replication, *J. Virol* **64**:6196.

Gard, H. B., Vezza, A. C., Bishop, D. H. L., and Compans, R. W., 1977, Structural proteins of Tacaribe and Tamiami virions, *Virology* **83**:84.

Ghiringhelli, P. D., Rivera Pomar, R. V., Baro, N. I., Rosas, M. F., Grau, 0., and Romanowski, V., 1989, Nucleocapsid protein gene of Junin arenavirus (cDNA sequence), *Nucleic Acids Res.* **17**:8001.

Ghiringhelli, P., Rivera Pomar, R. V., Lozano, M. E., Grau, 0., and Romanowski, V., 1991, Molecular organization of Junin virus S RNA: Complete nucleotide sequence, relationship with other members of the arenaviridae and unusual secondary structures, *J. Gen. Virol.* **72**:2129.

Howard, C. R., 1986, Physicochemical properties and chemical composition, in: *Arenaviruses—Perspectives in Medical Virology* (A. J. Zuckerman, ed.), Vol. 2, pp. 130–138, Elsevier, Amsterdam.

Iapalucci, S., Lopez, R., Rey, O., Lopez, N., Franze-Fernandez, M. T., Cohen, G., Lucero, M., Ochoa, A., and Zakin, M. M., 1989a, Tacaribe virus L. gene encodes a protein of 2210 amino acid residues, *Virology* **170**:40.

Iapalucci, S., Lopez, R., Rey, O., Lopez, N., and Franze-Fernandez, M. T., 1989b, The 5′ region of Tacaribe virus L RNA encodes a protein with a potential metal binding domain, *Virology* **173**:357.

Iapalucci, S., Lopez, N., and Franze-Fernandez, M. T., 1991, The 5′ end termini of the Tacaribe arenavirus subgenomic RNAs, *Virology* **182**:269.

Kamer, G., and Argos, P., 1984, Primary structural comparison of RNA-dependent polymerases from plant, animal and bacterial viruses, *Nucleic Acid Res.* **12**:7269.

Leung, W. C., Gosh, H. P., and Rawls, W. E., 1977, Strandedness of Pichinde virus RNA, *J. Virol.* **22**:235.

Miller, J., McLachlan, A. D., and Klug, A., 1985, Repetitive zinc-binding domains in the protein transcription factor IIIA from *Xenopus* oocytes, *EMBO J.* **4**:1609.

Platt, T., 1986, Transcription termination and the regulation of gene expression, *Annu. Rev. Biochem.* **55**:339.

Poch, O., Blumberg, B. M., Bougueleret, L., and Tordo, N., 1990, Sequence comparison of five polymerases (L proteins) of unsegmented negative-strand RNA viruses: theoretical assignment of functional domains, *J. Gen. Virol.* **71**:1153.

Raju, R., Raju, L., Hacker, D., Garcin, D., Compans, R., and Kolakofsky, D., 1990, Nontemplated bases at the 5′ ends of Tacaribe virus mRNAs, *Virology* **174**:53.

Salvato, M. S., and Shimomaye, E. M., 1989, The completed sequence of lymphocytic choriomeningitis virus reveals an unique RNA structure and a gene for a zinc finger protein, *Virology* **173**:1.

Salvato, M. S., Shimomaye, E. M., and Oldstone, M. B. A., 1989, The primary structure of the lymphocytic choriomeningitis virus L gene encodes a putative RNA polymerase, *Virology* **169**:377.

Southern, P. J., Singh, M. K., Riviere, Y., Jacoby, D. R., Buchmeier, M. J., and Oldstone, M. B. A., 1987, Molecular characterization of the genomic S RNA segment from lymphocytic choriomeningitis virus, *Virology* **157**:145.

Tinoco, I., Boer, P. N., Dengler, B., Leuine, M., Uhlenbeck, O., Crothers, D., and Gralla, J., 1973, Improved estimation of secondary structure in ribonucleic acids, *Nature New Biol.* **264**:40.

Tordo, N., Poch, O., Ermine, A., Keith, G., and Rougeon, F., 1988, Completion of the rabies virus genome sequence determination; highly conserved domains among the L (polymerase) proteins of unsegmented negative-strand RNA viruses, *Virology* **165**:565.

Vezza, A. C., Clewley, J. P., Gard, G. P., Abraham, N. Z., Compans, R. W., and Bishop, D. H. L., 1978, Virion RNA species of the arenaviruses Pichinde, Tacaribe and Tamiami, *J. Virol.* **26**:485.

Von Hippel, P. H., Bear, D. G., Morgan, W. D., and McSwiggen, J. A., 1984, Protein–nucleic acid interactions in transcription: A molecular analysis, *Annu. Rev. Biochem.* **53**:389.

Wickens, M. P., and Dahlberg, J. E., 1987, RNA protein interactions, *Cell* **51**:339.

Wilson, S. M., and Clegg, J. C. S., 1991, Sequence analysis of the S RNA of the African arenavirus Mopeia: An unusual secondary structure feature in the intergenic region, *Virology* **180**:543.

Yamashita, M., Krystal, M., and Palese, P., 1989, Comparison of the three large polymerase proteins of Inflenza A, B, and C viruses, *Virology* **171**:458.

Young, P. R. and Howard, C. R., 1983, Fine structure analysis of Pichinde virus nucleocapsids, *J. Gen. Virol.* **64**:833.

CHAPTER 8

Molecular Biology of the Prototype Arenavirus, Lymphocytic Choriomeningitis Virus

Maria S. Salvato

I. INTRODUCTION

Structural analysis of a well-characterized virus such as lymphocytic choriomeningitis virus (LCMV) is of particular importance in interpreting complex biological phenomena. Complete determination of the nucleotide sequence of LCMV establishes the structural features and coding capacity of the virus genome. It enables further studies on the gene expression and replication mechanisms of arenaviruses, with potential significance for our understanding of viral persistence and pathogenesis.

In this chapter, the structural features of LCMV are summarized, emphasizing those correlated with function through genetic analysis. In particular, a sequence comparison of the CTL^{+} and CTL^{-} variants of LCMV is described as a prerequisite for functional studies on the molecular basis of virus-mediated immunosuppression. New open reading frames are revealed at the 5′-end sequence of the L-genomic segment, including one that encodes a small zinc-binding protein. The presence of such a protein was verified and correlated with the presence of subgeno-

---

MARIA S. SALVATO • Department of Pathology and Laboratory Medicine, University of Wisconsin, Madison, Wisconsin 53706.

The Arenaviridae, edited by Maria S. Salvato. Plenum Press, New York, 1993.

mic packaged viral RNA. The ambisense coding arrangement of arenavirus genes on both the L-RNA segment and the S-RNA segment continues to evoke speculation as to its functional significance. Likewise, the panhandle termini of the LCMV genomic segments are discussed in terms of their importance for models of arenavirus replication and gene expression.

II. CURRENT STATE OF THE GENETIC ANALYSIS OF LCMV

A. Pathogenic Mechanisms in a Genetically Defined Host

In the mouse model system, the immune response plays a central role in controlling LCMV infection and pathogenicity (Oldstone and Dixon, 1970). For example, in persistent infections, virus-specific immune complexes may accumulate, leading to glomerulonephritis. In acute infections, the cytotoxic T-lymphocyte (CTL) response to virus-infected cells is the principal means of clearing infection, but may also cause sufficient tissue destruction to result in death. The study of infection in different murine MHC backgrounds revealed the role of host genetics in LCMV pathogenesis and led to the initial description of MHC restriction of the CTL response (Zinkernagel and Doherty, 1974).

In addition to the effect of LCMV on the immune response, its effect in several cell systems has been to diminish specialized cell functions. For example, the reduction of macrophage lymphokines, growth hormone, thyroid hormones, and acetylcholinesterase has been observed in persistently infected murine systems (Jacobs and Cole, 1976; Oldstone *et al.*, 1982, 1988; Rodriguez *et al.*, 1983; Klavinskis and Oldstone, 1987; Valsamakis *et al.*, 1987). The immune-mediated pathogenic effects of LCMV, as well as the ability of LCMV to abrogate differentiated functions of the immune, neural, and endocrine systems, depend on the genetic background of the host. (The contribution of host genetics is elaborated in the Chapters by Pfau and Thomsen.) In contrast to the emphasis on host genetics, the focus of the research discussed here is on viral mutations in the context of genetically defined host systems in order to elaborate the role of viral genes in disease.

B. Genetic Variants of LCMV with Altered Pathogenesis

Variants of LCMV with altered pathogenesis have been mapped to either the large (L) or small (S) viral genome segment by analysis of reassortants between the Armstrong, Pasteur, and WE strains of LCMV (reviewed by Riviere, 1987). The single-stranded, bisegmented RNA genome of LCMV is not likely to form intergenic recombinants and is, therefore, well suited to reassortant analysis.

Genetic characteristics of LCMV that map to the L RNA include plaque morphology (Kirk *et al.*, 1980), lethality of the WE strain for guinea pigs (Riviere *et al.*, 1985b), and a minor contribution to suppression of the CTL response (Matloubian *et al.*, 1990; Salvato *et al.*, 1991). The L segment encodes the L polymerase, leading to speculation that plaque morphology, lethality, and CTL suppression are due to altered polymerase function. However, two additional cistrons, "X" and "Z," have recently been detected by our sequence data and might be responsible for those phenotypes.

Genetic characteristics of LCMV that map to the S RNA include the target specificity of the CTL response (Riviere *et al.*, 1986) and the ability of the Armstrong strain to cause growth hormone deficiency in C3H mice (Riviere *et al.*, 1985a). The first has since been precisely mapped using LCMV-vaccinia recombinants and peptide-coated target cells (Whitton *et al.*, 1988, 1989; Joly *et al.*, 1989). Mapping the ability of LCMV to affect growth hormone will be considerably more difficult. The S segment encodes the glycoprotein (GP) and the nucleocapsid (NP) genes, either of which might be involved in abrogating growth hormone production. LCMV WE is not found in the anterior pituitary and does not cause growth hormone disease, whereas a variant of WE (Ahmed, Oldstone, and Salvato, unpublished) is found in the anterior pituitary and causes growth hormone disease. It is presumed that mutations within the NP or GP genes are determining viral tropism for the anterior pituitary and thus leading to the effect on growth hormone (Riviere *et al.*, 1985a). This is a case in which a more precise genetic analysis could identify the individual viral genes involved. Accordingly, we initiated sequence comparisons of the growth-suppressive and growth-neutral isolates of LCMV WE identified NP as the critical locus for this phenotype (Salvato and Oldstone, unpublished).

C. Sequence Analysis of LCMV Variants

The complete sequence of the S segment (WE strain) was first reported by Romanowski *et al.* (1985). It showed the nucleocapsid protein (NP) gene encoded in the negative sense at the 3′ end and the envelope glycoprotein (GP) gene encoded at the 5′ end of the RNA in the positive or mRNA sense. Sequence data for the Armstrong strain came from Southern and Bishop (1987) and Salvato *et al.* (1988) and established the structure of the S RNA. The relationship between sequence information and gene products was confirmed when antisera raised against predicted viral peptides reacted with known viral proteins (Riviere *et al.*, 1985a; Southern *et al.*, 1987). Sequence comparisons for the S RNA were instrumental in locating the differences in CTL epitope recognition between the Pasteur and Armstrong strains of LCMV (Joly *et al.*, 1989; Whitton *et al.*, 1989). In addition, the significance of certain amino acid changes to the CTL-suppressing phenotype of an Armstrong variant was elucidated by detailed sequence analysis and comparison of this data among numerous strains of LCMV (Salvato *et al.*, 1988, 1991).

Limited sequence information was reported initially for the L RNA segment (Romanowski *et al.*, 1985; Singh *et al.*, 1987). The negative coding sense of the L polymerase gene was established using antipeptide antisera based on the predicted amino acid sequence (Singh *et al.*, 1987). The sequence of the L polymerase gene at the 3′ end of the L RNA was then determined (Salvato *et al.*, 1989b), and the final 5′ end of the L RNA (Salvato and Shimomaye, 1989) will be described in this chapter.

The L and S genomic segments encompass 11 kb of RNA and encode five cistrons: NP (558 amino acids), GP (498 amino acids), L (2210 amino acids), Z (90 amino acids), and X (95 amino acids) (Fig. 1). The Z and X genes were previously unknown and their existence was revealed in the course of completing the nucleotide sequence of the L RNA. The sequence predicts single zinc-binding motifs in the NP and Z proteins (Salvato, 1989). Sequence comparisons highlight domains most likely to be functionally significant (Southern and Bishop, 1987; Salvato *et al.*, 1989b; Salvato and Shimomaye, 1989), and which are now being tested in defined assay systems, e.g., CTL assays with viral epitopes or expression assays in cell culture or transgenic animals.

D. Sequence Comparison of the CTL$^+$ and CTL$^-$ Variants of LCMV

To locate mutations responsible for the altered phenotypes of LCMV, it was necessary to obtain complete viral sequence . This proved to be a considerable task and required a combination of analytical ap-

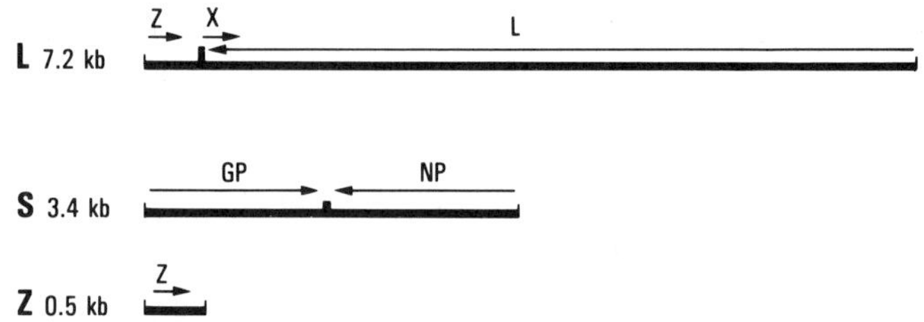

FIGURE 1. Virion-encapsidated RNA of LCMV. Three virus-specific RNAs are found in purified virions, L at 7.2 kb, S at 3.4 kb , and Z at 0.5 kb. Thin arrows denote coding regions of the five open reading frames: NP and L proteins are encoded in the negative sense with respect to the genomic RNA, whereas GP, Z, and X are encoded in the positive sense.

proaches. Attempts to map the CTL-suppressing phenotype of the Armstrong variant, Clone 13, illustrate the problems encountered. Direct sequence analysis of wild-type Armstrong and its variant revealed five mutational differences. One on the S RNA did not result in amino acid changes, and another caused an amino acid change that was also seen in strains of LCMV with wild-type phenotypes (Salvato *et al.*, 1988, 1989a). Reassortant analysis (Ahmed *et al.*, 1988) initially indicated that the mutations on the S RNA are not involved and that the CTL-suppressing phenotype maps solely to the L RNA. However, later reassortant analysis with homologous LCMV strains contradicted this conclusion and implicated the S RNA segment as the locus for the immunosuppressive phenotype (Matloubian *et al.*, 1990). Sequence analysis on the L RNA yielded one nucleotide change that affects amino acid sequence in the middle of the polymerase gene and two nucleotide changes that do not affect the coding capacity (Salvato *et al.*, 1991). Four revertants of the CTL-suppressing variant have the same three L RNA mutations and are presumed to be second site revertants. Accordingly, it was necessary to amass complete sequence information from several variants and their revertants, to ascertain which of the five mutations is indeed responsible for the CTL-suppressive phenotype. In conclusion, a single amino acid change (Phe to Leu) at the carboxyl-terminus of GP-1 was necessary, but not sufficient, for the immunosuppressive phenotype. The most likely candidates for the secondary mutations contributing to this phenotype are mutations in the polymerase or transcription-control regions on the L RNA (Salvato *et al.*, 1991).

E. Future Approaches to Genetic Analysis of LCMV

Infectious recombinants have been used in genetic mapping of positive-stranded RNA viruses such as polio (Racaniello and Baltimore, 1981; van der Werf *et al.*, 1986), human rhinovirus (Mitsutani and Colonno, 1985), Bromemosaic virus (Dreher *et al.*, 1984), and Sindbis (Rice *et al.*, 1987). Mutations are mapped in these systems using a wild-type version of the viral genome cloned in an RNA expression vector. Segments of the mutant virus are cloned and then substituted for wild-type sequences in the expression vector. This places the region potentially responsible for phenotypic variation into a clean genetic background, uncomplicated by mutations that may not contribute to the variant phenotype. The recombinant viral RNA is then expressed *in vivo* (or *in vitro* followed by reintroduction) to determine its contribution to the mutant phenotype.

It is now possible to obtain infectious RNA for several of the negative-stranded RNA viruses and to complement mutations in all of them by coexpression of recombinant genes (e.g., VSV, Li *et al.*, 1988). Nucleocapsids of most negative-strand RNA viruses are infectious (Strauss and Strauss, 1983), and LCMV is no exception (Salvato, unpublished). Infectious nucleocapsids of LCMV can be prepared by disruption of virus with detergent, gradient centrifugation through a 10–50% sucrose gradient, and electroporation of cores into BHK cells. It may be possible in the future to assemble the LCMV core proteins (L, Z, and NP) with recombinant RNA to make infectious cores.

An alternate approach, similar to one taken with VSV, is to express LCMV RNA from a T7 RNA polymerase promotor transfected into appropriate cells. Overproduction of one RNA segment in the cytoplasm might allow it to substitute for RNA segments of a coinfecting LCMV and form reassortants. Full-length cDNA for the S segment of LCMV (WE) is available (Romanowski *et al.*, 1985), as is the L RNA segment of the LCMV Armstrong strain (Salvato, unpublished). It is necessary to develop screens and selections for reassortment between LCMV and synthetic RNA segments based on the currently available markers of LCMV, i.e., differences in genomic sequence, plaque morphology, antibody epitopes, and CTL recognition. Reassortants of LCMV have been identified recently by screening Northern blots with sequence-specific cDNA probes, though it may be simpler to screen by hybridization to plaques lifted from infected Vero cell monolayers. In both cases, sequence data are used to develop probes that discriminate genetically distinct LCMV segments. Additional markers, such as the temperature-sensitive phenotypes of the related arenavirus Pichinde, should also be developed (Vezza *et al.*, 1978a).

Reporter genes expressing β-galactosidase, luciferase, or chloramphenicol acetyl transferase, flanked by arenavirus transcription-control signals, can be reintroduced into cell culture. If primary transcription is designed to yield a mRNA-complementary product, then subsequent transcription and expression is dependent on LCMV coinfection or addition of LCMV transcription components. Preinfection with LCMV sets up interference that prevents the functional expression of reporter genes; therefore, helper virus must be added after the transfection of the reporter genes. The availability of infectious recombinants of LCMV will streamline the ability to map mutations and would enhance the usefulness of this virus as a probe for pathogenic mechanisms affecting host biology.

III. SEQUENCE AT THE 5′ TERMINUS OF THE LARGE (L) SEGMENT OF LCMV

A. Approaches to RNA Sequence Analysis

The highly structured 5′ end of the L RNA segment has posed significant problems for the construction of full-length cDNA clones and has frustrated attempts to determine its sequence by the methods previously employed for the S RNA (Auperin *et al.*, 1984; Romanowski *et al.*, 1985). Accordingly, direct RNA sequencing was employed in which radiolabeled oligonucleotides were used to prime reverse transcription in the presence of dideoxy chain-terminators with either viral RNA or infected-cell RNA templates. This method fails to reveal minor heterogeneity, due to errors of viral transcription, in the RNA preparation. These errors, amounting to approximately one change per 10^3 molecules at each nucleotide position (Holland *et al.*, 1982), may be of some biological importance. However, our objective was to determine the nucleotide sequence of the most highly represented RNA in the preparation, presumably contributing most to the observable viral phenotype.

Sequence analysis of the L RNA proceeded from the 3′ end, with each new stretch of sequence dictating the oligonucleotides used as primers for the next stretch. For the initial 6.5 kb, the sequence data were relatively unambiguous. Reverse transcription came to a staggered halt at the end of the L-protein cistron, leading us to believe we had come to the 5′ terminus. Surprisingly, the use of reverse transcriptase at the elevated temperature of 47°–55°C (Shimomaye and Salvato, 1989) produced longer products and allowed the determination of additional sequence. Eventually, a more abrupt terminus was observed on sequencing gels; it also contained sequence complementarity to the 3′ end of the genomic segment (Salvato and Shimomaye, 1989).

After reaching the ultimate 5′ end of the L RNA, portions of the intergenic region (lying between the L gene and the newly defined Z cistron) remained ambiguous, and the available sequence information was used to design oligonucleotides for the production of cDNA clones. The cDNA clones contained the same staggered stops seen in the direct RNA sequence at the end of the L-protein cistron. Nonetheless, in the presence of dITP and the modified T7 DNA polymerase (Sequenase), it was possible to determine cDNA sequence unambiguously. Out of 15 cDNA clones representing four different cloning attempts from three stocks of LCMV Armstrong, we found two major types of variations: deletions and variations in the length of homopolymer tracts. First, the cloned cDNAs were consistently almost 30 bases shorter than the direct RNA sequence of the same region. This proved to be a deletion of a cytidylic acid–rich region upon passaging the cDNA clone in bacteria, and the entire sequence could only be observed by direct RNA sequence analysis.

Even by direct RNA sequence, ambiguity was observed in the intergenic region, and cDNA clones revealed sequence variation. The most notable variation occurs in a stretch of cytidylic acid (Cs) three bases away from the deleted region. Among 15 cDNA clones, one has four Cs, four have five Cs, five have six Cs, one has seven Cs, and four have eight Cs. Such phenomena are not unknown for the RNA viruses. The paramyxoviruses yield examples of nontemplated nucleotides, also stretches of cytidylic acid, which differentiate the messenger RNAs of the P gene from the genomic RNA (Thomas *et al.*, 1988). To determine whether the C-stretch is uniform in genomic LCMV RNA and only variable in messenger RNA, one round of cloning was performed with gradient-purified L RNA. The resulting cDNA clones still contained varied sequences but the size distribution was reduced; therefore, unlike the paramyxoviruses, the variability does not seem to be due to nontemplated additions during messenger transcription (Salvato and Shimomaye, 1989). Insertions and deletions in homopolymeric stretches have also been seen in VSV (DePolo *et al.*, 1987).

The intergenic region between the L and Z genes has an unusual structure. To illustrate the regions of the L RNA giving rise to the strong transcription stops observed during sequencing, a thermodynamically stable form of the RNA structure was obtained from the computer program FOLD (University of Wisconsin, Genetics Computer Group) and is presented in Fig. 2; this structure includes the 849 bases at the 5′ end of the L RNA. The 20 bases at the extreme 5′ end are found annealed to the intergenic region, although, in the presence of the entire L segment, it is more likely that they anneal to the 20 bases at the 3′ end of the L. The RNA structure is depicted with and without the region that is deleted in cDNA clones, showing that the deleted region is not involved in much base pairing, and hence its removal does not significantly affect the predicted structure of the surrounding RNA. This structure is a good candidate for pseudoknot formation (in which intramolecular pairing of bases in a hairpin loop forms an additional stem and loop region in three-dimensional space) because the stability of the structure to melting excedes that predicted by the thermodynamics of simple two-dimensional stem-loops (see Puglisi *et al.*, 1988).

Bishop and Auperin (1987) observed that the intergenic region of the S RNA probably serves to terminate the mRNA transcription for the abutting NP and GP genes and that an antiterminating mechanism would be necessary to allow accumulation of full-length replicative forms. Similarly, the intergenic region of the L and Z genes might function as a transcription terminator unless the block is overcome during replication. RNA-binding proteins with specificity for homopolymeric regions, like the L/Z intergenic C-rich regions, have been described (Dreyfuss *et al.*, 1988); so it is interesting that these regions are predicted to form prominent non-base-paired structures in LCMV. It is difficult to assess the biological significance of hypothetical RNA structures, yet it is encouraging that a bifurcated structure is predicted in which the L and Z genes form two separate arms. The predicted structural complexity of the intergenic region is also consistent with the problems encountered in analyzing these sequences.

B. Evidence for Expression of the Z Gene

Sequence information at the 5′ terminus of the L RNA segment reveals two small open reading frames, one for a 95-residue protein (X) encoded in the opposite sense, entirely within the L polymerase gene, and a second one for a 90-residue protein (Z) positively encoded 90 bases from the 5′ end of the genomic RNA (Fig. 1). Examination of purified virus by SDS polyacrylamide gel electrophoresis shows a small protein in the 10- to 11-kDa range (Salvato, 1989), the size expected for either the X or the Z protein. A 12-kDa protein was reported for Pichinde, which varied in amount from one preparation of virus to another (Vezza *et al.*, 1977). A 14-kDa LCMV protein was also reported by Buchmeier and Parekh (1987); however, its size was based on α-lactalbumin, now known to be an unreliable marker for that molecular weight, and it is likely that the 14-kDa product is, in fact, the 10- to 11-kDa protein.

The first evidence for the existence of the 90-residue Z protein comes from comparative sequence information of several viral strains. The predicted Z protein is conserved in Armstrong, Traub, Pasteur, and WE strains of LCMV (Fig. 3). A zinc-finger motif (Cys-X_2-Cys-X_3-Phe-X_2-Leu-X_5-His-X_4-His) is also conserved in all four strains. The direct demonstration of zinc-binding capacity for the 10- to 11-kDa viral protein is an indication that it may be the Z-gene product and not the X-gene product (which lacks a zinc-binding motif). Both the 11-kDa protein and the nucleocapsid protein (which has a single zinc finger at its carboxyl-terminus) bind $^{65}ZnCl_2$ in a Western blot overlay assay. In addition, the 10- to 11-kDa protein was subjected to N-terminal analysis and the results are consistent with the predicted Z sequence (Salvato, 1989; Salvato and Shimomaye, 1989).

Definitive biochemical and immunological evidence that the 11-kDa zinc-binding protein is a structural component of LCMV came from expressing its cDNA clone in *Escherichia coli*, purifying the gene product, and raising rabbit polyclonal sera to this protein. The antisera bind to p11 Z on Western blots of viral proteins, immunoprecipitate p11 from *in vitro* translation assays, and immunoprecipitate it from infected cell extracts (Salvato *et al.*, 1992). Subcellular fractionation localizes the Z protein to the postmicrosomal pellet and to LCMV transcription complexes. Whether it has a role in transcription is still unknown.

The 11-kDa protein is associated loosely with viral cores. Preparations of LCMV were lysed with 0.1% NP-40 as described by Howard and Buchmeier (1983), followed by fractionation on a 10–50% renografin gradient: it was found in undisrupted fractions containing untreated virus and in fractions containing disrupted virions stripped of glycoproteins, however, it was minimally present in the high-density fraction containing ribonucleoprotein viral cores (Salvato, unpublished). A similar fractionation has been described in which virus was treated with Triton X-100 and most of the p11 appeared at the top of a sucrose gradient whereas virus cores pelleted (Salvato *et al.*, 1992).

A hypothetical scheme for the association of p11 Z with the nucleoprotein and the glycoproteins is presented in Fig. 4. The stoichiometry of the viral proteins, L: NP: GP-1: GP-2: Z is approximately 30: 1500:

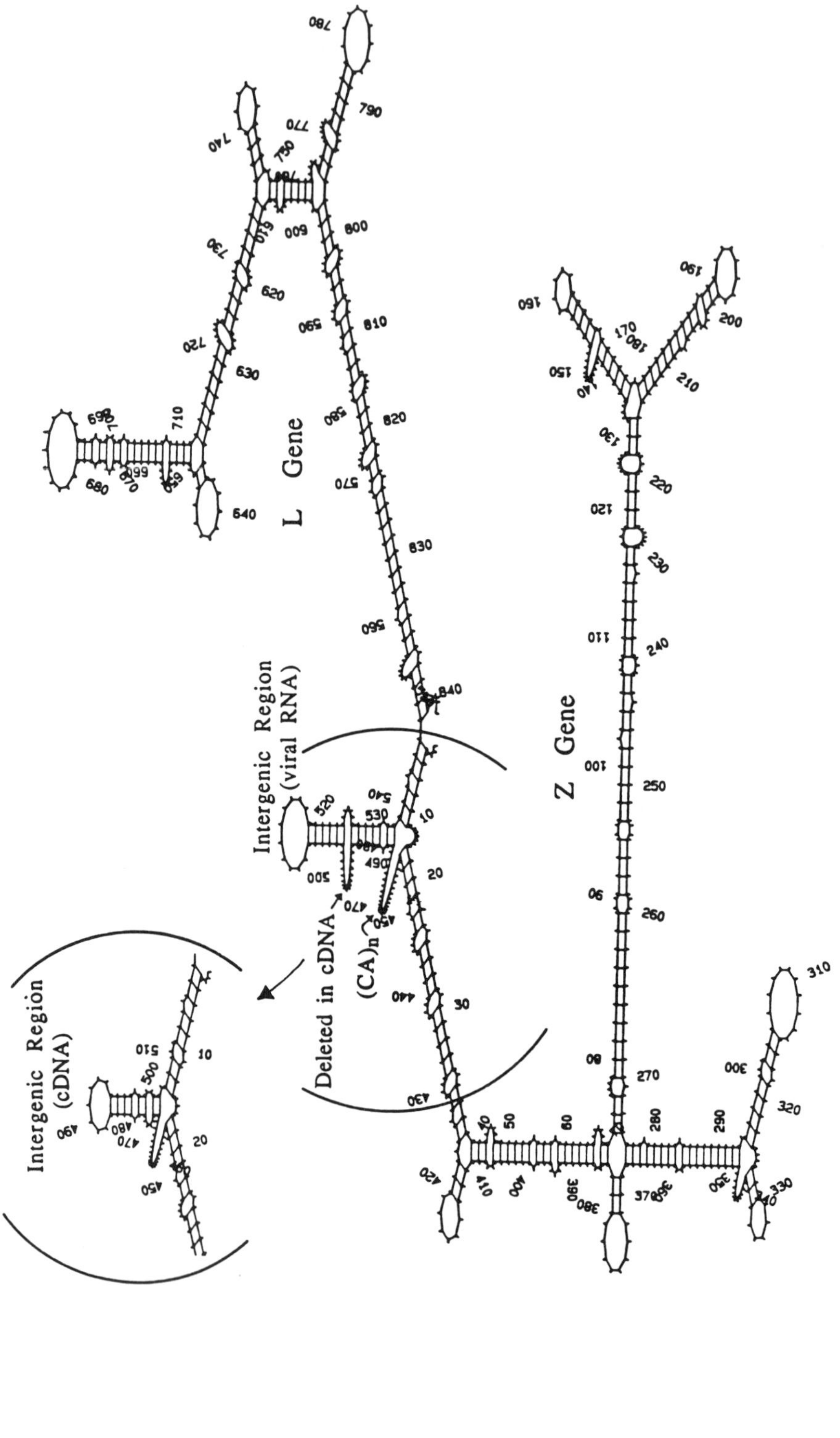
L Gene
Z Gene
Intergenic Region
(viral RNA)
Deleted in cDNA
(CA)n
Intergenic Region
(cDNA)

650: 650: 450, as estimated by metabolic labeling of virions with ^{35}S cysteine (Salvato *et al.*, 1992). The Z protein and GP-2 are likely to be associated with the viral membrane because they segregate as hydrophobic proteins upon Triton X-114 extraction. Association of Z protein with the nucleocapsid protein has been demonstrated by DMS cross-linking (Salvato *et al.*, 1992).

C. Structural Similarities of Z to Other Proteins

The predicted Z-gene product is structurally similar to several proteins having gene regulatory functions. The zinc-finger motif was first identified in TFIIIa, a protein from *Xenopus* oocytes that regulates the production of 5S RNA (Miller *et al.*, 1985). The multiple fingers of TFIIIa bind to both RNA and DNA (Klug and Rhodes, 1987). Many proteins have cysteine- or histidine-rich regions that bind zinc, but only the canonical "finger" motif, Cys-X_2-Cys-X_3-Phe-X_5-Leu-X_2-His-X_{3-4}-His, binds nucleic acid in a zinc-dependent manner; the Z and the NP proteins of LCMV each have a single zinc finger. The transcription regulatory proteins of cytomegalovirus, IE-1 and IE-2, each have a single finger as well, and it is postulated that they multimerize to perform their enhancing function (Nelson *et al.*, 1990). An example of a zinc-finger protein with RNA-only binding function is ribosome initiation factor eIF-2, which binds to mRNAs in a zinc-dependent manner and ensures correct translation initiation (Donahue *et al.*, 1988; Pathak *et al.*, 1988).

Some viral proteins contain cysteine-rich regions that bind zinc but are not "fingers." The sigma III protein of reovirus has separate RNA-binding and zinc-binding domains (Schiff *et al.*, 1988) and transregulates both viral and cellular translation (Fields and Greene, 1982). The P proteins of the paramyxoviridae have a cysteine-rich region (not yet shown to bind zinc) and function in concert with the polymerase and ribonucleoprotein during transcription (Thomas *et al.*, 1988). The E1A protein, which transactivates adenovirus-2 gene expression (Moran and Mathews, 1987), also has a cysteine-rich region but has not been able to footprint to DNA. Perhaps it is functioning at a post-transcriptional level, enhancing expression of viral proteins by stabilizing their mRNAs.

The HIV Tat protein has several structural features in common with the Z protein. It is similar in size to the predicted Z protein, it contains a single cysteine-rich region (not a finger), and an amino-terminal proline-

←

FIGURE 2. Predicted RNA structure of the 5′ end of the L-genome segment (bases 1–849). Paired bases, either G–C pairs or A–T pairs, are connected with short lines and nonpaired bases are represented as spikes on the looped structures. The 5′ terminal 20 bases are depicted as annealing to the intergenic region.

```
LCMV Strain                      10                  20                  30                  40
                                  |                   |                   |                   |
ARMSTRONG     M G Q G K S R E E K G T N S T N R A E I L P D T T Y L G P L S[C]K S[C]W Q K F D S L V R C
WE            . . . . . . K . . R D . S N . G . . . L . . . . . . . . . . N . . . . . . . . . . . . . .
TRAUB         . . . S . . K . . . . I S G . S . . . . . . . . . . . . . . N . . . . . . . . . . F S K .
PASTEUR                                             . . . . . . . . . N . . . . . . . . . . L V R .

                        50                  60                  70                  80                  90
                         |                   |                   |                   |                   |
ARMSTRONG     H D[H]Y L C R[H]C L N L L L S V S D R C P L C K Y P L P T R L K I S T A P S S P P P Y E E
WE            . . . . . . . . . . . . . . . . . . . . . . . . . . . . . K . . V . . V . . . L . . . . .
TRAUB         . . . . . . . . . . . . . . T S . . . . . . . . . . .
PASTEUR       . . . . . . . . . . . . . . T S . . . . . . . . . . .
```

FIGURE 3. Comparison of the predicted p11 Z protein sequence for four strains of LCMV. Armstrong, WE, Traub, and Pasteur conserve the zinc-finger motif (boxed residues). Alternate cysteines and histidines (not boxed) are available for coordinating zinc and may allow the protein to assume alternate conformations. The single-letter code for amino acids is: A, alanine; C, cysteine; D, aspartic acid; E, glutamic acid; F, phenylalanine; G, glycine; H, histidine; I, isoleucine; K, lysine; L, leucine; M, methionine; N, asparagine; P, proline; Q, glutamine; R, arginine; S, serine; T, threonine; V, valine; W, tryptophan; Y, tyrosine.

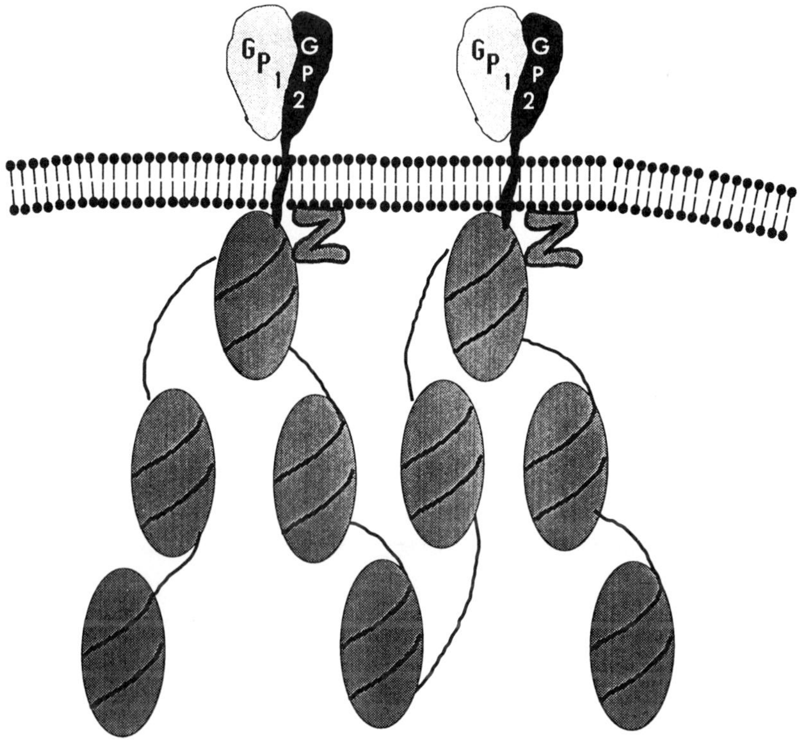

FIGURE 4. Hypothetical scheme for the association of proteins in the LCM virion. This scheme is derived from three pieces of information: (1) the stoichiometry of L: NP: GP-1: GP-2: Z is 30: 1500: 650: 650: 450, (2) GP-2 and Z are hydrophobic, and (3) NP can be cross-linked to Z (Salvato *et al.*, 1992).

rich region (the Z protein has a proline-rich carboxyl terminus). Tat functions as an antiterminator of transcription (Kao *et al.*, 1987) and possibly as a regulator of HIV translation by affecting Tar activation of the double-strand RNA-specific kinase (Edery *et al.*, 1989). It has been suggested that the *tat* gene product can affect cellular as well as viral functions in *trans* (Edery *et al.*, 1989). At present, we can only infer functions for the LCMV Z product by analogy to other viral proteins. Whereas all the other LCMV proteins have been assigned structural (NP and GP) or enzymatic (L) function, the predicted Z-gene product is the first candidate for transregulatory function. Inhibition of LCMV transcription by monoclonal antibodies to the Z protein is preliminary evidence that Z is essential for transcription (Salvato, unpublished). In view of the specific affect of LCMV on certain differentiated cell functions of the host, it is important to determine whether Z-gene expression can directly influence host cell functions.

D. Expression of a Message-Sense Subgenomic Z RNA

Subgenomic mRNAs have been identified for the nucleoprotein (NP) and glycoprotein (GP) genes of LCMV (Auperin *et al.*, 1984; Southern *et al.*, 1987). A messenger RNA for the L protein has been described only for Tacaribe virus (see chapter by Franze-Fernandez *et al.*), and it is expected to be 6.5 kb. We observed a small subgenomic RNA corresponding to the message sense of the Z gene in Northern blots of both infected cell RNA and purified virion RNA (Salvato, 1989; Salvato and Shimomaye, 1989). As only trace amounts of the mRNA for other viral genes can be found encapsidated in virions, the discovery of a message-sense Z RNA in large amounts, roughly equimolar to the genomic segment RNAs (Salvato and Shimomaye, 1989), provokes speculation as to the role of this RNA and the small protein it encodes.

Once it was determined that the structural genes on the S segment are encoded in an ambisense arrangement, a model was proposed for the expression of genes during the virus life cycle (Auperin *et al.*, 1984). It was predicted that the negatively encoded genes (NP and L) would be transcribed first, and that the positively encoded GP gene would await the production of full-length genome-complementary RNA before being transcribed and finally translated. Fuller-Pace and Southern (1988) noted that the NP mRNA appeared before the GP mRNA, but were unable to determine whether the onset of GP mRNA production was prior to or concomitant with the replication of genomic-sense RNA segments. As a positive-sense gene, the Z gene would be expected, as is the case for GP, to be transcribed following the negatively encoded genes. However, the substantial amount of Z-gene transcript carried within the virion might override the delay in transcription experienced by the other positively encoded gene (GP) and direct early expression of Z.

The Z-gene transcript might also function as a small satellite RNA, as seen in other viral systems. The VA_1 RNA of adenovirus, for example, associates with ribosome initiation factors and prevents interferon-induced changes in the translation machinery (Kitajewsky *et al.*, 1986). The fact that LCMV encapsidates ribosome-like particles suggests that one of its RNAs might act as a ribosome-associated factor. An alternative concept is that the Z RNA is an independent genetic element of LCMV and may, in fact, be a required viral chromosome. We know that its 5′ end contains the terminal repeat sequence; however, we have not yet characterized the 3′ end to determine whether it is complementary to the 5′ end. The Z gene is definitely found within the L RNA in a contiguous structure, so it would seem redundant to have a separate genomic segment for this gene. Should the Z transcript prove to be a separate genetic element, reassortment results would have to be reinterpreted in light of this possibility. A third possibility is that this transcript could be an early mRNA for the Z-gene product and is carried by the virion to facilitate rapid production of a protein with a required early function.

All these possibilities contribute to the supposition that the Z-gene product may have a regulatory function.

E. Future Approaches to Defining Z-Gene Function

The proposal that the Z protein and/or its RNA are regulatory elements suggests numerous experiments that will reveal the molecular details of interactions with other molecules. After establishing the relatively simple molecular interactions, more complex assay systems will be used to assess the role of Z in transcription, translation, and replication of LCMV.

Numerous systems are available for examining the consequences of any possible Z transregulatory function *in vivo*. Reintroduction of the Z gene into cells that are normally used for LCMV plaque assays may function to alter the plaque morphology of superinfecting LCMV. Reintroduction of the Z gene into mice as a transgene would reveal whether the Z gene alone can abrogate differentiated cell functions. The magnitude of research directed toward understanding the role of HIV Tat protein illustrates the importance of these regulatory factors in the regulation of RNA virus life cycles.

IV. THE GENOMIC TERMINAL COMPLEMENTARITY OF LCMV

A. Similarity to Other Viruses

The completed sequence of LCMV confirms that the 5′ and 3′ ends of the L RNA are complementary. Auperin *et al.* (1982) reported that the 5′ and 3′ ends of the S RNA are complementary, and that their sequence is conserved in a variety of arenaviruses. From the sequence conservation it was surmised that the termini might serve a common function, such as recognition elements for the initiation of transcription (Bishop and Auperin, 1987). Sequence analysis and electron micrographs show that all the negative-stranded RNA viruses circularize, and this physical structure reflects the complementary pairing of RNA termini (reviewed in Strauss and Strauss, 1983).

In some cases, such as influenza, the self-complementarity amounts to only 15 base pairs. The naked RNA does not circularize under conditions of spreading for electron microscopy, though the RNA–protein complexes are able to form circles. The circularized influenza ribonucleoprotein has been shown to be biologically significant; *in vivo* cross-linking studies place most of the viral RNA in circularized structures at the peak of viral transcription (Hsu *et al.*, 1987). Defective interfering particles of influenza conserve at least their 5′ and 3′ termini; thus it is supposed that these termini are needed for the replication or packaging of particles (Nakajima *et al.*, 1979). This contrasts with the defective interfering particles of VSV that conserve only the 5′ end (Keene *et al.*, 1979).

For the bunyaviruses (Porterfield *et al.*, 1975) and arenaviruses (Palmer *et al.*, 1977), circularized ribonucleoproteins are observed. The naked RNA does not circularize for the bunyavirus La Crosse (Dahlberg *et al.*, 1977), though it does for the arenaviruses [Tacaribe and Tamiami (Vezza *et al.*, 1978b); LCMV (Dutko, Griffith, and Oldstone, unpublished)]. There is no direct evidence yet that the arenavirus termini are joined during transcription or are conserved in defective particles.

B. Asymmetry of the Terminal Complementarity

The inverted terminal repeats of LCMV are not perfectly base-paired units (Fig. 5). The positions of mismatched bases may be important recognition elements: e.g., the ability of tRNA synthetases to see tRNAs and the ability of splicing complexes to recognize their cleavage sites depend on "bulged" or unpaired bases (Parker *et al.*, 1987). In addition to affecting recognition, unpaired bases in the terminal complexes might affect the stability of one complex over another. The 5′ and 3′ ends of the self-paired L RNA form a complex with a negative free energy (−40 kcal/mole) roughly equivalent to the free energy of the self-paired S RNA. In contrast, the association between the 5′ end of the L and the 3′ end of the S has fewer unpaired bases and a lower negative free energy (−60 kcal/mole), and is therefore predicted to be more stable. More stable association between L and S, as opposed to L self-association or S

```
5' end of L   C G C A C C G G G G A U C C U A G G C G U U U A G U U
              | | | | | | | * | | | | | | | | | | | | * | | | * | * |   –45 Kcal
3' end of L   G C G U G G C U C C U A G G A U C C G A A A A A C U A

5' end of S   C G C A C C G G G G A U C C U A G G C U U U U U G G A
              | | | | | * | * | | | | | | | | | | | | * | | | * * * *   –40 Kcal
3' end of S   G C G U G U C A C C U A G G A U C C G U A A A C U A A

5' end of S   C G C A C C G G G G A U C C U A G G C U U U U U G G A
              | | | | | | | * | | | | | | | | | | | | | | | | | * *   –60.8 Kcal
3' end of L   G C G U G G C U C C U A G G A U C C G A A A A A C U A

5' end of L   C G C A C C G G G G A U C C U A G G C G U U U A G U U
              | | | | | * | * | | | | | | | | | | | | * | | | * * | |   –41.2 Kcal
3' end of S   G C G U G U C A C C U A G G A U C C G U A A A C U A A
```

FIGURE 5. The termini of the LCMV Armstrong L and S genomic segments are complementary. The calculated free energy of the hybrid complexes (Tinoco *et al.*, 1973) is listed to the right of each one.

self-association, could well affect the relative production of of L and S replicative intermediates.

C. Model for Arenavirus Replication

Predicting that the most stable association of genomic termini is between one molecule of S and one molecule of L RNA leads to a model for the possible associations of L and S genomic segments during replication (Fig. 6). The model assumes that replication is initiated on viral RNA paired at the termini. Two types of the resulting panhandle structures are depicted. One type is monomolecular, in which the L and S undergo replication independently (Fig. 6a). The other type is a bimolecular structure in which the L and S termini are base-paired with each other (Fig. 6b). This structure would be favored thermodynamically and

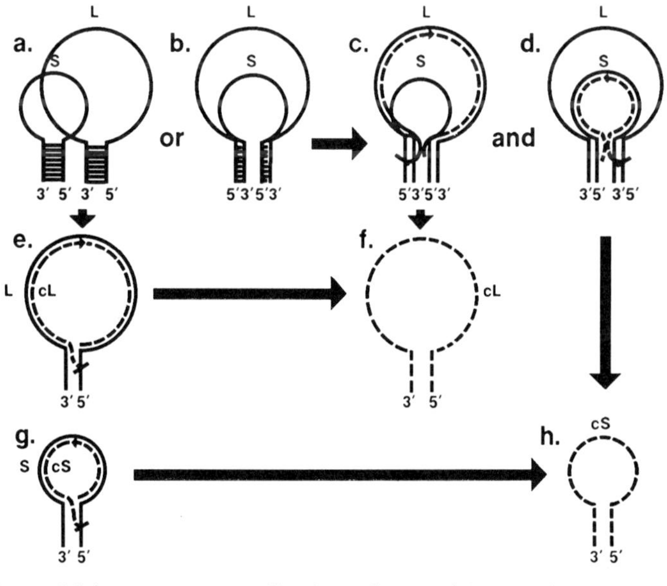

FIGURE 6. Model for arenavirus replication. The L and S genomic segments are represented with panhandle structures due to base pairing at their termini. The replicative intermediates are represented by dashed lines. Initiation of replication from the monomolecular (self-paired) structures in (a) would result in the independent replication events in (e) and (g). Initiation of replication from the bimolecular structure in (b) would result in simultaneous replication of L (c) and of S (d). Replication may be an unprimed event or primed by a fragment complementary to the 3′ end (such as could be derived by cleavage of five nucleotides from the 5′ end of either the L or the S). (The chapter by Kolakofsky and Garcin describes an *in vitro* transcription system that is greatly facilitated by specific dinucleotide primers.) The replicative intermediates cL (f) and cS (h) from monomolecular or bimolecular structures would be the same.

would ensure that each genomic segment would undergo an equal number of replication rounds.

The proportion of L to S genomic segments appears to diminish with long-term culture of LCMV in BHK cells (Southern *et al.*, 1984; Francis and Southern, 1988). Initially, LCMV Armstrong-infected cell RNA contains roughly one copy of L for every two to five copies of S, but in long-term BHK cell culture, the amount of L RNA in relation to S RNA is less. The gradual disappearance of L segment may well result from a disruption of bimolecular replication. Perhaps a viral product, such as accumulated Z RNA, causes this disruption. The fact that Z RNA contains the 5′ terminus and is found packaged in virions at high levels makes it likely that it associates with the 3′ termini of the other two segments. During replication, its 5′ end could insert itself into the initiation complex (assumed to be comprised of base-paired termini) and thus lead to abortive replication.

D. Future Approaches to Defining the Structure/Function of LCMV RNA

The structure of viral RNA during the viral life cycle can be probed with several techniques. An excellent study has been carried out to define the structure of influenza viral RNA during transcription (Hsu *et al.*, 1987). *In vivo* cross-linking with psoralen compounds at various stages of the infection, followed by analysis of RNA structures by gel electrophoresis, S1 nuclease analysis, and electron microscopy, identified circularized RNA in high proportions within virions and at the peak of transcription in the cell. Such a study with LCMV would determine the existence or biological significance of the hypothetical structures we have proposed.

Other types of cross-linkers and antibody coprecipitation could be used to explore the association of viral components with each other and with cellular components. Structural approaches such as these will be essential for determining whether certain processes, as well as components, are functionally interdependent. The interdependence of transcription and translation (described for Tacaribe by Boersma and Compans, 1985, and for the bunyaviruses by Raju and Kolakofsky, 1987 and Bellocq *et al.*, 1987) or assembly and replication might be elucidated via cross-linking and coprecipitation studies. The mechanism by which replicative processes are curtailed during persistent infection and the means for viral interference with differentiated cell functions may also be revealed in the course of these studies.

To solve the replication mechanism of LCMV we need to explore the capacities not only of the viral proteins involved, but also of the viral RNA. Recently there has been an explosion of information pertaining to the catalytic activities of RNAs and to the cellular components involved in RNA processing. For example, hepatitis delta virus RNA catalyzes its own cleavage and self-ligation (Wu and Lai, 1989). The discovery of a biased hypermutation in the measles Matrix (M) gene (Cattaneo *et al.*, 1988) was followed quickly by the observation that a cellular enzyme with unwinding-modifying activity (Bass and Weintraub, 1989) could account for localized changes of guanosine to inosine residues if the viral RNA was present in a double-stranded form at the (M) gene locus.

Cell functions actively attempt to eliminate double-stranded RNA, not only at the level of the unwinding-mutating activity, but also at the translational level; interferon production is the most recognized function in the latter category. Several viruses have protective mechanisms to deal with the interferon or dsRNA-induced kinase that curtails translation initiation (Farrell *et al.*, 1977). The VA_1 RNA of adenovirus inhibits eIF-2a kinase, which would otherwise block translation of adenovirus RNA (Kitajewsky *et al.*, 1986). Polio induces an inhibitor of eIF-2 phosphorylation by the dsRNA kinase (Ransone and Dasgupta, 1988). Several mRNAs, including the *tar* region of HIV RNA, contain a stem-loop structure that activates the dsRNA-induced kinase. Whereas the activation of kinase represses translation initiation, unwinding of the stem-loop structures (exposing the capped structure of the HIV messengers) increases the efficiency of initiation (Edery *et al.*, 1989). Therefore, any models for virus transcription and translation need to take the catalytic activities of the RNA and the processing activities of the cell into account. Such considerations should also dictate the design of reintroduction experiments, to include features essential to the viral RNA for its preservation and function.

V. CONCLUSION

The complete sequence of LCMV was a prerequisite for further advances in the study of this virus. By obtaining these data we have facilitated experimental approaches to problems as diverse as immune responses to LCMV infection or the structure of replicating RNA. Delineation of new components of LCMV, such as the Z protein, opens up research into the role of zinc-binding proteins for the structure and regulation of the virus as well as the activities of the infected cell. The RNA sequence enables the design of synthetic RNAs for reintroduction and binding assays; these experiments are crucial to understanding the biological effects of LCMV infection and other features of the host:pathogen interaction. Structural predictions from the RNA sequence have inspired a model for the role of RNA structures in replication, and this model will be examined for its potential impact on hypotheses concerning the mechanism of viral persistence. Accordingly, structural characterization of LCMV supplies a wealth of new experimental tools for the study of this virus and brings a bit more light to bear on the biology of this interesting pathogen.

ACKNOWLEDGMENTS. This is publication 5820-Imm from the Department of Immunology, Scripps Clinic and Research Foundation, La Jolla, California, 92037. This work was supported in part by USPHS Grants AI-09484 (M. S. and M. B. A. Oldstone) and AI-25522 (M. S.). The author is grateful to M. B. A. Oldstone under whose auspices and encouragement this work was completed, to Elaine Shimomaye and Karl Schweighofer for excellent technical assistance, and to Dave Pauza for critical and extensive revisions of the manuscript.

REFERENCES

Ahmed, R., Simon, R., Matloubian, M., Kolhekar, S., Southern, P., and Freedman, D., 1988, Genetic analysis of immunosuppressive viral variants causing chronic infection: Importance of mutation in the L RNA segment of lymphocytic choriomeningitis virus, *J. Virol.* **62**:3301.

Auperin, D. D., Compans, R. W., and Bishop, D. H. L, 1982, Nucleotide sequence conservation at the 3′ termini of the virion RNA species of New World and Old World arenaviruses, *Virology* **121**:200.

Auperin, D. D., Romanowski, V., Galinski, M., and Bishop, D. H. L, 1984, Sequence studies of Pichinde arenavirus S RNA indicate a novel coding strategy, ambisense viral S RNA, *J. Virol.* **52**:897.

Bass, B. L., and Weintraub, H., 1988, An unwinding activity that covalently modifies its double-stranded RNA substrate, *Cell* **55**:1089.

Bellocq, C., Raju, R., Patterson, J., and Kolakofsky, D., 1987, Translational requirement of La Cross Virus S-mRNA synthesis: *In vitro* studies, *J. Virol.* **61**:87.

Bishop, D. H. L., and Auperin, D. D., 1987, Arenavirus gene structure and organization, *Curr. Top. Microbiol. Immunol.* **133**:5–17.

Boersma, D. P., and Compans, R. W., 1985, Synthesis of Tacaribe virus polypeptides in an *in vitro* coupled transcription and translation system, *Virus Res.* **2**:261.

Buchmeier, M. J., and Parekh, B. S., 1987, Protein structure and expression among arenaviruses, *Curr. Top. Microbiol. Immunol.* **133**:41.

Cattaneo, R., Schmid, A., Eschle, D., Baczko, K., ter Meulen, V., and Billeter, M. A., 1988, Biased hypermutation and other genetic changes in defective measles viruses in human brain infections, *Cell* **55**:255.

Dahlberg, J. E., Obijeski, J. F., and Korb, J., 1977, Electron microscopy of the segmented RNA genome of La Crosse virus: Absense of circular molecules, *J. Virol.* **22**:203.

DePolo, N. J., Giachetti, C., and Holland, J. J., 1987, Continuing co-evolution of virus and defective interfering particles and of viral genome sequences during undiluted passages: virus mutants exhibiting nearly complete resistance to formerly dominant defective interfering particles, *J. Virol.* **61**:454.

Donahue, T. F., Cigan, A. M., Pabich, E. K., and Valavicius, B. C., 1988, Mutations at a Zn(II) finger motif in the yeast eIF-2β gene alter ribosomal start-site selection during the scanning process, *Cell* **54**:621.

Dreher, T. W., Bujarski, J. J., and Hall, T. C., 1984, Mutant viral RNAs synthesized *in vitro* show altered aminoacylation and replicase template activities, *Nature* **311**:171.

Dreyfuss, G., Swanson, M. S., and Pinol-Roma, S., 1988, Heterogeneous nuclear ribonucleoprotein particles and the pathway of mRNA formation, *TIBS* **13**:86.

Edery, I., Petryshyn, R., and Sonenberg, N., 1989, Activation of double-stranded RNA-dependent kinase (dsl) by the TAR region of HIV-1 mRNA: A novel translational control mechanism, *Cell* **56**:303.

Farrell, P. J., Balkow, K., Hunt, T., Jackson, R. J., and Trachsel, H., 1977, Phosphorylation of initiation factor eIF-2 and the control of reticulocyte protein synthesis, *Cell* **11**:187.

Fields, B. N., and Green, M. I., 1982, Genetic and molecular mechanisms of viral pathogenesis: implications for prevention and treatment, *Nature* (London) **300**:19.

Francis, S. J., and Southern, P. J., 1988, Deleted viral RNAs and lymphocytic choriomeningitis virus persistence *in vitro*, *J. Gen. Virol.* **69**:1893.

Fuller-Pace, F., and Southern, P. J., 1988, Temporal analysis of transcription and replication during acute infection with lymphocytic choriomeningitis virus, *Virology* **162**:260.

Holland, J., Spindler, K., Horodyski, F., Grabau, E., Nichol, S., and Vanderpol, S., 1982, Rapid evolution of RNA genomes, *Science* **215**:1577.

Howard, C. R., and Buchmeier, M. J., 1983, A protein kinase activity in lymphocytic choriomeningitis virus and identification of the phosphorylated product using monoclonal antibody, *Virology* **126**:538.

Hsu, M-T., Parvin, J. D., Gupta, S., Krystal, M., and Palese, P., 1987, Genomic RNAs of influenza virus are held in a circular conformation in virions and in infected cells by a terminal panhandle, *Proc. Natl. Acad. Sci.* **84**:8140.

Jacobs, R. P., and Cole, G. A., 1976, Lymphocytic choriomeningitis induced immunosuppression: A virus-induced macrophage defect, *J. Immunol.* **117**:1004.

Joly, E., Salvato, M., Whitton, J. L., and Oldstone, M. B. A., 1989, Polymorphism of CTL clones that recognize a defined nine amino acid immunodominant domain on lymphocytic choriomeningitis virus glycoprotein, *J. Virol.* **63**:1845.

Kao, S., Calman, A., Luciw, P., and Peterland, B. M., 1987, Anti-termination of transcription within the long terminal repeat of HIV-I by tat gene product, *Nature* **330**:489.

Keene, J. D., Schubert, M., and Lazzarini, R. A., 1979, Terminal sequences of vesicular stomatitis virus RNA are both complimentary and conserved, *J. Virol.* **32**:167.

Kirk, W., Cash, P., Peters, C. J., and Bishop, D. H. L., 1980, Formation and characterization of an intertypic lymphocytic choriomeningitis recombinant virus, *J. Gen. Virol.* **51**:213.

Kitajewski, J., Schneider, R. J., Safer, B., Munemitsu, S. M., Samuel, C. E., Thimmappaya, B., and Shenk, T., 1986, Adenovirus VA1 RNA antagonizes the antiviral action of interferon by preventing activation of the interferon-induced eIF2a kinase, *Cell* **45**:195.

Klavinskis, L. S., and Oldstone, M. B. A., 1987, Lymphocytic choriomeningitis virus can persistently infect thyroid epithelial cells and perturb thyroid hormone production, *J. Gen. Virol.* **68**:1867.

Klug, A., and Rhodes D., 1987, 'Zinc fingers': A novel protein motif for nucleic acid recognition, *Trends Biochem. Sci.* **12**:461.

Li, Y., Luo, L., Snyder, R. M., and Wagner, R. R., 1988, Expression of the M gene of vesicular stomatitis virus cloned in various vaccinia virus vectors, *J. Virol.* **62**:776.

Matloubian, M., Somasundaram, T., Kolhekar, S. R., Selvakumar, R., and Ahmed, R., 1990, Genetic basis of viral persistence: Single amino acid change in the viral glycoprotein affects ability of lymphocytic choriomeningitis virus to persist in adult mice, *J. Exp. Med.* **172**:1043.

Miller, J., McLachlan, A. D., and Klug, A., 1985, Repetitive zinc-binding domains in the protein transcription factor IIIA from *Xenopus* oocytes, *EMBO J.* **4**:1609.

Mitsutani, S., and Colonno, R. J., 1985, *In vitro* synthesis of an infectious RNA from cDNA clones of human rhinovirus type 14, *J. Virol.* **56**:628.

Moran, E., and Mathews, M. B., 1987, Multiple functional domains in the adenovirus E1A gene, *Cell* **48**:177.

Nakajima, K., Ueda, M., and Sugiura, A., 1979, Origin of small RNA in von Magnus particles in influenza virus, *J. Virol.* **29**:1142.

Nelson, J. A., Gnann, J. W., and Ghazal, P., 1990, Regulation and tissue specific expression of human cytomegalovirus, *Curr. Top. Microbiol. Immunol.* **154**:75.

Oldstone, M. B. A., and Dixon, F., 1970, Pathogenesis of chronic disease associated with persistent lymphocytic choriomeningitis virus. II. Relationship of anti-lymphocytic choriomeningitis immune response to tissue injury in chronic lymphocytic choriomeningitis disease, *J. Exp. Med.* **131**:1.

Oldstone, M. B. A., Sinha, Y. N., Blount, P., Tishon, A., Rodriguez, M., von Wedel, R., and Lampert, P. W., 1982, Virus induced alterations in homeostasis: Alterations in differentiated functions of infected cells *in vivo*, *Science* **218**:1125.

Oldstone, M. B. A., Salvato, M., Tishon, A., and Lewicki, H., 1988, Virus–lymphocyte interactions. III. Biologic parameters of a virus variant that fails to generate CTL and establishes persistent infection in immunocompetent hosts, *Virology* **164**:507.

Palmer, E. L., Obijeski, J. F., Webb, P. A., and Johnson, K. M., 1977, The circular segmented nucleocapsid of an Arenavirus, Tacaribe virus, *J. Gen. Virol.* **36**:541.

Parker, R., Siliciano, P. G., and Guthrie, C., 1987, Recognition of the TACTAAC box during mRNA splicing in yeast involves base pairing to the U2-like snRNA, *Cell* **49**:229.

Pathak, V. K., Nielson, P. J., Trachsel, H., and Hershey, J. W. B., 1988, Structure of the β subunit of translation initiation factor eIF-2, *Cell* **54**:633.

Porterfield, J. S., Casals, J., Chumakov, M. P., Gaidamovich, S. Y., Hannoun, C., Holmes, I. H., Horzinek, M. C., Mussgay, M., Okerblom, N., and Russell, P. K., 1975, Bunyaviruses and Bunyaviridae, *Intervirology* **6**:13.

Puglisi, J. D., Wyatt, J. R., and Tinoco, I., 1988, A pseudonotted RNA oligonucleotide, *Nature* **331**:283.

Racaniello, V. R., and Baltimore, D., 1981, Cloned poliovirus complementary DNA is infectious in mammalian cells, *Science* **214**:916.

Raju, R., and Kolakofsky, D., 1987, Translational requirement of La Cross virus S-mRNA synthesis: *In vivo* studies, *J. Virol.* **61**:96.

Ransone, L. J., and Dasgupta, A., 1988, A heat-sensitive inhibitor in poliovirus-infected cells which selectively blocks phosphorylation of the α subunit of eukaryotic initiation factor 2 by the double-stranded RNA-activated protein kinase, *J. Virol.* **62**:3551.

Rice, C. M., Levis, R., Strauss, J. H., and Huang, H. V., 1987, Production of infectious RNA transcripts from Sindbis virus cDNA clones: Mapping of lethal mutations, rescue of a temperature-sensitive marker, and *in vitro* mutagenesis to generate defined mutants, *J. Virol.* **61**:3809.

Riviere, Y., 1987, Mapping arenavirus genes causing virulence, *Curr. Top. Microbiol. Immunol.* **133**:59.

Riviere, Y., Ahmed, R., Southern, P. J., and Oldstone, M. B. A., 1985a, Perturbation of differentiated functions during viral infection *in vivo:* II. Viral reassortants map growth hormone defect at the S RNA of the lymphocytic choriomeningitis virus genome, *Virology* **142**:175.

Riviere, Y., Ahmed, R., Southern, P. J., Buchmeier, M. J., and Oldstone, M. B. A., 1985b, Genetic mapping of lymphocytic choriomeningitis virus pathogenicity: Virulence in guinea pigs is associated with the L RNA segment, *J. Virol.* **55**:704.

Riviere, Y., Southern, P. J., Ahmed, R., and Oldstone, M. B. A., 1986, Biology of cloned cytotoxic T-lymphocytes specific for lymphocytic choriomeningitis virus. V. Recognition is restricted to gene products encoded by the viral S RNA segment, *J. Immunol.* **136**:304.

Rodriguez, M., von Wedel, R. J., Garrett, R. S., and Lampert, P. S., 1983, Pituitary dwarfism in mice persistently infected with LCMV, *Lab. Invest.* **49**:48.

Romanowski, V., Matsuura, Y., and Bishop, D. H. L., 1985, Complete sequence of the S RNA of lymphocytic choriomeningitis virus (WE strain) compared to that of Pichinde arenavirus, *Virus Res.* **3**:101.

Salvato, M., 1989, Ambisense nature of the L genomic segment of LCMV, in: *Genetics and Pathogenicity of Negative Strand Viruses* (B. Mahy and D. Kolakofsky, eds.), pp. 168–174. Elsevier Press, New York, Amsterdam.

Salvato, M. S., and Shimomaye, E. M., 1989, The completed sequence of LCMV reveals a unique RNA structure and a new gene for a zinc-finger protein, *Virology* **173**:1.

Salvato, M. S., Shimomaye, E. M., Southern, P. J., and Oldstone, M. B. A., 1988, Virus, lymphocyte interactions. IV. Molecular characteristics of LCMV Armstrong (CTL^+) small genomic segment and that of its variant, clone 13 (CTL^-), *Virology* **164**:517.

Salvato, M. S., Shimomaye, E. M., Schweighofer, K. S., and Oldstone, M. B. A., 1989a, Mapping genes of lymphocytic choriomeningitis virus which determine the cytotoxic T-lymphocyte response, in: Cell biology of virus entry, replication, and pathogenesis, *UCLA Symp. Mol. Cell. Biol.* **90**:329.

Salvato, M. S., Shimomaye, E. M., and Oldstone, M. B. A., 1989b, The primary structure of the lymphocytic choriomeningitis L gene encodes a putative RNA polymerase, *Virology* **169**:377.

Salvato, M. S., Borrow, P. Shimomaye, E. M., and Oldstone, M. B. A., 1991, Molecular basis of viral persistence: as single amino acid change in the glycoprotein of lymphocytic choriomeningitis virus is associated with suppression of the antiviral cytotoxic T-lymphocyte response and establishment of persistence, *J. Virol.* **65**:1863.

Salvato, M. S., Schweighofer, K. J., Burns, J., Shimomaye, E. M., 1992, Biochemical and immunological evidence that the 11 kDa zinc-binding protein of lymphocytic choriomeningitis virus is a structural component of the virus, *Virus Research* **22**:185.

Schiff, L. A., Nibert, M. L., Co, M. S., Brown, E. G., and Fields, B. N., 1988, Distinct binding sites for zinc and double-stranded RNA in the reovirus outer capsid protein sigma 3, *Mol. Cell. Biol.* **8**:273.

Shimomaye, E. M., and Salvato, M. S., 1989, Use of AMV reverse transcriptase at high temperature for sequence analysis of highly structured RNA, *Gene Analysis Techniques* **6**:25.

Singh, M. K., Fuller-Pace, F. V., Buchmeier, M. J., and Southern, P. J., 1987, Analysis of the genomic L segment from lymphocytic choriomeningitis virus, *Virology* **161**:448.

Southern, P. J., and Bishop, D. H. L., 1987, Sequence comparison among arenaviruses, *Curr. Top. Microbiol. Immunol.* **133**:19–39.

Southern, P. J., Blount, P., and Oldstone, M. B. A., 1984, Analysis of persistent virus infections by *in situ* hybridization to whole-mouse sections, *Nature* **312**:555.

Southern, P. J., Singh, M. K., Riviere, Y., Jacoby, D. R., Buchmeier, M. J., and Oldstone, M. B. A., 1987, Molecular characterization of the genomic S RNA segment from LCMV, *Virology* **157**:145.

Strauss, E. G., and Strauss, J. H., 1983, Replication strategies of the single-stranded RNA viruses of eukaryotes, *Curr. Top. Microbiol. Immunol.* **105**:1.

Thomas, S., Lamb, R. A., and Patterson, R. G., 1988, Two mRNAs that differ by two nontemplated nucleotides encode the amino coterminal proteins P and V of the paramyxovirus SV5, *Cell* **54**:891.

Tinoco, I, Borer, P. N., Dengler, B., Levine, M. D., Uhlenbeck, O. C., Crothers, D. M., and Gralla, J., 1973, Improved estimation of secondary structure in ribonucleic acids, *Nat. New Biol.* **246**:40.

Valsamakis, A., Riviere, Y., and Oldstone, M. B. A., 1987, Perturbation of differentiated functions *in vivo* during persistent viral infection, *Virology* **156**:214.

van der Werf, S., Bradley, J., Wimmer, E., Studier, W., and Dunn, J. J., 1986, Synthesis of infectious poliovirus RNA by purified T7 RNA polymerase, *Proc. Natl. Acad. Sci. USA* **83**:2330.

Vezza, A. C., Gard, G. P., Compans, R. W., and Bishop, D. H. L., 1977, Structural components of the arenavirus Pichinde, *J. Virol.* **23**:776.

Vezza, A. C., Clewley, J. P., Gard, G. P., Abraham, N. Z., Compans, R. W., and Bishop, D. H. L., 1978a, Virion RNA species of the Arenaviruses Pichinde, Tacaribe, and Tamiami, *J. Virol.* **26**:485.

Vezza, A. C., Gard, G. P., Compans, R. W., and Bishop, D. H. L., 1978b, Genetic and molecular studies of arenaviruses, in: *Negative Strand Viruses and the Host Cell* (B. W. J. Mahy and R. D. Barry, eds.), pp. 73. Academic Press. New York.

Whitton, J. L., Southern, P. J., and Oldstone, M. B. A., 1988, Analysis of the cytotoxic T-lymphocyte response to glycoprotein and nucleoprotein components of lymphocytic choriomeningitis virus, *Virology* **16**:321.

Whitton, J. L., Tishon, A., Lewicki, H., Gebhard, J., Cook, T., Salvato, M. S., Joly, E., and Oldstone, M. B. A., 1989, Molecular analyses of a five-amino-acid cytotoxic T lymphocyte (CTL) epitope: An immunodominant region which induces nonreciprocal CTL cross-reactivity, *J. Virol.* **63**:4303.

Wu, H-N., and Lai, M. M. C., 1989, Reversible cleavage and ligation of hepatitis delta virus RNA, *Science* **243**:652.

Zinkernagel, R. M., and Doherty, P. C., 1974, Restriction of *in vitro* T cell–mediated cytotoxicity in lymphocytic choriomeningitis virus with a syngeneic or semiallogeneic system, *Nature* **248**:701.

CHAPTER 9

Arenavirus Replication

Molecular Dissection of the Role of Viral Protein and RNA

DELSWORTH G. HARNISH, STEPHEN J. POLYAK, AND WILLIAM E. RAWLS

I. INTRODUCTION

Arenaviruses have provided model systems in which to study host–virus relationships during persistent infections. Many of the members of this virus group, which includes Lassa virus, Junin virus, lymphocytic choriomeningitis virus (LCMV), and Machupo virus, have been associated with hemorrhagic fevers in humans, but more often than not, they do not appear to be a serious human health hazard. Interest in arenaviruses derives from the ability to persistently infect their natural hosts, and therefore much effort has been directed to an understanding of virus–cell interactions and immune system modulation of arenavirus infection.

Several excellent reviews on arenavirus biology have been published in recent years. Although it is the intention of this review to examine the essential features of arenavirus replication with respect to the function of viral protein and RNA in acute infection the reader is referred to complementary reviews on pathogenesis (Walker and Murphy, 1987) and molecular genetics (Bishop and Auperin, 1987) for comprehensive

---

DELSWORTH G. HARNISH, STEPHEN J. POLYAK, AND WILLIAM E. RAWLS • Molecular Virology and Immunology Programme, Department of Pathology, McMaster University, Hamilton, Ontario, L8N 3Z5, Canada.

The Arenaviridae, edited by Maria S. Salvato. Plenum Press, New York, 1993.

information. In addition, several chapters in this volume deal with specific aspects of gene organization and expression, coding strategies, glycoprotein function, host defense mechanisms, and pathogenesis. The information provided will be discussed with particular reference to Pichinde virus.

II. ARENAVIRUS REPLICATION

A. Molecular Components of Arenavirus Replication

Although in nature arenaviruses exhibit a highly restricted host range, a wide variety of animals have been infected experimentally or inadvertently (Johnson *et al.*, 1973). The hallmark of arenaviruses, persistent infection, is usually only observed with a specific member of the virus group and its natural host. It was demonstrated over 50 years ago that LCMV caused a persistent infection in *Mus musculus* (Traub, 1936), and this virus–host system is still the most explored model for host defense mechanisms.

All arenaviruses characterized to date encode genetic information in two virus-specific RNAs, which have been designated L (large) and S (small). S RNA encodes two structural proteins in ambisense orientation (Bishop, 1986). An intergenic hairpin separates nucleoprotein (NP) and glycoprotein (GPC) sequences in Lassa virus (Auperin *et al.*, 1986), LCMV (Romanowski and Bishop, 1985; Romanowski *et al.*, 1985), Tacaribe virus (Franze-Fernandez *et al.*, 1987), Junin virus (Ghiringhelli *et al.*, 1991), and Pichinde virus (Auperin *et al.*, 1984). Two S RNA intergenic hairpins have been suggested from the sequence of Mopeia virus (Wilson and Clegg, 1991). An ambisense coding strategy has now been determined for the L RNAs of LCMV (Salvato and Shimomaye, 1989) and Tacaribe virus (Iapalucci *et al.*, 1989a,b). The intergenic L hairpin separates sequences that encode a 200K polypeptide designated L (Harnish *et al.*, 1983; Iapalucci *et al.*, 1989a) and an 11K polypeptide, termed Z or p11 (Salvato and Shimomaye, 1989; Iapalucci *et al.*, 1989b; Singh *et al.*, 1987). The function of these proteins in virus replication is not known although it has been speculated that L protein may be part of the viral transcriptase complex (Harnish *et al.*, 1981). Arenavirus particles also contain host cell ribosomal RNAs (Farber and Rawls, 1975), which do not appear to be essential for virus replication (Leung and Rawls, 1977).

Three primary gene products have been identified in cells infected with LCMV, Tacaribe virus, Junin virus, Lassa virus, or Pichinde virus (reviewed in Buchmeier and Parekh, 1987). The most abundant protein is an RNP-associated nucleoprotein (NP). Mature virions contain two glycoproteins, GP1 and GP2, which are derived from a common cell-associated precursor (GPC) and can be found in association with the cell membrane. The L-derived proteins are thought to be primarily involved in transcription and replication of viral nucleic acid. Specifically, the L protein itself is believed to possess the RNA-dependent RNA polymerase activity initially characterized by Leung and co-workers (1979), while the Z or p11 protein is thought to provide transcriptional regulatory functions, based on its homology to zinc finger–binding proteins (Salvato and Shimomaye, 1989; Iapalucci *et al.*, 1989b). It is not known at present whether the 11K protein is part of the viral transcriptase complex, or whether it affects cellular transcription functions necessary for viral replication.

The discovery of the ambisense coding strategy of arenaviruses has led to a proposed model of arenavirus replication (Auperin *et al.*, 1986) that predicts discordant transcription of NP and L mRNA relative to GPC and 11K polypeptide mRNA. NP and L mRNAs are transcribed from viral RNA (vRNA) while the GPC and 11K messages are templated by viral complementary RNA (vcRNA). The events that occur during the 8-h eclipse period after infection of cells in culture and the nature of proposed autoregulatory processes that terminate the replicative cycle without obvious cytopathology are not known (Harnish, 1982). Clearly, an understanding of arenavirus replication will rely on a molecular characterization of the relative role of viral proteins in the replicative cycle and the precise contribution, if any, of host cell factor(s) to replicative events. A number of arenavirus variants are now available, including ts mutants of Junin virus (Ceriatti *et al.*, 1986) and Pichinde virus (Shivaprakash *et al.*, 1988) and attenuated strains of LCMV (Broomhall *et al.*, 1987), which should provide the necessary insight into arenavirus replication.

B. Glycoprotein-Dependent Events in Infection

It is clear that arenaviruses can infect a number of continuous cell lines *in vitro*, including cells of mouse, hamster, monkey, and human origin. The cellular receptor for virus binding has not been identified and the relative roles of GP1 and GP2 in binding and penetration are not known. Tunicamycin inhibitor studies with Junin virus–infected Vero cells suggest that virus maturation occurs normally when GP1 and GP2 are not glycosylated but that carbohydrate is important for infectivity of the mature virus (Daelli and Coto, 1982). Although the authors did not examine processing of the Junin virus cell-associated glycoprotein precursor (GPC) to GP1 and GP2, Wright *et al.* (1989) have shown that LCMV GPC cleavage to GP1 and GP2 is dependent on glycosylation. The same group also demonstrated that glycosylation of GPC was required for virus maturation (Wright *et al.*, 1990). The discrepancies between the two groups could be due to the time at which tunicamycin was added to the cells. In one study, tunicamycin was added immediately after virus adsorption (Daelli and Coto, 1982), while in the other study, tunicamycin was added to the cells 24 hr prior to virus adsorption (Wright *et al.*, 1990). At present, it is still not clear whether carbohydrate addition and/or GPC cleavage is required for virus infectivity.

Temperature-sensitive mutants of Pichinde virus that have defects in GPC have recently been characterized (Shivaprakash *et al.*, 1988). Two mutants (ts 13, ts 908) synthesize and accumulate increased amounts of GPC at the nonpermissive temperature. Although glycosylation appears normal, cleavage to GP1 and GP2 and translocation to the surface of infected BHK-21 cells were not observed. The mutation is unlikely to be in the conserved arenavirus GPC cleavage site (Buchmeier *et al.*, 1987) since cleavage appeared normal on shiftdown to the permissive temperature and viral glycoprotein was evident at the cell surface (authors' unpublished observations).

Recent studies have investigated the mechanism of arenavirus penetration into target cells. Using various lysosomotropic agents, it has been shown that Pichinde, Lassa, and Mopeia viruses enter cells via acidic lysosomes (Glushakova and Lukashevich, 1989; M. Godlewski, S. Polyak, W. Rawls, and D. Harnish, unpublished observations). Our own observations extend the initial report in that we were able to exclude late events in the viral life cycle as an explanation for inhibition of replication since both ammonium chloride and chloroquine abrogated viral protein synthesis as monitored by immunoprecipitation with hamster anti–Pichinde virus sera. Neither study can exclude secondary effects of the inhibitors as in the case of vesicular stomatitis virus, in which viral nucleic acid (as measured by interferon production) reaches the cell cytoplasm in the presence of lysosomotropic agents (Svitlik and Marcus, 1984).

C. Nucleoprotein Function

The major structural protein synthesized in arenavirus-infected cells has been termed the nucleoprotein. The physical properties of this protein have recently been reviewed (Buchmeier and Parekh, 1987); however, the molecular weight varies from 60,000 for Mobala (Gonzalez *et al.*, 1983), Lassa (Clegg and Lloyd, 1983), and Junin viruses (Martinez-Segovia and de Mitri, 1977) to 68,000 for Tacaribe virus (Gard *et al.*, 1977). A number of nucleoprotein-related polypeptides have been documented in Pichinde virus–infected BHK-21, Vero, and MDCK cells (Harnish *et al.*, 1981; Harnish, 1982) and in LCMV-infected Vero cells (Bruns *et al.*, 1983). The relative proportion of these in different cell types varies, and it is unclear whether they represent degradation products, products of premature termination of translation, or distinct mRNA-encoded polypeptides. Young *et al.* (1986a) have further characterized the intracellular location of NP-related polypeptides and have observed a distinct nuclear localization of the 28K molecule. This may suggest distinct functional roles for the NP-related molecules although the relative proportion of these compared to NP varies in different cells and with time postinfection (Harnish, 1982; Buchmeier and Parekh, 1987).

Very little information is available on the synthesis of viral polypeptides during the replicative cycle. NP synthesis is first detectable late (6 hr) in the eclipse period of viral replication as determined by L-[^{35}S]-methionine radiolabeling of LCMV and Pichinde virus–infected cells (Buchmeier *et al.*, 1978; Harnish *et al.*, 1981). Synthesis of new molecules as measured by pulse labeling precedes release of new virus, and NP declines as virus replication declines in Pichinde virus–infected BHK-21 cells (Dimock *et al.*, 1982). The half-life of NP in infected cells is approximately 12 hr in acute infections. It has been noted that the half-life of 17K, 16.5K, and 12K NP-related polypeptides is much longer, on the order of 4 days. The differences in half-life may be related to differential incorporation into virus, although all of the NP-related polypeptides have been observed in purified Pichinde virus.

The ambisense organization of S RNA suggests that NP synthesis is discordant with respect to GPC synthesis (Auperin *et al.*, 1984). The role of NP in the regulation of viral replication has not been directly determined although it is clearly the major component of viral ribonucleoprotein complexes. It has been suggested that NP and L protein are required for transcriptase activity and that NP may play a pivotal role in determining the balance between transcription of viral complementary RNA, genome replication transcription, and mRNA synthesis (Harnish, 1982; Young *et al.*, 1986b). Studies with influenza virus (Orthomyxoviridae), have revealed that antitermination (or conversion from transcription to replication) is mediated by the nucleoprotein (Beaton and Krug, 1986). It is conceivable that the nucleoprotein of arenaviruses has a similar function. The relative levels of NP in Pichinde virus–infected cells have also been proposed to regulate the apparent *in vitro* shutoff of viral replication in the absence of cytopathology (Harnish, 1982).

LCMV-infected cell RNP complexes containing L protein, NP and viral RNA will synthesize NP and GPC mRNAs *in vitro* (Fuller-Pace and Southern, 1988, 1989). The absolute requirement for NP, L protein, or both has not been assessed to date. It is interesting in this context that Howard and Buchmeier (1983) have described an LCMV RNP-associated protein kinase that preferentially phosphorylates serine and threonine of NP. It is not known whether the activity is host-cell-derived or virally encoded, but a putative role of phosphorylation in transcriptase activity or virion assembly has been envisioned. Previous attempts to demonstrate phosphorylation of arenavirus polypeptides have met with varying degrees of success. Pichinde virus polypeptides immunoprecipitated from infected cells after labeling with ^{32}P-ATP did not have detectable phosphorylation (Harnish, 1982), but LCMV-infected cells contain a soluble phosphorylated form of NP, which has been termed p63E (Bruns *et al.*, 1986).

D. Other Proteins

One other virus-encoded polypeptide has been consistently documented in arenavirus-infected cells by immunoprecipitation with anti-virus sera. A protein of approximately 200 kDa was first observed in Pichinde virus–infected BHK-21 cells. Tryptic mapping of L polypeptide in parental and reassortant viruses previously characterized with respect to viral RNA reassortment (Vezza *et al.*, 1980) showed that L polypeptide was encoded by L RNA (Harnish *et al.*, 1983). Similar high-molecular-weight L polypeptides have now been documented for LCMV (Bruns *et al.*, 1983), Lassa virus, Mobala virus, and Mopeia virus (Gonzalez *et al.*, 1983). The L protein may represent either part or all of the viral-specific RNA polymerase. In Pichinde virus (Leung *et al.*, 1979) and LCMV (Fuller-Pace and Southern, 1988, 1989) the polymerase activity has been shown to be associated with the nucleocapsid. Fuller-Pace and Southern (1988, 1989) have demonstrated *in vitro* synthesis of NP and GPC mRNA by viral RNP complexes, but further studies of arenavirus L protein have been hampered by the low quantities of virus and virus-infected cells and the apparent lability of the enzyme (F. V. Fuller-Pace, personal communication).

L RNA also appears to encode a P11 protein in LCMV (Salvato, this volume) and Tacaribe virus (Iapalucci *et al.*, 1989b). Although a function for this protein has not been determined, it retains sequence homology with zinc-binding proteins that have been implicated in transcriptional regulation in other systems (Salvato, this volume). We have, to date, been unable to demonstrate this protein in Pichinde virus or Pichinde virus–infected BHK-21 cells utilizing either Western blots or immunoprecipitation of L-[^{35}S] methionine– or [^{35}S] cysteine–labeled Pichinde virus polypeptides. A rabbit anti-P11 sera does not cross-react with a P11 protein of Pichinde virus, although it is likely that Pichinde virus does encode such a protein. The L RNA of Pichinde virus has not been cloned and therefore the putative P11 gene has not been characterized.

A number of other enzyme activities and minor polypeptides have been associated with purified arenavirus preparations or arenavirus-infected cells but these are of unknown significance. Poly A polymerase and poly U polymerase activities have been demonstrated in Pichinde virus (Leung *et al.*, 1979). It is clear, on the basis of association of these activities with ribosomes, that they are not activities intrinsic to the viral RNA polymerase but probably are of host cell origin; as yet there is no documented role for the enzymes in viral infection.

III. Molecular Dissection of Pichinde Virus Replication

A. Viral Nucleic Acid in Acute Infections

The genome of arenaviruses consists of two single-stranded RNA molecules designated L (large) and S (small). Several reviews deal with early observations on the physical properties of the RNA in arenaviruses (see, for example, Bishop and Auperin, 1987) and these will not be further documented here. The main features of the RNAs that may be of interest in terms of tropism and virulence derive from the ambisense coding strategy and more recent observations in Pichinde virus–infected BHK-21 cells, which document larger-than-unit-size (LUS) RNAs for at least the S RNA segment (Shivaprakash *et al.*, 1988).

Each of L and S RNAs appears to encode two polypeptides as documented for S RNAs of LCMV (Romanowski and Bishop, 1985), Pichinde virus (Auperin *et al.*, 1984), and Lassa virus (Auperin *et al.*, 1986) and L RNAs of Tacaribe virus (Iapalucci *et al.*, 1989a) and LCMV (Salvato, this volume). Viral mRNA for NP is templated directly by viral-sense S RNA; however, mRNA for GPC is transcribed from viral-complementary-sense RNA. The two open reading frames in viral-sense RNA are separated by a 42-nucleotide intergenic hairpin (Auperin *et al.*, 1984). Sequence comparisons of S RNAs of Lassa virus, Pichinde virus, and LCMV (Bishop and Auperin, 1987) as well as early RNA sequence data on Pichinde S RNA (Auperin *et al.*, 1982) have revealed complementarity of 3′ and 5′ termini. For Pichinde virus the complementarity extends for 19 nucleotides with two mismatches (Auperin *et al.*, 1984). This implies that viral RNAs may contain more extensive secondary structure, including circular molecules that are base-paired at the termini and in the intergenic hairpin. The same features of complementarity have been documented for LCMV (Salvato, this volume) and Tacaribe virus (Iapalucci *et al.*, 1989a,b) L RNAs, but the position of the intergenic hairpin is distinctly asymmetrical since L RNA encodes a 200-kDa and an 11-kDa polypeptide.

Recent studies have revealed that the 5′ end of Tacaribe virus genomic RNA contains an extra, nontemplated G nucleotide (Raju *et al.*, 1990; Garcin and Kolakofsky, 1990). This indicates that the proposed panhandle configuration has a one-base overhang. The function of the extra residue is not known, but it may provide a novel function in arenavirus replication with respect to a "slippage initiation" mechanism proposed by Kolakofsky and Garcin (this volume).

The synthesis of viral RNAs during the arenavirus replicative cycle has been examined for Pichinde virus (Shivaprakash *et al.*, 1988), Tacaribe virus (Franze-Fernandez *et al.*, 1987), and LCMV (Fuller-Pace and Southern, 1988, 1989). The following general observations have been made which support proposed ambisense models of replication (Auperin *et al.*, 1984). Messenger RNA for NP is observed as early as 2 hr after infection. NP mRNA is polyadenylated and synthesis is not sensitive to inhibitors of protein synthesis such as pactamycin (Franze-Fernandez *et al.*, 1987). Glycoprotein mRNA appears later and synthesis appears to be dependent on protein synthesis. Both RNAs accumulate but NP mRNA attains higher concentrations. The observed sensitivity of GPC mRNA synthesis to pactamycin inhibition of protein synthesis is also reflected by lack of synthesis of vcRNA (antigenomes) or vRNA (genomes). The quantity of genome RNA accumulating over 24 hr exceeds accumulations of viral complementary RNA. Northern blot analysis permits a relative comparison of RNA synthesis and accumulation but it is clearly insensitive with respect to determining the point of initial synthesis. L RNA accumulates 12–15 hr after infection in LCMV (Fuller-Pace and Southern, 1988, 1989) but a requirement for newly synthesized L mRNA and L protein might precede this by 10–12 hr.

The precise nature of molecular events involved in viral RNA synthesis, transcription of viral mRNAs, and replication (including accurate transcription of termini) is not known. Studies of viral mRNA transcription have revealed that the 5′ ends of Tacaribe virus mRNAs contain one to four nontemplated bases which have a methylated cap structure (Raju *et al.*, 1990; Garcin and Kolakofsky, 1990). This finding suggests that initiation of arenavirus transcription occurs by capsnatching from cellular mRNAs, similar to influenza virus (Kingsbury, 1990), although there is at present no direct evidence for this. The nontemplated bases are shorter than those observed in influenza virus, and it is not clear whether they contain the methylated cap or whether they are capped by a guanyl-transferase activity at a later point (Garcin and Kolakofsky, 1990). Other recent studies examining transcription termination by Tacaribe virus indicate that termination occurs within the intergenic hairpin region (Iapalucci *et al.*, 1991). The authors suggested that the mRNAs obtained stable hairpin structures when transcription entered the intergenic region of the template RNA, and this newly acquired configuration resulted in release of the nascent mRNA from its template.

We have speculated that viral RNA may replicate through circular intermediates due to complementarity of 3′ and 5′ termini but there is no evidence of this. Recently we have observed RNAs in Pichinde virus–infected cells that are not of unit length and hybridize with viral-sense and viral-complementary-sense cDNA probes to S RNA. The RNAs, termed larger than unit size (LUS), appear heterodisperse and have mobilities that differ from those of ribosomal RNA (Shivaprakash *et al.*, 1988). They accumulate late in infection and can exceed the quantities of S RNA in infected cells (Fig. 1). It is of interest that these observations are not restricted to Pichinde virus since S RNA–specific cDNA probes also appear to recognize anamolous RNAs synthesized late in infection

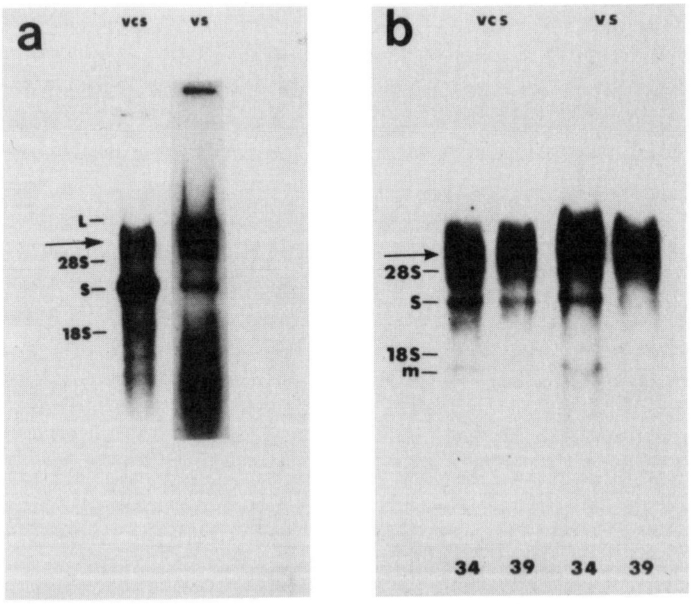

FIGURE 1. RNA in purified Pichinde virus and ts 538 infected BHK-21 cells. One microgram of viral RNA (a) or 20 μg of total infected cell RNA (b) was electrophoretically separated in formaldehyde–agarose gels, transferred to nitrocellulose, and hybridized to single-stranded cDNA probes for viral-sense (vs) or viral-complementary-sense sequences representing nucleotides 25–880 of Pichinde S RNA (NP sequences) (Shivaprakash *et al.*, 1988). VC and VCS refer to the sense of the probe such that VCS hybridizes to viral-sense RNA. RNA species represent viral S RNA (S), NP mRNA (m), LUS RNA (arrows), and ribosomal RNAs (28s, 18s). The position of L RNA (L) as determined by staining with acridine orange is indicated in (a). For ts 538 (b), RNA was extracted 48 hr after infection at a multiplicity of infection (moi) of 3 in permissive (34) and nonpermissive (39) temperature conditions.

with Tacaribe virus (Franze-Fernandez *et al.*, 1987) and LCMV (Southern *et al.*, 1987; Fuller-Pace and Southern, 1988, 1989). However, the precise nature of these RNAs is still a matter of speculation.

B. Reassortant Viruses

One of the consequences of infection of the same cell with two viral strains from the same family is genetic reassortment of the segmented RNA genome. Reassortment has been documented both *in vitro* and *in vivo*, and this has been useful in attempts to define the role of viral genes in pathogenesis (Webster *et al.*, 1982; Vezza and Bishop, 1977).

Variant viruses arising by reassortment of L and S RNAs after coinfection have been described for Pichinde virus (Vezza and Bishop, 1977) and for LCMV (Riviere and Oldstone, 1986). These variants have been used to map viral genes to L and S RNA (Vezza *et al.*, 1980; Harnish *et al.*, 1983) of Pichinde virus. Two strains of LCMV (Arm and WE) can be distinguished with respect to lethality in an adult guinea pig infection model; the Arm strain causes a subclinical infection and the WE strain is lethal (Lehmann-Grube, 1971). The pathogenesis of WE is not attributable to antiviral immune responses since immunosuppression of WE-infected animals does not alter the kinetics or outcome (Riviere *et al.*, 1985). Reassortants between ARM and WE yielded progeny that are distinctly avirulent (ARM/WE) and virulent (WE/ARM). Since virulent strains contained the L RNA of WE, the mortality has been attributed to the L RNA and, by extension, to the putative polymerase activity of L protein (Riviere *et al.*, 1985). The existence of a second L RNA–derived protein was not known until recently (Salvato, this volume) and therefore the precise role of L RNA–encoded proteins is uncertain. Similar experiments have implicated the L RNA of WE/ARM and Traub/ARM reassortants in lethal infections of neonatal mice. In this case, however, there may be an additional complexity introduced by the induction of high levels of interferon as well as high virus titers relative to nonpathogenic reassortants (Riviere and Oldstone, 1986). It should be noted that although this type of study is useful in defining putative changes in viral nucleic acid sequence or protein function that alter pathogenesis, pathogenesis is generally a multifactorial process that is dependent on the entire virus and the associated gene products.

C. Conditional Mutants

Arenavirus variants that are conditional mutants or differ in growth properties either *in vitro* or *in vivo* have not been extensively characterized at the molecular genetic level. Classically, conditional mutants have provided important information on the role of viral proteins in replication. Temperature-sensitive mutants have been reported for Junin virus (Ceriatti *et al.*, 1986) and Pichinde virus (Vezza and Bishop, 1977; Shivaprakash *et al.*, 1988), and these may serve as useful tools to define molecular events of virus replication.

Recently we have assigned temperature-sensitive mutants of Pichinde virus to five broad groups with respect to protein and RNA synthesis during acute infection of BHK-21 cells (Shivaprakash *et al.*, 1988). Our findings, as summarized in Table I, indicate that mutations that map to L RNA result in decreased synthesis of complementary- and genomic-sense S RNA as well as NP mRNA, with a consequent decrease in NP and GPC synthesis at the nonpermissive temperature. L RNA mutants further segregate with respect to synthesis of GPC (group 2 versus group 4). These mutants probably have temperature-sensitive lesions associated with viral polymerase activity and hence provide the most direct evidence for polymerase activity encoded by L RNA. The three remaining groups (Table I) synthesized viral complementary- and

TABLE I. Summary of Group Phenotypes of Pichinde Virus ts Mutants[a]

Group ts mutant	Titer[b] 34°	39°	ts locus[c] (RNA)	Protein at 39°C relative to 34°C	RNA at 39°
Wild type	2.8×10^8	2.3×10^8	—	Standard	Wt
Group 1					
ts 13	1.2×10^8	7.5×10^4	S	GPC	Wt
(ts 908)					
Group 2					
ts 538	1.0×10^8	1.0×10^5	L	NP	
(ts 1, 3, 5, 939)			(ts 3, 5)	no L, GPC	VS:[d] no S, 15S VCS: S LUS
Group 3					
ts 488	2.0×10^8	5.0×10^6	N.D.	Wt	Wt
(ts 474)					
Group 4					
ts 2	1.2×10^8	1.2×10^6	L	NP, GPC, L	LUS
Group 5					
ts 11	2.0×10^7	4.0×10^5	S	NP, GPC, L	S, 15S
(ts 454)				(not detectable at 24 hr)	cS, LUS

[a] Adapted from Shivaprakash *et al.* (1988).
[b] Titers provided for ts mutants that are not bracketed.
[c] ts locus for selected mutants was determined by Vezza and Bishop (1977).
[d] vs = viral sense; vcs = viral complementary sense.

genomic-sense S RNA as well as mRNAs at the nonpermissive temperature. Group 1 mutants appear to be mutants in GPC cleavage and translocation to the cell surface since cleavage is normal when the cultures are shifted to the permissive temperature (Table II).

Group 3 and 5 mutants have been much more difficult to reconcile

TABLE II. Effect of Temperature Shift on Expression on Viral Antigens on the Surface of Infected Cells

Virus	34°C	39.5°C	Shift[a] from 39.5°C to 34°C
ts 1	4+ (100%)[b]	4+ (1%)	2+ (1%)
ts 3	2+ (15%)	0	0
ts 908[c]	4+ (85%)	0	4+ (50%)
ts 454	4+ (15%)	3+ (1%)	4+ (10%)
ts 274	4+ (100%)	3+ (50%)	3+ (75%)

[a] Infected cells incubated 24 hr at nonpermissive temperature and then shifted to permissive temperature and incubated for an additional 24 hr before being tested for surface antigen.
[b] Intensity of fluorescence at cell surface assessed on scale of 0 (none) to 4+ (maximum). The percent of cells with surface fluorescence in parentheses was estimated by observing five separate microscopic fields with each containing at least 200 cells.
[c] Group 1 mutant (Table I).

with respect to currently proposed models of arenavirus replication (Bishop and Auperin, 1987). Group 3 mutants appear normal at either permissive or nonpermissive temperatures in the context of viral RNA and protein synthesis. The ts defect may relate to thermal stability, adsorption, or penetration, but in the absence of information on the ts protein, the nature of the defect is a matter of speculation. Group 5 mutants were characterized by reduced viral protein at both temperatures as detected by immunoprecipitation, and reduced cell numbers with surface antigen at the nonpermissive temperature (Table III) and increasing multiplicities of infection did not increase the quantity of detectable protein synthesized (Table IV).

Three explanations have been considered. First, the mutation could affect the antigenicity of a viral protein such that it would not be detected by immunoprecipitation. This appears unlikely since neither NP nor GPC was detected in cell lysates, although GPC appeared to be detectable with the antiserum on the cell surface. Second, there may be reduced synthesis of all viral proteins. Virus yields from cells infected with these mutants and incubated at the permissive temperature were comparable to those obtained with wild-type virus. For example, the titers of stocks for ts 454 and wild-type virus were 1.4×10^8 and 2.8×10^8 pfu/ml, respectively. This suggests that protein production is not reduced. Finally, the proteins could be antigenically intact and made in normal amounts but turn over at an increased rate. Little is known about the transport of viral proteins or assembly of Pichinde virus, and additional studies will be required to examine this possibility.

The findings imply several steps in the synthesis and processing of

TABLE III. Assay for Surface Antigens by Indirect Immunofluorescence

		Immunofluorescence[b] at	
Virus ts mutants	Group	34°C	39.5°C
13	1	3+ (10%)	0
908	1	3+ (5%)	0
1	2	2–3+[a] (40%)	3+ (1%)
3	2	3+ (10%)	0
5	2	3+ (50%)	0
538	2	3+ (15%)	0
939	2	4+ (20%)	1+ (<1%)
274	3	3+ (70%)	3+ (80%)
488	3	3+ (50%)	3+ (50%)
2	4	3+ (10%)	3+ (2%)
11	5	3+ (50%)	3+ (5%)
454	5	4+ (50%)	3+ (<1%)
Wild type		4+ (80%)	4+ (75%)

[a] Intensity of fluorescence was assigned on a scale from 0 to 4+, where 4+ was maximum.
[b] The number of cells (in parentheses) fluorescing is expressed as an approximate percentage of five microscopic fields. The maximum number of cells per field, based on observation, was 200.

TABLE IV. Effect of Multiplicity of Infection on Synthesis of Viral Antigens

		Antigen-positive cells standardized to pfu[a] (multiplicity of infection)			
Mutant	Group	0.1	0.3	1.0	3.0
1	1	NT	40	4.4	0.33
908	1	40	20	28	67
5	2	500	40	4	1
538	2	530	45	3.5	NT
939	2	214	NT	61	10.7
274	3	160	146	105	117
2	4	137	NT	88	13.3
454	5	NT	9	7	4.6
wt		380	130	95	NT

[a] Number of antigen-positive cells/200 cells divided by multiplicity of infection (moi).

Pichinde virus proteins and add further complexity to the ambisense model of arenavirus replication. Further analysis of the molecular events during acute cell infection with these viruses should enhance our understanding of arenavirus replication.

D. Host Cell Involvement in Virus Replication

Cytopathology is often absent in arenavirus infections (Pedersen, 1979; Dimock *et al.*, 1982), suggesting that the host cell is neither affected by nor affects viral replication. However, a role for nuclear function in arenavirus replication was suggested by Mims (1966) on the basis of nuclear fluorescence of LCMV-infected cells of virus-carrier mice. Similarly, the requirement of a cell nucleus in the replication of Pichinde virus in BHK-21 cells was suggested from the observation that viral replication was inhibited in cells enucleated with cytochalasin B (Banerjee *et al.*, 1976).

Further support for the concept of the requirement for host cell factors in arenavirus replication is derived from transcription inhibition studies using actinomycin D. Rawls *et al.* (1976) have reported inhibition of Pichinde virus by actinomycin D late in the replication cycle since virus protein synthesis and antigen on the cell surface were increased, with an apparent decrease in virus yield. This suggested the requirement of host cell protein(s) during virus budding from cells. LCMV replication is inhibited in L cells (Buck and Pfau, 1969), and Machupo virus titers are reduced 2 logs or more in the presence of actinomycin D (Lukashevich *et al.*, 1984). We have also demonstrated that host cell transcription is required for Pichinde virus replication in phorbol ester–treated THP-1 cells (Polyak *et al.*, 1991). However, Lopez *et al.*

(1986) did not observe any effect of actinomycin D on yields of Junin, Pichinde, or Tacaribe virus from virus-infected BHK-21 cells. The latter study limited actinomycin D treatments to 6–10 hr during the replication cycle. Therefore, differences in the cited studies may relate to the half-life of host cell components required for replication.

In terms of the effect of arenavirus infection on cellular function, two groups have demonstrated dose-dependent inhibitions of host cell DNA synthesis by Pichinde virus infection of mouse peritoneal macrophages (Friedlander *et al.*, 1984) and Tacaribe virus infection of Vero cells (Lopez and Franze-Fernandez, 1985). Young *et al.* (1986a) have carefully documented Pichinde virus–specific nuclear inclusions in Vero cells. Defined monoclonal antibodies further identified the antigen as an NP-related 28K polypeptide or a conformational variant of NP. The inclusions were observed between 10 and 20 days postinfection, but antigen could be detected as early as 10–15 hr after infection. However, the effect of such viral nuclear inclusions on nuclear function was not assessed.

To date, the evidence remains circumstantial for a role of the nucleus or nuclear transcription products in arenavirus replication. It is likely that host cell functions are required for early (Polyak *et al.*, 1991) and late (Rawls *et al.*, 1976) events in virus replication and that a demonstration of the requirement utilizing α amanitin, actinomycin D, or enucleation reflects the ability of these treatments to abrogate the function. Without knowledge of the precise nature of the function(s), the evidence will remain circumstantial and probably reflects the rate of turnover of host cell protein. However, our recent findings indicate that the differentiation or activation state of the host cell can be a major factor in viral tropism (Polyak *et al.*, 1991). The human promonocytic cell line THP-1 does not support Pichinde virus replication without treatment with phorbol ester. The inability of untreated THP-1 cells to support infection was not due to the absence of virus binding to the cells, nor was it due to absence of internalization of virus bound to the cell surface. These findings, when considered with observations on the accumulation of NP-related antigen in the nucleus (Young *et al.*, 1986a) and the late accumulation of LUS RNAs in Pichinde virus–infected cells (Shivaprakash *et al.*, 1988), remain to be explained in the context of the viral replication cycle.

IV. SUMMARY

We have attempted to summarize recent studies on the organization and expression of arenavirus genetic information in the belief that a clear understanding of virus replication during acute infection will lead to further insight into the determinants of tropism and virulence. Much remains to be investigated with respect to the role of the glycoproteins in infection and maturation, the nucleoprotein and L RNA–derived proteins in transcription and replication of viral RNA, and the role of the host cell in viral replication. The development of conditional mutants and molecular genetic approaches to characterize them will provide essential information on the interaction of arenaviruses with the host cell.

REFERENCES

Auperin, D. D., Dimock, K., Cash, P., Rawls, W. E., Leung, W-C., and Bishop, D. H. L., 1982, Analysis of the genomes of prototype Pichinde arenavirus and a virulent derivative Pichinde munchique: Evidence for sequence conservation at the 3′ termini of their viral RNA species, *Virology* **116**:363.

Auperin, D. D., Galinski, M., and Bishop, D. H., 1984, The sequences of the N protein gene and intergenic region of the S RNA of Pichinde arenavirus, *Virology* **134**:208.

Auperin, D. D., Sasso, D. R., and McCormick, J. B., 1986, Nucleotide sequence of the glycoprotein gene and intergenic region of the Lassa virus S genome RNA, *Virology* **154**:155.

Banerjee, S. N., Buchmeier, M., and Rawls, W. E., 1976, Requirement of the cell nucleus for the replication of an arenavirus, *Intervirology* **6**:190.

Beaton, A. R., and Krug, R. M., 1986, Transcription antitermination during influenza viral template RNA synthesis requires the nucleocapsid protein and the absence of a 5′ capped end, *Proc. Natl. Acad. Sci. USA* **83**:6282.

Bishop, D. H., 1986, Ambisense RNA genomes of arenaviruses and phleboviruses, *Adv. Virus Res.* **31**:1.

Bishop, D. H. L., and Auperin, D. D., 1987, Arenavirus gene structure and organization, *Curr. Top. Microbiol. Immunol.* **134**:5.

Broomhall, K. S., Morin, M., Pevear, D. C., and Pfau, C. J., 1987, Severe and transient pancytopenia associated with a chronic arenavirus infection, *J. Exp. Pathol.* **3**:259.

Bruns, M., Peralta, L. M., and Lehmann-Grube, F., 1983, Lymphocytic choriomeningitis virus. III. Structural proteins of the virion, *J. Gen. Virol.* **64**:599.

Bruns, M., Zeller, W., Rohdewohld, H., and Lehmann-Grube, F., 1986, Lymphocytic choriomeningitis virus. IX. Properties of the nucleocapsid, *Virology* **151**:77.

Buchmeier, M. J., and Parekh, B. S., 1987, Protein structure and expression among arenaviruses, *Curr. Top. Microbiol. Immunol.* **133**:41.

Buchmeier, M. J., Elder, J. H., and Oldstone, M. B. A., 1978, Protein structure of lymphocytic choriomeningitis virus: Identification of the virus structural and cell associated polypeptides, *Virology* **89**:133.

Buchmeier, M. J., Southern, P. J., Parekh, B. S., Wooddell, M. K., and Oldstone, M. B., 1987, Site-specific antibodies define a cleavage site conserved among arenavirus GPC glycoproteins, *J. Virol.* **61**:982.

Buck, L. L., and Pfau, C. J., 1969, Inhibition of lymphocytic choriomeningitis virus replication by actinomycin D and 6-azauridine, *Virology* **37**:698.

Ceriatti, F. S., Damonte, E. B., Mersich, S. E., and Coto, C. E., 1986, Partial characterization of two temperature-sensitive mutants of Junin virus, *Microbiologica* **9**:343.

Clegg, J. A. C., and Lloyd, G., 1983, Structural and cell associated proteins of Lassa virus, *J. Gen. Virol.* **64**:1127.

Daelli, M. G., and Coto, C. E., 1982, Inhibition of the production of infectious particles in cells infected with Junin virus in the presence of tunicamycin, *Rev. Argent. Microbiol.* **14**:171.

Dimock, K., Harnish, D. G., Sisson, G., Leung, W-C., and Rawls, W. E., 1982, Synthesis of virus-specific polypeptides and genomic RNA during the replicative cycle of Pichinde virus, *J. Virol.* **43**:273.

Farber, F. E., and Rawls, W. E., 1975, Isolation of ribosome-like structure from Pichinde virus, *J. Gen. Virol.* **26**:21.

Franze-Fernandez, M. T., Zetina, C., Iapalucci, S., Lucero, M. A., Bouissou, C., Lopez, R., Rey, O., Daheli, M., Cohen, G., and Zakin, M., 1987, Molecular structure and early events in the replication of Tacaribe arenavirus S RNA, *Virus Res.* **7**:309.

Friedlander, A. M., Jahrling, P. B., Merrill, J. P., and Tobery, S., 1984, Inhibition of mouse peritoneal macrophage DNA synthesis by infection with the arenavirus Pichinde, *Infect. Immun.* **43**:283.

Fuller-Pace, F. V., and Southern, P. J., 1988, Temporal analysis of transcription and replication during acute infection with lymphocytic choriomeningitis virus, *Virology* **162**:260.

Fuller-Pace, F. V., and Southern, P. J., 1989, Detection of virus-specific RNA-dependent RNA polymerase activity in extracts from cells infected with lymphocytic choriomeningitis virus: In vitro synthesis of full-length viral RNA species, *J. Virol.* **63**:1938.

Garcin, D., and Kolalofsky, D., 1990, A novel mechanism for the initiation of Tacaribe arenavirus genome replication, *J. Virol.* **64**:6196.

Gard, G. P., Vezza, A. C., Bishop, D. H. L., and Compans, R. W., 1977, Structural proteins of Tacaribe and Tamiami virions, *Virology* **83**:84.

Ghiringhelli, P. D., Rivera-Pomar, R. V., Lozano, M. E., Grau, O., and Romanowski, V., 1991, Molecular organization of Junin virus S RNA: Complete nucleotide sequence, relationship with other members of the arenaviridae and unusual secondary structures, *J. Gen. Virol.* **72**:2129.

Glushakova, S. E., and Lukashevich, I. S., 1989, Early events in arenavirus replication are sensitive to lysosomotropic compounds, *Arch. Virol.* **104**:157.

Gonzalez, J. P., McCormick, J. B., Saluzzo, J. F., and Herve, J. P., 1983, An arenavirus isolated from wild-caught rodents (*Pramys* species) in the Central African Republic, *Intervirology* **19**:105.

Harnish, D. G., 1982, Characterization of virus specific polypeptides during Pichinde virus infection in cell cultures, Doctoral thesis, McMaster University, Hamilton, Ontario, Canada, 218 pp.

Harnish, D. G., Leung, W-C., and Rawls, W. E., 1981, Characterization of polypeptides immunoprecipitable from Pichinde virus–infected BHK-21 cells, *J. Virol.* **38**:840.

Harnish, D. G., Dimock, K., Bishop, D. H. L., and Rawls, W. E., 1983, Gene mapping in Pichinde virus: Assignment of viral polypeptides to genomic L and S RNAs, *J. Virol.* **46**:638.

Howard, C. R., and Buchmeier, M. J., 1983, A protein kinase activity in lymphocytic choriomeningitis virus and identification of the phosphorylated product using monoclonal antibody, *Virology* **126**:538.

Iapalucci, S., Lopez, R., Rey, O., Lopez, N., Franze-Fernandez, M. T., Cohen, G., Lucero, M., Ochoa, A., and Zakin, M. M., 1989a, Tacaribe virus L gene encodes a protein of 2210 amino acid residues, *Virology* **170**:40.

Iapalucci, S., Lopez, N., Rey, O., Zakin, M. M., Cohen, G. N., and Franze-Fernandez, M. T., 1989b, The 5′ region of Tacaribe virus L RNA encodes a protein with a potential metal binding domain, *Virology* **173**:357.

Iapalucci, S., Lopez, N., and Franze-Fernandez, M. T., 1991, The 3′ end termini of the Tacaribe arenavirus subgenomic RNAs, *Virology* **182**:269.30.

Johnson, K. M., Webb, P. A., and Justines, G., 1973, Biology of the Tacaribe-complex viruses, in: *Lymphocytic Choriomeningitis Virus and Other Arenaviruses* (F. Lehmann-Grube, ed.), Springer-Verlag, New York, p. 241.

Kingsbury, D. W., 1990, Orthomyxoviridae and their replication, in: *Virology*, 2nd ed. (B. N. Fields, D. M. Knipe, *et al.*, eds.), Raven Press, New York, p. 1075.

Lehmann-Grube, F., 1971, Lymphocytic choriomeningitis virus, *Virol. Monogr.* **10**:1.

Leung, W-C., and Rawls, W. E., 1977, Virion-associated ribosomes are not required for the replication of Pichinde virus, *Virology* **81**:174.

Leung, W-C., Leung, F. K. L., and Rawls, W. E., 1979, Distinctive RNA transcriptase,

polyadenylic acid polymerase and polyuridylic acid polymerase activities associated with Pichinde virus, *J. Virology* **30**:98.
Lopez, R., and Franze-Fernandez, M. T., 1985, Effect of Tacaribe virus infection on host cell protein and nucleic acid synthesis, *J. Virol.* **66**:1753.
Lopez, R., Grau, O., and Franze-Fernandez, M. T., 1986, Effect of actinomycin D on arenavirus growth and estimation of the generation time for a virus particle, *Virus Res.* **5**:213.
Lukashevich, I. S., Lemeshko, N. N., and Shkolina, T. V., 1984, Effect of actinomycin D on the replication of the Machupo virus, *Vopr. Virusol.* **29**:569.
Martinez-Segovia, Z., and de Mitri, M. I., 1977, Junin virus structural proteins, *J. Virol.* **21**:579.
Mims, C. A., 1966, Immunofluorescence study of the carrier state and mechanisms of vertical transmission in LCM virus infection in mice, *J. Pathol. Bacteriol.* **91**:395.
Pedersen, I. R., 1979, Structural components and replication of arenaviruses, *Adv. Virus Res.* **24**:277.
Polyak, S. J., Rawls, W. E., and Harnish, D. G., 1991, Characterization of Pichinde virus infection of cells of the monocytic lineage, *J. Virol.* **65**:3575.
Raju, R., Raju, L., Hacker, D., Garcin, D., Compans, R., and Kolakofsky, D., 1990, Non-templated bases at the 5′ ends of Tacaribe virus mRNAs, *Virology* **174**:53.
Rawls, W. E., Banerjee, S. N., McMillan, C. A., and Buchmeier, M. J., 1976, Inhibition of Pichinde virus replication by actinomycin D, *J. Gen. Virol.* **33**:421.
Riviere, Y., and Oldstone, M. B. A., 1986, Genetic reassortants of lymphocytic choriomeningitis virus: unexpected disease and mechanism of pathogenesis, *J. Virol.* **59**:363.
Riviere, Y., Ahmed, R., Southern, P. J., Buchmeier, M. J., and Oldstone, M. B. A., 1985, Genetic mapping of lymphocytic choriomeningitis virus pathogenicity: Virulence in guinea pigs is associated with L RNA segment, *J. Virol.* **55**:704.
Romanowski, V., and Bishop, D. H. L., 1985, Conserved sequences and coding of two strains of lymphocytic choriomeningitis virus (We and Arm) and Pichinde Arenavirus, *Virus Res.* **2**:35.
Romanowski, V., Matsuura, Y., and Bishop, D. H., 1985, Complete sequence of the S RNA of lymphocytic choriomeningitis virus (WE strain) compared to that of Pichinde arenavirus, *Virus Res.* **3**:101.
Salvato, M. S., and Shimomaye, E. M., 1989, The completed sequence of lymphocytic choriomeningitis virus reveals a unique RNA structure and a gene for a zinc finger protein, *Virology* **173**:10.40.
Shivaprakash, M., Harnish, D. G., and Rawls, W. E., 1988, Characterization of temperature sensitive mutants of Pichinde virus, *J. Virol.* **62**:4037.
Singh, M. K., Fuller-Pace, F. V., Buchmeier, M. J., and Southern, P. J., 1987, Analysis of the genomic L RNA segment from lymphocytic choriomeningitis virus, *Virology* **161**:448.
Southern, P. J., Singh, M. K., Riviere, Y., Jacoby, D. R., Buchmeier, M. J., and Oldstone, M. B. A., 1987, Molecular characterization of the genomic S RNA segment from lymphocytic choriomeningitis virus, *Virology* **157**:145.
Svitlik, C., and Marcus, P. I., 1984, Interferon induction by viruses. XI. Early events in the induction process, *J. Interferon Res.* **4**:585.
Traub, E., 1936, Persistence of lymphocytic choriomeningitis virus in immune animals and its relation to immunity, *J. Exp. Med.* **63**:847.
Vezza, A. C., and Bishop, D. H. L., 1977, Recombination between temperature sensitive mutants of the arenavirus, Pichinde, *J. Virol.* **24**:712.
Vezza, A. C., Cash, P., Jahrling, P., Eddy, G., and Bishop, D. H. L., 1980, Arenavirus recombination: The formation of recombinants between prototype Pichinde and Pichinde-Munchique viruses and evidence that arenavirus S RNA codes for N polypeptide, *Virology* **106**:250.
Walker, D. H., and Murphy, F. A., 1987, Pathology and Pathogenesis of arenavirus infections, *Curr. Top. Microbiol. Immunol.* **133**:89.

Webster, R. G., Laver, W. G., Air, G. M., and Schild, G. S., 1982, Genetic and molecular mechanisms of viral pathogenesis: Implications for prevention and treatment, *Nature* **300:**19.

Wilson, S. M., and Clegg, J. C., 1991, Sequence analysis of the S RNA of the African arenavirus Mopeia: An unusual secondary structure feature in the intergenic region, *Virology* **180:**543.

Wright, K. E., Salvato, M. S., and Buchmeier, M. J., 1989, Neutralizing epitopes of lymphocytic choriomeningitis virus are conformational and require both glycosylation and disulfide bonds for expression, *Virology* **171:**417.

Wright, K. E., Spiro, R. C., Burns, J. W., and Buchmeier, M. J., 1990, Post-translational processing of the glycoproteins of lymphocytic choriomeningitis virus, *Virology* **177:**175.

Young, P. R., Chanas, A. C., Lee, S. R., Gould, E. A., and Howard, C. R., 1986a, Localization of an arenavirus protein in the nuclei of infected cells, *J. Gen. Virol.* **68:**2465.

Young, P. R., Lee, S. R., and Howard, C. R., 1986b, Regulation of Pichinde virus replication in vero and BHK-21 cells, *Med. Microbiol. Immunol. (Berl)* **175:**63.

CHAPTER 10

Molecular Phylogeny of the Arenaviruses and Guide to Published Sequence Data

J. CHRISTOPHER S. CLEGG

I. INTRODUCTION

The arenaviruses are primarily viruses of rodents, although a minority of them cause serious disease in humans. They were originally grouped together on the basis of their appearance in the electron microscope (Dalton *et al.*, 1968; Murphy and Whitfield, 1975) and serological cross-reactivity in complement fixation and immunofluorescence tests (Rowe *et al.*, 1970; Casals *et al.*, 1975; Wulff *et al.*, 1978). The production and use of monoclonal antibodies have reinforced and refined our ability to group these viruses and to recognize variation among strains of the same virus (Buchmeier *et al.*, 1981; Howard *et al.*, 1985; Sanchez *et al.*, 1989; Ruo *et al.*, 1991). Analysis of the structural components of several members of the family has confirmed their general similarity at the molecular level (see Bishop and Auperin, 1987; Buchmeier and Parekh, 1987, for reviews) and has provided, through nucleotide sequencing, a means by which we can begin to make quantitative estimates of the relationships among these viruses. Here I have collected together a directory of the arenavirus sequence data currently available in widely distributed databases, multiple alignments of nucleocapsid (N) and glycopro-

---

J. CHRISTOPHER S. CLEGG • Public Health Laboratory Service, Centre for Applied Microbiology and Research, Division of Pathology, Porton Down, Salisbury, Wiltshire, SP4 0JG, England.

The Arenaviridae, edited by Maria S. Salvato. Plenum Press, New York, 1993.

tein precursor (GPC) protein amino acid sequences, and present phylogenetic trees based on these sequences, which graphically summarize the relationships among these viruses.

II. GUIDE TO ARENAVIRUS SEQUENCE DATA

Databases and programs on the SEQNET facility at the Daresbury Laboratory of the Science and Engineering Research Council were used to search the GenBank and EMBL nucleic acid sequence databases, together with the daily updates available up to November 4, 1991. Entries in these primary databanks provided information on the translated amino acid sequences that are deposited in protein databases (e.g., PIR, SWISSPROT, OWL). In Tables I (nucleic acids) and II (proteins), the accession numbers of relevant entries in these databases are provided. These allow easy retrieval of these sequences from the relevant library, whether these are available on-line from a central computing facility, by electronic mail from a file server, or on CD-ROM disk.

III. SEQUENCE ANALYSIS

A. Multiple Alignments

The program CLUSTAL V (Higgins and Sharp, 1988, 1989; Higgins *et al.*, 1992) was used to construct multiple alignments of the nine arenavirus N protein sequences and eight GPC sequences currently available (Figs. 1 and 2). As in the progressive alignment procedure of Feng and Doolittle (1987, 1990), the program proceeds by first making all pairwise sequence comparisons and then aligning the sequences starting with the most similar pair, inserting gaps where necessary. Further sequences are added in order of decreasing similarity, treating the already aligned sequences as a unit.

All the structural proteins contain regions that are relatively conserved among all the arenaviruses analyzed and also regions where extensive divergence has taken place. In G1, the most variable of the structural proteins, the N-terminal 57-amino-acid residues are relatively well conserved, with only 25 nonconservative substitutions. This part of the molecule has been shown to be proteolytically cleaved from the mature protein found in LCM virions (Burns *et al.*, 1990). The remainder of this protein shows extensive variation, with the exception of blocks at residues 89–102 and 254–265 (numbering according to Fig. 1) and a conserved cysteine residue at residue 180. As has been noted previously (Franze-Fernandez *et al.*, 1987; Auperin and McCormick, 1988), the proteins G2 and N show a much greater proportion of identical residues or conservative substitutions distributed throughout their length. In

TABLE I. Arenavirus Nucleotide Sequences

Virus	Strain	Sequence	Length (bp)	Coding Regions	Accession no.	Ref.
LCM	WE	S RNA	3375	GPC and N	M22138	Romanowski *et al.* (1985)
		S RNA (3′-part)	2040	N	M22017	Romanowski and Bishop (1985)
		L RNA (3′-end)	1123	L (part)	M22016	
LCM	Armstrong	S RNA	3376	GPC and N	M20869	Salvato *et al.* (1988)
		S RNA (3′-terminus)	150	—	J02242	Auperin *et al.* (1982b)
		L RNA (3′-part)	6680	L	J04331	Salvato *et al.* (1989)
		L RNA (5′-part)	850	Z	M27693	Salvato and Shimomaye (1989)
		L RNA (3′-terminus)	50	—	J02241	Auperin *et al.* (1982b)
		L RNA (part)	697	L (part)	M18381	Singh *et al.* (1987)
		L RNA (part)	589	L (part)	M18382	Singh *et al.* (1987)
		L RNA (part)	786	L (part)	M18383	Singh *et al.* (1987)
Lassa	GA391	S RNA	3417	GPC and N	X52400	Clegg *et al.* (1990)
		S RNA (3′-part)	1830	N	K03362	Clegg and Oram (1985)
Lassa	Josiah	S RNA	3402	GPC and N	J04324	Auperin and McCormick (1988)
	(800593)	S RNA (5′-part)	1658	GPC	M15076	Auperin *et al.* (1986)
Mopeia	800150	S RNA	3419	GPC and N	M33879	Wilson *et al.* (1991)
Pichinde	3739	S RNA	3419	GPC and N	K02734	Auperin *et al.* (1984)
		S RNA (3′-part)	2040	N	K02735	
		S RNA (complementary sense)	3419	N and GPC	M16734	Bishop and Auperin (1987)
		S RNA	3419	GPC and N	M16735	
		S RNA (3′-terminus)	120	—	J02279	Auperin *et al.* (1982a)
		L RNA (3′-terminus)	50	—	J02277	
Pichinde	Munchique	S RNA (3′-terminus)	120	—	J02280	
		L RNA (3′-terminus)	50	—	J02278	
Tacaribe	T.RVL.II 573	S RNA	3432	GPC and N	M20304 M65834	Franze-Fernandez *et al.* (1987)
		S RNA (3′-terminus)	73	—	J02411	Auperin *et al.* (1982b)
		L RNA	7102	L and P11 (Z)	J04340 M33513	Iapalucci *et al.* (1989a and 1989b)
		Subgenomic RNAs (3′-end)	88	—	M65833	Iapalucci *et al.* (1991)
Junin	MC2	S RNA	3400	GPC and N	D10072	Ghiringhelli *et al.* (1991)
		S RNA (3′-end)	1800	N	X15827	Ghiringhelli *et al.* (1989)
Machupo	AA288-77	S RNA (3′-end)	1814	N	X62616	Griffiths *et al.* (1992)

TABLE II. Arenavirus Protein Sequences

Virus	Strain	Protein	Acession no.	Ref.
LCM	WE	N	P07400	Romanowski and Bishop (1985)
		GPC	P07399	Romanowski *et al.* (1985)
		L (part)	P14241	Romanowski and Bishop (1985)
LCM	Armstrong	N	P09992	Salvato *et al.* (1988)
		GPC	P09991	
		L (part)	P14240	Singh *et al.* (1987)
		Z	P18541	Salvato and Shimomaye (1989)
Lassa	GA391	N	P04935	Clegg and Oram (1985)
		GPC	P17332	Clegg *et al.* (1990)
Lassa	Josiah	N	P13699	Auperin and McCormick (1988)
		GPC	P08669	Auperin *et al.* (1986)
Mopeia	800150	N	P19239	Wilson and Clegg (1991)
	800150	GPC	P19240	
Pichinde		N	P03541	Auperin *et al.* (1984)
		GPC	P03540	
Tacaribe	T.RVL.II 573	N	P18140	Franze-Fernandez *et al.* (1987)
		GPC	P18141	
		L	P20430	Iapalucci *et al.* (1989a)
		P-11 (Z)	[a]	Iapalucci *et al.* (1989b)
Junin	MC2	N	P14239	Ghiringhelli *et al.* (1989)
		GPC	P26313	Ghiringhelli *et al.* (1991)
Machupo	AA288-77	N	P26578	Griffiths *et al.* (1992)

[a] Not in protein sequence databases.

these molecules it is perhaps more appropriate to point out regions of unusually high variation, e.g., the C-terminal 9–13 residues of G2, and regions 146–158, 339–358, and 548–563 of N. These regions also tend to contain gaps in one or more members of the set, suggesting relaxed structural requirements for a functional molecule, or even that these regions are dispensable. Potential N-linked glycosylation sites are conserved across all the viruses at positions 97 and 193/195 (allowing for some flexibility in the alignment of the divergent Old and New World sequences in this region) in G1, and at positions 399, 424, and 428 in G2.

B. Molecular Phylogeny

The use of serology to differentiate or classify viruses relies on differences only between antigenic sites. A more comprehensive and more readily quantifiable picture of the relationships within a group of viruses is to be obtained from direct analysis of nucleotide or amino acid sequence data. The three main approaches to phylogenetic tree building are those based on the calculation of distance matrices or on the principles of maximum parsimony or of maximum likelihood (reviewed in Swofford and Olsen, 1990). The CLUSTAL V program uses a distance matrix-based approach, implementing the "neighbor-joining" procedure of Saitou and Nei (1987). To try and rectify the apparent foreshortening of distances between widely divergent sequences caused by multiple changes at single sites, the empirical correction of Kimura (1982) was applied. The trees generated by this procedure are unrooted; that is, the connections and distances between sequences are specified but the direction of evolution through the network is not. In Fig. 3 the trees (calculated for each of the proteins G1, G2, and N independently) have been arbitrarily drawn as if the Old World and New World groups have evolved from some common ancestral virus and have subsequently diverged to give the present-day members of the groups. This seems a reasonable assumption in the absence of any evidence to the contrary, but it is important to keep in mind that it is not determined by the sequence data.

For each protein the tree has the same branching order. In the Old World group the two Lassa virus strains are quite closely related, while Mopeia virus is more divergent and the two closely related LCM virus strains are even more divergent. In the New World group, the N protein tree shows that Junin and Machupo viruses are the most closely related, with Tacaribe virus forming a third, less closely related member of this subgroup. In all three trees, Pichinde virus is rather distantly related to the other New World viruses. The similarity of the tree structures suggests there has been no recombination event between the structural proteins of this set of viruses in their evolutionary history, as has been observed in the case of the alphaviruses (Hahn *et al.*, 1988; Levinson *et al.*, 1990).

Branch lengths vary considerably between the trees. In particular, the extent of divergence in the G1 tree is strikingly larger, especially among the New World viruses. This reflects the known variability of this protein, a consequence of its role in forming the exterior surface of the virus and thus sites of interaction with the host organism, in particular with its immune system.

How reliable is the information about relationships portrayed by these trees? An attempt to assess this has been made in two ways. In a procedure known as bootstrap analysis (Felsenstein, 1985), the alignments are repeatedly sampled at random and reanalyzed. Thus each sample from an alignment of N residues consists of a random selection of N residues. If the tree structure is strongly supported by the data, i.e., all parts of the sequences being analyzed give rise to the same tree, we can be more confident that the topology of the tree does reflect the relationships among the sequences. This is in fact the case with each of the proteins considered here, where at least 998 of 1000 bootstrap trials for each protein resulted in trees of the topology shown in Fig. 3. Second, the N-protein sequences have been aligned and a tree constructed using

Lassa GA391 GPC	MGQIVTFFQE	VPHVIEEVMN	IVLIALSILA	ILKGLYNVAT	CGLIGLVTFL	50
Lassa Josiah GPC	MGQIVTFFQE	VPHVIEEVMN	IVLIALSVLA	VLKGLYNFAT	CGLVGLVTFL	
Mopeia GPC	MGQIVTFFQE	VPHILEEVMN	IVLMTLSILA	ILKGIYNVMT	CGIIGLITFL	
LCM Arm GPC	MGQIVTMFEA	LPHIIDEVIN	IVIIVLIVIT	GIKAVYNFAT	CGIFALISFL	
LCM WE GPC	MGQIVTMFEA	LPHIIDEVIN	IVIIVLIIIT	SIKAVYNFAT	CGILALVSFL	
Junín GPC	MGQFISFMQE	IPTFLQEALN	IALVAVSLIA	IIKGVVNLYK	SG----CSIL	
Tacaribe GPC	MGQFISFMQE	IPIFLQEALN	IALVAVSLIC	IVKGLVNLYR	CGLFQLMVFL	
Pichinde GPC	MGQIVTLIQS	IPEVLQEVFN	VALIIVSVLC	IVKGFVNLMR	CGLFQLVTFL	

LLSGRSCSL-	------IYKG	TYELQTLEL-	-NMETL**NMT**M	PLSCTK**NNS**H	100
LLCGRSCTT-	-----SLYKG	VYELQTLEL-	-NMETL**NMT**M	PLSCTK**NNS**H	
FLCGRSCS--	-----SIYKD	NYEFFSLDL-	-DMSSL**NAT**M	PLSCSK**NNS**H	
LLAGRSCGMY	GLKGPDIYKG	VYQFKSVEF-	-DMSHL**NLT**M	PNACSA**NNS**H	
FLAGRSCGMY	GLNGPDIYKG	VYQFKSVEF-	-DMSHL**NLT**M	PNACSV**NNS**H	
DLAGRSCPR-	AFK-IGLHTE	VPDCVLLQWW	VSFSNNPHDL	PLLCTL**NKS**H	
VLAGRSCSEE	TFK-IGMHTK	FQE-VSLSLS	ALLT**NQS**HEL	PMLCLA**NKT**H	
ILSGRSCD--	----SMMIDR	RH**NLT**HVEF-	-**NLT**RMFDNL	PQSCSK**NNT**H	

HYIRVG**NET**-	-GLELTLT**NT**	**S**ILNHK----	----------	FC**NLS**DAHKR	150
HYIMVG**NET**-	-GLELTLT**NT**	**S**IINHK----	----------	FC**NLS**DAHKK	
HYIQVG**NET**-	-GLELTLT**NT**	**S**IINHK----	----------	FC**NLS**DAHRR	
HYISMGT-S-	-GLELTFT**ND**	**S**IISHN----	----------	FC**NLT**SAFNK	
HYISMGS-S-	-GLEPTFT**ND**	**S**ILNHN----	----------	FC**NLT**SALNK	
LYIKGG**NAS**-	--FKISFDDI	AVL------L	PEYDV-----	IIQHPADMSW	
LYLKSGRSS-	--FKINIDSV	TVLTRSEV**NL**	**T**SINLTRSID	VFVHSPKLGS	
HYYKGPS**NTT**	WGIELTLT**NT**	**S**IA**NET**SG**NF**	**S**NIG---SLG	YG**NIS**NCDRT	

NLYDHSLMSI	ISTFHLSI-P	NFNQYEAMSC	D-FNGGKITV	QY**NLS**HSFAV	200
NLYDHALMSI	ISTFHLSI-P	NFNQYEAMSC	D-FNGGKISV	QY**NLS**HSYAG	
NLYDKALMSI	LTTFHLSI-P	DFNQYEAMSC	D-FNGGKISV	QY**NLS**HSNYV	
KTFDHTLMSI	VSSLHLSI-R	GNSNYKAVSC	D-FNNG-ITI	QY**NLT**FSDAQ	
KSFDHTLMSI	VSSLHLSI-R	GNSNYKAVSC	D-FNNG-ITI	QY**NLS**SSDPQ	
CSKSDDQIRL	SQWFMNAVGH	DWYLDPPFLC	R**NRT**KTEGFI	-FQV**NTS**KTG	
CFESDEE-WV	VAWWIEAIGH	RWDQDPGLLC	RNKTKTEGKL	-IQI**NIS**RAD	
REAGHTLKWL	LNELHFNV-L	HVTRHIGARC	KTVEGAGVLI	QY**NLT**VGDRG	

DAAGHCGTLA	NGVLQTF--M	RMAWGGSYIA	LDSGRGNWD-	----CIM-TS	250
DAANHCGTVA	NGVLQTF--M	RMAWGGSYIA	LDSGRGNWD-	----SIM-TS	
DAGNHCGTIA	NGIMDVF--R	RMYWSTS-LS	VASDISGTQ-	----CIQ-TD	
SAQSQCRTFR	GRVLDMF---	RTAFGGKYMR	SGWGWTGSDG	KTTWCSQ-TS	
SAMSQCRTFR	GRVLDMF---	RTAFGGKYMR	SGWGWTGSDG	KTTWCSQ-TS	
INENYAKKFK	TGMHHLY---	----------	--------RE	YPDSCLDGKL	
GNVHYGWRLK	NGLDHIY---	----------	--------RG	REEPCFEGEQ	
GEVGR--HLI	ASLAQIIGDP	KIAWVGKCFN	NCSGDTC---	RLTNCEGGTH	

YQYLIIQ**NTT**	WDDHCQFSRP	SPIGYLGLLS	QRTRDIYISR	RLLGTFTWTL	300
YQYLIIQ**NTT**	WEDHCQFSRP	SPIGYLGLLS	QRTRDIYISR	RLLGTFTWTL	
YKYLIIQ**NTS**	WEDHCMFSRP	SPMGFLSLLS	QRTKNFYISR	RLLGLFTWTL	
YQYLIIQ**NRT**	WENHCTY--A	GPFGMSRILL	SQEKTKFFTR	RLAGTFTWTL	
YQYLIIQ**NRT**	WENHCRY--A	GPFGMSRILF	AQEKTKFLTR	RLSGTFTWTL	
C-LMKAQPTS	WPLQCPLDHV	NTL---HFLT	RGKNIQLPRR	SLKAFFSWSL	
C-LIKIQPED	WPTDCKADHT	NTF---RFLS	RSQKSIAVGR	TLKAFFSWSL	
YNFLIIQ**NTT**	WENHCTY---	TPMATIRMAL	QRTAYSSVSR	KLLGFFTWDL	

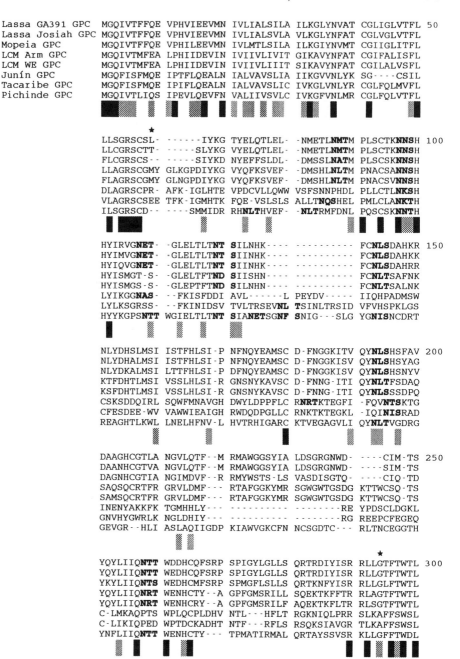

FIGURE 1. Alignment of the eight available arenavirus GPC proteins. Proteins were aligned using the program CLUSTAL V. A graphic indication of the degree of conservation is provided under each set of sequences. Positions that are exactly conserved in all the sequences are marked with a dark block, and those where conservative substitutions have taken place are marked with a hatched block. Potential N-linked glycosylation sites are shown in boldface type. The positions of the N-terminal residues of G1 and G2 as found in LCM virions (Burns *et al.*, 1990) are indicated by stars.

Lassa GA391 GPC	SDSEG**NET**PG	GYCLTRWMLI	EAELKCFGNT	AVAKCNEKHD	EEFCDMLRLF	350
Lassa Josiah GPC	SDSEGKDTPG	GYCLTRWMLI	EAELKCFGNT	AVAKCNEKHD	EEFCDMLRLF	
Mopeia GPC	SDSEGNDMPG	GYCLTRSMLI	GLDLKCFGNT	AIAKCNQAHD	EEFCDMLRLF	
LCM Arm GPC	SDSSGVENPG	GYCLTKWMIL	AAELKCFGNT	AVAKCNVNHD	AEFCDMLRLI	
LCM WE GPC	SDSSGVENPG	GYCLTKWMIL	AAELKCFGNT	AVAKCNVNHD	EEFCDMLRLI	
Tacaribe GPC	TDPLGNEAPG	GYCLEKWMLV	ASELKCFGNT	AIAKCNQNHD	SEFCDMLRLF	
Junín GPC	TDSSGKDTPG	GYCLEEWMLV	AAKMKCFGNT	AVAKCNLNHD	SEFCDMLRLF	
Pichinde GPC	SDSSGQHVPG	GYCLEQWAII	WAGIKCFDNT	VMAKCNKDHN	EEFCDTMRLF	
	DFNKQAIRRL	KTEAQMSIQL	INKAVNALIN	DQLIMKNHLR	DIMGIPYC**NY**	400
	DFNKQAIQRL	KAEAQMSIQL	INKAVNALIN	DQLIMKNHLR	DIMGIPYC**NY**	
	DFNKQAISKL	RSEVQQSINL	INKAVNALIN	DQLVMRNHLR	DLMGIPYC**NY**	
	DYNKAALSKF	KEDVESALHL	FKTTVNSLIS	DQLLMRNHLR	DLMGVPYC**NY**	
	DYNKAALSKF	KQDVESALHV	FKTTLNSLIS	DQLLMRNHLR	DLMGVPYC**NY**	
	DYNKNAIKTL	NEETKTRVNV	LSHTINALIS	DNLLMKNKIR	ELMSVPYC**NY**	
	DYNKNAIKTL	NDETKKQVNL	MGQTINALIS	DNLLMKNKIR	ELMSVPYC**NY**	
	DFNQNAIKTL	QLNVENSLNL	FKKTINGLIS	DSLVIRNSLK	QLAKIPYC**NY**	
	SRYWYL**NHT**S	TGKTSLPRCW	LIS**NGS**YL**NE**	**T**KFSDDIEQQ	ADNMITEMLQ	450
	SKYWYL**NHT**T	TGRTSLPKCW	LVS**NGS**YL**NE**	**T**HFSDDIEQQ	ADNMITEMLQ	
	SKFWYL**NDT**R	TGRTSLPKCW	LVT**NGS**YL**NE**	**T**QFSTEIEQE	ANNMFTDMLR	
	SKFWYLEHAK	TGETSVPKCW	LVT**NGS**YL**NE**	**T**HFSDQIEQE	ADNMITEMLR	
	SKFWYLEHAK	TGETSVPKCW	LVT**NGS**YL**NE**	**T**HFSDQIEQE	ADNMITEMLR	
	TRFWYV**NHT**L	SGQHSLPRCW	MIR**NNS**YL**NS**	**S**EFRNEWILE	SDFLISEMLG	
	TKFWYV**NHT**L	SGQHSLPRCW	LIK**NNS**YL**NI**	**S**DFRNDWILE	SDFLISEMLS	
	TKFWYI**NDT**I	TGRHSLPQCW	LVH**NGS**YL**NE**	**T**HFKNDWLWE	SQNLYNEMLM	
	KEYIDRQGKT	PLGLVDLFVF	STSFYLISIF	LHLVKIPTHR	HIVGKPCPKP	500
	KEYMERQGKT	PLGLVDLFVF	STSFYLISIF	LHLVKIPTHR	HIVGKSCPKP	
	KEYEKRQSTT	PLGLVDLFVF	STSFYLISVF	LHLIKIPTHR	HIKGKPCPKP	
	KDYIKRQGST	PLALMDLLMF	STSAYLVSIF	LHLVKIPTHR	HIKGGSCPKP	
	KDYIKRQGST	PLALMDLLMF	STSAYLISIF	LHFVRIPTHR	HIKGGSCPKP	
	KEYSERQGRT	PITLVDICFW	STVFFTSTLF	LHLIGFPTHE	HIRGEGCPLP	
	KEYSDRQGKT	PLTLVDICFW	STVFFTASLF	LHLVGIPTHR	HIRGEACPLP	
	KEYEERQGKT	PLALTDICFW	SLVFYTITVF	LHIVGIPTHR	HIIGDGCPKP	
	HRLNHMGICS	CGLYKQPGVP	VRWKR--			
	HRLNHMGICS	CGLYKQPGVP	VKWKR--			
	HRLNHMAICS	CGFYKQPGLP	TQWKR--			
	HRLTNKGICS	CGAFKVPGVK	TVWKRR-			
	HRLTNKGICS	CGAFKVPGVK	TIWKRR-			
	HRLNSMGGCR	CGKYLPLKKP	TIWHRRH			
	HRLNSLGGCR	CGKYPNLKKP	TVWR---			
	HRITRNSLCS	CGYYK---YQ	RNLTNG-			

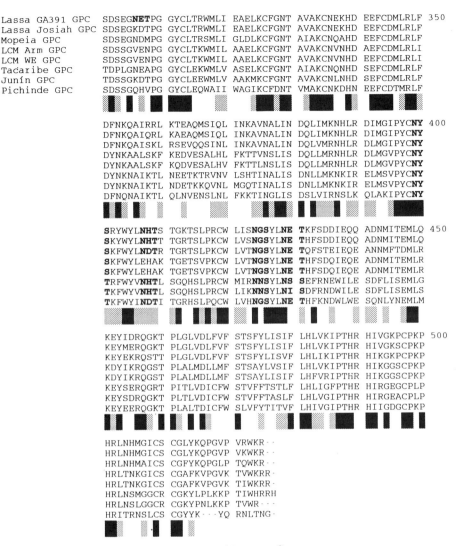

FIGURE 1. *(Continued)*

the programs of Feng and Doolittle (1987, 1990). This tree is practically identical to the N-protein tree shown in Fig. 3 (Griffiths *et al.*, 1992).

An accurate understanding of relationships among different members of a virus family may have important practical consequences, as well as taxonomic significance. The relationship between the three New World viruses Junin, Machupo, and Tacaribe, indicated in the N-protein tree in Fig. 3, is a case in point. It is known that prior infection by the nonpathogenic Tacaribe virus will protect experimental animals against challenge with Junin virus (reviewed in Peters *et al.*, 1987; Weissenbacher *et al.*, 1987), and also that animals that have recovered from

```
Lassa GA391 N     MSASKEVRSF LWTQSLRREL SGYCSNIKLQ VVKDAQALLH GLDFSEVSNV  50
Lassa Josiah N    MSASKEIKSF LWTQSLRREL SGYCSNIKLQ VVKDAQALLH GLDFSEVSNV
Mopeia N          MSNSKEVKSF LWTQSLRREL SGYCSNIKIQ VIKDAQALLH GLDFSEVANV
LCM Arm N         MSLSKEVKSF QWTQALRREL QSFTSDVKAA VIKDATNLLN GLDFSEVSNV
LCM WE N          MSLSKEVKSF QWTQALRREL QGFTSDVKAA VIKDATSLLN GLDFSEVSNV
Junín N           MAHSKEVPSF RWTQSLRRGL SQFTQTVKSD VLKDAKLIAD SIDFNQVAQV
Machupo N         MAHSKEIPSF RWTQSLRRGL SQVHPTVKTD VLKDAKLIAD SIDFNQVSQV
Tacaribe N        MAQSKEVPSF RWTQSLRKGL SQFTQTVKSD ILKDAKLIAD SIDFNQVAQV
Pichinde N        MSDN--IPSF RWVQSLRRGL SNWTHPVKAD VLSDTRALLS ALDFHKVAQV

                  QRLMRKQKRD DGDLKRLRDL NQAVNNLVEL KSTQQKSVLR VGTLSSDDLL 100
                  QRLMRKERRD DNDLKRLRDL NQAVNNLVEL KSTQQKSILR VGTLTSDDLL
                  QRLMRKEKRD DSDLKRLRDL NQAVNNLVEL KSVQQKNVLR VGTLTSDDLL
                  QRIMRKEKRD DKDLQRLRSL NQTVHSLVDL KSTSKKNVLK VGRLSAEELM
                  QRIMRKERRD DKDLQRLRSL NQTVHSLVDP KSTSKKNVLK VGRLSAEELM
                  QRALRKTKRG EEDLNKLRDL NKEVDRLMSM RSVQRNTVFK AGDLGRVERM
                  QRALRKNKRG EEDLNKLRDL NKEVDRLMSM KSIQKNTIFK IGDLGRDELM
                  QRVLRKTKRT DDDLNKLRDL NIEVDRLMSM KSVQKNTIFK VGDLARDELM
                  QRMVRKDKRT DSDLTKLRDM NKEVDALMNM RSVQRDNVLK VGGLAKEELM

                  ILAADLEKLK SKVTRTERPL SSGVYMGNLS SQQLDQRRAL LNMIGMTGVS 150
                  ILAADLEKLK SKVIRTERPL SAGVYMGNLS SQQLDQRRAL LNMIGMSGGN
                  VLAADLDRLK AKVIRGERPL AAGVYMGNLT AQQLEQRRVL LQMVGMGGGF
                  SLAADLEKLK AKIMRSERPQ ASGVYMGNLT TQQLDQRSQI LQIVGMRKPQ
                  TLAADLEKLK AKIMRTERPQ ASGVYMGNLT AQQLDQRSQI LQMVGMRRPQ
                  ELASGLGNLK TKFRRAETGS Q-GVYMGNLS QSQLAKRSEI LRTLGF--QQ
                  ELASDLEKLK NKIKRTESGP Q-GLYMGNLS QLQLTKRSEI LKTLGF--QQ
                  ELASDLEKLK DKIKRTESNG T-NAYMGNLP QSQLNRRSEI LRTLGF--AQ
                  ELASDLDKLR KKVTRTEGLS QPGVYEGNLT NTQLEQRAEI LRSMGF--AN

                  GGGKGASDGI VRVWDVKNAE LLNNQFGTMP SLTLACLTKQ GQVDLNDAVQ 200
                  QGARAGRDGV VRVWDVKNAE LLNNQFGTMP SLTLACLTKQ GQVDLNDAVQ
                  RAGNTLGDGI VRVWDVRNPE LLNNQFGTMP SLTIACMCKQ GQADLNDVIQ
                  QGAS----GV VRVWDVKDSS LLNNQFGTMP SLTMACMAKQ SQTPLNDVVQ
                  QGAS----GV VRVWDVKDSS LLNNQFGTMP SLTMACMAKQ SQTPLNDVVQ
                  QGTGGN--GV VRVWDVKDPS KLNNQFGSVP ALTIACMTVQ GGETMNSVIQ
                  QRGAGN--GV VRIWDVSDPS KLNNQFGSMP ALTIACMTVQ GGETMNSVVQ
                  QGGRPN--GI VRVWDVKDSS KLNNQFGSMP ALTIACMTVQ GGETMNNVVQ
                  ARPAGNRDGV VKVWDIKDNT LLINQFGSMP ALTIACMTEQ GGEQLNDVVQ

                  ALTDLGLIYT AKYPNSSDLD RLSQSHPILN MIDTKKSSLN ISGYNFSLGA 250
                  ALTDLGLIYT AKYPNTSDLD RLTQSHPILN MIDTKKSSLN ISGYNFSLGA
                  SLSDLGLVYT AKYPNMSDLD KLSQTHPILG IIEPKKSAIN ISGYNFSLSA
                  ALTDLGLLYT VKYPNLNDLE RLKDKHPVLG VITEQQSSIN ISGYNFSLGA
                  ALTDLGLLYT VKYPNLSDLE RLKDKHPVLG VITEQQSSIN ISGYNFSLGA
                  ALTSLGLLYT VKYPNLSDLD RLTQEHDCLQ IVTKDESSIN ISGYNFSLSA
                  ALTSLGLLYT VKYPNLNDLD KLTLEHECLQ IVTKDESSIN ISGYNFSLSA
                  ALTSLGLLYT VKYPNLSDLD KLIPNHECLQ IITKEESSIN ISGYNLSLLA
                  ALSALGLLYT VKFPNMTDLE KLTQQHSALK IISHEPSALN ISGYNLSLSA

                  AVKAGACMLD GGNMLETIKV SPQTMDGILK SILKVKKSLG MFVSDTPGER 300
                  AVKAGACMLD GGNMLETIKV SPQTMDGILK SILKVKKALG MFISDTPGER
                  AVKAGACLID GGNMLETIKV TKSNLEGILK AALKVKRSLG MFVSDTPGER
                  AVKAGAALLD GGNMLESILI KPSNSEDLLK AVLGAKRKLN MFVSDQVGDR
                  AVKAGAALLH GGNMLESILI KPSNSEDLLK AVLGAKKKLN MFVSDQVGDR
                  AVKAGASILD DGNMLETIRV TPDNFSSLIK STIQVKRREG MFIDEKPGNR
                  AVKAGASILD GGNMLETIRV TPDNFSSLIK STLQVKRKEG MFIDEKPGNR
                  AVKAGASILD GGNMLETIRV SPDNFSSLIK NTLQVKRREG MFIDDRPGSR
                  AVKAAACMID GGNMLETIQV KPSMFSTLIK SLLQIKNREG MFVSTTPGQR
```

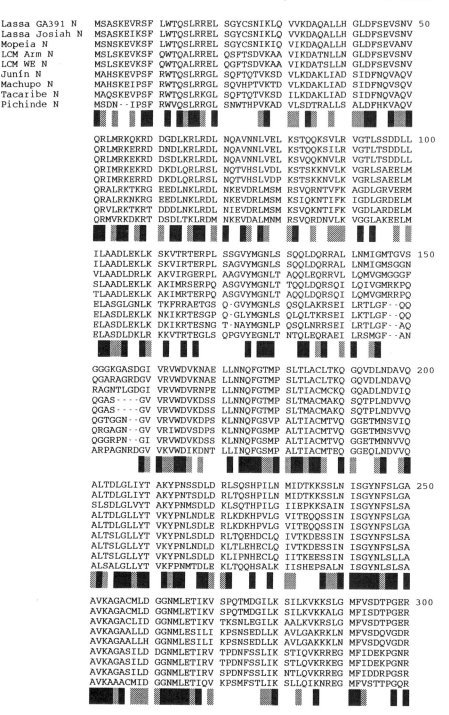

FIGURE 2. Alignment of the nine available arenavirus N proteins. Conserved positions are indicated as in Fig. 1.

Lassa GA391 N	NPYENILYKI	CLSGDGWPYI	ASRTSIVGRA	WENTVVDLEQ	DNKPQ-KIGN	350
Lassa Josiah N	NPYENILYKI	CLSGDGWPYI	ASRTSITGRA	WENTVVDLES	DGKPQ-KADS	
Mopeia N	NPYENLLYKL	CLSGEGWPYI	ASRTSIVGRA	WDNTTVDLSG	DVQQNAKPDK	
LCM Arm N	NPYENILYKV	CLSGEGWPYI	ACRTSIVGRA	WENTTIDLTS	E-----KPAV	
LCM WE N	NPYENILYKV	CLSGEGWPYI	ACRTSVVGRA	WENTTIDLTN	E-----KLVA	
Junín N	NPYENLLYKL	CLSGDGWPYI	GSRSQIIGRS	WDNTSIDLTR	KPVA------	
Machupo N	NPYENLLYKL	CLSGDGWPYI	GSRSQILGRS	WDNTSVDLTK	KPQV------	
Tacaribe N	NPYENLLYKL	CLSGDGWPYI	GSRSQIMGRS	WDNTSVDLTK	KPDAVPEPGA	
Pichinde N	NPYENLLYKI	CLSGDGWPYI	GSRSQVQGRA	WDNTTVDLDS	KPSAI-----	
	GGSNKSLQSA	GFAA-GLTYS	QLMTLKDFKC	FNLIPNAKTW	MDIEGRPEDP	400
	NNSSKSLQSA	GFTA-GLTYS	QLMTLKDAM-	LQLDPNAKTW	MDIEGRPEDP	
	GNSNRLAQAQ	GMPA-GLTYS	QTMELKDSM-	LQLDPNAKTW	IDIEGRPEDP	
	NSPRPAPGAA	GPPQVGLSYS	QTMLLKDLMG	-GIDPNAPTW	IDIEGRFNDP	
	NSSRPVPGAA	GPPQVGLSYS	QTMLLKDLMG	-GIDPNAPTW	IDIEGRFNDP	
	GPRQPEKNGQ	NLRLANLTEI	QEAVIREAVG	-KLDPTNTLW	LDIEGPATDP	
	GPRQPEKNGQ	NLRLANLTEM	QEAVIKEAVK	-KLDPTNTLW	LDIEGPPTDP	
	APRPAERKGQ	NLRLASLTEG	QELIVRAAIS	-ELDPSNTIW	LDIEDLQLDP	
	--QPPVRNGG	SPDLKQIPKE	KEDTVVSSIQ	-MLDPRATTW	IDIEGTPNDP	
	VEIALYQPSS	GCYVHFFREP	TDLKQFKQDA	KYSHGIDVTD	LFAAQPGLTS	450
	VEIALYQPSS	GCYIHFFREP	TDLKQFKQDA	KYSHGIDVTD	LFATQPGLTS	
	VEIAIYQPNN	GQYIHFYREP	TDIKQFKQDS	KHSHGIDIQD	LFSVQPGLTS	
	VEIAIFQPQN	GQFIHFYREP	VDQKQFKQDS	KYSHGMDLAD	LFNAQPGLTS	
	VEIAIFQPQN	GQFIHFYREP	TDQKQFKQDS	KYSHGMDLAD	LFNAQAGLTS	
	VEMALFQPAG	SKYIHCFRKP	HDEKGFKNGS	RHSHGILMKD	IEDAMPGVLS	
	VELALYQPAN	KHYIHCFRKP	HDEKGFKNGS	RHSHGILMQD	IEDAMPGVLS	
	VELALYQPAK	KQYIHCFRKP	HDEKGFKNGS	RHSHGILMKD	IEDAVPGVLS	
	VEMAIYQPDT	GNYIHCYRFP	HDEKSFKEQS	KYSHGLLLKD	LADAQPGLIS	
	AVIEALPRNM	VITCQGSEDI	RKLLESQGRR	DIKLIDITLS	KADSRKFENA	500
	AVIDALPRNM	VITCQGSDDI	RKLLESQGRK	DIKLIDIALS	KTDSRKYENA	
	AVIESLPKNM	VLSCQGADDI	RKLLDSQNRR	DIKLIDVSMQ	KDDARKFEDK	
	SVIGALPQGM	VLSCQGSDDI	RKLLDSQNRK	DIKLIDVEMT	REASREYEDK	
	SVIGALPQGM	VLSCQGSDDI	RKLLDSQNRR	DIKLIDVEMT	KEASREYEDK	
	YVIGLLPPDM	VVTTQGSDDI	RKLFDLHGRR	DLKLVDVRLT	SEQARQFDQQ	
	YVIGLLPQDM	VITTQGSDDI	RKLLDIHGRK	DLKLVDVKLT	SDQARLYDQQ	
	YVIGLLPPNM	VITTQGSDDI	RKLLDIHGRK	DLKLIDVKFT	SDQARLFEHQ	
	SIIRHLPQNM	VFTAQGSDDI	IRLFEMHGRR	DLKVLDVKLS	AEQARTFEDE	
	VWDQFKDLCH	MHTGVVVEKK	KRGGKEEIT-	----PHCALM	DCIMFDAAVS	550
	VWDQYKDLCH	MHTGVVVEKK	KRGGKEEIT-	----PHCALM	DCIMFDAAVS	
	IWDEYKHLCR	MHTGIVTQKK	KRGGKEEVT-	----PHCALL	DCLMFEAAVI	
	VWDKYGWLCK	MHTGIVRDKK	KK----EIT-	----PHCALM	DCIIFESA-S	
	VWDKYGWLCK	MHTGVVRDKK	KK----EIT-	----PHCALM	DCIIFESAAS	
	VWEKFGHLCK	HHNGVVVSKK	KRDKDAPFKL	ASSEPHCALL	DCIMFQSVLD	
	IWEKFGHLCK	HHNGVVVNKK	KREKDSPFKL	SSGEPHCALL	DCIMYQSVMD	
	VWDKFGHLCK	QHNGVIISKK	NKSKDSPPSP	SPDEPHCALL	DCIMFHSAVS	
	IWERYNQLCT	KHKGLVIKKK	KKGAVQ----	TTANPHCALL	DTIMFDATVT	
	GG--LDAKVL	RVVLPRDMVF	RTSTPKVVL			
	GG--LNTSVL	RAVLPRDMVF	RTSTPRVVL			
	GS--PQIPTP	RPVLSRDLVF	RTGPPRVVL			
	KARLPDLKTV	HNILPHDLIF	R-GPNVVTL			
	KARLPDLKTV	HNILPHDLIF	R-GPNVVTL			
	G-KLYEEELT	P-LLPPSLLF	LPKAAY-AL			
	G-KMVDEEPV	A-LLPLSLLF	LPKAAF-AL			
	G-ELPKEEPI	P-LLPKEFLF	FPKTAF-AL			
	G-WVRDQKPM	R-CLPIDTLY	RNNTDLINL			

FIGURE 2. *(Continued)*

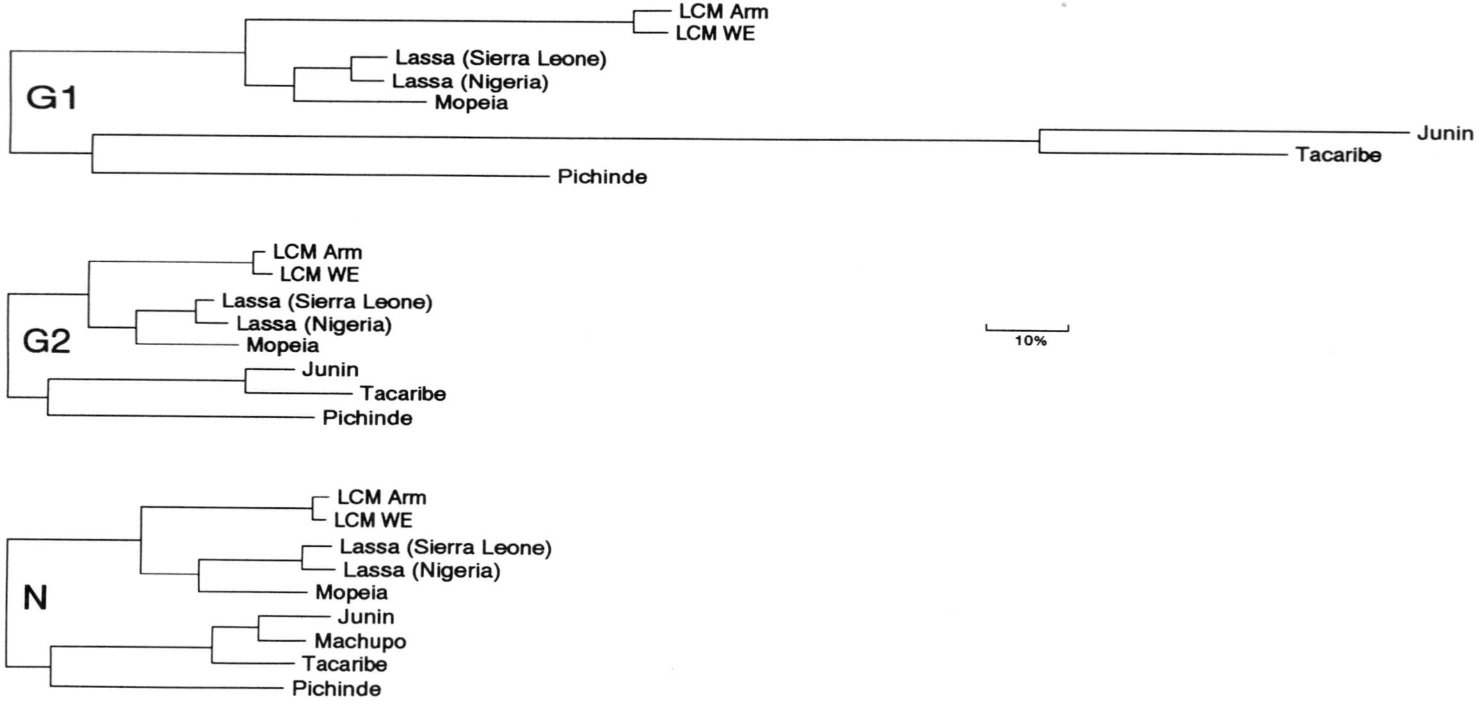

FIGURE 3. Phylogenetic trees for arenavirus structural proteins. Trees were calculated by the program CLUSTAL V using the neighbor-joining method of Saitou and Nei (1987). Regions of the alignments where gaps occur in one or more sequences were not considered, and distances were empirically corrected for multiple substitutions by the method of Kimura (1983). Alignment lengths used for G1, G2, and N were 214, 230, and 537 residues, respectively. Horizontal lines are proportional in length to divergence as indicated by the scale bar. Vertical distances are for clarity only.

Machupo virus infection are protected against a subsequent Junin virus challenge (Weissenbacher *et al.*, 1976). We have suggested on molecular phylogenetic grounds (Griffiths *et al.*, 1992) that the vaccine strain of Junin virus, currently undergoing clinical trial in the human at-risk population in Argentina (Maiztegui *et al.*, 1990) may well also protect against Machupo virus infection. If it could be confirmed that a vaccine for human use at this advanced stage of development was effective in this way, it would be of great benefit for the control of any future outbreak of Bolivian hemmorhagic fever.

REFERENCES

Auperin, D. D., and McCormick, J. B., 1988, Nucleotide sequence of the Lassa Virus (Josiah strain) S genome RNA and amino acid sequence comparison of the N and GPC proteins to other arenaviruses, *Virology* **168**:421.

Auperin, D., Dimock, K., Cash, P., Rawls, W. E., Leung, W. C., and Bishop, D. H. L., 1982a, Analyses of the genomes of prototype Pichinde arenavirus and a virulent derivative of Pichinde Munchique: Evidence for sequence conservation at the 3'-termini of their viral RNA species, *Virology* **116**:363.

Auperin, D. D., Compans, R. W., and Bishop, D. H. L., 1982b, Nucleotide sequence conservation at the 3' termini of the viron RNA species of New World and Old World arenaviruses, *Virology* **121**:200.

Auperin, D. D., Romanowski, V., Galinski, M., and Bishop, D. H. L., 1984, Sequencing studies of Pichinde arenavirus S RNA indicate a novel coding strategy, an ambisense viral S RNA, *J. Virol.* **52**:897.

Auperin, D. D., Sasso, D. R., and McCormick, J. B., 1986, Nucleotide sequence of the glycoprotein gene and intergenic region of the Lassa virus S genome RNA, *Virology* **154**:155.

Bishop, D. H. L., and Auperin, D. D., 1987, Arenavirus gene structure and organization, *Curr. Top. Microbiol. Immunol.* **13**:5.

Buchmeier, M. J., and Parekh, B. S., 1987, Protein structure and expression among arenaviruses, *Curr. Top. Microbiol. Immunol.* **133**:41.

Buchmeier, M. J., Lewicki, H. A., Tomori, O., and Oldstone, M. B. A., 1981, Monoclonal antibodies to lymphocytic choriomeningitis and Pichinde viruses: Generation, characterization and cross-reactivity with other arenaviruses, *Virology* **113**:73.

Burns, J. W., Salvato, M. S., and Buchmeier, M. J., 1990, Molecular architecture of lymphocytic choriomeningitis virus, in: *Proc. VIIIth Int. Cong. Virol.*, International Union of Microbiological Societies, Berlin, Abstr. W3-001.

Casals, J., Buckley, S. M., and Cedeno, R., 1975, Antigenic properties of the arenaviruses, *Bull. WHO* **52**:421.

Clegg, J. C. S., and Oram, J. D., 1985, Molecular cloning of Lassa virus RNA: Nucleotide sequence and expression of the nucleocapsid protein gene, *Virology* **144**:363.

Clegg, J. C. S., Wilson, S. M., and Oram, J. D., 1990, Nucleotide sequence of the S RNA of Lassa virus (Nigerian strain) and comparative analysis of arenavirus gene products, *Virus Res.* **18**:151.

Dalton, A. J., Rowe, W. P., Smith, G. H., Wilsnack, R. E., and Pugh, W. E., 1968, Morphological and cytochemical studies on lymphocytic choriomeningitis virus, *J. Virol.* **2**:1465.

Felsenstein, J., 1985, Confidence limits on phylogenies: an approach using the bootstrap, *Evolution* **39**:783.

Feng, D-F., and Doolittle, R. F., 1987, Progressive sequence alignment as a prerequisite to correct phylogenetic trees, *J. Mol. Evol.* **25**:351.

Feng, D-F., Doolittle, R. F., 1990, Progressive alignment and phylogenetic tree construction of protein sequences, *Meth. Enzymol.* **183**:375.

Franze-Fernandez, M. T., Zetina, C., Iapalucci, S., Lucero, M. A., Bouissou, C., Lopez, R., Rey, O., Daheli, M., Cohen, G. N., and Zakin, M. M., 1987, Molecular structure and early events in the replication of Tacaribe arenavirus S RNA, *Virus Res.* **7**:309.

Ghiringhelli, P. D., Rivera Pomar, R. V., Baro, N. I., Rosas, M. F., Grau, O., and Romanowski, V., 1989, Nucleocapsid protein gene of Junin arenavirus (cDNA sequence), *Nucleic Acids Res.* **17**:8001.

Ghiringhelli, P. D., Rivera-Pomar, R. V., Lozano, M. E., Grau, O., and Romanowski, V., 1991, Molecular organization of Junin virus S RNA: Complete nucleotide sequence, relationship with other members of the Arenaviridae and unusual secondary structures, *J. Gen. Virol.* **72**:2129.

Griffiths, C. M., Wilson, S. M., and Clegg, J. C. S., 1992, Sequence of the nucleocapsid protein gene of Machupo virus: close relationship with another South American pathogenic arenavirus, Junin, *Arch. Virol.* **124**:371.

Hahn, C. S., Lustig, S., Strauss, E. G., and Strauss, J. H., 1988, Western equine encephalitis virus is a recombinant virus, *Proc. Natl. Acad. Sci. USA* **85**:5997.

Higgins, D. G., and Sharp, P. M., 1988, CLUSTAL: A package for performing multiple sequence alignments on a microcomputer, *Gene* **73**:237.

Higgins, D. G., and Sharp, P. M., 1989, Fast and sensitive multiple sequence alignments on a microcomputer, *Comput. Appl. Biosci.* **5**:151.

Higgins, D. G., Bleasby, A. J., and Fuchs, R., 1992, CLUSTAL V: Improved software for multiple sequence alignment, *Comput. Appl. Biosci.* **8**:189.

Howard, C. R., Lewicki, H., Allison, L., Salter, M., and Buchmeier, M. J., 1985, Properties and characterization of monoclonal antibodies to Tacaribe virus, *J. Gen. Virol.* **66**:1383.

Iapalucci, S., Lopez, R., Rey, O., Lopez, N., Franze-Fernandez, M. T., Cohen, G. N., Lucero, M., Ochoa, A., and Zakin, M. M., 1989a, Tacaribe virus L gene encodes a protein of 2210 amino acid residues, *Virology* **170**:40.

Iapalucci, S., Lopez, N., Rey, O., Zakin, M. M., Cohen, G. N., and Franze-Fernandez, M. T., 1989b, The 5′ region of Tacaribe virus L RNA encodes a protein with a potential metal binding domain, *Virology* **173**:357.

Iapalucci, S., Lopez, N., and Franze-Fernandez, M. T., 1991, The 3′ end termini of the Tacaribe virus sub-genomic RNAs, *Virology* **182**:269.

Kimura, M., 1983, *The Neutral Theory of Molecular Evolution*, Cambridge University Press, Cambridge.

Levinson, R. S., Strauss, J. H., and Strauss, E. G., 1990, Complete sequence of the genomic RNA of O′Nyong-nyong virus and its use in the construction of alphavirus phylogenetic trees, *Virology* **175**:110.

Maiztegui, J. I., Barrera Oro, J. G., Feuillade, M. R., Peters, C. J., Vallejos, D., and McKee, K. T., 1990, Inoculation of human volunteers with Candid 1, a live attenuated Junin virus vaccine candidate, in: *Proc. VIIIth Int. Cong. Virol.*, International Union of Microbiological Societies, Berlin, Abstr. W70-005.

Murphy, F. A., and Whitfield, S. G., 1975, Morphology and morphogenesis of arenaviruses, *Bull. WHO* **52**:479.

Peters, C. J., Jahrling, P. B., Liu, C. T., Kenyon, R. H., McKee, K. T., and Barrera Oro, J. G., 1987, Experimental studies of arenaviral hemorrhagic fevers, *Curr. Top. Microbiol. Immunol.* **134**:5.

Romanowski, V., and Bishop, D. H. L., 1985, Conserved sequences and coding of two strains of lymphocytic choriomeningitis virus (WE and ARM) and Pichinde arenavirus, *Virus. Res.* **2**:35.

Romanowski, V., Matsuura, Y., and Bishop, D. H. L., 1985, Complete sequence of the S RNA of lymphocytic choriomeningitis virus (WE strain) compared to that of Pichinde arenavirus, *Virus Res.* **3**:101.

Rowe, W. P., Pugh, W. E., Webb, P. A., and Peters, C. J., 1970, Serological relationship of

the Tacaribe complex of viruses to lymphocytic choriomeningitis virus, *J. Virol.* **5**:289.
Ruo, S. L., Mitchell, S. W., Kiley, M. P., Roumillat, L. F., Fisher-Hoch, S. P., and McCormick, J. B., 1991, Antigenic relatedness between arenaviruses defined at the epitope level by monoclonal antibodies, *J. Gen. Virol.* **72**:549.
Saitou, N., and Nei, M., 1987, The neighbor-joining method: a new method for reconstructing phylogenetic trees, *Mol. Biol. Evol.* **4**:406.
Salvato, M., and Shimomaye, E. M., 1989, The completed sequence of lymphocytic choriomeningitis virus reveals a unique RNA structure and a gene for a zinc-finger protein, *Virology* **173**:1.
Salvato, M., Shimomaye, E., Southern, P., and Oldstone, M. B. A., 1988, Virus-lymphocyte interactions. IV. Molecular characterization of LCMV Armstrong (CTL+) small genomic segment and that of its variant, clone 13 (CTL−), *Virology* **164**:517.
Salvato, M. S., Shimomaye, E. M., and Oldstone, M. B. A., 1989, The primary structure of the lymphocytic choriomeningitis virus L gene encodes a putative RNA polymerase, *Virology* **169**:377.
Sanchez, A., Pifat, D. Y., Kenyon, R. H., Peters, C. J., McCormick, J. B., and Kiley, M. P., 1989, Junin virus monoclonal antibodies: characterization and cross-reaction with other viruses, *J. Gen. Virol.* **70**:1125.
Singh, M. K., Fuller-Pace, F. V., Buchmeier, M. J., and Southern, P. J., 1987, Analysis of the genomic L RNA segment from lymphocytic choriomeningitis virus, *Virology* **161**:448.
Swofford, D. L., and Olsen, G. J., 1990, Phylogeny reconstruction, in: *Phylogeny Reconstruction* (D. M. Hillis and C. Moritz, eds.), Sinauer Associates, Sunderland, MA.
Weissenbacher, M. C., Coto, C. E., and Calello, M. A., 1976, Cross-protection between Tacaribe complex viruses: Presence of neutralizing antibodies against Junin virus (Argentine hemorrhagic fever) in guinea pigs infected with Tacaribe virus, *Intervirology* **6**:42.
Weissenbacher, M. C., Laguens, R. P., and Coto, C. E., 1987, Argentine hemorrhagic fever, *Curr. Top. Microbiol. Immunol.* **134**:79.
Wilson, S. M., Clegg, J. C. S., and Oram, J. D., 1991, Sequence analysis of the S RNA of the African arenavirus Mopeia: An unusual secondary structure feature in the intergenic region, *Virology* **180**:543.
Wulff, H., Lange, J. V., and Webb, P., 1978, Interrelationship among arenaviruses measured by immunofluorescence, *Intervirology* **9**:344.

PART II

IMMUNOPATHOLOGY, VACCINES, AND EPIDEMIOLOGY

CHAPTER 11

Lymphocytic Choriomeningitis Virus

History of the Gold Standard for Viral Immunobiology

CHARLES J. PFAU AND
A. RANDRUP THOMSEN

I. ARENAVIRUSES—THE STARTING POINT

Of all the families of viruses, it is tempting to say that none but the arenaviruses has brought to light so many principles on which modern-day viral immunobiology and viral immunopathology are based. The touchstone, and prototype of the Arenaviridae, is murine lymphocytic choriomenigitis virus (LCMV). As strikingly demonstrated in the following chapters, the LCMV murine model has proven to be without equal for the study of acute and persistent infections, including related disease, in a natural host. For this reason, as well as to give the reader some feeling of the excitement and frustration of everyday science as we know it, we recount the history of the discovery of LCMV and its long-awaited place and convergence in a newly emerging taxon. We also briefly discuss the epidemiology of LCMV, since it has been so troublesome for researchers, especially nonvirologists working in cancer and immunobiology fields.

CHARLES J. PFAU • Department of Biology, Rensselaer Polytechnic Institute, Troy, New York 12181. A. RANDRUP THOMSEN • Institute of Medical Microbiology and Institute of Experimental Immunology, University of Copenhagen, DK-2200 Copenhagen N, Denmark.

The Arenaviridae, edited by Maria S. Salvato. Plenum Press, New York, 1993.

II. HISTORY

In 1925, the Swedish pediatrician Arvid Wallgren laid down standards for defining a clinical entity that he termed acute aseptic meningitis. The outstanding features were an abrupt onset of fever with symptoms of meningeal involvement and a rapid recovery. Other criteria to satisfy the definition were a slight or moderate increase in the number of cells in bacteria-free cerebrospinal fluid (initially polymorphonuclear leukocytes followed by replacement with lymphocytes); a benign course with no complications; the absence of a focus of acute or chronic infection in the vicinity of the brain, for example sinusitis; and the absence from the community of a disease known to be capable of producing irritation of the meninges. Later this was also termed benign lymphocytic choriomeningitis.

Wallgren's description would eventually cause a great deal of confusion in the medical literature for a number of years because, in the light of hindsight, it encompassed conditions associated with a number of viruses, including some members of the Picornaviridae (polio, Coxsackie, ECHO), and mumps (Baird and Rivers, 1938; Spink, 1978). However, the first virus associated with acute aseptic meningitis in humans was LCMV. Working at the National Institute of Health in Washington, D.C., Armstrong and Lillie published the initial description in 1934. The virus was found after repeated passage in monkeys of human brain tissue homogenates from an apparent case of St. Louis encephalitis. While the monkeys presented symptoms similar to those elicited by other encephalitis virus isolates, the presence of another agent was suspected when the 6th-passage material unexpectedly produced clinical disease in a monkey with proven immunity to St. Louis encephalitis. This new isolate produced symptoms, faithfully recapitulated by repeated passage in monkeys or mice, that were quite distinct from those induced by any other known encephalitis virus. The true source of the virus was never identified. It might have been the patient, who could have been infected with two viruses concurrently, or the virus might have originated in one of the monkeys, despite the fact that the other monkeys used at that time were free from the infection.

As so often happens in science (and art), the same discovery was made almost simultaneously in different surroundings. During the winter of 1934, a virologist, Richard E. Shope, and a maintenance-staff painter, both working at the Division of Animal Pathology of the Rockefeller Institute for Medical Research at Princton, New Jersey, were admitted to the Rockefeller Institute Hospital in New York City. Their complaints brought them to the ward of Thomas M. Rivers, who put them under the care of Thomas F. McNair Scott. These patients underwent illnesses of several weeks' duration, resembling severe influenza with symptoms of meningitis (Corner, 1964). Rivers and McNair Scott obtained a "filterable" infectious agent from the spinal fluid of both patients, with which they transmitted the disease to mice. In a prolonged investigation that was a model of logic and precision, Rivers and Scott tested their newly isolated virus against every procurable virus known to have similar characteristics. It differed from all except the one reported by Armstrong and Lillie a year earlier (Rivers and Scott, 1935; Benison, 1967).

At about the same time, a close friend of Shope's (Robert Shope, personal communication, 1990), Eric Traub, also at the division of Animal Pathology in Princeton, was trying to adapt hog cholera virus to mice. He was injecting hog cholera material into the brains of white Swiss mice and noted that many of them contracted encephalitis. At first he thought he had succeeded, but then he noted that his control mice, which had received sterile broth (used to dilute the virus), also displayed the same kind of disease. A careful examination revealed that practically all the mice in the Princeton laboratories, including those used by Shope, were carrying the contaminant virus in a latent form and that it was passed from mother to offspring (Traub, 1935; Benison, 1967). Both the Traub and Rivers–McNair Scott isolates produced symptoms in mice identical to those reported for the Armstrong and Lillie LCMV. This feature, which 55 years later is still the hallmark, is the characteristic illness after intracerebral inoculation of adult animals. Rivers and Scott's description (1936) is preeminent in conciseness and clarity.

> During the first 5 days after inoculation the mice appear well. Occasionally on the 5th day, but more commonly on the 6th day, symptoms appear, at which time some of the mice may be found dead although none of them were obviously sick on the preceding day, while others with dirty, ruffled fur, half-closed eyes, and hunched backs remain motionless. When disturbed they occasionally leap up and down in the jar and fall over backwards; but the characteristic reaction, especially when the animals are suspended by the tail, is for them to exhibit coarse tremors of the head and extremities frequently going on to a series of clonic convulsions terminating in a tonic extension of the hind legs. In male mice an erection sometimes occurs during the convulsions. The convulsions, often the cause of death, may also occur spontaneously either in sick mice or even in those that appear to be normal. As a rule, the animals either die within 1 to 3 days after the onset of symptoms or quickly recover after 5 or 6 days. Paralyses have never been observed.

In short order, viruses were exchanged among the three laboratories (Rivers and Scott, 1936), and each confirmed that all isolates were identical based on *in vitro* neutralization and *in vivo* cross-protection. In the mid-1930s there seemed to be quite a bit of lymphocytic choriomeningitis in humans. In fact, since 11% (n=1149)* of sera from patients (in a veterans' hospital in Baltimore) with no history of CNS disease contained neutralizing antibodies to the virus, it was thought that the most common infections in humans were asymptomatic or produced febrile,

* 4.7% (n = 1,149) of sera collected between 1986 and 1988 from a sexually transmitted disease clinic in Baltimore was positive, by ELISA test, for LCMV (Childs *et al.*, 1991).

grippe-like illness (Farmer and Janeway, 1942). In more modern times, sera collected between 1986 and 1988 from a sexually transmitted disease clinic in Baltimore was 4.7% (n=1149) positive for LCMV by ELISA (Childs *et al.*, 1991).

The epidemiological pattern was revealed with the recovery of LCMV from peridomestic gray mice trapped in infected households (Armstrong and Sweet, 1939). Interestingly, although it is unquestionable that LCMV has proven potential to cause serious illness in humans, we hardly hear of anyone being infected today (see Epidemiology). However, the unusual pathogenesis of the disease in mice has never ceased to excite investigators from a number of disciplines.

III. TAXONOMY

For more than three decades LCMV remained a morphological and biological orphan, albeit an important orphan. Today, LCMV is considered the prototype virus of the family Arenaviridae. The road to taxonomic classification was both long and tortuous. The first tile in the taxonomic mosaic was set in place between 1956 and 1958 when workers at the Trinidad Regional Virus Laboratory, using tissues of fructivorous bats, recovered 11 apparently identical strains of a virus that could cause disease in mice. These isolates could not be linked to any of the then-recognized viruses. The new agent was given the name Tacaribe virus and was consigned to the freezers of the Rockefeller Foundation laboratories to await its future (Downs *et al.*, 1963). It did not wait long. The etiological agent of a new human disease called Argentine hemorrhagic fever was isolated in 1958 and designated Junin virus (Parodi *et al.*, 1958).

How Junin and Tacaribe virus were shown to share complement fixing (CF) antigens (Mettler *et al.*, 1963) beautifully illustrates those ever so important chance events that all too often go unrecorded in the literature. In the late 1950's when Robert Shope (the son of Richard) was working for the Rockefeller Foundation in Belem, Brazil, an illness occurred in his family that forced them to return to the United States. During the time of hospitalization and protracted recovery, Shope was asked by Max Theiler, then Director of the Rockefeller Foundation Virus Laboratories in New York, to try to identify some viruses. One of these was Tacaribe virus. When Shope set up the first CF test for Tacaribe virus, he asked Norma Mettler, a young Argentine scientist who had recently been awarded a fellowship to work with Jordi Casals on the identification and classification of Junin virus, to let him have some serum from one of her guinea pigs that was convalescent from a Junin virus infection. A further footnote to their publication, which does not do justice to the string of events that led to this CF relationship, was that Shope had previously interviewed Mettler (in Buenos Aires) in connection with the eventual award of the Rockefeller Foundation fellowship (Robert Shope, personal communication, 1990).

When Machupo virus, the causative agent of Bolivian hemorhaggic fever (a disease remarkably similar to Argentine hemorrhagic fever) was encountered in 1963, it was found to be related to Tacaribe and Junin viruses (Johnson *et al.*, 1965). Thus, the so-called "Tacaribe complex" of viruses was born (Johnson *et al.*, 1973).

As early as 1965, Patricia Webb and Karl Johnson, both working at the Middle America Research Unit in Panama, called attention to the similarity in chronic infection patterns between LCMV in mice and Machupo virus in hamsters, as well as in its natural host, *Calomy callosus* (Murphy, 1975). Things began to move rapidly now! Webb and Johnson's thoughts were right on target as shown by the finding that, under laboratory conditions, infant *C. callosus* infected with Machupo virus would become chronically infected, just as in the neonatal mouse–LCMV model (Justines and Johnson, 1969). In the meantime, Wallace Rowe at the National Institutes of Health in Washington was looking for an immunological link between Machupo virus (supplied by Johnson) and LCMV. None could be found at the level of CF antibodies (Murphy, 1975), and it was only much later that relatedness was shown between the nucleoproteins of these viruses.

Johnson also sent virus samples to Gernot Bergold at the Venezuelan Scientific Research Institute for electron microscopic (EM) examination. Early studies, by Bergold and untold others working with LCMV, were thwarted because of the lability of the virus under standard EM-processing techniques, such as centrifugation or negative staining with phosphotungstic acid (Johnson *et al.*, 1973). However, related studies gave Bergold the credit for seeing an Arenavirus (Tacaribe) for the first time (Bergold *et al.*, 1969). Another problem that faced the electron microscopist was the low yield of virus from infected cells. Today we would explain this by the fact that defective-interfering (DI) virus are readily generated after high multiplicity infection of tissue culture cells, and these drastically reduce the total number of virions produced (Pevear and Pfau, 1989). Eventually this problem was overcome, more or less empirically. Morphologically, LCMV was found to be very pleomorphic (50–300 nm in diameter) and spherical to cup-shaped (Dalton *et al.*, 1968). The virus particles appeared to have spikes and to be formed by budding from the plasma membrane. The Latin-derived areno, meaning sand, was to become the name of the emerging taxon (later changed to arena), because all the LCM virions contained one to eight ribonuclease-sensitive, electron-dense granules.

Confirmation of the Dalton and Rowe discovery came within a couple of months. It was based on collaborative studies undertaken in late 1968 between Federick Murphy at the Center for Disease Control in Atlanta and Webb and Johnson at the Middle America Research Unit (Murphy, 1975). Murphy's initial photomicrographs established structural similarities between Machupo and LCMV, which were quickly extended to other Tacaribe complex viruses. Soon, sufficient serological, physiochemical, and morphological data were available to become the basis for a formal proposal for the new Arenavirus group (Rowe *et al.*, 1970). The merit of this taxon was soon borne out. In 1970, Lassa fever was first described, and both CF and EM studies quickly placed it in the new family of viruses (Murphy, 1975). This rapid classification contributed significantly to the selection of laboratory methodologies as well as the choice of rodent most likely to be the reservoir of the virus. With the possible exception of Tacaribe virus, all the arenaviruses are carried as silent infections in murids or cricetids.

IV. EPIDEMIOLOGY

LCMV is of particular interest as a zoonotic infection in which humans acquire the virus from a mouse or hamster. It is the only member of the arenavirus family where infections are multifocal over much of Europe and the Americas. Aside from laboratory-associated infection, Scandinavia and Australia appear uniquely free of the virus. In mice the virus does not persist as a silent integrated nucleic acid, as is the case with herpes viruses. Quite the contrary, viremia is maintained throughout life (cells in *in vitro* "carrier" cultures also produce large amounts of virus). Such mice are "walking timebombs" excreting virus in respiratory droplets, feces, and urine. The virus appears to be transmitted to humans in three ways: by direct contact; by contact with excreta; and probably by infectious aerosols (Johnson, 1989). With this in mind, it could be argued that the general decline in clinical LCMV infection in recent years is due to improved hygiene and differing degrees of urbanization that limit human contact with mice. It is worthwhile noting, however, that, even today, when mouse sightings are high, as in multiunit residential sections of inner-city Baltimore, the incidence of human seroconversion is about half that reported for that general area 50 years ago (Childs *et al.*, 1991). Consequently, one wonders how much of the 50-year-old serological data were due to nonspecific neutralization of the virus (see I. History, in Johnson, 1989).

Although the mouse has been the traditional source of human infection, pet owners and laboratory workers coming into contact with infected hamsters represent a new population at risk. In 1974 tumor research workers in Rochester, New York, suffered an outbreak of respiratory disease, including some cases of meningitis and encephalomyelitis. An unrecognized LCMV infection in the hamster colony was the cause. All but 1 of the 48 infected persons had entered the room where the hamsters were housed (Hinman *et al.*, 1975). At the same time, pet hamsters in New York State were found to be infected. In 1974 12 cases of meningitis and encephalomyelitis, 34 cases of flu-like illness, and 13 other paired sera showed evidence of LCMV infection. Of these, 55 patients had pet hamsters and four were employees of wholesalers or retailers of hamsters (Deibel *et al.*, 1975).

ACKNOWLEDGMENTS. We thank Dr. O. Marker for constructive criticism and helpful discussions during the preparation of the manuscript. We are also grateful to Dr. Y. L. Eyler for her numerous redactic suggestions.

REFERENCES

Armstrong, C., and Lillie, R. D., 1934, Experimental lymphocytic choriomeningitis of monkeys and mice produced by a virus encountered in studies of the 1933 St. Louis encephalitis epidemic, *Pub. Health Rep. (Washington)* **49:**1019.

Armstrong, C., and Sweet, L. K., 1939, Lymphocytic choriomeningitis. Report of two cases, with recovery of virus from gray mice (*Mus musculus*) trapped in two infected households, *Pub. Health Rep. (Washington)* **54:**673.

Baird, R. D., and Rivers, T. M., 1938, Relationship of lymphocytic choriomeningitis to acute aseptic meningitis (Wallgren), *Amer. J. Public Health* **28:**47.

Benison, S., 1967, *Tom Rivers: Reflections on a Life in Medicine and Science,* MIT Press, Cambridge, MA.

Bergold, G. H., Graf, T., and Munz, K., 1969, Structural differences among arboviruses, in: *Arboviruses of the California Complex and the Bunyamwera Group* (V. Bardos, ed.), Slovak Academy of Sciences, Bratislava.

Childs, J. E., Glass, G. E., Ksiazek, T. G., Rossi, C. A., Barrera Oro, J. G., and Leduc, J. W., 1991, Human-rodent contact and infection with lymphocytic choriomeningitis and Seoul viruses in an inner-city population, *Am. J. Trop. Med. Hyg.* **44:**117.

Corner, G. W., 1964, *A History of the Rockefeller Institute 1901–1953: Origins and Growth,* Rockefeller Institute Press, New York.

Dalton, A. J., Rowe, W. P., Smith, G. H., Wilsnack, R. E., and Pugh, W. E., 1968, Morphological and cytochemical studies on lymphocytic choriomeningitis virus, *J. Virol.* **2:**1465.

Deibel, R., Woodall, J. P., Decker, W. J., and Schryver, G. D., 1975, Lymphocytic choriomeningitis in man. Serological evidence of association with pet hamsters, *JAMA* **232:**501.

Downs, W. G., Anderson, C. R., Spence, L., Aitken, T. H. G., and Greenhall, A. H., 1963, Tacaribe virus, a new agent isolated from Artibeus bats and mosquitoes in Trinidad, West Indies, *Am. J. Trop. Med. Hyg.* **12:**640.

Farmer, T. W., and Janeway, C. A., 1942, Infections with the virus of lymphocytic choriomeningitis, *Medicine (Baltimore)* **21:**1.

Hinman, A. R., Fraser, D. W., Douglas, R. G., Bowen, G. S., Kraus, A. L., Winkler, W. G., and Rhodes, W. W., 1975, Outbreak of lymphocytic choriomeningitis virus infections in medical center personnel, *Am. J. Epidemiol.* **101:**10.

Johnson, K. M., 1989, Arenaviruses, in: *Viral infections of humans: Epidemiology and Control,* 3rd ed. (A. S. Evans, ed.), pp. 133–152, Plenum Press, New York.

Johnson, K. M., Wiebenga, N. H., Mackenzie, R. B., Kuns, M. L., Tauraso, N. M., Shelokov, A., Webb, P. A., Justines, G., and Beye, H. K., 1965, Virus isolations from human cases of haemorrhagic fever in Bolivia, *Proc. Soc. Exp. Biol. Med.* **118:**113.

Johnson, K. M., Webb, P. A., and Justines, G., 1973, Biology of the Tacaribe complex of viruses, in: *Lymphocytic Choriomeningitis Virus and Other Arenaviruses* (F. Lehmann-Grube, ed.), pp. 241–258, Springer-Verlag, Vienna.

Justines, G., and Johnson, K. M., 1969, Immune tolerance to *Calomys callosus* infected with machupo virus, *Nature* **222**:1090.

Mettler, N. E., Casals, J., and Shope, R. E., 1963, Study of the antigenic relationships between Junin virus, the aetiologic agent of Argentine haemorrhagic fever and other arthropod-borne viruses, *Am. J. Trop. Med.* **12**:647.

Murphy, F. A., 1975, Arenavirus taxonony: A review, *Bull. W.H.O.* **52**:389.

Parodi, A. S., Greenway, D. J., Rugiero. H. R., Rivero, E., Frigerio, M. J., Mettler, W. E., Garzon, F., Boxaca, M., Guerrero, L. B., and Nota, N. R., 1958, Sobre la etiologia del bute epidemico de Junin, *Dia Med.* **30**:2300.

Pevear, D. C., and Pfau, C. J., 1989, Lymphocytic choriomeningitis virus, in: *Clinical and Molecular Aspects of Neurotropic Virus Infections* (D. H. Gilden and H. L. Lipton, eds.), pp. 141–172, Kluwer Academic Publishers, New York.

Rivers, T. M., and Scott, T. F. M., 1935, Meningitis in man caused by a filterable virus, *Science* **81**:439.

Rivers, T. M., and Scott, T. F. M., 1936, Meningitis in man caused by a filterable virus. II. Identification of the etiological agent, *J. Exp. Med.* **63**:415.

Rowe, W. P., Murphy, F. A., Bergold, G. H., Casals, J., Hotchin, J., Johnson, K. M., Lehmann-Grube, F., Mims, C. A., Traub, E., and Webb, P. A., 1970, Arenoviruses: proposed name for a newly-defined virus group, *J. Virol.* **5**:651.

Spink, W. W., 1978, *Infectious Diseases: Prevention and Treatment in the Nineteenth and Twentieth Centuries,* University of Minnesota Press, Minneapolis.

Traub, E., 1935, A filterable virus recovered from white mice, *Science* **81**:298.

Wallgren, A, 1925, Une nouvelle maladie infectieuse du system nerveux central? *Acta Pediatr.* **4**:158

CHAPTER 12

Influence of Host Genes on the Outcome of Murine Lymphocytic Choriomeningitis Virus Infection

A Model for Studying Genetic Control of Virus-Specific Immune Responses

A. RANDRUP THOMSEN AND CHARLES J. PFAU

I. INTRODUCTION

One of the most important challenges in pathogenesis is to understand the influence of host genes on disease susceptibility. It is an attractive hypothesis to geneticists that the existence of extensive polymorphisms within populations implies that various alleles confer certain selective advantages for survival under particular evolutionary pressures. Thus, it is not surprising that when the extensive polymorphism at HLA loci (the designation for major histocompatibility complex of genes, MHC, in

A. RANDRUP THOMSEN • Institute of Medical Microbiology and Institute of Experimental Immunology, University of Copenhagen, DK-2200 Copenhagen N, Denmark. CHARLES J. PFAU • Department of Biology, Rensselaer Polytechnic Institute, Troy, New York 12181.

The Arenaviridae, edited by Maria S. Salvato. Plenum Press, New York, 1993.

humans) was discovered, geneticists examined a variety of diseases for possible associations with one or more alleles at this complex of loci. Further impetus for these studies was the observation by Lilly and coworkers (1964) that in mice the H-2 complex (the designation for MHC in mice) controlled susceptibility to viral leukemogenesis. Unfortunately, the most striking finding in the majority of ensuing studies on resistance to infectious disease, in both men and mice, was how little disease resistance is influenced by MHC (Brinton and Nathanson, 1981; Clatch *et al.*, 1987; Klein, 1986). A rare exception is the finding of two HLA types frequent among West Africans being associated with resistance to severe malaria (Hill *et al.*, 1991). This is in contrast to the well-known associations between HLA and many autoimmune diseases (Tiwari and Terasaki, 1985). This correlation is quite understandable since until recently there was virtually no selection against noninfectious diseases, probably because people died before the diseases could develop (Klein, 1986). Although this is not the case today, autoimmune diseases still do not appear to exert any selective pressure on the population because most of them become apparent after the age of reproduction. Even at this time, the development of disease is basically a statistical risk. This could be due to multigenic effects (see below) or the need for exposure to a triggering factor (even in monozygous twins there is discordance, Tiwari and Terasaki, 1985). In addition, in almost no case do we know the etiological agent(s) responsible for an HLA-associated disease. As Klein (1986) states, "In fact, at the present stage, the study of HLA-associated diseases has about the same value to medicine as beetle counting had to zoology 200 years ago."

What, then, fuels the continued interest in establishing patterns of genetic predisposition to disease? To be sure, there are a number of advantages. Such knowledge could be applied to disease diagnosis, prevention, prognosis, pathogenesis, and, finally, genetic counseling (Klein, 1986). Furthermore, keeping in mind the fact that the prevalence of autoimmune diseases increase with age, we can anticipate this category of diseases being an ever-increasing problem as life expectancy rises in the Western world. One must also remember that the classic infectious diseases (smallpox, polio, measles, etc.) are being, or have been, vaccinated out of existence. What we are increasingly left with are disease-producing organisms that have found ways to evade or subvert the immune system. Specifically, we are currently seeing a rise in persistent viral infections in the world (Haywood, 1986), and new methodologies will be needed to identify and control these infections. With regard to control, new synthetic vaccine technology, such as viral peptides, will require that we know much more about the regulation of the immune response to infectious agents. This in turn will require better understanding of the role of the MHC in immune regulation, as well as the role of infectious agents in the evolution of MHC polymorphism. Aside from persistent viral infections, one should not entirely dismiss the case for acute lethal virus infections. Humans have probably been subject to major epidemics for 5000 years and during this time selection has been weeding out unsuitable genes (Klein, 1986). Thus, it is not surprising that there is little correlation between specific genes and and resistance to acute infectious diseases. This picture could, however, change dramatically in a population exposed to a "new" infectious agent. For example, an epidemic could arise from an interspecies transfer of a virus from animal to human (e.g., HIV-1, Gallo, 1987).

As already exemplified, the study of genetically determined resistance to the development of a particular disease in humans is exceedingly complex. One is dealing with mostly outbred populations, and one cannot easily dissociate genetic effects from environmental factors (nutrition for example). Mice are the species, par excellence, for this type of study. There are a large number of inbred strains; an immense source of genetic information already exists; and new congenic lines can easily be developed. We will attempt to show in this review that a better understanding of genetic predisposition to disease can be achieved if the murine "genetic pool" is considered in its simplest form as two compartments. We will treat the MHC as one compartment and the non-MHC or "background" genes as the other compartment. We have chosen to make this uneven division of the genome as a natural consequence of the central role of MHC genes in providing the antigen presentation elements for immune recognition. As has rarely been achieved before in a single mouse-virus model (Brinton and Nathanson, 1981; Clatch *et al.*, 1987), we will demonstrate the separate and combined effects of MHC and background genes on the immune response against a viral pathogen, lymphocytic choriomeningitis virus (LCMV).

II. THE MURINE LCMV MODEL

Just as polio has served as the classic paradigm for the biochemical and molecular virologist, so has the murine LCMV model for the viral immunopathologist. It would be hard to bring to mind any other single model that has brought to light so many principles on which modern-day immunology/virology is based (Pevear and Pfau, 1989). Basically, the route of infection has an effect on the outcome. The textbook picture is as follows: injection of adult mice by the intracerebral (i.c.) route results in acute lethal disease, whereas peripheral infection is usually abortive and immunizing. On the other hand, virus introduced *in utero* or into the neonatal mouse by any route of injection results in lifelong persistence of fully mature infectious virus. However, as is often the case with textbooks, this description is too simplistic—in certain mouse-virus strain combinations no fatal disease is induced by i.c. inoculation, and more chronic infection is observed. Initially considered merely vexing in attempting to bring order to the field, these chance combinations of virus and mouse strain have had a seminal effect on the research we are about to describe.

III. PIVOTAL ROLE OF T CELLS ON THE COURSE OF MURINE LCMV INFECTION

The role of humoral and cell-mediated immunity in LCM virus clearance, as well as virus-induced lethal CNS disease, has been the subject of intense investigation for a number of years. The very beginning of this search can be traced to the observation that preirradiation of mice with X-rays protected against lethal infection (Rowe, 1956), while having no effect on virus replication (Hotchin and Weigand, 1961b). On the other hand, there was severe leukocytopenia and an absence of histological lesions routinely associated with the infection in unirradiated control mice (Hotchin and Weigand, 1961a). The early observations of Rowe (1954), showing lack of correlation between serum antibody level and immunity, followed by the finding that neonatally thymectomized mice could not be killed by i.c. infection (Rowe *et al.*, 1963) were the first indications of the central role of T cells in critical responses to the infection. Later studies by Hirsch and co-workers (1967) using antithymocyte serum also supported the role of the thymus-derived cells in the development of disease. Such treatment spared the mice even though they mounted an antiviral antibody response. Following the pioneering work of Volkert (1962), Gilden and colleagues (1972) established a model system for adoptive transfer of lymphocytes into adult i.c. infected mice. In these elegant studies, Cole and co-workers (1972) demonstrated, by selective reconstitution of the immune system of cyclophosphamide-spared mice, that only immune lymphocytes bearing Thy-1 antigen (the pan-T-cell marker of the mouse) caused rapid convulsive death. Following the discovery in 1973 of LCMV-specific cytotoxic T (T_c) cells (Marker and Volkert, 1973), much interest was focused on this population. Since T-cell-mediated lysis *in vitro* was found to require MHC compatibility between effector T cells and targets (Zinkernagel and Doherty, 1974), it was tested, and established, that similar restrictions applied to adoptive transfer of lethal disease (Doherty and Zinkernagel, 1975). Evidence continued to mount with the observation that congenitally athymic mice were resistant to LCMV-induced lethal CNS disease (Christoffersen *et al.*, 1976). Furthermore, Johnson and colleagues (1978), and later Cerny and co-workers (1986), conclusively demonstrated that B cells and their secreted products played no role in pathogenesis by showing that mice markedly depleted of Ig-bearing lymphocytes were just as susceptible as normal mice to LCMV-induced lethal disease. Most recently, adoptive transfer of cloned MHC class I restricted T_c cells specific for LCMV have been found capable of inducing lethal immunopathological disease (Baenzinger *et al.*, 1986).

Not only have T cells been found to be central to the development of immune-mediated pathology, they have also been shown to play a major role in virus elimination. Early experiments (Mims and Blanden, 1972; Volkert *et al.*, 1974) demonstrated that the antiviral activity of "primed" donor lymphocytes in recipients infected with LCMV was significantly diminished, when depleted of cells bearing Thy-1 antigen. Later, more conclusive studies (Volkert *et al.*, 1975) showed that donor cell suspensions so devoid of B cells (by passage though affinity columns) that all antibody-producing capacity was abrogated still possessed full antiviral activity in recipients. Unequivocal evidence has come from the finding that class I restricted T cells cultured and cloned *in vitro* possess antiviral activity when adoptively transferred into acutely infected mice (Byrne and Oldstone, 1984).

Several lines of evidence indicate that NK cells do not influence LCMV synthesis *in vivo*. There is no genetic basis for sensitivity in mice to LCMV that correlates with low NK-cell activity (Welsh and Kiessling, 1980), and virus synthesis is not enhanced in mice depleted of NK-cell activity by treatment with antibody against the asialo GM1 epitope (Bukowski *et al.*, 1983).

IV. IMPORTANCE OF MHC AND NON-MHC GENES IN REGULATION OF THE T_c RESPONSE

It is well established that T cells can only recognize antigen in combination with cell membrane molecules coded for by MHC. This phenomenon is known as MHC restriction. One of the first observations indicating that this is so was actually obtained in the LCMV model. While trying to correlate susceptibility to lethal LCM disease with T_c activity *in vitro*, Doherty and Zinkernagel (1974) noted that only lymphocytes from mouse strains sharing at least part of their MHC region with the target cells they used (L cells, H-2^k) caused significant lysis *in vitro*. In the following years the all-important role of MHC molecules as restriction elements not only for LCMV-specific T_c cells, but for T cells in general, was established (Katz and Benacerraf, 1975; Shearer *et al.*, 1976; Doherty *et al.*, 1976). As a rule, T_c cells are restricted by class I molecules (K, D, and L in the mouse), whereas class II molecules restrict T-helper (T_H) cells (I-A and I-E molecules in the mouse). To this division of the T-cell population corresponds a difference in cell surface markers in that class I restricted cells are CD4−CD8+ whereas class II restricted T cells are CD4+CD8−.

However, even before the importance of MHC in antigen presentation was known, it had been established (McDevitt and Chinitz, 1969; Ellman *et al.*, 1970; McDevitt *et al.*, 1972) that genes located in the MHC region were involved in regulating immune responses to simple antigens (Ir genes). As it became clear that the Ir genes were in fact the same genes that coded for the restriction elements, a search was begun for MHC regulation of T_c responses. Originally no H-2 influence on virus-specific T_c-cell responses was detected because all mouse strains tested at the time gave significant lysis on completely MHC-compatible targets (Zinkernagel *et al.*, 1976). But when a sufficient number of reliable H-2 recombinant target cell lines became available so that MHC allele-specific cytotoxicity could be evaluated, it was discovered that there might be great differences in the level of K and D restricted activity in the same animal (Doherty *et al.*, 1978; Zinkernagel *et al.*, 1978). Because all mice tested had significant T_c activity restricted to at least one class I element, it was deduced that the control did not operate at the T_H level, but rather at the T_c-cell level. Mapping of the controlling genes showed these to be located in the K and D region. Thus the virus-specific T_c-cell response was not only class I restricted, but also class I gene regulated.

Two types of class I gene regulation of T_c responsiveness have been observed (Zinkernagel and Doherty, 1979), both of which seem to be operative in the LCMV system:

1. Stimulation with virus in the context of a given class I allele does not result in generation of T_c cells restricted to the same class I allele. Although this type of Ir-gene effect was not found initially in the LCMV system, it has been described recently: In BALB/c mice* infected with LCMV virtually all T_c activity is restricted by L^d and none by K^d or D^d (Orn *et al.*, 1982). Since BALB/c mutants with a nonfunctional L-gene–BALB/c-H-2^{dm2} (see Table I) were subsequently found to be poor T_c responders to LCMV (Frelinger *et al.*, 1983), the low response associated with K^d and D^d seems to be of this type.

2. The response is restricted by one class I gene product but is influenced by a second class I allele; this type of regulation may be seen regardless of whether the regulating gene is in a *cis* or *trans* arrangement with the restricting gene. Our studies indicate that haplotype preference within the LCMV system may reflect this type of regulation (Thomsen and Marker, 1989b). Comparing ($K^dD^dL^dxK^kD^k$) F_1 and ($K^dD^dL^-xK^kD^k$) F_1 hybrids, we found a marked haplotype preference for H-2^d in the LCMV-specific T-cell response of the former animals whereas this was not the case in the latter. Since the only difference between the two types of hybrids is the defective L gene in the latter, this finding maps the regulating element for haplotype preference to a class I locus.

Independent of the underlying mechanism, the Ir-gene effects are generally virus-specific, and the low response linked to a particular class I allele has a dominant character (Zinkernagel and Doherty, 1979), although phenotypically responsiveness is the dominant character. (Thomson and Marker, 1989b).

* All the mouse strains mentioned in this review can be found in Table I, with details on their H-2 haplotype and background.

TABLE I. Mouse Strains; H-2 Haplotype and Background

Mouse strain	H-2 K	H-2 I	H-2 D	H-2 L	Non-H-2 background
BALB/c	d	d	d	d	BALB/c
BALB/c-H-2^{dm2}	d	d	d	—	BALB/c
B10 congenics, e.g.,		Variable			C57BL/10
B10.AKM	k	k	q	q	C57BL/10
B10.BR	k	k	k		C57BL/10
C3H	k	k	k		C3H[a]
CBA	k	k	k		CBA
DBA/2	d	d	d	d	DBA/2
D2.GD	d	d/b	b		DBA/2
DBA/1	q	q	q	q	DBA/1[b]
D1.C	d	d	d	d	DBA/1
SWR	q	q	q	q	Swiss

[a] Note that substrains of C3H, e.g., C3H/HeJ and C3HeB/FeJ, differ markedly in background genes.
[b] DBA/1 and DBA/2 were split from common origin about 60 years ago.

No case of class II regulation of a class I–restricted, virus-specific T_c response has been described, although such an influence may be seen in T_c responses directed against minor histocompatibility antigens (minor H ags) (Hurme *et al.*, 1978; von Boehmer and Haas, 1979). Although this could be a coincidence, recent data suggest that it is more likely to reflect a true difference in the dependence on CD4+ T_H cells in the generation of CD8+ T_c cells directed toward the two classes of antigen. Thus it has been shown that generation of virus-specific T_c cells and T_c memory is possible in the absence of CD4+ T cells (Buller *et al.*, 1987; Ahmed *et al.*, 1988; Mizuochi *et al.*, 1989), whereas the available evidence indicates that this is not the case for minor H ag-specific T-cell responses (Roopenian and Anderson, 1988). Probably this again reflects differences in the frequency of CD8+ T_H cells directed toward viral antigens and toward minor H ags.

Non-MHC genes also exert an important influence on the magnitude of the T_c response, and although this has been described in several antigen systems (Arora and Shearer, 1981; Fiertz *et al.*, 1982; Mullbacher *et al.*, 1983; Epstein *et al.*, 1986), the importance tends to be overlooked. In the LCMV model we have found that C3H mice will respond with a T_c response upon challenge with Docile strain LCMV whereas in H-2 identical B10.BR mice the T_c response tends to be marginal (Zinkernagel *et al.*, 1985a; Thomsen *et al.*, 1987). Similarly, mice with a DBA/2 background (DBA/2 and D2.GD) are T_c low responders to most strains of LCMV compared to mice of other strains with matched H-2 type (Zinkernagel *et al.*, 1976; Thomsen *et al.*, 1987; Thomsen and Marker, 1989a). In the latter case low responsiveness is recessive and most pronounced at high doses of virus.

V. INTERACTION BETWEEN MHC AND BACKGROUND GENES IN DETERMINING SUSCEPTIBILITY TO LETHAL LCM DISEASE

A. Early Indications

Although it has been known for some time that the response to LCMV could vary markedly among different mouse strains (Hotchin and Weigand, 1961a; Volkert and Hannover-Larsen, 1965; Oldstone and Dixon, 1968), the link between specific genes and disease outcome has been elusive. Attention was first focused on the role of MHC genes in determining susceptibility to LCMV (Oldstone *et al.*, 1973). This was done some 20 years ago, at the time when the role of MHC genes in controlling immune responses was emerging, and when it had been only recently discovered that the T-cell response to LCMV was central to, if not the mediator of, the disease process (Cole *et al.*, 1972). Oldstone and colleagues (1973), exploiting their earlier findings (1968) that significantly less LCMV was needed to kill adult SWR/J (H-2^q) mice than C3H/HeJ (H-2^k) mice, found that increased susceptibility was the dominant trait in all the F_1 hybrids. The F_1 backcross to the relatively resistant parent (C3H/HeJ) showed that the H-$2^{k/q}$ mice were relatively susceptible whereas the H-$2^{k/k}$ mice were relatively resistant. Unfortunately, the phenomenology was lost (Oldstone, 1975), and indeed the link to H-2 could not be verified in other laboratories with what were considered to be closely related or identical strains of virus and mice (Lehmann-Grube, 1975; Neustadt *et al.*, 1978).

More recently, however, with the isolation of two plaque variants of the UBC strain that differed markedly in their disease-inducing capacities (Jacobson and Pfau, 1980), the subject of host gene influence came alive again. It was found, namely, that whereas mice of most strains died from i.c. inoculation of the so-called Aggressive variant, susceptibility to the so-called Docile substrain was influenced markedly by host genotype (Zinkernagel *et al.*, 1985a). By compartmentalizing the influence of MHC and non-MHC genes, we have shown that both types of genes are involved in determining susceptibility to LCM disease.

B. Evidence for Influence of Class I Genes in Pathogenesis

By using a number of strains belonging to the B10 series of congenic mice, i.e., mice that have the same background (in this case C57BL/10) but differ in their H-2 type (see Table I), one of us found disease susceptibility to vary with H-2 (Zinkernagel *et al.*, 1985a). The susceptibility of B10.AKM mice ($K^kD^qL^q$) compared to the resistance of B10.BR mice (K^kD^k) seemed to map the crucial gene or gene complex to the D-end of H-2. However, since B10 congenics are derived not by mutation, but by selective inbreeding, the possibility existed that genes unrelated to MHC class I could be involved (Datta *et al.*, 1978; Melvold, 1986). We therefore considered it essential to confirm the mapping by segregation analysis of linkage between H-2D phenotype and susceptibility. Challenge of F_1 hybrids produced between these two mouse strains (AKM; BR), revealed that susceptibility was the dominant quality (Eyler *et al.*, 1989), as would be expected if a high T-cell response was dependent on the presence of H-2D^qL^q (it should be remembered that LCM disease is immune-mediated, so susceptibility is equivalent to a high T-cell response). In three of the four possible combinations, the backcross progeny of the F_1 (D^qxD^k or D^kxD^q) to the resistant recessive parental strain (BR; D^k) demonstrated approximately 50% mortality (55%, 45%, 51%). This fits completely with the 1:1 segregation ratio between resistant and susceptible individuals that one would expect if a single gene or gene complex was involved. In the fourth combination ((AKMxBR)xBR), however, a mortality rate of 70% was observed, which is significantly higher than expected. Experiments involving H-2 typing of individual backcross mice followed by i.c. infection revealed an absolute linkage between the H-2D^q phenotype and death (Eyler *et al.*, 1989). A few backcross animals with the H-2D^k phenotype also died, but the mortality rate was not different from that in the H-2^k parental strain. Furthermore, the observed skewed distribution of the (AKMxBR)xBR animals toward susceptibility was correlated with a similar skewed distribution of H-2D phenotypes in that progeny. Thus, the results strongly indicate that the D region contains a gene (or set of genes) crucially involved in determining whether lethal disease will occur or not.

We have not proven that this gene is D^q itself. Any gene close to the D locus could be the one involved. However, Doherty and co-workers have also presented evidence supporting a role for class I genes. In their study (Allan and Doherty, 1985) they found a moderate but significant difference in susceptibility between BALB/c mice, which were susceptible, and BALB/c-H-2^{dm2} mutants, which were more resistant. Since a deletion in the L locus is the only difference between mutants and ordinary BALB/c, the evidence for class I regulation is quite convincing.

C. Evidence for Influence of Background Genes in Pathogenesis

Similar to the influence of H-2 genes, background genes were also found to affect disease outcome (Zinkernagel *et al.*, 1985a); e.g., as already mentioned, B10.BR mice were resistant whereas the majority of H-2 identical CBA mice died following i.c. infection with the Docile strain of LCMV. Based on this observation, a similar backcross analysis of the influence of non-H-2 genes was done using these two mouse strains as parental prototypes (Eyler *et al.*, 1989). The purpose of these experiments was: (a) to provide formal proof of the genetic basis of background influence and (b) to determine the number of loci involved.

As was the case with the MHC influence, susceptibility was observed to be dominant. In the segregating backcross population produced from B10.BR and CBA parental strains, 81% of the animals succumbed to i.c infection with the Docile strain of LCMV. This is close to what is expected if two independently assorting loci are involved (75%), both of which have to be homozygous for the allele of the resistant parental strain in order to avoid lethal disease. It is of interest that a similar estimate of the number of non-H-2 genes involved is reached when analyzing a segregating backcross generation produced between two parental strains differing both in H-2 type and in background (C3H; SWR) (Eyler *et al.*, 1989).

However, not all mouse–virus strain combinations yield backcross data that easily fit with a pattern predictable on the basis of simple Mendelian inheritance (Thomsen, unpublished). This has led us to the assumption that multiple non-H-2 loci influence susceptibility, and that the particular importance of a given gene depends on a number of factors, such as MHC type, other background genes, and virus strain. This prediction is supported by our finding that in one case F_1 hybrids between two H-2 identical resistant strains (C3H; B10.BR) are fully susceptible (Eyler *et al.*, 1989), suggesting either that complementing alleles exist or, as seems more likely, that different sets of non-H-2 loci confer resistance in these two strains. Therefore, we conclude that even in association with a single H-2 haplotype, one cannot predict the genetic basis for the resistance to lethal disease. It is noteworthy that we have previously suggested (based on analysis of T_c response profiles in H-2 compatible strains) that different mechanisms could be involved in preventing a lethal outcome in different mouse strains (Thomsen *et al.*, 1987). Thus, the genetic data presented strongly support this hypothesis and point to the necessity of analyzing the virus–host interaction in several mouse–virus strain combinations, even of the same H-2 haplotype, in order to fully understand the complexity involved.

D. Possible Mechanisms for Cooperative Effects of H-2 and Background Genes

As a whole, the results presented above, together with previously published data (Zinkernagel *et al.*, 1985b) indicate that knowledge of neither H-2 nor non-H-2 genes individually is sufficient to predict whether or not i.c. infection will result in fatal disease. Rather, disease outcome seems to be determined by the combined action of genes located both inside and outside of MHC. For example, the H-2^k haplotype, when in combination with the B10 background, is not conducive to disease, yet this same haplotype on the CBA background is associated with high mortality. The combinatorial effect of H-2 and non-H-2 genes is further illustrated by our recent finding that a pattern of increasing susceptibility to fatal disease is observed comparing the outcome of infection with the Aggressive strain of LCMV in the mouse strains DBA/2 ($H\text{-}2^d$, DBA/2 background), D1.C ($H\text{-}2^d$, DBA/1 background-separated from DBA/2 about 60 years ago), and DBA/1 ($H\text{-}2^q$, DBA/1 background) (Thomsen *et al.*, unpublished). The level at which the involved MHC and non-MHC genes interact is not known with certainty, but the available data point to regulation of antiviral T-cell responsiveness as a key event.

Although other possibilities exist (Price *et al.*, 1990), MHC genes are likely to affect susceptibility by directly controlling T-cell responder status. This assumption is supported by recent data (Speiser and Zinkernagel, 1990) showing that in thymectomized $(B10.BRxB10.AKM)F_1$ hybrids reconstituted with fetal thymus of either parental strain, disease outcome is controlled by the H-2 type of the thymus-graft; i.e., the selection of the T-cell repertoire markedly affects susceptibility. Adoptive transfer experiments (Allan *et al.*, 1987; Baenzinger *et al.*, 1986) and, more recently, experiments involving *in vivo* depletion of T-cell subsets with monoclonal antibodies (Leist *et al.*, 1987b) have clearly shown that the effector cells mediating LCM disease belong to the class I restricted, CD8+ T-cell subpopulation. Thus, an effect of class I genes on LCM disease is not surprising considering the evidence for class I genes acting as Ir genes for T_c-cell activity (see above).

With respect to the role of background genes, these could exert their effect in two principally different ways: (1) by direct or indirect influence on T-cell responder status and (2) by modulation of the effector phase. The latter could take place through effects on, e.g., virus distribution, ability of activated T cells to home to the target organ, susceptibility of target organ to the immune attack. The first data to suggest that both levels of regulation might be relevant was our finding that in certain mouse strains i.c. infection was associated with low mortality, despite a potent antiviral T-cell response, measured as LCMV-specific T_c activity (Thomsen *et al.*, 1987). Recently, however, evidence has been obtained that more directly demonstrates this point. Thus, in adoptive transfer experiments between BALB/c and DBA/2 mice representing H-2 identical susceptible and resistant mouse strains, respectively, we found that LCMV-primed BALB/c cells induced substantial mortality in i.c.-infected, immunosuppressed recipients of either strain whereas DBA/2 donor cells had no effect at all. However, the BALB/c cells worked with different efficiency in recipients of the two strains: BALB/c recipients were killed significantly faster than DBA/2 recipients (5 vs. 7 days after cell transfer; $p < 0.05$) of which not all died (Thomsen *et al.*, manuscript in preparation). Therefore, these two strains seem to differ not only in their capacity to generate an LCMV-specific T-cell response, but also with regard to some as yet unknown factor that modulates the efferent phase of the immune response. It should be noted that the development of virus titers in the brain followed the same pattern in both strains, suggesting that there is no lack of virus-infected target cells in the brains of DBA/2 mice.

One obvious way in which non-H-2 background could directly influence T-cell responder status is through the T-cell-receptor variable (V) region genes, which have been shown to map outside of MHC and to differ between mouse strains (Epstein *et al.*, 1986, Marrack and Kappler, 1987). With this type of genetic influence, high responsiveness would be dominant, as is the case in the LCMV system. Minor histocompatibility antigens have also been found to affect the T-cell repertoire based on the mandatory requirement for self-tolerance (Vidovic and Matzinger, 1988). Thus, exogenous antigen mimicking a minor self-antigen would not elicit an immune response. This explanation is not easily applicable to the present system, however, since F_1 mice from crosses between H-2 identical resistant and susceptible mice would then need to be resistant (self tolerance → no T-cell response → no immune pathogenesis), and this is clearly not the case.

Indirectly, non-H-2 genes could possibly affect the ability to raise a virus-specific T-cell response through an influence on the susceptibility to virus-induced nonspecific immunosuppression (Roost *et al.*, 1988). However, no data have been obtained that support this possibility in our model (Thomsen, unpublished). An indirect effect could also be exerted through host genes regulating the kinetics of initial virus multiplication. No data directly point to the relevance of this mechanism (Leist *et al.*, 1987a), and significant differences in the level of T-cell priming may be found without any obvious difference in the development of spleen virus titers (Thomsen and Marker, 1989a; Thomsen *et al.*, manuscript in preparation). However, this mechanism could be relevant in certain cases, as it has been found that treatment with anti-interferon antibody, which substantially enhances initial virus multiplication, also reduces susceptibility to fatal meningitis (Pfau *et al.*, 1983). Furthermore, it is well known that an increase in inoculum size may markedly affect the LCMV-specific T-cell response (see further below), and if virus spread is more rapid in one mouse strain than in another, this would correspond to challenge with a higher virus dose.

E. Does Genetic Influence on Pathogenesis Correlate with Regulation of Virus-Specific T_c Activity?

Since LCM disease is mediated by CD8+ effector T cells and may be transferred by cloned LCMV-specific T_c cells to immunodeficient recipients (Baenzinger *et al.*, 1986), it is natural to ask how well regulation of disease susceptibility correlates with regulation of T_c activity. In an extensive analysis Zinkernagel and co-workers (1985b) found that early and high T_c activity, and the rapidity and extent of virus-specific delayed-type hypersensitivity, directly correlated with susceptibility.

However, using other mouse strains (Pfau *et al.*, 1982a, 1985; Thomsen *et al.*, 1987), we observed that while the correlation holds true in many cases, it is not absolute. For example, in C3HeB/FeJ mice infected with Docile strain LCMV, the T_c response follows the same kinetics as in mice infected with Aggressive virus, although only the latter virus variant is lethal to this strain of mice (Thomsen *et al.*, 1987). Furthermore, T_c activity of similar magnitude was also detected directly in meningeal exudate cells (Pfau *et al.*, 1985), indicating that there was no failure of virus-specific T_c cells to home to the relevant target organ, i.e. the meninges, in mice infected with Docile virus.

Exactly how this observation can be reconciled with the relatively strong evidence that T_c clones will induce lethal disease (Baenzinger *et al.*, 1986) is not clear at present. One possibility is that although T_c clones will transfer the disease, it is not their lytic activity which is the critical factor. T_c clones are often plurifunctional (Prystowsky *et al.*, 1982; MacDonald *et al.*, 1983; Conta *et al.*, 1985; Taylor *et al.*, 1985), and it could be another function of the cloned cells that actually brings about the lethal meningitis. In this context it is pertinent to mention that lethal disease is associated with a diffuse breakdown of the normal blood–brain barrier (Marker *et al.*, 1984; Andersen *et al.*, 1991). This is more consistent with a role for a soluble factor than with the supposed kiss of death by T_c cells (Schwendemann *et al.*, 1983).

Alternatively, cytolysis could constitute the underlying mechanism for lethal LCM disease, but in the cases where no direct correlation is observed, either (1) inhibitory mechanisms block activity *in situ*, or (2) optimal lysis by primary effector T cells *in vivo* requires interaction with another cell subset not present. The reason that no need for this putative cell subset is observed in recipients given cloned T_c cells could be that any requirement for cooperation was bypassed in this case either by the sheer number of T_c cells introduced directly into the brain or by the cloned cells possessing such qualities that no cooperation is needed; note that cloned cells are often plurifunctional and are likely to have T-cell receptors with a higher affinity for antigen. The possibility that something in addition to cytolytic activity may be required for primary effector cells to work *in vivo* is strongly supported by the finding that whereas C3HeB/FeJ mice infected with either Aggressive or Docile virus have similar splenic T_c activity (see above), only cells from the former mice, i.e., the lethally infected animals, will kill in an adoptive transfer to X-irradiated, i.c.-infected recipients (Pfau *et al.*, 1985).

VI. INTERACTION BETWEEN MHC AND BACKGROUND GENES IN REGULATION OF VIRUS CLEARANCE

One of the main conclusions of the preceding section is that the outcome of i.c. infection with LCMV is influenced by MHC as well as non-MHC genes. While studying this aspect of the LCMV infection, we noted that survivors of i.c. infection always carried the virus in their organs for a prolonged period of time (Thomsen *et al.*, unpublished). This suggested to us that virus clearance, which is another T-effector-cell–dependent parameter (see above), would also vary with the genetic makeup of the host. Since T-cell-mediated virus clearance clearly represents a more universal phenomenon than immune mediated meningitis, we decided to investigate: (1) whether the influence of MHC and non-MHC genes also applied with regard to virus elimination following intravenous (i.v.) infection, (2) if a genetic influence was indeed important, what the underlying mechanism would be, i.e., whether dissimilarities would primarily reflect differences in intrinsic virus growth characteristics in the mouse strain and/or differences in T-cell responder capability, and (3) whether variations in *in vivo* antiviral T-cell efficiency would correlate with variations in other parameters of LCMV-specific T-cell activity, i.e., T_c activity and delayed-type hypersensitivity (T_d activity).

A. Evidence for Influence of Class I Genes and Background on Regulation of Virus Clearance

Comparing the rate of virus clearance in BALB/c and H-2 identical DBA/2 mice (Thomsen and Marker, 1989a), we found a marked difference in ability to eliminate LCMV (Traub strain). Thus, using a challenge dose of 2000 LD_{50} i.v., BALB/c mice had cleared the infection in about 2 weeks whereas similarly infected DBA/2 mice were still haboring virus at high titers in their organs at this time. Moreover, BALB/c mice effectively cleared virus doses of up to 10^4 LD_{50} whereas DBA/2 mice had difficulty clearing even 10 LD_{50} doses. Initial virus replication prior to appearance of the immune response was not found to differ significantly between the two strains, indicating that the difference in clearance did not reflect a difference in virus load. Consequently, we concluded that a difference in the antiviral immune response formed the basis for the difference in virus clearance. Challenge of F_1 hybrids produced by crossing the two strains revealed that rapid clearance was the dominant quality.

Similar results were obtained when comparing the rate of virus clearance in BALB/C-H-2^{dm2} mutants with that in matched BALB/c mice (Thomsen and Marker, 1989a). In the dose range 200–20,000 LD_{50} i.v., LCMV was cleared more slowly by the mutants. Further support for a role of class I genes in influencing virus clearance has been obtained by comparing clearance of Docile strain virus in B10.BR mice and B10.AKM (cf. the difference in outcome of i.c. infection); at 10 days postinfection significantly higher spleen titers were measured in B10.BR mice (Thomsen *et al.*, unpublished).

To confirm that at least two independent loci, one in the MHC and one or more outside of MHC, were involved in regulating virus clearance, we produced F_1 hybrids between DBA/2 and BALB/c-H-2^{dm2} and challenged those with LCMV. As expected if independent recessive genes formed the basis for the low responsiveness of these two strains, these hybrids cleared LCMV as rapidly as did (BALB/cxDBA/2) F_1's (Thomsen and Marker, 1989a).

Analysis of other parameters of the antiviral immune response revealed that delayed virus clearance correlated with poor T-cell responsiveness as measured in terms of virus-specific T_c and T_d activity. No correlation was observed between NK-cell activity and antibody response on the one hand and clearance rate on the other (Thomsen and Marker, 1989a).

B. Possible Mechanisms of Gene Control

Theoretically, the mechanisms by which host genes may influence clearance of LCMV are the same as those described for the influence of host genes on LCMV-induced immunopathology: (1) host genes may directly or indirectly affect the generation of antiviral effector T cells, and (2) host genes may modulate the biological effect of a given effector cell response.

In another viral model, the murine cytomegalovirus infection, MHC genes have been found to affect the course of infection by regulating virus binding or entry into the host cells (Price *et al.*, 1990). This may lead to an increased virus load and also affect the level of T-cell responsiveness. However, in the LCMV model no evidence suggesting this kind of mechanism for the influence of MHC genes have been obtained. On the contrary, we found that initial virus replication appeared to progress at similar speed in BALB/cH-2^{dm2} and wild-type BALB/c controls (Thomsen and Marker, 1989a). Furthermore, since BALB/c-H-2^{dm2} represent a loss mutation, it is hard to visualize how such a mechanism could be working here. Also, considering that class I restricted, CD8+ T cells are the mediators of virus clearance in the LCMV system (Zinkernagel and Welsh, 1976; Moskophidis *et al.*, 1987), the observed class I influence most likely reflects direct regulation of T-cell responder status as already described for LCMV-induced immunopathology.

One may wonder why a single deletion mutation may so profoundly delay virus clearance when there are several class I genes and several potential epitopes available. However, studies in several viral systems indicate that the frequency of low-responder alleles at least for T_c responsiveness is relatively high (Bennink and Yewdell, 1988; Gomez *et al.*, 1989). Moreover, a recent study indicates that the immune response to a natural whole protein is primarily determined by responsiveness to a single immunodominant determinant, even though other epitopes may be recognized (Kojima *et al.*, 1988). Therefore, considering that LCMV replicates very fast and to high titers in the murine host, and by itself has immunosuppressive qualities, it not difficult to envision that lack of an immunodominant determinant might have marked consequences (Thomsen and Marker, 1989b).

With regard to the influence of non-MHC genes, it has previously been argued that these genes work by affecting preimmune virus spread. Thus more rapid dissemination would lead to an increased virus load and a depletion of the spleen of effector T cells through exhaustive recruitment to other organs (Zinkernagel *et al.*, 1976). Our studies have failed to support this explanation (Thomsen and Marker, 1989a). Organ titers in low-responder (DBA/2) mice were not found to be higher than in high responders (BALB/c) prior to appearance of effector T cells. Furthermore, immunosuppression postponed until just prior to the start of T-cell-mediated clearance (day 5 of infection) completely abolished the T-cell response and resulted in equally high titers being obtained in both strains (Thomsen, unpublished). Together these observations strongly indicate that the delayed clearance in DBA/2 versus BALB/c mice is the consequence of an inefficient immune response in the former rather than a reflection of an intrinsic difference in ability to support viral replication.

To further analyze the mechanism underlying the influence of non-MHC genes on virus clearance, adoptive transfer experiment were carried out (Thomsen *et al.*, manuscript in preparation). BALB/c cells were effective in mediating virus clearance in nude H-2 compatible recipients whereas DBA/2 cells had little or no effect. Similar results were obtained using recipients only compatible with the donors at D-end class I loci. This, together with the fact that depletion of CD8+ cells abolished antiviral activity, provides direct evidence that non-MHC genes may influence the ability to clear virus through an effect on the induction of class I restricted, CD8+ effector cells. However, the environment in which the primed T cells are working also appears to be important: BALB/c donor cells were much less effective in clearing virus from spleens of DBA/2 recipients ($\ll$O.25 log) than from nude syngeneic or D-end class I compatible recipients (>2 logs). Since virus loads did not differ between these groups of recipients, and no such difference evolved in untransplanted controls during the assay period (day 3->5 p.i.), it seems reasonable to conclude that something in the DBA/2 environment modifies the effector capacity of appropriately primed virus-specific T cells. Therefore non-MHC genes may also modulate the rate of virus clearance through an influence on the expression of antiviral activity.

The fact that poor clearance always correlated with poor T-cell responsiveness measured in terms of T_c and T_d activity adds further support to the assumption that the same, or closely related, subset(s) are mediating all three effects (Marker and Thomsen, 1987). It should be mentioned that in another report this correlation between T_c activity and virus clearance was not found (Lehmann-Grube *et. al.*, 1985). In view of our finding that clearance rate may be influenced by other factors than generation of effector T cells, this is perhaps not surprising, and our own recent results seem to support this interpretation. Thus, using a segregating backcross population to study the linkage between level of LCMV specific T_c/T_d activity and ability to clear the virus, we observed a strong, but not absolute, correlation between the responsiveness measured with regard to these three parameters (Thomsen *et al.*, unpublished observation). It should be added that all parameters showed a continuous distribution, which is consistent with our impression of a complex polygenic regulation.

VII. OTHER VARIABLES MODIFYING THE OUTCOME OF LCMV INFECTION

It should be clear from the preceding sections that host genotype markedly influences the outcome of LCMV infection in adult immunocompetent mice. However, not only the genetic makeup of the host but also the dose and strain of virus, and to some extent the external environment, are crucial in determining whether a potent T-cell response will be induced. This is a relevant fact that should kept in mind not only when trying to compare results from different laboratories. But perhaps more important, the modifying effect of factors unrelated to host genotype has clear bearing on the implications of the findings we obtain in the LCMV model (see further below).

It is a classic observation in the LCMV system that high doses of virus induce a state of T-cell anergy to the virus (Hotchin, 1971). This is perhaps most markedly illustrated by our observation that C3H mice infected i.c. with 10^2 LD_{50} of Traub strain virus all die; whereas only 10–15% of mice inoculated with 10^4 LD_{50} succumb to the infection (Marker *et al.*, 1985). The underlying mechanism is not really clear, but analysis of the T_c and T_d response suggests that following an initial burst of activity, differentiation from primed precursor to mature effector cell is insufficient for the upkeep of effector cell capacity (Lehmann-Grube *et al.*, 1982; Marker and Thomsen, 1987; Thomsen and Marker, 1989a). Unresponsiveness is rapidly reversible by transfer of the cells to an environment less heavily loaded with antigen, and generally anergy is temporary, as evidenced by the gradual elimination of virus from the host. Although active suppression has been invoked as an explanation for this phenomenon (Lehmann-Grube *et al.*, 1985), no evidence implicating a specific suppressor cell subset has been obtained (Marker and Thomsen, 1987). It is of interest that a major difference between mice infected with high and low doses of virus is the high titers of virus that are very rapidly attained in the spleen of high-dose-primed mice (Lehmann-Grube *et al.*, 1982). Since there is evidence that susceptibility to the suppressive influence of high-virus titers decreases as the immune response matures (as the specific T-cell clones expand?) (Dunlop and Blanden, 1977; Marker and Thomsen, 1986), one could imagine a race between immune response and virus replication as the important determining factor underlying high-dose unresponsiveness.

Besides dose, the virus strain is also important for the outcome of LCMV infection. This is very clearly illustrated by the marked difference in susceptibility of many mouse strains to two plaque variants of the UBC strain (Zinkernagel *et al.*, 1985a,b). Thus, whereas mice of most strains succumb to i.c. infection with the Aggresive variant, only mice of some strains die with typical LCM disease when infected with Docile virus (as already mentioned above). The influence of virus strain on outcome of infection may also, in part, be a reflection of the influence of dose. At least in one case clear correlation has been established between ability of the virus isolate to multiply rapidly in the host and ability to reduce T-cell responsiveness (Pfau *et al.*, 1982b). Furthermore, it seems to be a general observation that at low doses of virus, differences between virus strains may more or less disappear (Pfau *et al.*, 1982b; Zinkernagel *et al.*, 1985b). In this context it is also of interest that we have noted a tendency to find the most clear-cut differences between mouse strains using relatively high doses of virus (e.g., Thomsen *et al.*, 1987). This observation parallels the fact that Docile LCMV, which multiplies very rapidly in the host, appears better in revealing an MHC influence on the immune response than does the same dose of Aggressive virus (Zinkernagel *et al.*, 1985a). Furthermore, using the Armstrong strain, which replicates only minimally in the viscera, Allan and Doherty (1985) found no more than a marginal difference in the susceptibility to fatal LCM disease of BALB/c ($K^dD^dL^d$) versus BALB/c-H-2^{dm2} ($K^dD^dL^-$) mice. On the other hand, they observed a more pronounced difference using the WE strain, which multiplies to high titers in many organs. A difference in the ability to multiply in the new host may also be part of the reason why spleen-derived plaque variants are more effective in causing immunosuppression and persistent infection than are plaque variants isolated from the brain of the same carrier animals (Ahmed and Oldstone, 1988; Ahmed *et al.*, 1984). Thus ability to replicate well in cells of the lymphoid tissues (lymphocytes or macrophages) *in vitro* has recently been associated with ability to cause persistent infection *in vivo* (King *et al.*, 1990).

Finally, even external environmental factors appear to exert some influence on the responses induced. Using the B10 congenics B10.AKM and B10.BR one of us (A.R.T.) finds 100% and about 60% mortality, respectively, when mice are infected i.c. with 600 p.f.u. of Docile LCMV. In contrast, 100% and 10–20% mortality was observed in same mouse strains when tested in another laboratory using the identical virus stock. Since the mice in both cases were obtained from the same source (the Jackson laboratories) and only bred locally for 1 or 2 generations, genetic drift appears an unlikely explanation for this phenomenon. Therefore, this finding suggests that environmental factors may significantly modulate the influence of MHC genes, and probably of host genes in general.

VIII. IMPLICATIONS FOR UNDERSTANDING OF MHC–DISEASE ASSOCIATION

The most intriguing feature of MHC genes is their uniquely high degree of polymorphism. Other polymorphic gene systems possess only a few alleles and the products differ only by one or a couple of amino acids. This is in contrast to the MHC system, where there may be 50–100 alleles identified for each locus. Further, individual alleles may differ by as much as one-fifth of their amino acid sequence (Steinmetz and Hood, 1983). Considering that the differences are found mainly in and around the site that serves to present peptide fragments to the T cells, it appears very likely that this polymorphism has been established for a purpose (Nagy *et al.*, 1989). An adaptive significance is also suggested by the fact that extensive polymorphism occurs in birds as well as mammals and thus seems to have been maintained for a long period of evolution (Klein, 1986). Finally, the distribution of allele frequencies at the HLA-A and -B locus deviates significantly from that expected if the alleles were neutral (Hedrick and Thomson, 1983).

Of the functions ascribed to the MHC, an influence on immune responses to pathogenic microorganisms seems the most likely to explain the polymorphism. Certainly infectious diseases tend to strike at an early age and therefore represent a strong selection pressure since they may eliminate individuals prior to or during reproductive age. At the level of T-cell recognition *in vitro* there is ample evidence to assume that both class I and II genes may serve as Ir genes (Katz and Benacerraf, 1975; Zinkernagel and Doherty, 1979; Nagy *et al.*, 1981; de Waal *et al.*, 1983). However, in order to extrapolate from such *in vitro* findings to the *in vivo* situation, information is required about the consequences of Ir-gene defects *in vivo*. In general, it has been the experience that although genetically determined differences in susceptibility to various pathogens may readily be found between strains of laboratory mice, the differences rarely correlate with MHC (Klein, 1986). This, together with the fact that only rare and weak HLA-disease associations have been described when it comes to infectious diseases, has led to some doubt about the importance of pathogens as a major evolutionary force in maintaining MHC polymorphism (Andersson *et al.*, 1987). Therefore, what has been lacking is clear evidence linking together MHC genes, acting in their capacity as Ir genes, and a changed biological outcome. Furthermore, previous failures needs to be explained in order to revive the subject.

In our analysis of the murine LCMV model we have found that a number of factors interact in determining the outcome of infection. With regard to host genes, non-MHC genes play a dominant role in affecting the outcome of LCMV infection, but given well-defined conditions a significant influence of H-2 was observed. The influence of MHC genes was, however, modulated by environmental factors and virus strain and dose. This seems to be the rule whether lethal immune-mediated disease or rapid virus clearance is taken as end-point. Considering that in both cases the outcome of infection is believed to reflect rather directly the action of class I restricted CD8+ effector T cells, and still this complex picture appears, it is not surprising that most models, where the pathogenesis may be even more complicated, may not provide useful information. Furthermore, looking at the human population there are two major obstacles to finding an MHC influence on susceptibility to infectious diseases. First, it would be impossible in a human study to control all the variables, based on what we have learned in the model system, that might modulate the influence of MHC. For that reason alone susceptibility genes would be extremely difficult to identify. Second, humans have been subject to major epidemics for at least 5000 years. This means that low-responder alleles have been selected against for at least the same amount of time. Consequently, we are now left with genes that have withstood the test, and only under exceedingly rare sets of conditions (special background genes, a special virus variant, etc.) may all the alleles in an individual be associated with a poor T-cell response. Statistically it would be almost impossible to establish the influence of MHC on susceptibility to infectious diseases (Serjeantson, 1983). This is in contrast to the situation with autoimmune diseases, where sometimes quite strong HLA–disease associations may be found. But, in this context it should be remembered that high-responder genes for autoimmune responses are few, since obviously there has been no selection eliminating low-responder alleles, so that they stand out much more clearly.

The LCMV model may therefore provide us with an almost unique tool for studying the importance of Ir genes for T-cell function in the *in vivo* setting. The data indicate that Ir genes do function at the level of effector T-cell generation *in vivo* and in this way may affect the outcome of infection; i.e., Ir genes are biologically relevant. Thus, our findings give credence to the idea that MHC genes, because of this function, may be subject to selection pressure. Although this does not prove that pathogens are a primary evolutionary force in maintaining MHC polymorphism (Potts *et al.*, 1991), it certainly provides us with the conceptual framework needed to explain the mechanism that could be at work.

ACKNOWLEDGMENTS. The authors thank Dr. O. Marker for constructive criticism of the manuscript and Dr. Y. Eyler for suggesting redactic improvements. This work was supported in part by the Danish Medical

Research Council, the Gerda and Aage Haensch Foundation, and the Novo Foundation.

REFERENCES

Ahmed, R., and Oldstone, M. B. A., 1988, Organ-specific selection of viral variants during chronic infection, *J. Exp. Med.* **167**:1719.

Ahmed, R., Salmi, A., Butler, L. D., Chiller, J. M., and Oldstone, M. B. A., 1984, Selection of genetic variants of lymphocytic choriomeningitis virus in spleens of persistently infected mice, *J. Exp. Med.* **60**:521.

Ahmed, R., Butler, L. D., and Bhatti, L., 1988, T4 T helper cell function *in vivo:* Differential requirement for induction of antiviral cytotoxic T-cell and antibody responses, *J. Virol.* **62**:2102.

Allan, J. E., and Doherty, P. C., 1985, Consequences of a single Ir-gene defect for the pathogenesis of lymphocytic choriomeningitis, *Immunogenetics* **21**:581.

Allan, J. E., Dixon, J. E., and Doherty, P. C., 1987, Nature of the inflammatory process in the central nervous system of mice infected with lymphocytic choriomeningitis virus, *Curr. Top. Microbiol. Immunol.* **134**:131.

Andersen, I. H., Marker, O., and Thomsen, A. R., 1991, Breakdown of blood–brain barrier function in the murine lymphocytic choriomeningitis virus infection mediated by virus-specific CD8+ T cells, *J. Neuroimmunol.* **31**:155.

Andersson, L., Paabo, S., and Rask, L., 1987, Is allograft rejection a clue to the mechanism promoting MHC polymorphism, *Immunol. Today* **8**:206.

Arora, P. K., and Shearer, G. M., 1981, Non-MHC-linked genetic control of murine cytotoxic T lymphocyte responses to hapten-modified syngeneic cells, *J. Immunol.* **127**:1822.

Baenzinger, J., Hengartner, H., Zinkernagel, R. M., and Cole, G. A., 1986, Induction or prevention of immunopathological disease by cloned cytotoxic T cell lines specific for lymphocytic choriomeningitis virus, *Eur. J. Immunol.* **16**:387.

Bennink, J. R., and Yewdell, J. W., 1988, Murine cytotoxic T lymphocyte recognition of individual influenza virus proteins. High frequency of nonresponder MHC class I alleles, *J. Exp. Med.* **168**:1935.

Brinton, M. A., and Nathanson, N., 1981, Genetic determinants of virus susceptibility: Epidemiologic implications of murine models, *Epidemiol. Rev.* **3**:115.

Byrne, J. A., and Oldstone, M. B. A., 1984, Biology of cloned cytotoxic T lymphocytes specific for lymphocytic choriomeningitis virus: clearance of virus *in vivo, J. Virol.* **51**:682.

Bukowski, J. F., Woda, B. A., Habu. S., Okumura, K., and Welsh, R. M., 1983, Natural killer cell depletion enhances virus synthesis and virus-induced hepatitis *in vivo, J. Immunol* **131**:1531.

Buller, R. M. L., Holmes, K. L., Hagin, A., Frederickson, T. N., and Morse, H. C., 1987, Induction of cytotoxic T-cell responses *in vivo* in the absence of CD4 helper cells, *Nature* **328**:77.

Cerny, A., Huegin, A. W., Sutter, S., Bazin, H., Hengartner, H. H., and Zinkernagel, R. M., 1986, Immunity to lymphocytic choriomeningitis virus in B cell-depleted mice: evidence for B cell and antibody independent protection by memory T cells, *Eur. J. Immunol.* **16**:913.

Christoffersen, P. J., Volkert, M., and Rygaard, J., 1976, Immunological unresponsiveness of nude mice to LCM virus infection, *Acta Pathol. Microbiol. Scand. Sec. B* **84**:520.

Clatch, R. J., Melvold, R. W., DalCanto, M. C., Miller, S. D., and Lipton, H. L., 1987, The Theiler's murine encephalomyelitis virus (TMEV) model for multiple sclerosis shows a strong influence of the murine equivalents of HLA-A,B, and C, *J. Neuroimmunol.* **15**:121.

Cole, G. A., Nathanson, N., and Prendergast, R. A., 1972, Requirement for theta-bearing cells in lymphocytic choriomeningitis virus-induced central nervous system disease, *Nature* **238**:335.

Conta, B. S., Powell, M. B., and Ruddle, N.H., 1985, Activation of Lyt-1+ and Lyt-2+ T cell cloned lines: stimulation of proliferation, lymphokine production and self-destruction, *J. Immunol.* **134**:2185.

Datta, S. K., Tsichlis; P., Schwartz, R. S., Chattophadhyay, S. K., and Melief, C. J. M., 1978, Genetic difference unrelated to H-2 in H-2 congenic mice, *Immunogenetics* **7**:359.

de Waal, L. P., Kast, W. M., Melvold, R. W., and Melief, C. J. M., 1983, Regulation of the cytotoxic T-lymphocyte response against Sendai virus analyzed with H-2 mutants, *J. Immunol.* **130**:1090.

Doherty, P. C., and Zinkernagel, R. M., 1974, T-cell-mediated immunopathology in viral infections, *Transplant. Rev.* **19**:89.

Doherty, P. C., and Zinkernagel, R. M., 1975, Capacity of sensitized thymus-derived lymphocytes to induce fatal lymphocytic choriomeningitis is restricted by the H-2 complex, *J. Immunol.* **114**:30.

Doherty, P. C., Blanden, R. V., and Zinkernagel, R. M., 1976, Specificity of virus-immune effector T cells for H-2K or H-2D compatible interactions, *Transplant. Rev.* **29**:89.

Doherty, P. C., Biddison, W. E., Bennink, J. R., Knowles, B. B., 1978, Cytotoxic T cell responses in mice infected with influenza and vaccinia virus vary in magnitude with H-2 type, *J. Exp. Med.* **148**:534.

Dunlop, M. B. C., and Blanden, R. V., 1977, Mechanisms of suppression of cytotoxic T cell responses in murine lymphocytic choriomeningitis virus infection, *J. Exp. Med.* **145**:1131.

Ellman, L., Green, I., Martin, W. J., and Benacerraf, B., 1970, Linkage between the PLL gene and the locus controlling the major histocompatibility antigen in strain 2 guinea pigs, *Proc. Natl. Acad. Sci. USA* **66**:322.

Epstein, R., Sham, G., Womack, J., Yague, J., Palmer, E., and Cohn, M., 1986, The cytotoxic T cell response to malespecific histocompatibility antigen (H-Y) is controlled by two dominant immune response genes, one in the MHC, the other in the Tar α-locus, *J. Exp. Med.* **163**:759.

Eyler, Y. L., Pfau, C. J., Broomhall, K. S., and Thomsen, A. R., 1989, The combination of MHC and non-MHC genes influence murine LCM virus pathogenesis, *Scand. J. Immunol.* **29**:527.

Fiertz, W., Brenan, M., Mullbacher, A., and Simpson, E., 1982, Non-H-2 and H-2-linked immune response genes control the cytotoxic T-cell response to H-Y, *Immunogenetics* **15**:261.

Frelinger, J. A., Orn, A., Brayton, P. R., and Hood, L., 1983, Use of cloned H-2 genes for study of H-2-restricted cytotoxicity: L^d is the LCMV restriction element for $H\text{-}2^d$, *Transplant. Proc.* **15**:2024.

Gallo, R. C., 1987, The AIDS Virus, *Sci. Am.* **256**:47.

Gilden, D. H., Cole, C. A., and Nathanson, N., 1972, Immunopathogenesis of acute central nervous system disease produced by lymphocytic choriomeningitis virus. II. Adoptive immunization of virus carriers, *J. Exp. Med.* **135**:874.

Gomez, A., Bourgault, I., Gomard, E., Picard, F., and Levy, J-P., 1989, Role of different lymphocyte subsets in human antiviral T cell cultures, *Cell. Immunol.* **118**:312.

Haywood, A., 1986, Patterns of persistent virus infections, *N. Engl. J. Med.* **315**:939.

Hedrick, P. W., and Thomson, G., 1983, Evidence for balancing selection at HLA, *Genetics* **104**:449.

Hill, A. V. S., Allsopp, C. E. M., Kwiatkowski, D., Anstey, N. M., Twumasi, P., Rowe, P. A., Bennett, S., Brewster, D., McMichael, A. J., and Greenwood, B. M., 1991, Common West African HLA antigens are associated with protection from severe malaria, *Nature* **352**:595.

Hirsch, M. S., Murphy, F. A., Russe, H. P., and Hicklin, M. D., 1967, Effects of antithymocyte serum on lymphocytic choriomeningitis (LCM) virus infection in mice, *Proc. Soc. Exp. Biol. Med.* **125**:980.

Hotchin, J., 1971, Persistent and slow virus infections, in: *Monographs in Virology* Vol. 3, (J. L. Melnick, ed.), S. Karger, Basel.

Hotchin, J., and Weigand, H., 1961a, Studies on lymphocytic choriomeningitis virus in mice. I. The relationship between age at inoculation and outcome of infection, *J. Immunol.* **86**:392.

Hotchin, J., and Weigand, H., 1961b, The effects of pretreatment with X-rays on the pathogenesis of lymphocytic choriomeningitis in mice. I. Host survival, virus multiplication and leukocytosis, *J. Immunol.* **87**:675.

Hurme, M., Chandler, P. R., Hetherington, C. M., and Simpson, E., 1978, Cytotoxic T cell responses to H-Y: Mapping of the Ir genes, *J. Exp. Med.* **147**:758.

Jacobson, S., and Pfau, C. J., 1980, Viral pathogenesis and resistance to defective interfering particles, *Nature* **238**:311.

Johnson, E. D., Monjan, A. A., and Morse III, H.C., 1978, Lack of B-cell participation in acute lymphocyte choriomeningitis disease of the central nervous system, *Cell. Immunol.* **36**:143.

Katz, D. H., and Benacerraf, B., 1975, The function and interrelationship of T-cell receptors, Ir genes and other histocompatibility gene products, *Transplant. Rev.* **22**:175.

Klein, J., 1986, *Natural History of the Major Histocompatibility Complex*, Wiley, New York.

King, C-C., de Fries, R., Kolhekar, S. R., and Ahmed, R., 1990, *In vivo* selection of lymphocyte-tropic and macrophage-tropic variants of lymphocytic choriomeningitis virus during persistent infection, *J. Virol.* **64**:5611.

Kojima, M., Cease, K. B., Buckenmeyer, G. K., and Berzofsky, J. A., 1988, Limiting dilution comparison of high and low responder MHC-restricted T cells, *J. Exp. Med.* **167**:1100.

Lehmann-Grube, F., 1975, Discussion, *Bull. WHO* **52**:485.

Lehmann-Grube, F., Cihak, J., Varho, M., and Tijerina, R., 1982, The immune response of the mouse to lymphocytic choriomeningitis virus. II. Active suppression of cellmediated immunity by infection with high virus doses, *J. Gen. Virol.* **58**:223.

Lehmann-Grube, F., Assmann, U., Loliger, C., Moskophidis, D., and Lohler, J., 1985, Mechanism of recovery from acute virus infection. I. Role of T lymphocytes in clearance of lymphocytic choriomeningitis virus from spleens of mice, *J. Immunol.* **134**:608.

Leist, T. P., Aguet, M., Hassig, M., Pevear, D. C., Pfau, C. J., and Zinkernagel, R. M., 1987a, Lack of correlation between serum titres of interferon α, β, natural killer cell activity and clinical susceptibility in mice infected with two isolates of lymphocytic choriomeningitis virus, *J. Gen. Virol.* **68**:2213.

Leist, T. P., Cobbold, S. P., Waldmann, H., Aguet, M., and Zinkernagel, R. M., 1987b, Functional analysis of T lymphocyte subsets in antiviral host defense, *J. Immunol.* **138**:2278.

Lilly, F., Boyse, E. A., and Old, L. J., 1964, Genetic basis of susceptibility to viral leukemogenesis, *Lancet* **2**:1207.

MacDonald, H. R., Ceredig, R., Cerottini, J.-C., Kelso, A., and Glasebrook, A. L., 1983, Heterogeneity of lymphokine production by T lymphocytes: Analysis of established clones and primary limiting dilution microcultures, in: *Progress of Immunology*, Vol. V (Y. Todaro and T. Tada, eds.), pp. 247–258, Academic Press, Japan.

Marker, O., and Thomsen, A. R., 1986, T-cell effector function and unresponsiveness in the murine lymphocytic choriomeningitis virus infection. I. On the mechanism of a selective suppression of the virus-specific delayed-type hypersensitivity response, *Scand. J. Immunol.* **24**:127.

Marker, O., and Thomsen, A. R., 1987, Clearance of virus by T lymphocytes mediating delayed type hypersensitivity, *Curr. Top. Microbiol. Immunol.* **134**:145.

Marker, O., and Volkert, M., 1973, Studies on cell-mediated immunity to lymphocytic choriomeningitis virus in mice, *J. Exp. Med.* **137**:1511.

Marker, O., Nielsen, M. H., and Diemer, N. H., 1984, The permeability of the blood–brain barrier in mice suffering from fatal lymphocytic choriomeningitis virus infection, *Acta Neuropathol. (Berl.)* **63**:229.

Marker, O., Thomsen, A. R., Volkert, M., Hansen, B. L., and Clemmensen, I. N., 1985, High-dose survival in the lymphocytic choriomeningitis virus infection is accompanied by suppressed DTH but unaffected T-cell cytotoxicity, *Scand. J. Immunol.* **21**:81.

Marrack, P., and Kappler, J., 1987, The T-cell receptor, *Science* **238**:1073.

McDevitt, H. O., and Chinitz, A., 1969, Genetic control of antibody response: Relationship between immune response and histocompatibility (H-2) type, *Science* **163**:1207.

McDevitt, H. O., Deak, B. D., Shreffler, D. C., Klein, J., Stimpfling, J. H., and Snell, G. D., 1972, Genetic control of the immune response. Mapping of the Ir-1 locus, *J. Exp. Med.* **135**:1259.

Melvold, R. W., 1986, Inbred, congenic, recombinant-inbred and mutant mouse strains, in: *Handbook of Experimental Immunology*, Vol. 3. *Genetics and Molecular Immunology* (D. M. Weir, ed.), pp. 106.1–106.20, Blackwell Scientific Publications, Oxford.

Mims, C. A., and Blanden, R. V., 1972, Antiviral action of immune lymphocytes in mice infected with lymphocytic choriomeningitis virus, *Infect. Immun.* **6**:695.

Mizuochi, T., Hugin, A. W., Morse, H. C., Singer, A., and Buller, R. M. L., 1989, Role of lymphokine-secreting CD8+ T cells in cytoxic T lymphocyte responses against vaccinia virus, *J. Immunol.* **142**:270.

Moskophidis, D., Cobbold, S. P., Waldmann, H., and LehmannGrube, F., 1987, Mechanism of recovery from acute virus infection: treatment of lymphocytic choriomeningitis virus-infected mice with monoclonal antibodies reveals that Lyt-2+ T lymphocytes mediate clearance of virus and regulate the antiviral antibody response, *J. Virol.* **61**:1867.

Mullbacher, A., Brenan, M., and Bowern, N., 1983, The influence of non-MHC genes on the cytotoxic T-cell response to modified self, *Aust. J. Exp. Biol. Med. Sci.* **61**:57.

Nagy, Z. A., Baxevanis, C. N., Ishii, N., and Klein, J., 1981, Ia antigens as restriction molecules in Ir-gene-controlled T-cell proliferation, *Immunol. Rev.* **60**:59.

Nagy, Z. A., Lehmann, P. V., Falcioni, F., Muller, S., and Adorni, L., 1989, Why peptides? Their possible role in the evolution of MHC-restricted T-cell recognition, *Immunol. Today* **10**:132.

Neustadt, P. M., Cody, T. S., and Monjan, A. A., 1978, Failure to find H-2-associated susceptibility to LCM disease, *J. Immunogenetics* **5**:397.

Oldstone, M. B. A., 1975, Relationship between major histocompatibility antigens and disease, *Bull. WHO* **52**:479.

Oldstone, M. B. A., and Dixon, F. J., 1968, Susceptibility of different mouse strains to lymphocytic choriomeningitis virus, *J. Immunol.* **100**:355.

Oldstone, M. B. A., Dixon, F. J., Mitchell, G. F., and McDevitt, H. O., 1973, Histocompatibility-linked genetic control of disease susceptibility, *J. Exp. Med.* **137**:1201.

Orn, A., Goodenow, R. S., Hood, L., Brayton, P. R., Woodward, J. G., Harmon, R. C., and Frelinger, J. A., 1982, Product of a transferred H-2L^d gene acts as a restriction element for LCMV-specific killer T cells, *Nature* **297**:415.

Pevear, D. C., and Pfau, C. J., 1989, Lymphocytic choriomeningitis virus, in: *Clinical and Molecular Aspects of Neurotropic Virus Infections* (D. H. Gilden and H. L. Lipton, eds.), pp. 141–172, Kluwer Academic Publishers, New York.

Pfau, C. J., Valenti, J. K., Jacobson, S., and Pevear, D. C., 1982a, Cytotoxic T cells are induced in mice infected with lymphocytic choriomeningitis virus strains of markedly different pathogenicities, *Infect. Immun.* **36**:598.

Pfau, C. J., Valenti, J. K., Pevear, D. C., and Hunt, K. D., 1982b, Lymphocytic choriomeningitis virus killer cells are lethal only in weakly disseminated murine infections, *J. Exp. Med.* **156**:79.

Pfau, C. J., Gresser, I., and Hunt, K. D., 1983, Lethal role of interferon in lymphocytic choriomeningitis virus-induced encephalitis, *J. Gen. Virol.* **64**:1827.

Pfau, C. J., Saron, M-F., and Pevear, D. C., 1985, Lack of correlation between cytotoxic T lymphocytes and lethal murine lymphocytic choriomeningitis, *J. Immunol.* **135**:597.

Potts, W. K., Manning C. J., and Wakeland, E. K., 1991, Mating patterns in seminatural populations of mice influenced by MHC genotype, *Nature* **352**:619.

Price, P., Gibbons, A. E., and Shellam, G. R., 1990, H-2 class I loci determine sensitivity to MCMV in macrophages and fibroblast, *Immunogenetics* **32**:20.

Prystowsky, M. B., Ely, J. M., Beller, D. I., Eisenberg, L., Goldman. J., Goldman, M., Goldwasser, E., Ihle, J., Quitans, J., Remold, H., Vogel, S. N., and Fitch, F. W., 1982, Alloreactive cloned T cell lines. VI. Multiple lymphokine activities secreted by helper and cytolytic cloned T lymphocytes, *J. Immunol.* **129**:2337.

Roopenian, D. R., and Anderson, P. S., 1988, Generation of helper cell-independent cytotoxic T lymphocytes is dependent upon L3T4 helper T cells, *J. Immunol.* **141**:391.

Roost, H., Charan, S., Gobet, R., Ruedi, E., Hengartner, H., Althage, A., and Zinkernagel, R. M., 1988, An acquired immune suppression in mice caused by infection with lymphocytic choriomeningitis virus, *Eur. J. Immunol.* **18**:511.

Rowe, W. P., 1954, Studies on pathogenesis and immunity in lymphocytic choriomeningitis infection of the mouse, *Res. Rep. Naval Med. Res. Inst., Bethesda, Md.* **12**:167.

Rowe, W. P., 1956, Protective effect of pre-irradiation on lymphocytic choriomeningitis infection in mice, *Proc. Soc. Exp. Biol. Med.* **92**:194.

Rowe, W. P., Black, P. H., and Levey, R. H., 1963, Protective effect of neonatal thymectomy on mouse LCM infection, *Proc. Soc. Exp. Biol. Med.* **114**:248.

Schwendemann, G., Lohler, J., and Lehmann-Grube, F., 1983, Evidence for cytotoxic T-lymphocyte–target cell interaction in brains of mice infected with lymphocytic choriomeningitis virus, *Acta Neuropathol.* **61**:183.

Serjeantson, S. W., 1983, HLA and susceptibility to leprosy, *Immunol. Rev.* **70**:89.

Shearer, G. M., Rehn, T. G., and Schmitt-Verhulst, A. M., 1976, Role of the murine major histocompatibility complex in the specificity of *in vitro* T cell-mediated lympholysis against chemically-modified autologous lymphocytes, *Transplant. Rev.* **29**:222.

Speiser, D. E., and Zinkernagel, R. M., 1990, Thymic MHC class I gene regulation of susceptibility to lymphocytic choriomenigitis, *Thymus* **16**:187.

Steinmetz, M., and Hood, L., 1983, Genes of the major histocompatibility complex in mouse and man, *Science* **222**:727.

Taylor, P. M., Wraith, D. C., Askonas, B. A., 1985, Control of immune interferon release by cytotoxic T-cell clones specific for influenza, *Immunology* **54**:607.

Thomsen, A. R., and Marker, O., 1989a, MHC and non-MHC genes regulate elimination of lymphocytic choriomeningitis virus and antiviral cytotoxic T lymphocyte and delayed-type hypersensitivity mediating T lymphocyte activity in parallel, *J. Immunol.* **142**:1333.

Thomsen, A. R., and Marker, O., 1989b, Class I gene regulation of haplotype preference may influence antiviral immunity *in vivo, Cell. Immunol.* **122**:365.

Thomsen, A. R., Marker, O., and Pfau, C. J., 1987, Different T_c response profiles are associated with survival in the murine lymphocytic choriomeningitis virus infection, *Scand. J. Immunol.* **25**:637.

Tiwari, J. L., and Terasaki, P. I., 1985, *HLA and Disease Associations*, Springer-Verlag, New York.

Vidovic, D., and Matzinger, P., 1988, Unresponsiveness to a foreign antigen can be caused by self-tolerance, *Nature* **336**:222.

Volkert, M., 1962, Studies on immunological tolerance to LCM virus. A preliminary report on adoptive immunization of virus carrier mice, *Acta Pathol. Microbiol. Scand.* **56**:305.

Volkert, M., and Hannover-Larsen, J., 1965, Immunological tolerance to viruses, *Prog. Med. Virol.* **7**:160.

Volkert, M., Marker, O., and Bro-Jorgensen, K., 1974, Two populations of T lymphocytes immune to the lymphocytic choriomeningitis virus, *J. Exp. Med.* **139**:1329.

Volkert, M., Bro-Jorgensen, K., Marker, O., Rubin, B., and Trier, L., 1975, The activity of T and B lymphocytes in immunity and tolerance to the lymphocytic choriomeningitis virus in mice, *Immunology* **29**:455.

von Boehmer, H., and Haas, W., 1979, Distinct Ir genes for helper and killer cells in the cytotoxic response to H-Y antigens, *J. Exp. Med.* **150**:1134.

Welsh, R. M., and Kiessling, R. W., 1980, Natural killer cell response to lymphocytic choriomeningitis virus, *Scand. J. Immunol.* **11**:363.

Zinkernagel, R. M., and Doherty, P. C., 1974, Restriction of *in vitro* lymphocytic choriomeningitis within a syngeneic or semiallogeneic system, *Nature (Lond.)* **248**:701.

Zinkernagel, R. M., and Doherty, P. C., 1979, MHC-restricted cytotoxic T cells: Studies on the biological role of polymorphic major transplantation antigens determining T cell restriction-specificity, function, and responsiveness, *Adv. Immunol.* **27**:51.

Zinkernagel, R. M., and Welsh, R. M., 1976, H-2 compatibility requirements for virus-specific T cell-mediated effector function *in vivo.* I. Specificity of T cells conferring antiviral protection against lymphocytic choriomeningitis virus is associated with H-2K and H-2D, *J. Immunol.* **17**:1495.

Zinkernagel, R. M., Dunlop, M. B. C., Blanden, R. V., and Doherty, P. C., and Shreffler, D. C., 1976, H-2 compatibility requirements for virus-specific T-cell-mediated cytolysis. Evaluation of the role of H-2I region and non-H-2 genes in regulating immune response, *J. Exp. Med.* **144**:519.

Zinkernagel, R. M., Althage, A., Cooper, S., Kreeb, G., Klein, P. A., Sefton, B., Flaherty, L., Stimpfling, J., Shreffler, D., and Klein, J., 1978, Ir-genes in H-2 regulate generation of antiviral cytotoxic T cells: Mapping to K or D and dominance of unresponsiveness, *J. Exp. Med.* **148**:592.

Zinkernagel, R. M., Pfau, C. J., Hengartner, H., and Althage, A., 1985a, Susceptibility to murine lymphocytic choriomeningitis maps to class I MHC genes—A model for MHC/disease associations, *Nature* **316**:814.

Zinkernagel, R. M., Leist, T., Hengartner, H., and Althage, A., 1985b, Susceptibility to lymphocytic choriomeningitis virus isolates correlates directly with early and high cytotoxic T cell activity, as well as with footpad swelling reaction, and all three are regulated by H-2D, *J. Exp. Med.* **162**:2125.

CHAPTER 13

Molecular Anatomy of the Cytotoxic T-Lymphocyte Responses to Lymphocytic Choriomeningitis Virus

LINDA S. KLAVINSKIS, J. LINDSAY WHITTON, AND MICHAEL B. A. OLDSTONE

I. INTRODUCTION

Cell-mediated immunity (CMI) plays a cardinal role in the control and elimination of several viral infections, including the arenavirus lymphocytic choriomeningitis virus (LCMV) (Zinkernagel and Doherty, 1974, 1979; Buchmeier *et al.*, 1980). One manifestation of this response is the generation of cytotoxic T lymphocytes (CTL) that kill virus-infected target cells, thereby containing virus production and limiting the spread of infection. Although both humoral and cell-mediated responses are readily induced by the virus, it is clear that the induction of T cells is central in the recovery from the acute infection (Byrne and Oldstone, 1984; Cerny *et al.*, 1988; Moskophidis *et al.*, 1987) and for virus clearance in persistent/chronic infection (role of lymphoid cell: Volkert and Hannover Larsen, 1964, 1965; role of CTL: Oldstone *et al.*, 1986; Ahmed *et al.*, 1987).

LINDA S. KLAVINSKIS, J. LINDSAY WHITTON, AND MICHAEL B. A. OLDSTONE • Department of Neuropharmacology, Division of Virology, Research Institute of Scripps Clinic, La Jolla, California 92037. *Present address of L.S.K.:* Department of Immunology, Guy's Hospital Medical School, London, SE1 9RT, England.

The Arenaviridae, edited by Maria S. Salvato. Plenum Press, New York, 1993.

Much of the evidence for the role of CMI in arenavirus infections has come from studies of LCMV infection of mice. In this respect, the variety of inbred and congenic mouse strains has been invaluable in studying the MHC restriction of the CTL response and, in conjunction with recombinant DNA technology, has enabled a molecular analysis of the antiviral T-cell response to be carried out.

In this review the major emphasis will be a molecular description of the CTL response in LCMV infection. An overview of CMI in arenavirus infection is presented elsewhere. We will survey the CTL epitopes selected by the MHC in different strains of inbred mice; assess the biological function of these CTL epitopes, their role in antiviral immunity; and, finally, discuss the prospects for designing vaccines based on a molecular approach of identifying immunodominant T-cell epitopes.

II. ROLE OF CTL RESISTANCE IN LCMV INFECTION

A. Historical Background

The role of T-cell immunity in the pathogenesis of LCMV infection stems from the early observations of Traub (1936, 1960), Rowe (1954), and Hotchin (Hotchin and Benson, 1963; Hotchin and Weigand, 1961), who noted that peripheral inoculation of adult mice by the intraperitoneal (ip) or intravenous (iv) routes, resulted in an acute self-limiting infection with lasting immunity. Paradoxical at the time, intracerebral (ic) inoculation resulted in a fatal lymphocytic choriomeningitis and was thought to be associated with a "strong" immunological response to infection of the central nervous system. Lundstedt (1969) and Oldstone *et al.* (1969) were the first to demonstrate independently that lymphocytes taken from mice 6–9 days after infection with LCMV were cytotoxic for virus-infected target cells *in vitro.* Subsequent experiments showed that these cells were thymic in origin (Cole *et al.*, 1972; Gilden *et al.*, 1972) or eliminate virus from immunosuppressed mice infected by the peripheral route (Mims and Blanden, 1972; Zinkernagel and Welsh, 1976).

Following the observations of Zinkernagel and Doherty (1974) that T-cell-dependent functions are restricted by histocompatibility (H-2) antigen determinants, it was shown that effective clearance of LCMV by immune spleen cells *in vivo* required absolute H-2 compatibility between the donor and recipient (Zinkernagel and Welsh, 1976). Phenotypically, these cells were Thy-1^+, Lyt-2^+, and L3T4^- (Zinkernagel and Welsh, 1976; Moskophidis *et al.*, 1987). The conclusive evidence that virus-specific, H-2-restricted T cells with cytotoxicity *in vitro* could eliminate and maintain virus clearance from the spleens of acutely infected mice came finally from the adoptive transfer experiments of Byrne and Oldstone (1984) with cloned LCMV-specific CTL. These experiments, in conjunction with those of Cerney *et al.* (1988) examining virus clearance in B-cell-depleted mice, indicate that T cells alone efficiently clear infection with LCMV. However agammaglobulinemic or B-cell depleted mice clear LCMV a little less rapidly, suggesting that virus-specific antibodies are not mandatory, but may accelerate virus clearance in conjunction with a brisk CTL response.

B. CTL Response to Acute Infection

Infection of adult immunocompetent mice with 1×10^5 plaque-forming units (pfu) of LCMV Armstrong (LCMV ARM) induces a brisk virus-specific CTL response peaking between days 7 and 9 postinfection (Marker and Volkert, 1973; Cole *et al.*, 1973; Welsh and Zinkernagel, 1977). The appearance of the CTL response coincides with virus clearance (Marker and Volkert, 1973; reviewed by Buchmeier *et al.*, 1980). Adoptive transfer of immune spleen cells taken from syngeneic donors primed 7 days previously with LCMV effectively eliminates virus from acutely infected recipients (Mims and Blanden, 1972; Zinkernagel and Welsh, 1976; Lehmann-Grube *et al.*, 1985). The critical role played by CTL is underlined by the ablation studies of Moskophidis *et al.* (1987), using monoclonal antibodies to deplete CD4+ (helper/inducer) or CD8+ (cytotoxic/suppressor) T cell-subsets from mice acutely infected with LCMV. Whereas treatment with anti-CD4+ antibody had essentially no effect on virus clearance, inoculation with anti-CD8+ antibody impaired virus elimination (Moskophidis *et al.*, 1987). Adoptive transfer studies with cloned LCMV-specific H-2-restricted CTL have further demonstrated that CTL alone are sufficient to eliminate and terminate acute infection with LCMV (Byrne and Oldstone, 1984).

The use of recombinant vaccinia virus as an expression vector for foreign minigenes, and synthetic peptides, has allowed the epitope specificity of LCMV-specific CTL clones to be mapped (Whitton *et al.*, 1988a,b, 1989; Oldstone *et al.*, 1988). This has led to the finding that cloned CTL specific for either the surface glycoprotein (GP) or internal nucleoprotein (NP) of LCMV can either eliminate the acute infection or induce fatal choriomeningitis when adoptively transferred to LCMV-infected immunosuppressed mice (Klavinskis *et al.*, 1988, 1989a,b, 1990).

C. CTL Response to Persistent Infection

In contrast to the brisk CTL response during acute infection with LCMV, persistent LCMV infection is accompanied by low or undetectable levels of LCMV-specific CTL activity. Indeed, the establishment of persistent infection by LCMV can occur only in the absence of the primary CTL response. The virus can evade the generation of functional

CTL during acute infection (and consequently become persistent) in several ways. In nature, the most common mode is, presumably, congenital/neonatal infection (occurring while the CMI system is immature) establishing a state of "immunological T-cell tolerance" to the virus. Experimentally, immunosuppressed mice may be used as hosts; alternatively, viral variants that fail to elicit CTL can establish persistence in immunocompetent adult mice (Ahmed *et al.*, 1984b).

Mice infected *in utero* or neonatally with LCMV establish a lifelong persistent infection (Hotchin and Cinits, 1958). Such mice express, at best, low levels of functional LCMV-specific CTL (Cihak and Lehmann-Grube, 1978; Zinkernagel and Doherty, 1974; Oldstone *et al.*, 1986; Ahmed *et al.*, 1984b, 1987). CTL responses to several other viruses can be generated (Ahmed *et al.*, 1984b). The basis for this T-cell unresponsiveness to the virus is still unclear (Ahmed *et al.*, 1984b, 1987; Jamieson and Ahmed, 1988, 1989) but can be reconstituted by adoptive transfer of LCMV-specific immune memory cells of the Lyt-2^+, L3T4(CD8+) phenotype, i.e., CTL (Ahmed *et al.*, 1987; Jamieson and Ahmed, 1988). As a consequence, adoptive immunotherapy with LCMV immune spleen cells completely cures the persistent infection (Volkert and Hannover Larsen, 1964, 1965; Oldstone *et al.*, 1986; Ahmed *et al.*, 1987). In view of the critical role played by CTL in the outcome of LCMV infection, we have set out to study the molecular anatomy of the host CTL response. The following sections will discuss our understanding of T-cell recognition of LCMV; which virus proteins are involved; how they are restricted by the host MHC; and finally, whether we can manipulate these requirements to generate vaccines to confer protection from lethal infection.

III. T-CELL RECOGNITION

LCMV has a bisegmented RNA genome comprising a long (L) and short (S) segment (Bishop and Auperin, 1987). By the use of segmental reassortants between the Armstrong (ARM) and Pasteur (PAST) strains of LCMV both induction of CTL (in mice of the H-2^d haplotype) and target-cell recognition and lysis by CTL have been mapped to the virus S segment and the genes it encodes (Riviere *et al.*, 1986). The S segment of LCMV ARM has been cloned and sequenced (Southern *et al.*, 1987; Salvato *et al.*, 1988) and encodes two proteins: a 558-amino-acid NP and a 498-amino-acid precursor glycoprotein GP-C, which undergoes post-translational cleavage to generate the two mature structural glycoproteins, GP-1 (residues 1–262) and GP-2 (residues 263–498) (Buchmeier and Oldstone, 1979).

To estimate the relative CTL responses to the GP and NP and examine the role of anti-GP- and anti-NP-specific CTL in antiviral immunity to LCMV, we have expressed these proteins in vaccinia virus (VV). Using the transfer plasmid pSC11 (Chakrabarti *et al.*, 1985), the coding sequences for full-length NP and GP residues 1–363 followed by six plasmid-encoded amino acid residues preceding a termination codon have been cloned into vaccinia virus (Whitton *et al.*, 1988a). The full-length NP and GP residues 1–363 vectors were used as the parental plasmids, from which a family of serial carboxy-terminal truncations of the LCMV GP- or NP-specific probes reveals bands of approximately the expected sizes (Fig. 1).

A. Influence of the MHC on the Fine Specificity of the CTL Response

Ahmed *et al.* (1984a) first noted that CTL recognition of several LCMV strains varied in a host-dependent manner, suggesting that the variability mapped to the MHC locus. It was also inferred that the degree of cross-protective immunity induced by different strains of LCMV was associated with the H-2 haplotype (Ahmed *et al.*, 1984a).

The availability of recombinant VVs expressing either the GP or NP of LCMV has allowed us to evaluate the precise character of the anti-LCMV ARM CTL response in a variety of inbred strains of mice. Indeed on the H-2^b background, target cells expressing either the GP or NP are efficiently lysed by anti-LCMV ARM CTL, and thus on this haplotype, LCMV infection induces CTL against both these moieties (Whitton *et al.*, 1988a). However, on the H-2^d or H-2^q background we have found a dramatically altered pattern of CTL reactivity. On each of these haplotypes, anti-NP CTL are readily demonstrable, though anti-GP CTL are undetectable (Fig. 2), suggesting that anti-GP CTL are best, a minor component of the primary anti-LCMV ARM CTL response in H-2^d and H-2^q mice. Analysis of the primary LCMV WE specific CTL response also indicates that CTL from the H-2^d, H-2^q, and H-2^s haplotypes are almost exclusively NP-specific (Hany *et al.*, 1989). On the H-2^k background, recognition by primary anti-LCMV ARM- or anti-LCMV WE-specific CTL does not appear to be directed to the NP or GP, suggesting that H-2^k CTL recognize an epitope(s) encoded by the L RNA, expressing the polymerase or the newly discovered Z protein (Salvato and Shimomaye, 1989). These findings highlight the role of the MHC in selecting virus epitopes. Further they council caution in the administration of a single subunit vaccine. One would predict that vaccination with LCMV GP would induce a CTL response in H-2^b mice, whereas H-2^d, H-2^k, and H-2^q may remain unstimulated (see Section V of this chapter).

B. Target Epitopes of GP-Specific CTL

Analysis of the anti-LCMV CTL response in mice of the H-2^b haplotype, indicates that at least one-third of the CTL activity induced by

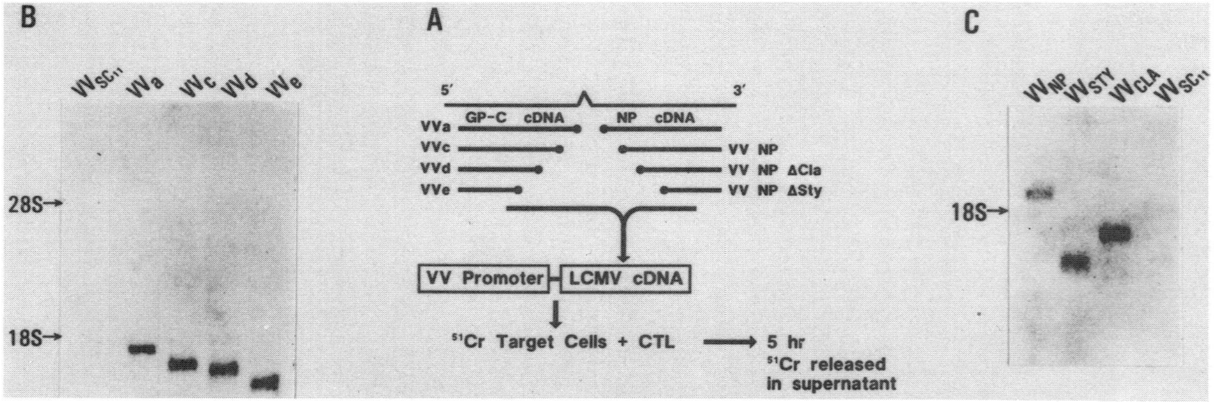

FIGURE 1. (A) Cartoon illustrating the generation of truncated LCMV ARM GP and NP minigenes expressed in VV. The GP gene was serially truncated from the 3′ end and the NP gene from the S′ end and attached to translational termination sequences (●). C-terminal truncated GP molecules comprising residues 1–363 (VVa), 1–293 (VVc), 1–271 (VVd), or 1–218(VVe) and C-terminal truncated NP molecules comprising residues 1–558 (VVNP), 1–321 (VVNPΔCla), or 1–201 (VVNPΔSty) were introduced into vaccinia virus by homologous recombination using the pSC11 transfer plasmid (Whitton *et al.*, 1988a,b, 1989). (B and C) Northern blot analysis of total cytoplasmic RNA from cells infected with VVSC11 (a control that contains no LCMV sequences) or with recombinant VVs expressing either truncated LCMV GP (B) or LCMV NP (C) minigenes and hybridized with either a GP-specific (B) or NP-specific probe (C), revealing bands of approximately the expected sizes.

Effector (anti LCMV ARM CTL)	E:T	LCMV ARM	VV Recomb GP	VV Recomb NP	VV_{wt}	UN	LCMV ARM
	Target:	$H2^{bb}$	$H2^{bb}$	$H2^{bb}$	$H2^{bb}$	$H2^{bb}$	$H2^{dd}$
$H2^{bb}$	50:1	[71]	[51]	[43]	15	15	13
	25:1	[49]	[32]	[28]	6	3	10
	Target:	$H2^{dd}$	$H2^{dd}$	$H2^{dd}$	$H2^{dd}$	$H2^{dd}$	$H2^{bb}$
$H2^{dd}$	50:1	[67]	4	[62]	4	2	6
	25:1	[55]	4	[54]	3	1	2
	Target:	$H2^{qq}$	$H2^{qq}$	$H2^{qq}$	$H2^{qq}$	$H2^{qq}$	$H2^{bb}$
$H2^{qq}$	50:1	[43]	5	[20]	5	2	2
	25:1	[29]	5	[13]	3	0	1

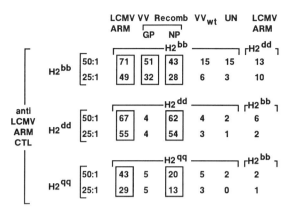

FIGURE 2. The host MHC determines which LCMV proteins will induce a CTL response. Effector cells were induced on three MHC backgrounds ($H\text{-}2^{bb}$, $H\text{-}2^{dd}$, and $H\text{-}2^{qq}$) by infection with LCMV Armstrong (anti-LCMV ARM CTL). Each effector population was assayed for cytoxicity against both syngeneic and allogeneic target cells at effector to target ratios of 50:1 and 25:1. Target cells were either uninfected (UN) or infected with LCMV ARM, VVGP (aa1–363), VVNP (aa1–558), or wild-type vaccinia virus (VV_{wt}) and labeled with ^{51}Cr prior to incubation with effector cells. Percentage ^{51}Cr release (i.e., cell lysis) considered significantly above background is boxed. For details see Byrne and Oldstone (1984) and Whitton *et al.* (1988a).

LCMV is directed to the GP (Whitton *et al.*, 1988a). This contrasts with the CTL response to viruses such as influenza, vesicular stomatitis (VSV), or respiratory syncytial (RSV), where the majority of the response is directed against "internal" viral proteins, with anti-GP responses either undetectable (e.g., against RSV; Bangham *et al.*, 1986) or a minor component. It is interesting to note that cross-reactive T-cell responses are generally associated with internal viral antigens, yet at the bulk splenocyte (Whitton *et al.*, 1988a) and cloned CTL (Whitton *et al.*, 1988b) levels, the anti-LCMV GP-specific CTL response restricted by $H\text{-}2^b$ appears to be broadly cross-reactive. CTL induced by serologically distinct strains of LCMV (PAST, Traub, and E-350) elicit CTL that efficiently lyse target cells infected with recombinant VV expressing LCMV ARM GP (Whitton *et al.*, 1988b).

The epitopes recognized by the $H\text{-}2^b$-restricted anti-GP CTL have been established from studies examining the recognition by primary and cloned CTL against a family of carboxy-terminal truncated LCMV GP molecules expressed by vaccinia (Whitton *et al.*, 1988b). At the clonal level, 17 of 18 independent $H\text{-}2^b$-specific CTL clones generated by Whitton *et al.* (1988b) were directed to the GP, with only a single clone specific for the NP. Using a library of recombinant VVs expressing truncated LCMV GP genes, two patterns of CTL recognition were found. The vast majority of these clones (15 or 17) mapped to an epitope in GP-2 positioned between residues 272 and 293 (Fig. 3; Oldstone *et al.*, 1988) defining a major CTL epitope at the clonal level restricted by $H\text{-}2^b$. The two residual clones recognized each of the serially truncated GP

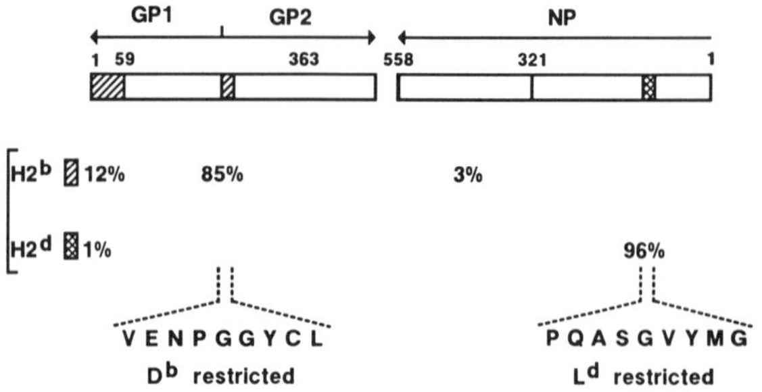

FIGURE 3. Cartoon illustrating the position and aa sequences of the immunodominant LCMV ARM epitope restricted by H-2^b (VENPGGYCL) or H-2^d (PQASGVYMG). The numbers represent the percent of CTL clones that recognize each epitope (17/18 for H-2^b, 48/50 for H-2^d). A single CTL clone derived from H-2^d mice consistently lysed LCMV ARM-infected targets but failed to lyse VVGP or VVNP targets. The specificity of this clone is unclear at present, but presumably recognizes a gene product encoded by the L RNA segment.

molecules, positioning at least one epitope between residues 1 and 59 in GP-1 (Fig. 3; Klavinskis *et al.*, 1990). In contrast, an analysis of 50 LCMV-specific CTL clones generated on the H-2^d background indicates that only 1 of 50 clones is directed to the GP and localized to a probable transmembrane region of GP-1 (Whitton *et al.*, 1989, unpublished observations). Thus, at both the clonal and primary CTL levels, the anti-GP response represents a very minor component of the overall H-2^d-restricted anti-LCMV CTL response.

C. Fine Specificities of GP Restricted CTL

Dissection of the GP recognition sites in mice of the H-2^b haplotype using a series of truncated LCMV GP molecules indicated that a major epitope at the clonal level required presentation of GP residues 272–293 for CTL recognition and target cell lysis (Whitton *et al.*, 1988b). Using a synthetic peptide encompassing the LCMV ARM GP aa272–293 sequence, it was directly demonstrated that this region contained a CTL epitope that directs killing by the appropriate H-2^b-restricted clones. This observation, in conjunction with those of Townsend *et al.* (1985, 1986) in the influenza virus system, and of Koszinowski and colleagues (DelVal *et al.*, 1988; Reddehase *et al.*, 1989) in the murine cytomegalovirus system, argue against any requirement for native antigen in CTL recognition and indicate that endogenous synthesis is not an absolute requirement for sensitization of target cells to CTL recognition and lysis *in vitro.*

Using a library of truncated peptides from GP aa272–293 to sensitize uninfected H-2^b target cells, the minimal viral epitope recognized by a family of H-2^b-restricted cloned LCMV ARM-specific CTL (Whitton *et al.*, 1988b) has been defined as GP278–286 (VENPGGYCL; Oldstone *et al.*, 1988). Analysis of target cells recombinant at the class I and class II MHC loci indicates that CTL recognition of the GP aa278–286 epitope is restricted to H-$2D^b$, and not to H-$2K^b$ or H-$2I^b$ (Oldstone *et al.*, 1988). Using the same approach, a second epitope on the H-2^b background has been minimally identified by amino acid residues GP34–40 (AVYNFAT) and is restricted by H-$2D^b$ (Klavinskis *et al.*, 1990).

Detailed study of the H-$2D^b$ clones specific for the 278–286 epitope indicate that individual CTL clones can distinguish single amino acid substitutions at position 278 of the sequence (Oldstone *et al.*, 1988; Fig. 4). For example, substitution of valine at position 278 with either large aromatic amino acids (phenylalanine or tyrosine) or amino acids with polar side chains (serine or threonine) can be used to segregate the clones into those that recognize only the latter (Fig. 4). This suggests that class I restricted CTL most likely recognize/view the same minimal epitope in a slightly different way, relating to different affinities of their T-cell receptors.

Side Chain Structure of Amino Acids Substituted at aa 278

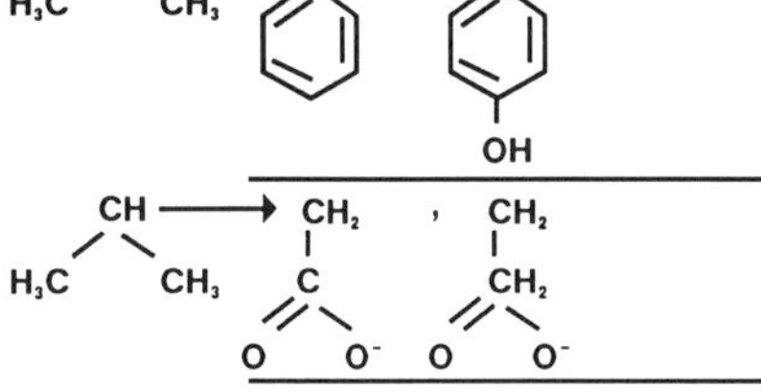

Sequence of Substituted Peptides (278)	% Specific Lysis by Clones 232	228	39	31	RG-9
...GVENPGGYCL..	100	100	100	100	100
L	100	87	99	97	80
I	100	100	61	75	84
A	100	100	100	78	67
G	38	38	ND	ND	ND
S	62	1	53	100	95
T	83	2	75	82	91
F	70	1	<1	79	67
Y	96	2	<1	34	23
D	16	2	13	21	10
E	43	1	10	20	21

FIGURE 4. CTL clones specific for LCMV GP aa278–286 (VENPGGYCL) can distinguish single amino acid substitutions at position 278. A series of peptides 272–293 were synthesized with a single amino acid substitution in position 278 where Val (V) was substituted for Leu (L), Ile (I), Ala (A), Gly (G), Ser (S), Thr (T), Phe (F), Tyr (Y), Asp (D), or Glu (E). Individual peptides at a final concentration of 200 μg/ml were added to 2×10^4 ^{51}Cr-labeled H-2^b or H-2^d uninfected target cells. Sensitized cells were incubated with either CTL clones 232, 228, 39, 31, or RG-9 (effector to target ratio 2.5:1) for 4–5 hr, and specific ^{51}Cr was quantitated. Specific lysis by CTL clones reacted with the native aa sequence (i.e., Val in position 278) on H-2^b targets was 40–60%.

D. Target Epitopes of NP-Specific CTL

The target epitopes of NP-specific CTL have been mapped in a similar manner, using recombinant VVs expressing truncated regions of the NP. In mice of the H-2^b haplotype, the NP determinant(s) recognized by bulk anti-LCMV ARM CTL have been localized to the carboxy end of the molecule, encoding residues 301–558 (Klavinskis *et al.*, 1988, 1989b; Whitton *et al.*, 1989). We have previously shown that H-2^b CTL specific for LCMV ARM NP can efficiently cross-react with other strains of LCMV (WE, PAST) and would hence predict that, were similar studies to be undertaken in the light of our findings, CTL epitopes would be found in the same region of these LCMV strains. These observations are in agreement with those in other viral systems, such as influenza (Townsend and Skehel, 1984; Yewdell *et al.*, 1985) and vesicular stomatitis virus (Yewdell *et al.*, 1986), wherein cross-reactive CTL responses are directed to internal viral proteins.

As mentioned previously (see Section III.A and III.B), the bulk of the CTL response on the H-2^d background is directed to the NP (Whitton *et al.*, 1988a, 1989). Using expression vectors encoding deleted LCMV NP molecules, a CTL epitope has been localized between LCMV NP residues 1 and 201 (Whitton *et al.*, 1989). The sequence comprising the core epitope (GVYMG) suggests that it is neither amphipathic nor an alpha helix and thus would not have been identified from an algorithm considering peptide conformation to predict antigenicity (DeLisi and Berzofsky, 1985; Margalit *et al.*, 1987). Similarly, the sequence lacks the linear motif of hydrophobic residues flanked by glycine, charged, or polar moieties, proposed by Rothbard and Taylor (1988) to predict antigenic sites in proteins. Thus simple sequence analysis would have failed to identify NP122–127 as a potential CTL epitope. Clearly, additional rules need to be incorporated to develop universally predictive schemes, though epitopes in other viral systems have been identified from predictive motifs (Gotch *et al.*, 1987; Reddehase *et al.*, 1989).

Both at the bulk splenic CTL and cloned CTL level, the H-2^d-restricted CTL response appears to be almost monospecific and directed to the NP122–127 epitope (Fig. 3). Analysis of a library of H-2^d-restricted LCMV ARM-specific CTL clones indicated that 48 of 50 clones were NP-specific (Whitton *et al.*, 1989). Finer analysis of a random selection of these clones demonstrated that they were all directed to the same region encompassing residues 116–129 (Whitton *et al.*, 1989). Although one cannot exclude the possibility that some degree of selection in favor of certain clones may have occurred during the cloning procedure, these results, in conjunction with those observed at the primary CTL level, indicated that on the H-2^d background > 95% of anti-LCMV ARM CTL are directed to a single domain of the NP. This is in contrast to the polyclonal B-cell response found on most proteins, including LCMV. As discussed in further detail elsewhere (see Section V), the epitope NP122–127 restricted by H-2^d is biologically relevant *in vivo*, conferring protection from lethal infection with LCMV (Klavinskis *et al.*, 1989b).

IV. CTL INDUCTION

Several studies have been directed toward identifying the minimal viral sequence requirements directing CTL recognition of LCMV (Klavinskis *et al.*, 1990; Oldstone *et al.*, 1988; Schulz *et al.*, 1991a; Whitton *et al.*, 1989). Currently four CTL recognition epitopes have been described, consisting of a minimum of five to nine amino acid residues. The critical issues that now need to be addressed are whether the minimal sequences identified for T-cell recognition are sufficient to direct CTL induction either as peptides or encoded as minigenes in expression vectors such as VV. Studies by Aichele *et al.* (1990), and Schulz *et al.* (1991b) with a 15-amino-acid peptide LCMV-WE NP118–132 that includes and extends from the immunodominant NP118–126 recognition site have shown that this peptide elicits CTL *in vivo* when administered in Freund's incomplete adjuvant. While this suggests that antiviral cytotoxic responses can be induced by synthetic peptides encoding CTL recognition sequences, it is unclear why only certain synthetic peptides have been shown to induce CTL successfully (e.g., Carbone and Bevan, 1989) while others do not. This dichotomy in ability of CTL recognition sequences to induce CTL is highlighted by the failure of the LCMV ARM GP278–286 epitope when contained as a minigene in VV to induce CTL *in vivo*, while when the epitope is incorporated in the normal LCMV ARM GP "backbone," it successfully elicits CTL (Klavinskis *et al.*, 1990). Since the epitope is readily recognized by LCMV ARM-specific CTL when contained as a minigene in VV GP272–293 (Whitton and Oldstone, 1989), this suggests that differences may exist in viral sequence requirements for CTL induction and recognition. Thus the "minimal" recognition sequences alone may not always be sufficient to induce a CTL response *in vivo*. Understanding the sequence constraints for inducing T cells may be critical when considering the design of subunit "CTL vaccines" to confer protective immunity. Furthermore, one may be able to maintain facets of the immune response and reconstitute the responsiveness to LCMV in the persistent infection.

A. Use of Expression Vectors to Induce CTL

Recombinant VV has been used as an expression vector to examine the ability of the NP and GP to induce CTL *in vivo*. In both H-2^b and

H-2[d] mice, a recombinant VV containing the full-length NP of LCMV ARM (VVNP1–558) has been shown to prime splenocytes for a secondary LCMV-specific CTL response following restimulation with LCMV ARM *in vitro* (Fig. 5; Klavinskis *et al.*, 1988, 1989b). In contrast, inoculation with VVGP1–363 (expressing GP-1 and the amino terminal end of GP-2 from LCMV ARM) fails to prime H-2[d] mice for a detectable secondary CTL response (Klavinskis *et al.*, 1988), though H-2[b] mice are primed to generate a vigorous secondary GP-specific CTL response (Klavinskis *et al.*, 1990; Fig. 5). These studies indicate that expression vectors encoding the NP or GP induce LCMV-specific CTL that mirror CTL induced during infection with LCMV. Additionally, these findings highlight the possibility of engineering LCMV sequences encoding CTL epitopes in VV, as a potential "subunit vaccine." With this objective in mind, a truncated LCMV ARM NP (VVNP1–201) encoding the immunodominant H-2[d]-restricted NP122–127 epitope has been shown to prime BALB/c (H-2[d]) mice for class I MHC-restricted memory CTL to the NP122–127 epitope (Klavinskis *et al.*, 1989b). Similarly, a truncated LCMV ARM GP (VV GP1–59) encoding an H-2[d]-restricted GP34–40

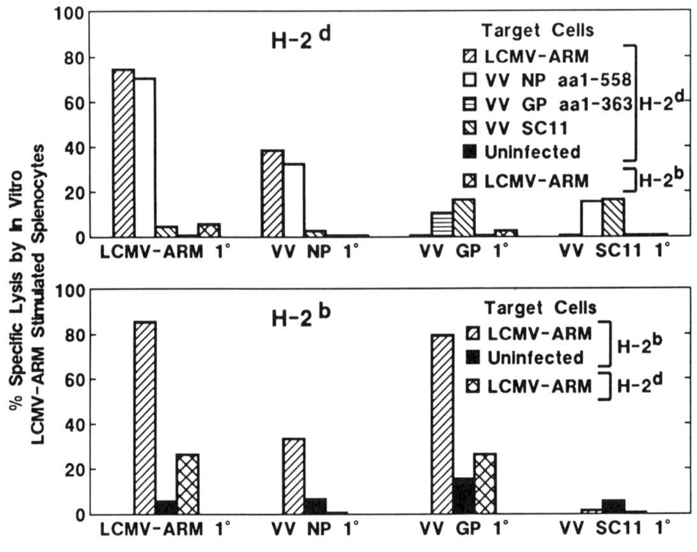

FIGURE 5. Immunization with expression vectors encoding LCMV NP (VVNPaa1–558) or LCMV GP (VVGPaa1–363) induces LCMV-specfic CTL that mirror CTL induced during infection with LCMV ARM. Secondary CTL cultures were established from H-2[d] mice (top) and H-2[b] mice (bottom) primed *in vivo* with either LCMV ARM, VVNPaa1–558, VVGPaa1–363, or VVSC11 and restimulated 6 weeks later *in vitro* with LCMV ARM-infected syngeneic peritoneal exudate cells. CTL were harvested after 5 days and tested by ^{51}Cr release assay as described (Klavinskis *et al.*, 1989b). Target cells were either uninfected or infected with LCMV ARM, VVNPaa1–558, VVGPaa1–363, or VVSC11 as described (Whitton *et al.*, 1988b). Percent ^{51}Cr release for each effector group primed *in vivo* (LCMV-ARM, VVNP, VVGP, and VVSC 11) is illustrated at effector/target ratios 25:1 (top) and 50:1 (bottom).

epitope has also been shown to elicit CTL to the GP34–40 epitope (Klavinskis *et al.*, 1990).

B. Use of Synthetic Peptides to Induce CTL

Soluble peptides or fragments of proteins have occasionally been documented to elicit class I MHC-restricted CTL *in vivo*, for example, ovalbumin (Carbone and Bevan, 1989). This is despite the usual requirements that class I MHC-restricted CTL are induced against endogenously synthesized proteins, which are processed and presented by a separate pathway from exogenously introduced soluble proteins (reviewed in Braciale and Braciale, 1991). Thus noninfectious virus or peptides do not usually stimulate a class I MHC-restricted CTL response. However, it has been suggested that when cell debris is phagocytosed by antigen-presenting cells, the resulting antigenic peptides may enter the class I presentation system (Bevan, 1987). According to this theory, there is no absolute requirement for a protein to be endogenously synthesized in order to induce class I MHC-restricted CTL. The only requirement may be the form of the endocytosed antigen and its ability to enter the correct subcellular location, determining whether it is presented by class-I or class-II MHC molecules.

The NP sequence 118–132 from the WE strain of LCMV (encompassing an immunodominant H-2^d restricted NP CTL epitope) has been shown to induce class-I restricted CTL *in vivo* following repeated subcutaneous priming with the free synthetic peptide emulsified in incomplete Freund's adjuvant (Aichele *et al.*, 1990). However, not all T-cell-recognition sequences are sufficient to induce CTL *in vivo* (Klavinskis *et al.*, 1990). Immunization with an H-2^b-restricted CTL epitope LCMV WE GP275–288 delivered as a synthetic peptide in incomplete Freund's adjuvant was unsuccessful (Schulz *et al.*, 1991b). Therefore, there may be specific rules regarding which synthetic peptides induce CTL, how they gain access to the appropriate class I presentation pathway, and whether the "minimal" CTL recognition sequence needs to be attached to an accessory signal (such as a T-helper epitope) required for the transition of a virgin CTL *in vivo* to an active effector/memory cell.

Recently it has become possible to isolate naturally processed peptide antigens derived from the MHC class I–restricted processing pathway by acid hydrolysis and reverse-phase high performance liquid chromatography (Falk *et al.*, 1990). Such approaches offer new avenues to explore the rules governing which peptide motifs bind to each MHC molecule. The possibility of exactly predicting and identifying CTL epitopes, which are then delivered as synthetic peptides in adjuvant (such as alum acceptable to humans) or as synthetic peptides coupled to lipids (Deres *et al.*, 1989) or mixed with immunostimulating complexes (Takahashi *et al.*, 1990), will be important in the development of a subunit vaccine for arenaviruses such as Lassa.

V. VACCINE DESIGN

Vaccination strategies have traditionally focused on raising humoral responses. In view of the pivotal role played by CTL in the control and elimination of LCMV during acute infection (Byrne and Oldstone, 1984; Moskophidis *et al.*, 1987; Klavinskis *et al.*, 1989b), the design of a vaccine containing LCMV sequences directing CTL induction and target-cell recognition represents an alternative approach to induce protective immunity.

A. CTL Vaccines Confer Protection *In Vivo*

Mice of the H-2^b, H-2^d, and H-2^q haplotype vaccinated with a recombinant VV expressing full-length NP are completely protected from lethal intracerebral challenge with LCMV ARM (Klavinskis *et al.*, 1989b; Klavinskis and Oldstone, unpublished observations). Identification of a T-cell epitope (NP122–127) associated with >95% of the virus-specific CTL response in H-2^d mice led to the design of a "CTL vaccine" (VVNP1–201) expressing the immunodominant epitope. A single immunization with VVNP1–201 completely protects H-2^d (and also H-2^q) mice from lethal infection with LCMV, but does not protect H-2^b mice (Klavinskis *et al.*, 1989b; Klavinskis and Oldstone, unpublished observations; Fig. 6). The difference in outcome between mice of these haplotypes is consistent with the presence or absence in the immunizing sequences of an epitope for CTL recognition and is correlated with the induction of LCMV-specific H-2 restricted CTL in H-2^d mice (Klavinskis *et al.*, 1989b). Presentation of the epitope is restricted to the L-MHC locus of the H-2^d haplotype (Whitton *et al.*, 1989), and indeed recombinant H-2^b mice expressing L^d (BlOA.18R) are completely protected by vaccination with VVNP1–201, though C57BL/6 congenic at the L locus are susceptible to lethal infection with LCMV (Klavinskis *et al.*, 1989b; Fig. 7). The mechanism of protection does not appear to be via a humoral response. LCMV-specific antibodies are not detected by ELISA in sera from VVNP1–201 primed mice, though mice vaccinated with VVNP1–558 do develop an anti-LCMV-specific antibody response (Klavinskis *et al.*, 1989b; Table I). Passive transfer of sera from VVNP1–201 immune mice does not confer protection on naïve recipients challenged with LCMV (Klavinskis *et al.*, 1989b). In fact, no evidence exists for neutralizing antibody directed to the NP of LCMV; the major neutralizing epitope of LCMV maps to the GP (Parekh and Buchmeier, 1986). The data indicate that a protective "CTL vaccine" has been constructed.

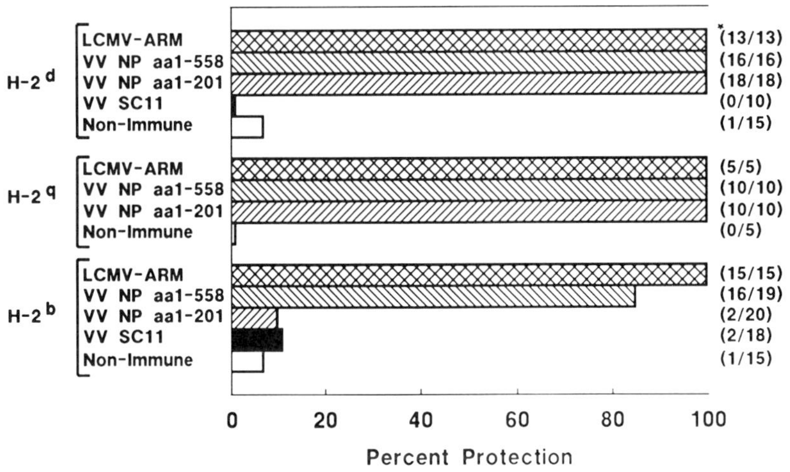

FIGURE 6. Immunization of H-2^d and H-2^q mice with a single inoculation of a recombinant VV expressing a truncated LCMV NPaa1–201 (containing the epitope for CTL recognition) protects against lethal LCMV infection. BALB/c (H-2^d), SWR/J (H-2^q), and C57BL/6 (H-2^b) mice were primed by intraperitoneal injection with 2×10^5 pfu of LCMV ARM or with 2×10^7 pfu of VVNPaa1–558, VVNP1–201, or VVSC11 (which does not contain LCMV-specific sequences). Six weeks later, these mice were challenged intracerebrally with 10 pfu LCMV ARM (20 LD_{50}). *(Number of mice protected/total immunized).

Identification of a major T-cell-recognition epitope in GP-1 restricted by H-2^b has also led to the construction of a recombinant VV/LCMV vaccine encoding as few as 60 residues of GP-1, which induces a vigorous CTL response and confers protection from lethal challenge (Klavinskis *et al.*, 1990). These experiments collectively show that the molecular dissection of LCMV glycoproteins can uncover immunodominant CTL epitopes that can be engineered into vaccines that elicit CTL.

B. Prospects of Subunit Vaccines

The development of VV expressing foreign genes has made the prospect of live subunit vaccines a reality. Our work has clearly demonstrated the efficacy of single virus proteins, indeed of single CTL epitopes, in protecting against lethal infection. We have also, however, uncovered a critical problem with subunit vaccination. Identification of the CTL recognition epitopes of LCMV has shown that there is an extremely limited number of antigenic sites on a molecule recognized by class I MHC-restricted CTL. Thus the less viral amino acid sequences contained in a vaccine, the greater the risk that certain MHC types will be unable to present any sequences for CTL recognition. We have now demonstrated the reality of this problem with subunit vaccination (Klavinskis *et al.*, 1989b), since H-2^b mice are not protected by a vaccine that confers protection on H-2^d mice. Recognition of these sites, as men-

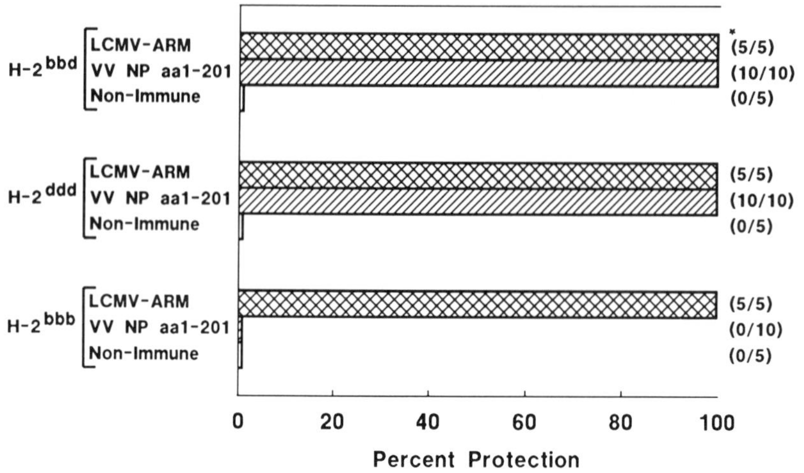

FIGURE 7. H-2^b recombinant mice expressing L^d are protected from a lethal intracerebral infection with LCMV. H-2^{bbd} (BlOA18R), H-2^{ddd} (BALB/c), and H-2^{bbb} (C57BL/6) mice were immunized with a single intraperitoneal inoculation of 2×10^5 pfu LCMV ARM or with 2×10^7 pfu VVNPaa1-201. Six weeks later, mice were challenged intracerebrally with 10 pfu LCMV ARM (20 LD_{50}). Percent protection at 14 days following lethal virus challenge is shown. *(Number of mice protected/total immunized).

tioned previously, is determined by the MHC. Incorporation of only one epitope is fully protective against lethal infection in H-2^d mice (Klavinskis *et al.*, 1989b). However, the polymorphism of MHC proteins in an outbred population suggests that multiple CTL epitopes may be required to minimize the chance that an individual will be unable to respond to the vaccine.

Another factor to consider in the design of subunit "CTL vaccines" is the incorporation of an epitope(s) that induces protection against serologically distinct strains of an arenavirus. In general, CTL responses to internal antigens are cross-reactive between different strains of a virus, such as the NP of LCMV in H-2^b mice (Whitton *et al.*, 1988a). However, it is interesting to note that, on the H-2^d background, a nonreciprocal CTL cross-reactivity to the NP exists between the ARM and Pasteur strains of LCMV (Whitton *et al.*, 1989). CTL induced by LCMV ARM fail to efficiently lyse target cells infected with LCMV Pasteur, yet the latter induce CTL that cross-react and kill LCMV ARM-infected target cells (Ahmed *et al.*, 1984a). Molecular analysis of the cross-reactivity indicates that despite local sequence divergence at the recognition epitope between ARM (PQASGVYMG) and Pasteur (LKTSGVYMG), the same minimal epitope NP122–127 (GVYMG) is used to induce a CTL response (Whitton *et al.*, 1989). Thus we predict a heirarchy of similar epitope sequences, all of which can induce CTL, but which induce CTL that differ in their abilities to cross-react. If this hypothesis is correct,

TABLE I. Immune Responses of H-2^{dd} Mice Immunized with a Truncated LCMV NP (Amino Acids 1–201)[a]

Immunization	LCMV-specific[b] ELISA antibody titer	LCMV-neutralizing[c] antibody titer	Percentage protection from lethal challenge
LCMV ARM	163,840	20	100
VV NPaal–558	10,240	<10	100
VV NPaal–201	<40	<10	100
None	<40	n.t.	0

[a] H-2^{dd} mice were immunized by intraperitoneal injection with 2×10^5 pfu LCMV ARM or with 2×10^7 pfu or VV NPaal–201. At 5 weeks postimmunization, LCMV-specific antibody titers were determined. At 6 weeks postimmunization, mice were challenged intracerebrally with 20 LD_{50} LCMV ARM.
[b] ELISA titers are expressed as the reciprocal of the end-point dilution.
[c] Neutralizing activity is expressed as the reciprocal of the final dilution neutralizing 200 pfu of LCM ARM *in vitro.*

vaccination with the ARM NP116–129 sequence would protect against ARM challenge, but not against PAST challenge, whereas vaccination with PAST NP116–129 would protect against both LCMV strains. We are currently testing this hypothesis, which may establish another important concept in subunit vaccination. Since we are eliciting CTL responses of limited variety, we shall have to consider not simply whether a vaccine induces CTL, but also whether these CTL will be effective against all strains of the virus. These observations highlight the importance of selecting immunodominant CTL epitopes that encode sequences with the ability to induce cross-reactivity with other virus strains.

Finally, it is important to establish that incorporation of an epitope(s) in a subunit "CTL vaccine" induces a sufficiently high-level (or early) CTL response against virus challenge. H-2^d mice vaccinated with a VV–NP recombinant (containing an immunodominant CTL epitope) are fully protected from intracerebral challenge with LCMV (Klavinskis *et al.*, 1989b; Oehen *et al.*, 1991). However, vaccination with a VV–GP recombinant that induces a low-level CTL response in this haplotype did not protect and resulted in accelerated disease (Oehen *et al.*, 1991). Effective vaccination against noncytopathic viruses such as LCMV requires early containment of viral spread to avoid immunopathological damage by a "late" CTL response when virus is widely disseminated. This form of disease enhancement is generally not a problem with whole-virus vaccines containing multiple protective CTL epitopes. However, if only one or a limited number of CTL epitopes are incorporated in a subunit vaccine, conditions may exist in which subsequent infection is associated with T-cell-mediated immunopathological disease.

VI. CONCLUSIONS AND SUMMARY

Work with LCMV has provided a useful model for examining the role and function of CTL in the regulation of arenavirus infections. The availability of cloned cDNA's encoding viral glycoproteins and primary sequence data for serologically distinct virus strains has allowed analysis of the molecular nature of the CTL response to LCMV in a variety of inbred strains of mice. This has allowed us to (1) examine the role of the host MHC in selecting the viral determinant used to induce a CTL response, (2) dissect the fine specificities of the induced CTL across serologically diverse strains, and (3) examine the immunodominance of particular CTL epitopes of a protein. This approach has successfully led to the design of a recombinant VV/LCMV subunit vaccine containing viral sequences that confer protection against lethal infection through the induction of CTL (Whitton *et al.*, 1989; Klavinskis *et al.*, 1989b, 1990). The identification of an immunodominant CTL epitope (LCMV WE 118–132) has also led to the demonstration that synthetic peptides delivered in Freund's adjuvant can also successfully prime for LCMV-specific CTL responses *in vivo* and enhance virus elimination from low-level infection (Schulz *et al.*, 1991b).

In view of the relatedness of LCMV with the human pathogen Lassa virus, both in sequence homology and in the ineffectual neutralizing antibody response noted in both infections (Jahrling and Peters, 1986; Peters *et al.*, 1987), parallels may be drawn from understanding the molecular basis and significance of the CTL response to LCMV in the design of an effective vaccine to Lassa. Additionally, knowledge derived from identifying the viral sequence requirements for CTL induction to LCMV should provide insights into manipulating synthetic peptides to induce CTL *in vivo*. With the development of alternative adjuvants for licensing in humans, it is conceivable that cocktails of peptides containing CTL epitopes will be administered to vaccinate against an outbred population and may thus make it possible to bypass the requirement of immunization with live vaccines in the future.

ACKNOWLEDGMENTS. This is Publication No. 5834-IMM from the Department of Neuropharmacology, Scripps Clinic and Research Foundation, La Jolla, CA 92037. This work was supported in parts by USPHS Grants NS-12428, AG-04342, and AI-09484 (MBAO) and AI-27028 (JLW). LSK was supported by a Fulbright Fellowship and Wellcome Grant. We thank Gay Schilling and Tracey Beck for excellent secretarial assistance.

REFERENCES

Ahmed, R., Byrne, J. A., and Oldstone, M. B. A., 1984a, Virus specificity of cytotoxic T lymphocytes generated during acute lymphocytic choriomeningitis virus infection:

Role of the H-2 region in determining cross-reactivity for different lymphocytic choriomeningitis virus strains, *J. Virol.* **51**:34.
Ahmed, R., Salmi, A., Butler, L. D., Chiller, J. M., and Oldstone, M. B. A., 1984b, Selection of genetic variants of lymphocytic choriomeningitis virus in spleens of persistently infected mice: Role in suppression of cytotoxic T lymphocyte response and viral persistence, *J. Exp. Med.* **60**:521.
Ahmed, R., Jamieson, B. D., and Porter, D. D., 1987, Immune therapy of a persistent and disseminated viral infection, *J. Virol.* **61**:3920.
Aichele, P., Hengartner, H., Zinkernagel, R. M., and Schulz, M., 1990, Antiviral cytotoxic T cell response induced by *in vivo* priming with a free synthetic peptide, *J. Exp. Med.* **171**:1815.
Bangham, C. R., Openshaw, P. J. M., Ball, L. A., King, A. M. Q., Wertz, G. W., and Askonas, B. A., 1986, Human and murine cytotoxic T cells specific to respiratory syncytial virus recognise the viral nucleoprotein N but not the major glycoprotein G expressed by vaccinia virus recombinants, *J. Immunol.* **137**:3973.
Bevan, M. J., 1987, Class discrimination in the world of immunology, *Nature* **325**:192.
Bishop, D. H., and Auperin, D. D., 1987, Arenavirus gene structure and organisation, *Curr. Top. Microbiol. Immunol.* **133**:5.
Braciale, T. J., and Braciale, V. L., 1991, Antigen presentation: Structural themes and functional variations, *Immunol. Today* **12**:124.
Buchmeier, M. J., and Oldstone, M. B. A., 1979, Protein structure of lymphocytic choriomeningitis virus: Evidence for a cell associated precursor of the virion glycopeptides, *Virology* **99**:111.
Buchmeier, M. J., Welsh, R. M., Dutko, F. J., and Oldstone, M. B. A., 1980, The virology and immunobiology of lymphocytic choriomeningitis virus infection, *Adv. Immunol.* **30**:275.
Byrne, J. A., and Oldstone, M. B. A., 1984, Biology of cloned cytotoxic T lymphocytes specific for lymphocytic choriomeningitis virus: Clearance of virus *in vivo*, *J. Virol.* **51**:682.
Carbone, F. R., and Bevan, M. J., 1989, Induction of ovalbumin-specific cytotoxic T cells by *in vivo* peptide immunisation, *J. Exp. Med.* **169**:603.
Cerny, A., Sutter, S., Bazin, H., Hengartner, H., and Zinkernagel, R. M., 1988, Clearance of lymphocytic choriomeningitis virus in antibody and B-cell deprived mice, *J. Virol.* **62**:1803.
Chakrabarti, S., Brechling, K., and Moss, B., 1985, Vaccinia virus expression vector: Coexpression of β-galactosidase provides visual screening of recombinant virus plaques, *Mol. Cell. Biol.* **5**:3403.
Cihak, J., and Lehmann-Grube, F., 1978, Immunologic tolerance to lymphocytic choriomeningitis virus in neonatally infected virus carrier mice: evidence supporting a clonal inactivation mechanism, *Immunology* **34**:265.
Cole, G. A., Nathanson, N., and Prendergast, R. A., 1972, Requirement for O bearing cells in lymphocytic choriomeningitis virus-induced central nervous system disease, *Nature* **238**:335.
Cole, G. A., Prendergast, P. A., and Henney, C. S., 1973, *In vitro* correlates of LCM virus-induced immune response, in: *Lymphocytic Choriomeningitis Virus and Other Arenaviruses* (F. Lehmann-Grube, ed.), pp. 60, Springer-Verlag, Berlin and New York.
DeLisi, C., and Berzofsky, J., 1985, T-cell antigenic sites tend to be amphipathic structures, *Proc. Natl. Acad. Sci. USA* **82**:7048.
DelVal, M., Volkmer, H., Rothbard, J. B., Jonjic, S., Messerle, M., Schickedanz, J., Reddehase, M., and Koszinowski, U. H., 1988, Molecular basis for cytolytic T-lymphocyte recognition of the murine cytomegalovirus immediate early protein pp 89, *J. Virol.* **62**:3965.
Deres, K., Schild, H., Wiesmuller, K-H., Jung, G., and Rammensee, H-G., 1989, *In vivo* priming of virus-specific cytotoxic T lymphocytes with synthetic lipopeptide vaccine, *Nature* **342**:561.

Falk, K., Rotzschke, 0., Stevanovic, S., Jung, G., and Rammensee, H-G., 1990, Allele-specific motifs revealed by sequencing of self-peptides eluted from MHC molecules, *Nature* **351**:290.
Gilden, D. H., Cole, G. A., and Nathanson, N., 1972, Immunopathogenesis of acute central nervous system disease produced by lymphocytic choriomeningitis virus. II. Adoptive immunisation of virus carriers, *J. Exp. Med.* **135**:874.
Gotch, F., McMichael, A., Smith, G., and Moss, B., 1987, Identification of viral molecules recognized by influenza-specific cytotoxic T lymphocytes, *J. Exp. Med.* **165**:408.
Hany, M., Oehen, S, Schulz, M, Hengartner, H, Mackett M, Bishop, D. H. L., Overton, H., and Zinkernagel, R. M., 1989, Anti-viral protection and prevention of lymphocytic choriomeningitis virus or of the local footpad swelling reaction in mice by immunisation with vaccinia recombinant virus expressing LCMV-WE nucleoprotein or glycoprotein, *Eur. J. Immunol.* **19**:417.
Hotchin, J., and Benson, L., 1963, The pathogenesis of lymphocytic choriomeningitis in mice: The effects of different inoculation routes and the footpad response, *J. Immunol.* **91**:460.
Hotchin, J., and Cinits, M., 1958, Lymphocytic choriomeningitis infection of mice as a model for the study of latent virus infection, *Can. J. Microbiol.* **4**:149.
Hotchin, J., and Weigand, H., 1961, Studies of lymphocytic choriomeningitis in mice. I. The relationship between age at inoculation and outcome of infection, *J. Immunol.* **86**:392.
Jahrling, P. B., and Peters, C. J., 1986, Serology and virulence diversity among Old World arenaviruses and the relevance to vaccine development, *Med. Microbiol. Immunol.* **175**:165.
Jamieson, B. D., and Ahmed, R., 1988, T-cell tolerance: Exposure to virus *in utero* does not cause a permanent deletion of specific T cells, *Proc. Natl. Acad. Sci. USA* **85**:2265.
Jamieson, B. D., and Ahmed, R., 1989, T-cell memory: Long-term persistence of virus specific cytotoxic T cells, *J. Exp. Med.* **169**:1993.
Klavinskis, L. S., Oldstone, M. B. A., and Whitton, J. L., 1988, Designing vaccines to induce cytotoxic T lymphocytes: Protection from lethal virus infection, in: *Vaccines 89* (R. Chanock, R. Lerner, and H. Ginsburg, eds.), pp. 485, Cold Spring Harbor Laboratory, Cold Spring Harbor, NY.
Klavinskis, L. S., Tishon, A., and Oldstone, M. B. A., 1989a, Efficiency and effectivness of cloned cytotoxic T lymphocytes *in vivo*, *J. Immunol.* **143**:2013.
Klavinskis, L. S., Whitton, J. L., and Oldstone, M. B. A., 1989b, Molecularly engineered vaccine expressing an immunodominant T cell epitope induces cytotoxic T lymphocytes that confer protection from lethal virus infection, *J. Virol.* **63**:4311.
Klavinskis, L. S., Whitton, J. L., Joly, E., and Oldstone, M. B. A., 1990, Vaccination and protection from a lethal viral infection: Identification, incorporation, and use of a cytotoxic T lymphocyte glycoprotein epitope, *Virology* **178**:393.
Lehmann-Grube, F., Assmann, V., Löliger, C., Moskophidis, D., and Löhler, J., 1985, Mechanism of recovery from acute virus infection. I. Role of T lymphocytes in the clearance of lymphocytic choriomeningitis virus from spleens of mice, *J. Immunol.* **134**:608.
Lundstedt, C., 1969, Interaction between antigenically different cells: Virus induced cytotoxicity by immune lymphocytes in vitro, *Acta Pathol. Microbiol. Scand.* **75**:134.
Margalit, H., Sponge, J. L., Cornette, J. L., Cease, K. B., DeLisi, C., and Berzofsky, J. A., 1987, Prediction of immunodominant helper T cell antigenic sites from the primary sequence, *J. Immunol.* **138**:2213.
Marker, O., and Volkert, M., 1973, Studies on cell-mediated immunity to lymphocytic choriomeningitis virus in mice, *J. Exp. Med.* **137**:1511.
Mims, C. A., and Blanden, R. V., 1972, Antiviral action of immune lymphocytes in mice infected with lymphocytic choriomeningitis virus, *Infect. Immun.* **6**:695.
Moskophidis, D., Cobbold, S. P., Waldmann, H., and Lehmann-Grube, F., 1987, Mechanism of recovery from acute virus infection: Treatment of lymphocytic choriomenin-

gitis virus-infected mice with monoclonal antibodies reveals that Lyt2+ T lymphocytes mediate clearance of virus and regulate the antiviral antibody response, *J. Virol.* **61**:1867.

Oehen, S., Hengartner, H., and Zinkernagel, R. M., 1991, Vaccination for disease, *Science* **251**:195.

Oldstone, M. B. A., Habel, K., and Dixon, F. J., 1969, The pathogenesis of cellular injury associated with persistent LCM viral infection, *Fed. Proc.* **28**:429.

Oldstone, M. B. A., Blount, P., Southern, P. J., and Lampert, P. W., 1986, Cytoimmunotherapy for persistent virus infection: Unique clearance pattern from the central nervous system, *Nature* **321**:239.

Oldstone, M. B. A., Whitton, J. L., Lewicki, H., and Tishon, A., 1988, Fine dissection of a nine amino acid glycoprotein epitope, a major determinant recognized by lymphocytic choriomeningitis virus-specific class I restricted H-2D^b cytotoxic T lymphocytes, *J. Exp. Med.* **168**:559.

Parekh, B. S., and Buchmeier, M. J., 1986, Proteins of lymphocytic choriomeningitis virus: Antigenic topography of the viral glycoproteins, *Virology* **153**:168.

Peters, C. J., Jahrling, P. B., Liu, C. T., Kenyon, R. H., McKee, K. T., and Barrera-Oro, J. G., 1987, Experimental studies of arenaviral hemorrhagic fever, *Curr. Top. Med. Microbiol.* **134**:5.

Reddehase, M. J., Rothbard, B., and Koszinowski, U. H., 1989, A pentapeptide as minimal antigenic determinant for MHC class-I restricted T lymphocytes, *Nature* **337**:651.

Riviere, Y., Ahmed, R., Southern, P. J., Buchmeier, M. J., Dutko, F. J., and Oldstone, M. B. A., 1986, The S RNA segment of lymphocytic choriomeningitis virus codes for the nucleoprotein and glycoproteins 1 and 2, *J. Virol.* **53**:966.

Rothbard, J. B., and Taylor, W. R., 1988, A sequence pattern common to T cell epitopes, *EMBO J.* **7**:93.

Rowe, W. P., 1954, Studies on pathogenesis and immunity in lymphocytic choriomeningitis infection of the mouse, *Res. Rep. Naval. Med. Res. Inst.* **12**:167.

Salvato, M. S., and Shimomaye, E. M., 1989, The completed sequence of LCMV reveals a unique RNA structure and a new gene for a zinc-finger protein, *Virology* **173**:1.

Salvato, M. S., Shimomaye, E. M., Southern, P. J., and Oldstone, M. B. A., 1988, Virus, lymphocyte interactions. IV. Molecular characteristics of LCMV Armstrong (CTL+) small genomic segment and that of its variant, Clone 13 (CTL−), *Virology* **164**:517–522.

Schulz, M., Aichele, P., Schneider, R., Hansen, T. H., Zinkernagel, R. M., and Hengartner H., 1991a, Major histocompatibility complex binding and T cell recognition of a viral nonepeptide containing a minimal tetrapeptide, *Eur. J. Immunol.* **21**:1181.

Schulz, M., Zinkernagel, R. M., and Hengartner, H., 1991b, Peptide-induced antiviral protection by cytotoxic T cells, *Proc. Natl. Acad. Sci. USA* **88**:991.

Southern, P. J., Singh, M. K., Riviere, Y., Jacoby, D. R., Buchmeier, M. J., and Oldstone, M. B. A., 1987, Molecular characterization of the genomic S RNA segment from lymphocytic choriomeningitis virus, *Virology* **157**:145.

Takahashi, H., Takeshita, R., Morein, B., Putney, S., Germain, R. N., and Berzofsky, J. A., 1990, Induction of CD8+ cytotoxic T cells by immunisation with purified HIV-1 envelope protein in ISCOMS, *Nature* **344**:873.

Townsend, A. R. M., and Skehel, J. J., 1984, The influenza A virus nucleoprotein gene controls the induction of both subtype specific and cross-reactive cytotoxic T cells, *J. Exp. Med.* **160**:552.

Townsend, A. R. M., Gotch, F. M., and Davey, J., 1985, Cytotoxic T cells recognise fragments of the influenza nucleoprotein, *Cell* **42**:457.

Townsend, A. R. M., Rothbard, J., Gotch, F., Bahadur, G., Wraith, D. C., and McMichael, A. J., 1986, The epitopes of influenza nucleoprotein recognised by cytotoxic T lymphocytes can be defined with short synthetic peptides, *Cell* **44**:959.

Traub, E., 1936, The epidemiology of lymphocytic choriomeningitis virus in white mice, *J. Exp. Med.* **64**:183.

Traub, E., 1960, Uber die immunologische toleranz beider lympocytaren choriomeningitus der mause, *Zentralb./Bacteriol. (Stuttgart)* **177**:472.

Volkert, M., and Hannover Larsen, J., 1964, Studies on immunological tolerance to LCM virus. 3. Duration and maximal effect of adoptive immunisation of virus carriers, *Acta Pathol. Microbiol. Immunol. Scand.* **60**:577.

Volkert, M., and Hannover Larsen, J., 1965, Immunological tolerance to viruses, *Prog. Med. Virol.* **7**:160.

Welsh, R. M., and Zinkernagel, R. M., 1977, Hetero-specific cytotoxic cell activity induced during the first three days of acute lymphocytic choriomeningitis virus infection in mice, *Nature* **268**:648.

Whitton, J. L., Southern, P. J., and Oldstone, M. B. A., 1988a, Analyses of the cytotoxic T lymphocyte responses to glycoprotein and nucleoprotein components of lymphocytic choriomeningitis virus, *Virology* **162**:321.

Whitton, J. L., Gebhard, J. R., Lewicki, H., Tishon, A., and Oldstone, M. B. A., 1988b, Molecular definition of a major cytotoxic T lymphocyte epitope in the glycoprotein of lymphocytic choriomeningitis virus, *J. Virol.* **62**:687.

Whitton, J. L., Tishon, A., Lewicki, H., Gebhard, J., Cook, T., Salvato, M., Joly, E., and Oldstone, M. B. A., 1989, Clonally restricted virus specific cytotoxic T lymphocyte response: Molecular analyses Of Ld class I restricted minimal five amino acid nucleoprotein epitope of LCMV, *J. Virol.* **63**:4303.

Whitton, J. L., and Oldstone, M. B. A., 1989, Class I MHC can present an endogenous peptide to cytotoxic T lymphocytes, *J. Exp. Med.* **170**:1033.

Yewdell, J. W., Bennink, J. R., Smith, G. L., and Moss, B., 1985, Influenza A virus nucleoprotein is a major target antigen for cross-reactive anti-influenza A virus cytotoxic T lymphocytes, *Proc. Natl. Acad. Sci. USA* **82**:1785.

Yewdell, J. W., Bennink, J. R., Mackett, M., Lefrancois, L., Lyles, D. S., and Moss, B., 1986, Recognition of cloned vesicular stomatitis virus internal and external gene products by cytotoxic T lymphocytes, *J. Exp. Med.* **163**:1529.

Zinkernagel, R. M., and Doherty, P. C., 1974, Restriction of *in vitro* T cell mediated cytotoxicity in lymphocytic choriomeningitis within a syngeneic or semiallogeneic system, *Nature* **248**:701.

Zinkernagel, R. M., and Doherty P. C., 1979, MHC-restricted cytotoxic T cells: Studies on the biological role of polymorphic major transplantation antigens determining T cell restriction-specificity, function and the responsiveness, *Adv. Immunol.* **27**:52.

Zinkernagel, R. M., and Welsh, R. M., 1976, H-2 compatibility requirement for virus-specific T-cell mediated effector functions *in vivo*. I. Specificity of T cells conferring antiviral protection against lymphocytic choriomeningitis virus is associated with H-2^K and H-2^D, *J. Immunol.* **117**:1495.

CHAPTER 14

Virus-Induced Acquired Immunosuppression by T-Cell-Mediated Immunopathology and Vaccine Strategies

ROLF M. ZINKERNAGEL

I. INTRODUCTION

T cells dominate early cellular immune responses against many viruses and are potentially protective (Zinkernagel and Doherty, 1979; Blanden, 1974). There is some evidence that cytotoxic T cells destroy virus-infected cells before viral progeny are assembled (Zinkernagel and Doherty, 1979; Blanden, 1974), thus eliminating virus during the eclipse phase of virus replication. Virus elimination via immunological host cell destruction is, in the case of cytopathic viruses, an efficient way to prevent virus spread and the resulting more extensive virus-mediated cell and tissue damage. In the case of noncytopathic viruses, this immunological defense mechanism becomes less attractive because host cells are not destroyed by virus but only by the T-cell immune response. Because T cells apparently cannot distinguish cytopathic from noncytopathic viruses, immune-mediated cell and tissue damage often re-

ROLF M. ZINKERNAGEL • Department of Pathology, Institute of Experimental Immunology, University of Zurich, Zurich, Switzerland.

The Arenaviridae, edited by Maria S. Salvato. Plenum Press, New York, 1993.

sults, in the latter infections, in immunopathology (Zinkernagel and Doherty, 1979).

Examples of infections with noncytopathic viruses are lymphocytic choriomeningitis (LCM) in mice (Hotchin, 1962; Lehmann-Grube, 1971) and hepatitis B in humans (Bianchi, 1981; Mondelli and Eddleston, 1984). Lymphocytic choriomeningitis in mice develops only in immunocompetent animals after intracerebral injection of LCM virus (LCMV). Mice vertically infected as embryos by their infected mothers or mice lacking T cells or those immunosuppressed by irradiation or cytostatic drugs do not develop inflammatory reactions or LCM disease. These mice fail to eliminate the virus and as a result become LCMV carriers (Traub, 1936; Lehmann-Grube, 1971; Zinkernagel and Doherty, 1979); they are tolerant to LCMV at least at the level of cytotoxic T cells (Oldstone and Dixon, 1967; Doherty and Zinkernagel, 1974). LCM disease as well as elimination of LCMV after acute infection has been carefully analyzed and has been clearly shown to be cytotoxic T-cell-mediated (Cole *et al.*, 1972; Byrne and Oldstone, 1984; Baenziger *et al.*,1986). Thus, LCMV-induced disease in mice is an immunopathologically mediated disease caused by T-cell-mediated destruction of infected host cells.

II. VIRUS-TRIGGERED ACQUIRED IMMUNOSUPPRESSION

A. Pathogenesis of Immunosuppression

It has been known for some time that LCMV causes immunosuppression in mice (Mims and Wainwright, 1968; Lehmann-Grube, 1984). When reevaluating this in various mouse strains using different LCMV isolates, we found that an LCMV infection of mice suppresses their capacity to mount an IgM or IgG response to vesicular stomatitis virus (VSV) (Roost *et al.*, 1988; Leist *et al.*, 1988). Mice preinfected with LCMV are considerably more susceptible to disease caused by VSV, which usually is nonpathogenic for mice if injected subcutaneously or intravenously.

The extent of immune suppression by LCMV depends on the following parameters: Different LCMV virus isolates influence immune reactivity differently. LCMV WE and some other LCMV isolates, such as LCMV AGG or LCMV DOC, cause immunosuppression, whereas LCMV ARM only rarely does so. Mouse strains differ considerably with respect to susceptibility to this immunosuppression; the MHC plays some as yet poorly defined role, but non-MHC genes also have a major influence. The kinetics of induction of the described impairment of mice to respond with T-cell-independent IgM and/or a strictly T-cell-dependent IgG response to a subsequent virus infection parallel that usually characteristic for the induction of an anti-LCMV T-cell response starting on day 6 after LCMV infection and reaching maximum levels around day 8–10. In C57BL/6 mice, suppression is seen earlier by 2–3 days when compared with DBA/2 mice; this correlates with the shifted later kinetics of the latter mice probably owing to their low percentage of CD8+ T cells (Via and Shearer, 1988). The impairment for mounting an IgM and/or IgG anti-VSV response after LCMV infection of mice can be transient or of rather long duration (up to 4–5 months, the longest period measured to date in DBA/2 mice), again depending on the LCMV isolate and doses used and on the mouse strain that was infected.

The following experimental results (Leist *et al.*, 1988) suggested that, similar to LCMV hepatitis (Zinkernagel *et al.*, 1986), the antiviral T-cell response was responsible for immunosuppression. When neonatally infected LCMV carrier mice were evaluated with respect to their immune responsiveness, they were found to mount anti-VSV, IgM, and IgG responses comparable to those of normal control mice. T-cell-deprived nude mice infected with LCMV also made normal IgM responses. This indicated that LCMV alone is not immunosuppressive and that the observed immunosuppression is not caused by the action of interferons on VSV. In contrast, LCMV-infected nude mice that received LCMV immune cytotoxic T cells exhibited suppressed antibody responses (Leist *et al.*, 1988). Also, while LCMV infected mice failed to mount an antibody response, similarly infected mice that were treated with anti-CD8 antisera some days before they received the VSV infection mounted normal IgM and IgG responses (Table I).

These results are compatible with the view that antiviral cytotoxic T cells are responsible for immunosuppression in this model infection. Accordingly, LCMV may infect lymphocytes and/or antigen-presenting cells, which are involved in antibody responses; these infected cells are

TABLE I. LCMV-Induced Immune Suppression[a]

Mouse strain	Infection with LCMV	Treatment	Serum antibody titer to subsequent infection with VSV: IgM	IgG
C57BL/6 +/+	10^6 pfu i.v.-8d	—	<1/50	<1/50
C57BL/6 nu/nu	10^6 pfu i.v.-8d	—	1/1000	<1/50
C57BL/6 +/+	Neonatal LCMV carrier	—	1/1000	1/10000
C57BL/6 +/+	10^6 pfu i.v.-8d	Anti-CD8 d + 2, +4	1/1000	1/5000
C57BL/6 nu/nu	10^6 pfu i.v.-8d	Adoptive transfer of CD8 8d LCMV immune T cells	1/100	<1/50

[a] Summarized from Leist *et al.* (1988) and Roost *et al.* (1988).

then in turn destroyed by anti-LCMV-specific cytotoxic T cells (Odermatt *et al.*, 1991).

This immune suppression has been analyzed further by monitoring the lymphocyte subset composition in spleens of mice acutely infected with LCMV by using FACS analysis. We did not find drastic reduction in the absolute numbers of T-helper cells or B cells but did see an increase in the number of cytotoxic T cells (Odermatt *et al.*, 1991; Moskophidis *et al.*, 1992).

Several alternative possible mechanisms of LCMV-caused immunosuppression have been discussed. It has been argued that LCMV may infect CD4+ T cells and either make them less or nonfunctional by marginal cytopathogenicity (McChesney and Oldstone, 1989; Borrow *et al.*, 1991); alternatively, infection may render them susceptible to lysis by CD8+ effector T cells. Because only very small numbers of T-helper cells have been found to be infected (i.e., 0.05–2%) even in LCMV carrier mice, this explanation is not yet supported experimentally. The argument that LCMV specific CD4+ T cells might be infected preferentially is a theoretical possibility that is, however, not supported by the fact that immunosuppression is rather general, i.e., that it includes anti-VSV-specific T-helper and antibody responses (Roost *et al.*, 1988; Moskophidis *et al.*, 1992). There is so far no evidence either that pre-B or B cells get LCMV-infected and thus become subject to the above-discussed mechanisms. There have been studies of several model situations where the possibility of specific B cells becoming infected or presenting antigen after uptake by their specific receptors has been analyzed (Shinohara *et al.*, 1988; Barnaba *et al.*, 1990). Such B cells may become functionally impaired or targets for CD8+ effector T cells. This possibility may explain some of the differential kinetics of hepatitis B virus or HIV-specific antibody responses in humans (Mondelli and Eddleston, 1984; Rosenberg and Fauci, 1990).

Studies of the lymph follicles and germinal centers in spleens of T-cell-competent ICR mice showed the following characteristic changes during an acute LCMV infection: By day 3–4 most of the typical distribution of cells expressing CD4, CD8, IgM, or macrophage markers was comparable to that in controls. However, in immunocompetent mice the follicular structure was massively destroyed between day 6 and 14 after LCMV infection (Odermatt *et al.*, 1991). The usual architecture of germinal centers in spleens reappeared slowly after day 12. Thus the orderly structure of germinal centers is drastically altered, and this may be responsible for the failure of LCMV-infected mice to mount an IgM and IgG response.

These findings may be relevant to our understanding of acquired immunodeficiency that is caused by the human immunodeficiency viruses (HIV) (Fauci, 1988). As in LCMV infections, recent evidence suggests that cytotoxic T cells may actually be instrumental in controlling HIV replication, but may also be responsible for destroying lymphocytes and macrophages and thus causing immune suppression (Walker *et al.*, 1986). The example of HIV infections in humans may therefore, in a way, represent the ultimate perversity in the balance between the host immune system and the infectious agent: noncytopathic viruses infect lymphocytes and macrophages, the essential partners of an immune response, which are then destroyed by the immune response; i.e., the virus infection misleads the immune system to partially destroy itself, and the virus may thereby be enabled to persist.

B. Role of IFN-γ in Protecting Immunological Cells against Virus Infection

If virus-infected lymphohemopoietic cells can be destroyed by virus-specific cytotoxic T cells, or alternatively by the cytopathic effects of other viruses, their protection against infection should be of utmost importance. The role of γ-interferon (IFN-γ) induced during a viral infection in the ability of the host to acquire antiviral immunity was studied in mice (Leist *et al.*, 1989; Wille *et al.*, 1989; Klavinskis *et al.*, 1989a). Treatment with an anti-IFN-γ-specific antibody preparation had no influence on the ability of mice to generate anti-vaccinia virus- or anti-vesicular stomatitis virus (VSV)-specific cytotoxic T-cell responses or T-helper-dependent immunoglobulin G responses to VSV. In contrast, in mice infected with LCMV and treated with sheep anti-IFN-γ the cytotoxic T-cell responses against LCMV were impaired. In addition, under the experimental conditions used, it prevented lethal LCM. Cytotoxic T-cell activity measured in the spleens of anti-IFN-γ-treated mice was comparable to that found in mice initially infected with a 100-fold larger dose of LCMV. Evaluation of the effects of treatment on the kinetics of virus replication revealed that in both euthymic and athymic nude C57BL/6 mice, anti-IFN-γ treatment led to an increase of virus titers up to 100-fold compared with control mice. Therefore, IFN-γ may play an essential role in controlling viruses with a tropism for lymphocytes and monocytes/macrophages, such as LCMV.

III. POSSIBILITIES AND LIMITATIONS OF T-CELL VACCINATION AGAINST VIRUS

A. Antiviral Protection by T Cells: Role of Viral Peptide and Class I MHC Alleles

To evaluate the protective potential of vaccinia recombinant viruses in an acute virus infection, we have studied T-cell responses to LCMV and to vaccinia recombinant viruses expressing the LCMV GP or

NP (Whitton *et al.*, 1988a,b, 1989; Oldstone *et al.*, 1988; Hany *et al.*, 1989; Schulz *et al.*, 1989; Klavinskis *et al.*, 1989b) both *in vitro* and *in vivo*.

The viral antigen specificity of primary cytotoxic T-cell responses of H-2^b, H-2^k, H-2^q, H-2^s, H-2^f, and some H-2-recombinant mice against LCMV (WE isolate), as well as the specificity of some T-cell clones and T-cell lines, was defined on target cells infected with vaccinia–LCMV–NP or vaccinia–LCMV–GP. NP was recognized together with H-2^q (L^q), H-2^d (L^d), H-2^s, and H-2^b (D^b). GP specificity was restricted to H-2^f and H-2^b (K^b and D^b); H-2^k-restricted anti-LCMV responses were neither GP nor NP specific.

The antiviral protective immunity induced by vaccinia-GP or vaccinia-NP recombinants was evaluated in mice. T-cell-mediated protection correlated well with the cytotoxic T-cell specificity defined *in vitro*. Some of the H-2 alleles plus NP or H-2 plus GP combinations that were found to be nonresponder combinations *in vitro* were, however, protected to variable and low degrees by vaccinia recombinant viruses *in vivo* (Hany *et al.*, 1989), indicating that antiviral protection is a more sensitive readout than cytotoxicity *in vitro*. After immunization with vaccinia–NP recombinants, H-2^d mice had, for example, 10^4 times lower LCMV titers in spleens than in vaccinia-primed controls. Although vaccinia-GP-immunized (H-2^d) mice revealed no cytotoxic T-cell activity *in vitro*, they nevertheless had 10^2 times lower LCMV titers in spleens than controls. Interestingly, antiviral protection, particularly in low-responder combinations, was usually short-lived and diminished after 3 weeks, whereas in a high-responder situation, protection was of a longer duration (>8 weeks).

Vaccination with vaccinia-NP or -GP recombinants protected mice against lethal T-cell-mediated LCM or prevented the local footpad swelling reaction; these *in vivo* effects were H-2 dependent and followed the identical rules established for CTL recognition *in vitro*. These experiments document for LCMV NP that one viral protein may exhibit several protective antigenic determinants recognized by T cells in a H-2 K or D allele-specific manner. These results also show that distinct single T-cell epitopes of an internal viral protein efficiently protected mice against various manifestations of viral disease. The location of these epitopes within amino- and carboxy-terminal regions of the nucleoprotein and their H-2 restriction fine specificity previously defined *in vitro* were confirmed *in vivo*. So far, the protective potential against mortality of a complete internal viral protein has been documented for cytomegalovirus immediate early protein (Jonjic *et al.*, 1988) and possibly the nucleoprotein of influenza A virus (Andrew *et al.*, 1987). In respiratory syncytial virus a slight reduction in virus titer (0.7–0.8 $\log_{10}$ pfu) was documented by vaccination with an internal protein expressed in vaccinia virus (King *et al.*, 1987).

In a corresponding responder mouse the protective capacity of the

LCMV NP epitopes expressed in vaccinia virus is comparable to that of wild-type LCMV. The implication of the data is that the defined fragments only protect mice possessing one or a very limited number of MHC alleles (Bennink and Yewdell, 1988; Hany *et al.*, 1989). To protect the entire mouse population against LCMV it would be necessary to introduce all possible fragments into a potential recombinant vaccinia virus.

The same limitation applies to the use of relevant peptides as vaccines. Studies by Aichele *et al.* (1990), Schulz *et al.* (1991), and then by Kast *et al.* (1991) have shown that T-cell-epitope peptides may be used as efficient protective vaccines. Again, for general use, a cocktail of peptides will have to be used in an outbred population. In addition, questions of peptide stability and peptide persistence and of the duration of peptide-induced protection have to be evaluated before the usefulness of this approach can be judged.

B. Vaccination for Disease: An Interesting Exception to the Rule

Effective vaccination protects a host against cytopathogenic viral infection generally by triggering an appropriate cellular or humoral immune response. In infections involving noncytopathogenic viruses, the host cell is not destroyed by the virus directly, but rather by the T-cell immune response. The severity of T-cell-mediated immunopathological disease, such as hepatitis B virus in humans, and LCMV in mice, is determined by the balance between the kinetics of virus spread and the kinetics of the T-cell immune response. The experiments summarized here (Oehen *et al.*, 1991) demonstrate that vaccination against a virus may influence this balance by triggering a T-cell response, not only to prevent disease but in some instances to cause or aggravate disease, resulting in vaccination for rather than against disease.

In general, mice primed with an inoculum of LCMV are protected from subsequent intracerebral (i.c.) challenge. If mice are primed with vaccinia recombinant viruses expressing the glycoprotein or the nucleoprotein of LCMV we find instances where unexpectedly and surprisingly mice are more susceptible to lethal challenge than when not primed. This finding is heavily influenced by virus titer: e.g., virus isolates that cause immune suppression at high dose and lethal disease at low dose are even more virulent upon i.c. low-dose challenge when the mice have been primed with certain vaccinia recombinants. As discussed in the previous paragraphs, virulence seems to be enhanced by vaccinia recombinants that leave an immunopathological immune response still developing so that LCMV causes lethal disease; for example, H-2^d mice, which make a predominant CTL response to NP, are protected from lethal challenge upon priming with vaccinia–NP recombinants but are more susceptible to lethal challenge upon being primed with vaccinia–GP recombinants. Mice primed with wild-type LCMV will always be protected so efficiently that lethal LCM does not develop (Oehen *et al.*, 1991).

This probably rare example of vaccination for disease is nevertheless interesting and merits a cautionary note: Vaccination with recombinant vaccinia vaccines not inducing neutralizing antibodies and expressing only some T-cell epitopes (but not wild-type virus) may enhance T-cell-mediated immunopathology triggered by a noncytopathic virus under some conditions. This should certainly not discredit the development of recombinant vaccines. "New" types of vaccines are very promising as long as the minimal requirements (multiple T- and/or B-cell epitopes, induction of high neutralizing antibody titers, or induction of constantly high levels of CTL-P frequencies) are fulfilled. Recombinant hepatitis B vaccine may serve as a good example. Our animal model demonstrates that apparently, unless these requirements are met, sometimes even if the vaccine expresses a neutralizing determinant (e.g., vaccinia–LCMV–GP recombinant virus presents the neutralizing determinant), a vaccine may not work the way we would like it to work.

IV. CONCLUSION

The summarized experiments illustrate that during an LCMV infection antiviral T cells may cause immunosuppression; the virus alone is not immunosuppressive since nude mice or LCMV carrier mice respond normally. In this murine model infection, antiviral cytotoxic T cells seem to destroy LCMV-infected cells involved in immune responses; one part of the immune system destroys another essential part because it is virus-infected. Since IFN-γ seems to control LCMV infection of lymphohemopoietic cells, it may play a crucial role in preventing the infection of immunologically important cells. These parameters may be involved in the pathogenesis of acquired immunodeficiency syndrome.

Vaccination against immunopathologically mediated disease is possibly not always easy, because its efficacy depends on many parameters defining the equilibrium between infectious virus and immune defense mechanisms. The probably rare example of a vaccination causing aggravation of a disease illustrates some of the possible limitations, which may have to be taken into account in planning vaccines against virus-induced and T-cell-mediated diseases. The presented results and concepts may impinge on requirements for pre- or postexposure vaccination inducing highly efficient protective CTLs (as shown here) and/or neutralizing antibodies (which are probably in most cases protective also against T-cell-mediated immunopathological disease). For postexposure vaccination to be beneficial, enhanced T-cell responses should prevent disease by eliminating cytopathic virus efficiently, such as is the case for rabies virus infection (Wiktor *et al.*, 1984; Macfarlan *et al.*, 1984). After infection with a noncytopathic virus, improved T-cell responses may enhance immunopathology and disease. It has to be kept in mind that even in this latter case enhancement of neutralizing antibodies by any vaccine should be beneficial for a pre- or postexposure vaccination, because reduction of viral titers reduces potential damage by direct cytopathic effects of a virus but also against immunopathological consequences.

REFERENCES

Aichele, P., Hengartner, H., Zinkernagel, R. M., and Schulz, M., 1990, Antiviral cytotoxic T cell response induced by *in vivo* priming with a free synthetic peptide, *J. Exp. Med.* **171**:1815.

Andrew, M. E., Coupar, B. E. H., Boyle, D. B., and Ada, G. L., 1987, The roles of influenza virus hemagglutinin and nucleoprotein in protection: Analysis using vaccinia virus recombinants, *Scand. J. Immunol.* **25**:21.

Baenziger, J., Hengartner, H., Zinkernagel, R. M., and Cole, G. A., 1986, Induction or prevention of immunopathological disease by cloned cytotoxic T cell lines specific for lymphocytic choriomeningitis virus, *Eur. J. Immunol.* **16**:387–393.

Barnaba, V., Franco, A., Alberti, A., Benvenuto, R., and Balsano, F., 1990, Selective killing of hepatitis B envelope antigen-specific B cells by class I-restricted, exogenous antigen-specific T lymphocytes, *Nature* **345**:258.

Bennink, J. R., and Yewdell, J. W., 1988, Murine cytotoxic T lymphocyte recognition of individual influenza virus proteins. High frequency of nonresponder MHC class I alleles, *J. Exp. Med.* **168**:1935.

Bianchi, L., 1981, The immunopathology of acute type B hepatitis. *Springer Semin. Immunopathol.* **3**:421.

Blanden, R. V., 1974, T cell response to viral and bacterial infection, *Transplant. Rev.* **19**:56.

Borrow, P., Tishon, A., and Oldstone, M. B. A., 1991 Infection of lymphocytes by a virus that aborts cytotoxic T lymphocyte activity and establishes persistent infection, *J. Exp. Med.* **174**:203.

Byrne, J. A., and Oldstone, M. B. A., 1984, Biology of cloned cytotoxic T lymphocytes specific for lymphocytic choriomeningitis virus: Clearance of virus *in vivo*, *J. Virol.* **51**:682.

Cole, G. A., Nathanson, N., and Prendergast, R. A., 1972, Requirement for theta-bearing cells in lymphocytic choriomeningitis virus-induced central nervous system disease, *Nature* **238**:335.

Doherty, P. C., and Zinkernagel, R. M., 1974, T cell-mediated immunopathology in viral infections, *Transplant. Rev.* **19**:89.

Fauci, A. S., 1988, The human immunodeficiency virus: Infectivity and mechanisms of pathogenesis, *Science* **239**:617.

Hany, M., Oehen, S., Schulz, M., Hengartner, H., Mackett, M., Bishop, D. H. L., and Zinkernagel, R. M., 1989, Anti-viral protection and prevention of lymphocytic choriomeningitis or of the local footpad swelling reaction in mice by immunisation with vaccinia-recombinant virus expressing LCMV-WE nucleoprotein or glycoprotein, *Eur. J. Immunol.* **19**:417.

Hotchin, J., 1962, The biology of lymphocytic choriomeningitis infection: Virus induced immune disease, *Cold Spring Harbor Symp. Quant. Biol.* **27**:479.

Jonjic, S., Val del, M., Keil, G. M., Reddehase, M. J., and Koszinowski, U. H., 1988, A nonstructural viral protein expressed by a recombinant vaccinia virus protects against cytomegalovirus infection, *J. Virol.* **62**:1653.

Kast, W. M., Roux, L., Curran, J., Blom, H. J. J., Voordouw, A. C., Meloen, R. H., Kolakofsky, D., and Melief, C. J. M., 1991, Protection against lethal Sendai virus infection by *in vivo* priming of virus-specific cytotoxic T lymphocytes with a free synthetic peptide, *Proc. Natl. Acad. Sci. USA* **88**:2283.

King, A. M., Stott, E. J., Langer, S. J., Young, K. K. Y., Ball, L. A., and Wertz, G. W., 1987, Recombinant vaccinia virus carrying the N gene of human respiratory syncytial virus: studies of gene expression in cell culture and immune response in mice, *J. Virol.* **61**:2885.

Klavinskis, L. S., Whitton, J. L., and Oldstone, M. B. A., 1989a, Molecularly engineered vaccine which expresses an immunodominant T-cell epitope induces cytotoxic T lymphocytes that confer protection from lethal virus infection, *J. Virol.* **63**:4311.

Klavinskis, L. S., Geckler, R., and Oldstone, M. B., 1989b, Cytotoxic T lymphocyte control of acute lymphocytic choriomeningitis virus infection: Interferon γ, but not tumour necrosis factor alpha, displays antiviral activity *in vivo*, *J. Gen. Virol.* **70**:33.

Lehmann-Grube, F., 1971, Lymphocytic choriomeningitis virus, *Virol. Monogr.* **10**:1.

Lehmann-Grube, F., 1984, *Bacterial and Viral Inhibition and Modulation of Host Defences*, Academic Press, London.

Leist, T. P., Ruedi, E., and Zinkernagel, R. M., 1988, Virus-triggered immune suppression in mice caused by virus-specific cytotoxic T cells, *J. Exp. Med.* **167**:1749.

Leist, T. P., Eppler, M., and Zinkernagel, R. M., 1989, Enhanced virus replication and inhibition of lymphocytic choriomeningitis virus disease in anti-IFN-γ-treated mice, *J. Virol.* **63**:2813.

Macfarlan, R. I., Dietzschold, B., Wiktor, T. J., Kiel, M., Houghten, R., Lerner, R. A., Sutcliffe, J. G., and Koprowski, H., 1984, T cell responses to cleaved rabies virus glycoprotein and to synthetic peptides, *J. Immunol.* **133**:2748.

McChesney, M. B., and Oldstone, M. B. A., 1989, Virus-induced immunosuppression: Infections with measles virus and human immunodeficiency virus, *Adv. Immunol.* **45**:335.

Mims, C. A. and Wainwright, S., 1968, The immunosuppressive action of lymphocytic choriomeningitis virus in mice, *J. Immunol.* **101**:717.

Mondelli, M., and Eddleston, A. L. W. F., 1984, Mechanisms of liver cell injury in acute and chronic hepatitis B, *Semin. Liver Dis.* **4**:47.

Moskophidis, D., Pircher, H. P., Ciernik, I., Odermatt, B., Hengartner, H., and Zinkernagel, R. M., 1992, Supression of virus specific antibody production by CD8+ class I-restricted antiviral cytotoxic T cells *in vivo*, *J. Virol.* **66**:3661.

Odermatt, B., Eppler, M., Leist, T. P., Hengartner, H., and Zinkernagel, R. M., 1991, Virus-triggered acquired immunodeficiency by cytotoxic T-cell dependent destruction of antigen-presenting cells and lymph follicle structure, *Proc. Natl. Acad. Sci. USA* **88**:8252.

Oehen, S., Hengartner, H., and Zinkernagel, R. M., 1991, Vaccination for disease, *Science 251*:195.

Oldstone, M. B. A., and Dixon, F. J., 1967, Lymphocytic choriomeningitis: Production of anti-LCM antibody by "tolerant" LCM-infected mice, *Science* **158**:1193.

Oldstone, M. B. A., Whitton, J. L., Lewicki, H., and Tishon, A., 1988, Fine dissection of a nine amino acid glycoprotein epitope, a major determinant recognized by lymphocytic choriomeningitis virus-specific class-1-restricted H-2D^b cytotoxic T lymphocytes, *J. Exp. Med.* **168**:559.

Ottenhoff, T. H., and de Vries, R. R., 1990, Antigen reactivity and autoreactivity: two sides of the cellular immune response induced by mycobacteria, *Curr. Top. Microbiol. Immunol.* **155**:111.

Roost, H. P., Charan, S., Gobet, R., Ruedi, E., Hengartner, H., Althage, A., and Zinkernagel, R. M., 1988, An acquired immunodeficiency in mice caused by infection with lymphocytic choriomeningitis virus, *Eur. J. Immunol.* **18**:511.

Rosenberg, Z. F., and Fauci, A. S., 1990, Immunopathogenic mechanisms of HIV infection: cytokine induction of HIV expression, *Immunol. Today* **11**:176.

Schulz, M., Aichele, P., Vollenweider, M., Bobe, F. W., Cardinaux, F., Hengartner, H., and Zinkernagel, R. M., 1989, MHC dependent T cell epitopes of LCMV nucleoprotein and their protective capacity against viral disease, *Eur. J. Immunol.* **19**:1657.

Schulz, M., Aichele, P., Schneider, R., Hansen, T. H., Zinkernagel, R. M., and Hengartner, H., 1991, Major histocompatibility complex binding and T cell recognition of a viral nonapeptide containing a minimal tetrapeptide, *Eur. J. Immunol.* **21**:1181.

Shinohara, N., Watanabe, M., Sachs, D. H., and Hozumi, N., 1988, Killing of antigen-reactive B cells by class II-restricted, soluble antigen-specific CD8+ cytolytic T lymphocytes, *Nature* **336**:481.

Traub, E., 1936, Persistence of lymphocytic choriomeningitis virus in immune animals and its relation to immunity, *J. Exp. Med.* **63**:847.

Via, C. S., and Shearer, G. M., 1988, T-cell interactions in autoimmunity: Insights from a murine model of graft-versus-host disease, *Immunol. Today* **9**:207.

Walker, C. M., Moody, D. J., Stites, D. P., and Levy, J. A., 1986, CD8+ lymphocytes can control HIV infection *in vitro* by suppressing virus replication, *Science* **234**:1563.

Whitton, J. L., Gebhard, J. R., Lewicki, H., Tishon, A., and Oldstone, M. B., 1988a, Molecular definition of a major cytotoxic T-lymphocyte epitope in the glycoprotein of lymphocytic choriomeningitis virus, *J. Virol.* **62**:687.

Whitton, J. L., Southern, P. J., and Oldstone, M. B. A., 1988b, Analyses of the cytotoxic T lymphocyte responses to glycoprotein and nucleoprotein components of lymphocytic choriomeningitis virus, *Virology* **162**:321.

Whitton, J. L., Tishon, A., Lewicki, H., Gebhard, J., Cook, T., Salvato, M., Joly, E., and Oldstone, M. B. A., 1989, Molecular analyses of a five-amino-acid cytotoxic T lymphocyte (CTL) epitope: An immunodominant region which induces nonreciprocal CTL cross-reactivity, *J. Virol* **63**:4303.

Wiktor, T. J., Macfarlan, R. I., Reagan, K. J., Dietzschold, B., Curtis, P. J., Wunner, W. H., Kieny, M-P., Lathe, R., Lecocq, J-P., Mackett, M., Moss, B., and Koprowski, H., 1984, Protection from rabies by a vaccinia recombinant containing the rabies glycoprotein, *Proc. Natl. Acad. Sci. USA* **81**:7194.

Wille, A., Gessner, A., Lother, H., and Lehmann-Grube, F., 1989, Mechanism of recovery from acute virus infection. VIII. Treatment of lymphocytic choriomeningitis virus infected mice with interferon-γ monoclonal antibody blocks generation of virus-specific cytotoxic T lymphocytes and virus elimination, *Eur. J. Immunol.* **19**:1283.

Zinkernagel, R. M., and Doherty, P. C., 1979, MHC-restricted cytotoxic T cells: Studies on the biological role of polymorphic major transplantation antigens determining T cell restriction-specificity, function and responsiveness, *Adv. Immunol.* **27**:52.

Zinkernagel, R. M., Haenseler, E., Leist, T. P., Cerny, A., Hengartner, H., and Althage, A., 1986, T cell mediated hepatitis in mice infected with lymphocytic choriomeningitis virus, *J. Exp. Med.* **164**:1075.

CHAPTER 15

Construction and Evaluation of Recombinant Virus Vaccines for Lassa Fever

DAVID D. AUPERIN

I. PERSPECTIVE

A. Lassa Fever in West Africa

The first clinical and pathological descriptions of Lassa fever were reported in 1970 as a result of a program to investigate the prevalence of certain virus infections among missionaries in West Africa (Frame *et al.*, 1970). The index patient was a nurse working in a mission hospital in Lassa, Nigeria. She developed a persistent fever, was transferred to Bingham Memorial Hospital in Jos, Nigeria, and died on the 14th day of illness. A second nurse, who cared for the index patient on the day of admission to the hospital in Jos, developed a febrile illness 9 days later. The clinical course of this illness paralleled that of the index case in that the patient maintained a high fever, developed a severe sore throat, facial edema, and macular rash, and died on the 11th day of illness. A third nurse who had contact with the first two patients became ill 1 week after the second patient died. As her condition deteriorated similarly to the first two patients, she was transferred to Presbyterian Hospital in New York City. Upon the administration of supportive therapy, including fluid and electrolyte replacement and regimens of tetracycline, chloramphenicol, and chloroquine, this patient gradually recovered over the

DAVID D. AUPERIN • Special Pathogens Branch, Division of Viral and Rickettsial Diseases, Centers for Disease Control, Atlanta, Georgia 30333. *Present address:* Department of Molecular Genetics, Pfizer Central Research, Groton, Connecticut 06340.

The Arenaviridae, edited by Maria S. Salvato. Plenum Press, New York, 1993.

next several weeks. The clinical and epidemiological observations associated with these three cases suggested a new viral hemorrhagic fever not previously reported in Africa.

Lassa virus was isolated from Vero cell tissue cultures inoculated with serum samples from all three patients at the Yale Arbovirus Research Unit (YARU) (Buckley and Casals, 1970). Infectivity was sensitive to sodium deoxycholate, and replication was not inhibited by 5-bromo-2-deoxyuridine (BUdR), which suggested an enveloped RNA virus. The diameter of the virus was estimated to be 70–150 nm by filtration studies. Serological cross-reactivity in complement fixation tests indicated that Lassa virus was related to lymphocytic choriomeningitis virus (LCMV) and the Tacaribe group viruses. Electron microscopic examination of Vero cell cultures infected with Lassa virus revealed pleomorphic particles of variable size, with surface projections and electron-dense granules within the particles resembling those previously described for LCM, Machupo, and Tacaribe viruses (Speir *et al.*, 1970).

The first epidemiological study of an outbreak of Lassa fever occurred during January–February 1970 on the Jos plateau in northern Nigeria (Troup *et al.*, 1970). There were 26 suspected cases, seven of which were confirmed as Lassa fever by complement-fixation test; 10 patients died. The observation of mild forms of illness associated with Lassa virus infection suggested a spectrum of disease ranging from subclinical infection to serious, even fatal illness. Subsequently, several investigators identified Lassa fever in widespread areas of West Africa, particularly Nigeria, Liberia, and Sierra Leone (Monath *et al.*, 1973, 1974a; Carey *et al.*, 1972; Fraser *et al.*, 1974). A more recent epidemiological study indicates that the prevalence of antibody to Lassa virus in the human population of Sierra Leone ranges from 8% to 52% (McCormick *et al.*, 1987a). Roughly 100,000–300,000 Lassa virus infections occur annually in West Africa, resulting in approximately 5000 fatalities. In a study of two hospitals in rural eastern Sierra Leone, Lassa fever accounted for 10–16% of all adult medical admissions and for 30% of adult deaths (McCormick *et al.*, 1987b). Patient outcome was directly proportional to the titers of virus in blood; survival rates declined precipitously when virus titers exceeded $10^{3.0}$ $TCID_{50}$/ml (Johnson *et al.*, 1987).

The reservoir for Lassa virus is the multimammate rat *Mastomys natalensis* (Monath *et al.*, 1974b; Murphy and Walker, 1978). The rats become persistently infected *in utero* or neonatally, continually excreting virus in urine. *Mastomys natalensis* is ubiquitous in West African villages, where human infection occurs by contact with urine-contaminated food, water, or soil, all resulting from rodent infestation.

B. Treatment

The apparent success of Lassa virus immune plasma in the treatment of a patient with a laboratory-acquired infection from the YARU in 1969 offered hope that this would provide an effective therapy for Lassa fever (Leifer *et al.*, 1970). Twenty-four hours after the administration of plasma, the patient's condition improved dramatically, and virus was undetectable in serum and throat washings, which previously had titers of virus of $10^{4.0}$ and $10^{2.2}$ $TCID_{50}$/ml, respectively. The immune plasma, donated by the missionary nurse who had recovered from Lassa fever in Presbyterian Hospital, New York, neutralized 10^2 $TCID_{50}$ of Lassa virus and had a complement fixation titer of 32–64. Subsequent applications of immune plasma therapy in Lassa fever patients, however, were much less successful, casting doubt on the validity of this approach (Clayton, 1977; Keane and Gilles, 1977; McCormick *et al.*, 1986).

More recent reports on the use of Lassa virus immune plasma in laboratory animals have improved our understanding of its limitations (Jahrling, 1983; Jahrling and Peters, 1984). The effectiveness of immune plasma therapy is well documented when the plasma has a log neutralization index (LNI) > 2.0 and is administered on days 0, 3, and 6 following infection. However, the effectiveness of immune plasma therapy that is not initiated until after the onset of illness—the more likely scenario—is unknown. Unlike immunofluorescent and complement-fixing antibodies (IFA and CFA), which develop early during acute infection, neutralizing antibodies tend to develop late after convalescence in guinea pigs, and not uniformly in humans. High titers of IF or CF antibodies are not indicative of neutralizing antibodies, thereby accounting for the failure of immune plasma noted in some earlier reports. Only 14% of the serum samples obtained from convalescent-phase patients in Liberia contained an LNI > 2.0 (Jahrling, 1983). Strain specificity is also an important consideration, as it has been shown that the LNI of immune plasma is highest when measured against the homologous strain. These restrictions, coupled with the difficulties of obtaining, testing, transporting, and storing the serum in West Africa, severely limit its use.

The antiviral drug ribavirin has proved effective in the treatment of Lassa fever in laboratory animals (Jahrling *et al.*, 1980, 1984) and in humans in Sierra Leone (McCormick *et al.*, 1986). The drug is most effective when administered during the first 6 days after the onset of illness, significantly decreasing case fatality rates from approximately 50% to 5%–9%. It is the preferred method of treatment for patients diagnosed with Lassa fever today and is also recommended as a prophylactic agent in cases of possible exposure to Lassa virus (e.g., close personal contact with a Lassa fever patient).

C. Implications for Cell-Mediated Immunity

Although very little direct evidence exists for the involvement of cell-mediated immunity in the prevention of and recovery from Lassa fever, several observations are indicative of its role. Humans and laboratory animals recover from Lassa virus infections without developing neutralizing antibodies, presumably owing to the development of a cellular immune response. Whereas the passive transfer of immune plasma from Mopeia- or Mobala- (African arenaviruses serologically related to Lassa virus) infected animals fails to protect guinea pigs from a lethal Lassa virus infection, vaccination of guinea pigs with live LCM (ARM), Mopeia, or Mobala viruses does confer protection against a Lassa virus challenge (Jahrling and Peters, 1986). Furthermore, spleen effector cells taken from guinea pigs infected 15 days earlier with an avirulent Lassa virus strain, LCMV (ARM), or Mopeia virus recognize and effectively lyse target cells infected with the homologous or cross-protective virus.

All these observations are consistent with and supported by numerous studies on the cellular immune response to LCMV infection of mice. Adult immunocompetent mice generate a potent cytotoxic T-lymphocyte (CTL) response to LCMV infection (Cole *et al.*, 1973; Marker and Volkert, 1973). The response is virus specific (Marker and Volkert, 1973) and H-2 restricted in its activity (Zinkernagel and Doherty, 1974; 1979). Spleen cells obtained from these mice 7–8 days after LCMV infection mediate virus clearance when transferred into acutely infected mice (Mims and Blanden, 1972; Zinkernagel and Welsh, 1976). The population of cells responsible for virus clearance and the viral-induced immune response disease seen in adult mice inoculated intracerebrally with LCMV are T lymphocytes (Cole *et al.*, 1972; Doherty and Zinkernagel, 1975; Zinkernagel and Welsh, 1976). A single CTL clone presumably representing one viral determinant in the context of a single histocompatibility epitope can mediate and maintain viral clearance *in vivo* (Byrne and Oldstone, 1984). Adoptive transfer of CTL cells into persistently infected mice results in virus clearance from the visceral organs in 15 days and from the central nervous system in 120 days (Oldstone *et al.*, 1986). A major CTL epitope corresponding to amino acid residues 272–293, near the amino-terminus of the GP-2 glycoprotein of LCMV, has been identified and is recognized by mice with an H-2b MHC background (Whitton *et al.*, 1988). Assuming there are some basic similarities between the host immune response to LCM and Lassa virus infections, an effective vaccine for Lassa fever must be capable of eliciting an effective cell-mediated immune response.

D. Prospects for Vaccine Development

The development of a safe and efficacious vaccine for Lassa fever has proved difficult. A classic killed virus vaccine prepared from gamma-irradiated virus injected into rhesus monkeys elicits a substantial humoral immune response to the major structural proteins of Lassa virus (N, G1, and G2) but fails to protect the animals upon challenge with live virus (J. B. McCormick, personal communication). The development of fatal disease in vaccinated monkeys closely parallels that in unvaccinated controls: a high persistent fever, high viremia, and shock resulting in death 12–16 days postinfection. These results are consistent with the observation that humoral antibodies elicited in response to Lassa antigen are, for the most part, nonneutralizing and that what is needed to offer protection against a Lassa fever is a cell-mediated immune response.

Live attenuated virus vaccines offer an alternative to killed vaccines. Yellow fever, poliomyelitis, and measles are examples of viral diseases for which effective live attenuated vaccines are being used today (Theiler and Smith, 1937; Sabin, 1955; Enders *et al.*, 1960). In addition, a cold-adapted strain of influenza virus (Maassab, 1967; Kendal *et al.*, 1981) and an attenuated strain of varicella zoster virus (Takahashi *et al.*, 1974) are under development to provide vaccines for influenza and chicken pox, respectively. The advantages of live vaccines include lower cost and the possibility of stimulating a more effective immune response because of their ability to establish an infection and thereby present viral antigens to the host's immune system in a manner resembling that of the parent virus. The disadvantages include the possibility of reversion to virulence, which is well documented for polio virus vaccine (Kew *et al.*, 1981; Nkowane *et al.*, 1987), and adverse side affects, which occurred during the smallpox eradication program (Arita and Fenner, 1985). To evaluate the benefit of any vaccine, the risks associated with the disease and those associated with vaccination must be compared.

Experimental evidence indicates that a live virus vaccine may effectively prevent Lassa fever. Laboratory animals infected with various avirulent viruses serologically related to Lassa virus, including LCM (ARM), Mopeia, and Mobalia viruses, survive a subsequent challenge with virulent Lassa virus and experience only mild, if any, symptoms of disease (Jahrling and Peters, 1986; Walker *et al.*, 1982). The drawback to the development of any of these viruses into a live virus vaccine originates from the propensity for arenaviruses to establish persistent infections in rodents and tissue culture. In addition, Mopeia and Mobala viruses were isolated from rodents in Mozambique and the Central African Republic, respectively, and their inability to cause human disease is not assured. According to one report, liver and renal pathology has been associated with Mopeia virus infection of rhesus monkeys (Lange *et al.*, 1985).

The recently acquired ability to generate recombinant viruses, including poxviruses, adenoviruses, herpesviruses, and baculoviruses, which are capable of expressing heterologous genes, provides an attractive alternative to attenuated vaccines. It is beyond the scope of this chapter to review the literature on each of these expression systems. Suffice it to say there are advantages and disadvantages to each (e.g., abundant levels of antigen production and limited host range for baculoviruses; the ability to accept and express very large fragments of DNA and broad host range for poxviruses) that make them appropriate for specific applications.

The recombinant vaccinia virus system (Mackett *et al.*, 1982; Panicali and Paoletti, 1982) was chosen for application to Lassa fever primarily because of the well-documented safety and success of vaccinia virus in the smallpox eradication program. In addition, it has been used widely during the last 6 years to produce prototype vaccines that have proven effective in laboratory tests against hepatitis B (Smith *et al.*, 1983; Paoletti *et al.*, 1984), rabies (Kieny *et al.*, 1984; Wiktor *et al.*, 1984), herpes simplex (Paoletti *et al.*, 1984; Cremer *et al.*, 1985), respiratory syncytial (Olmstead *et al.*, 1986), and vesicular stomatitis (Mackett *et al.*, 1985) viruses. Of particular interest to its application to a Lassa fever vaccine is the demonstrated ability of recombinant vaccinia viruses to prime the host's immune system for secondary proliferative, and in one case, CTL responses (Bennink *et al.*, 1984; Earl *et al.*, 1986; Zarling *et al.*, 1986).

The recent discovery that the cell-mediated immune response to viral infection is directed at internal as well as external viral proteins (Bangham *et al.*, 1986; Bennink *et al.*, 1987; Puddington *et al.*, 1986; Townsend *et al.*, 1984; Yewdell *et al.*, 1985) and the demonstration of at least one epitope of the LCMV nucleoprotein on the surface of infected cells and mature virions (Zeller *et al.*, 1988) mean that all viral proteins (structural and nonstructural, internal and external) should be considered and evaluated in the design and development of a vaccine. To date, only recombinant vaccinia viruses expressing the nucleoprotein and envelope glycoproteins of Lassa virus have been constructed and evaluated as possible vaccine candidates (Auperin *et al.*, 1988; Clegg and Lloyd, 1987; Fisher-Hoch *et al.*, 1989; Morrison *et al.*, 1989).

II. RECOMBINANT VIRUS CONSTRUCTION

A. Lassa Virus Genome Structure

A complete description of the ambisense organization and coding strategies of the arenavirus L and S genomic RNAs has recently been reviewed (Bishop and Auperin, 1987). Molecular cloning and nucleotide sequence analyses have been performed on the S-genome RNA segment of two virulent Lassa virus isolates: one from Nigeria and the other from Sierra Leone (Auperin and McCormick, 1989; Auperin *et al.*, 1986; Clegg and Oram, 1985; Clegg *et al.*, 1990). These studies confirm that the genetic organization of the Lassa virus S RNA is similar to that of other members of the Arenaviridae for which sequences are available. The amino acid sequences of the N, G1, and G2 proteins of these two viruses are greater than 90% homologous. This is not surprising con-

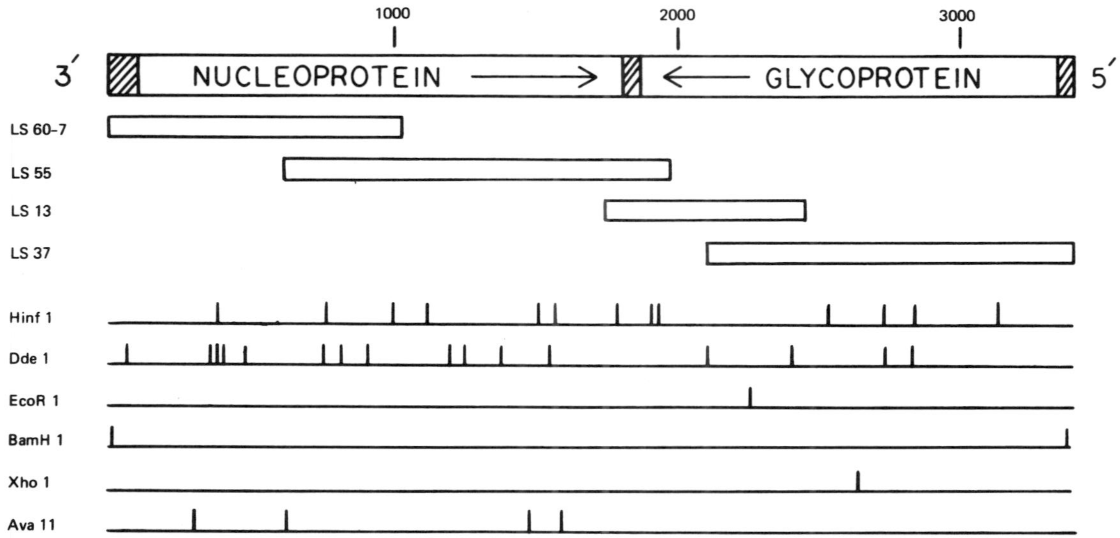

FIGURE 1. Schematic diagram of the Lassa virus S-genome RNA. The location and 5′ to 3′ orientation of the nucleoprotein and glycoprotein genes are indicated. Diagonally lined areas denote noncoding regions of the molecule. Below the RNA are positioned the four overlapping cDNA clones that have been isolated, sequenced, and assembled to express the N and GPC genes. The locations of several restriction enzyme cut sites are also included.

sidering the similarities of their biological properties and the proximity of their isolation.

A schematic diagram of the Lassa virus (Josiah strain) S-genome RNA, its coding strategy, and the four cDNA clones that span the sequence is shown in Fig. 1. The nucleoprotein and glycoprotein genes are located at the 3′ and 5′ regions of the S RNA, respectively, and arranged in ambisense organization. Assembly of the N and GPC genes for expression was accomplished by ligating the two appropriately overlapping cDNAs at a unique restriction site within the region of overlap.

B. Generation of Recombinant Viruses

The basic strategy for constructing recombinant vaccinia viruses that express heterologous genes is based on the demonstration that foreign DNA fragments can be specifically inserted into nonessential regions of the vaccinia virus genome by homologous recombination between flanking vaccinia DNA sequences (Mackett *et al.*, 1982; Panicali and Paoletti, 1982). A schematic diagram illustrating the process by which recombinant viruses are generated is shown in Fig. 2. A transfer vector, such as pSC11 (Chakrabarti *et al.*, 1985), is constructed to contain an expression cassette consisting of one or two vaccinia virus promoter elements positioned between the 5′ and the 3′ regions of a vaccinia virus DNA fragment corresponding to a nonessential part of the ge-

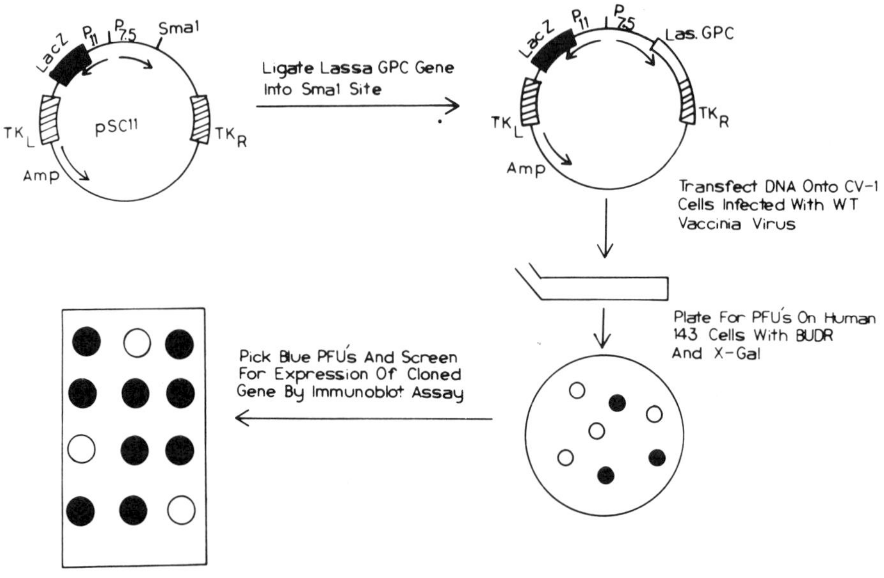

FIGURE 2. Schematic diagram of the process used to construct and isolate recombinant vaccinia viruses expressing Lassa virus genes. TK_L and TK_R refer to the 5′ (left) and 3′ (right) portions, respectively, of the vaccinia virus thymidine kinase gene. P_{11} and $P_{7.5}$ denote the promoter elements of vaccinia virus genes encoding 11-kDa and 7.5-kDa proteins, respectively.

nome. In this example, two promoter elements, P_{11}, and $P_{7.5}$, are positioned back to back within the vaccinia virus TK gene. The Lac Z gene has been inserted under the transcriptional control of the P_{11} promoter to provide a means for visual selection of recombinant virus. A foreign gene, in this case the Lassa virus glycoprotein (GPC) gene, is inserted into the unique Sma I site under the transcriptional control of the early-late $P_{7.5}$ promoter. Transfection of this plasmid onto CV-1 cells previously infected with wild-type vaccinia virus results in the production of recombinant virus with a TK phenotype containing the expression cassette inserted into the TK locus (insertional inactivation of the TK gene). Recombinant viruses are selected from wild-type viruses by plaque cloning in the presence of BUdR and visually identified by staining with X-Gal. Blue plaques are picked and screened for expression of the Lassa gene by reactivity with monoclonal antibodies. Once recombinant viruses are isolated and plaque-cloned, Southern blot analyses may be performed to confirm that homologous recombination has occurred between the targeting and the vaccinia DNA sequences.

Many factors, including the promoter, the parental poxvirus strain, the size and character of the 5′ and 3′ untranslated sequences, and the nature of the foreign protein, undoubtedly influence the levels of gene expression obtained by recombinant vaccinia viruses. The Lassa virus work to date has used only the $P_{7.5}$ promoter and two human-approved vaccine strains of vaccinia virus: Lister (Clegg and Lloyd, 1987) and New York Board of Health (NYBH) (Auperin *et al.*, 1988; Fisher-Hoch *et al.*, 1989; Morrison *et al.*, 1989). Examples of Lassa virus protein synthesis by NYBH recombinant vaccinia viruses containing the N or GPC genes are shown in Fig. 3. Abundant quantities of the Lassa nucleoprotein and more moderate quantities of the Lassa glycoproteins (GPC, G1, G2) are synthesized by recombinant virus-infected cells. The proteins appear identical to authentic Lassa virus proteins in electrophoretic mobility (Clegg and Lloyd, 1983) and react with a panel of monoclonal antibodies representing multiple distinct epitopes on each protein in a manner indistinguishable from authentic Lassa proteins (Morrison *et al.*, 1989). The appearance of the glycoprotein precursor (GPC) and mature glycoproteins, G1 and G2, indicate that these proteins are being translated and processed normally by vaccinia virus. Expression of the Lassa virus glycoproteins on the surface of V-LSGPC-infected cells has also been observed by using a pool of monoclonal antibodies specific for the G1 and G2 proteins.

III. VACCINE EFFICACY TRIALS

A. Guinea Pigs

The guinea pig has provided an effective small-animal model for a number of fatal arenavirus infections, including Pichinde, Junin, LCM,

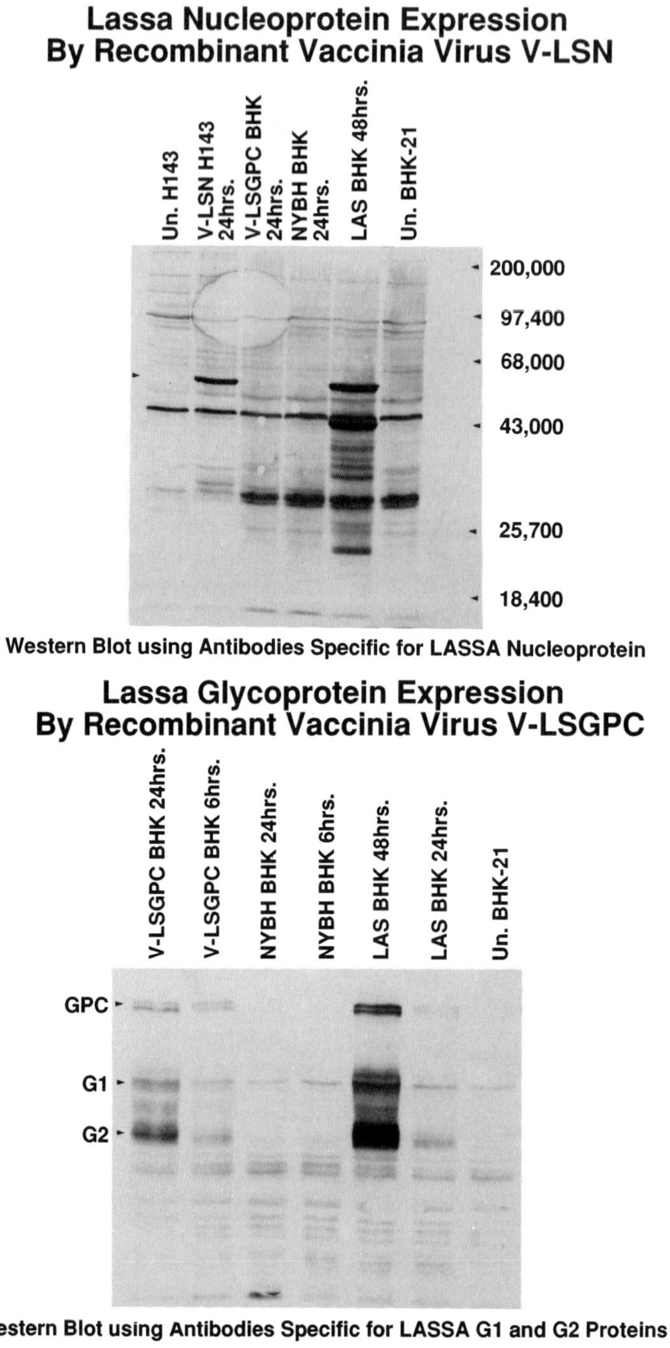

FIGURE 3. Western blot analyses of Lassa virus protein synthesis by recombinant vaccinia viruses: (A) vaccinia virus V-LSN; (B) vaccinia virus V-LSGPC. The positions of the N, GPC, G1, and G2 proteins are marked. H143 and BHK refer to human 143 cells, which exhibit a TK- phenotype, and BHK-21 cells, respectively.

and Lassa viruses. This material has been comprehensively reviewed by Peters *et al.* (1987). The pathogenesis and the development of disease in Lassa virus–infected guinea pigs has been described by Walker *et al.* (1975) and Jahrling *et al.* (1982). Inbred strain-13 guinea pigs are uniformly susceptible to Lassa virus infection. Death occurs in 100% of the animals 11–22 days postinfection with as few as two plaque-forming units (pfu) of virus. High titers of Lassa virus, 10^5 to 10^7 pfu/ml, can be isolated from the liver, the spleen, the pancreas, the adrenal gland, the kidney, the lymph node, the salivary gland, and the lung during acute infection. Lower titers of virus, usually 10^3 to 10^4 pfu/ml, are found in the plasma of lethally infected animals 9–11 days postinfection.

Lower mortality rates, approximately 30%, were observed upon Lassa virus infection of outbred Hartley-strain animals. Repeated passage of a Lassa virus inoculum in guinea pigs, however, produces a more virulent stock of virus capable of inducing fatal disease in 70–90% of outbred animals (J. V. Lange, personal communication). To date, only outbred Hartley-strain guinea pigs have been used in the preliminary evaluations of recombinant vaccinia virus vaccines for Lassa fever. Although they are not uniformly susceptible to Lassa virus infection, as are strain-13 animals, they offer the genetic diversity inherent in outbred populations, thereby permitting a more realistic assessment of vaccine efficacy.

Clegg and Lloyd (1987) reported that a recombinant vaccinia virus vaccine that expressed the Lassa virus (Nigerian strain) nucleoprotein gene successfully protected guinea pigs from a lethal Lassa virus infection. They used the Lister strain of vaccinia virus as the vector and expressed the Lassa virus N gene under the transcriptional control of the early–late vaccinia $P_{7.5}$ promoter inserted into the TK gene. Three groups of six guinea pigs each received no vaccination or vaccination with $10^{7.5}$ pfu of wild-type Lister virus or TK^- recombinant vaccinia virus by subcutaneous injection at several sites on the back of each animal. Antibodies to vaccinia virus proteins and Lassa virus N protein were detected by enzyme-linked immunosorbent assay (ELISA) following vaccination. All six guinea pigs receiving the vaccinia-vectored Lassa virus N protein vaccine survived the challenge infection in which 11 of 12 control animals died. No symptoms of Lassa fever were observed in the vaccine-protected animals. Lassa virus was undetectable in their blood, and their body temperatures remained normal after Lassa virus infection. In contrast to this, pyrexia, anorexia, and elevated serum aspartate aminotransferase (AST) levels—all typical of Lassa fever—were observed in the unprotected controls. Similar experiments conducted in our laboratory at the Centers for Disease Control (CDC) have confirmed the protective efficacy of a vaccinia-vectored Lassa nucleoprotein and extended the analyses to include the Lassa virus glycoproteins as well (Auperin *et al.*, 1988; Morrison *et al.*, 1989). The vaccinia-expressed N and GPC genes of the Josiah strain of Lassa virus each provide a significant level of protection from a lethal Lassa virus infection in guinea pigs. For these experiments, guinea pigs were vaccinated with a single intradermal injection, on the back, of wild-type NYBH or recombinant vaccinia virus (10^8 pfu). In addition, 24 animals received a single vaccination with each recombinant virus at separate sites.

Table I shows the survival of control and recombinant-virus-vaccinated guinea pigs following challenge infection with virulent Lassa virus. A single vaccination with either V-LSN or V-LSGPC protected 94% and 79%, respectively, of the animals from fatal Lassa fever. Vaccine failures did, however, occur in each group. Contrary to our expectations, a double vaccination with V-LSGPC and V-LSN resulted in higher mortality, 42%, than either single vaccination alone. Although we do not know the reasons for the higher mortality, it does not seem to be related to the presence or the quantity of postvaccination antibodies to Lassa virus. The proportion of animals with and without antibodies was approximately the same in those that survived and those that died, according to a Chi-square test for trends and proportions, p value of 0.68 (Morrison *et al.*, 1989).

The level of protection afforded by vaccination was not complete: 52 of 60 vaccinated animals became febrile after infection with Lassa virus, and all but one VLSN-vaccinated animal developed a detectable viremia. Although not prevented by vaccination, Lassa virus replication was limited, as mean virus titers in the blood of VLSN- and V-LSGPC-vaccinated animals were approximately 10-fold lower than in control animals 10–11 days postinfection. Mean viremia titers in animals receiving a double vaccination were intermediate, falling between those of the unprotected controls and the singly vaccinated guinea pigs. Recombinant-virus-vaccinated animals compared with control animals demonstrated an enhanced ability to clear virus. The percentage of recombinant-virus-vaccinated animals that were viremic 10 days postinfection and beyond declined rapidly in comparison to the control groups. All surviving guinea pigs were aviremic by 25 days postinfection; attempts

TABLE I. Lassa Fever Mortality in Unvaccinated Guinea Pigs and Guinea Pigs Vaccinated with NYBH and Recombinant Vaccinia Viruses Expressing the Lassa Virus Envelope Glycoproteins (V-LSGPC) and Nucleoprotein (V-LSN)

Vaccination	No. tested	No. died	% mortality
None	14	12	86
NYBH	23	14	61
V-LSGPC	19	4	21
V-LSN	17	1	6
V-LSGPC + V-LSN[a]	24	10	42

[a] Dual vaccination with V-LSGPC and V-LSN at separate sites.

TABLE II. Antibody Titers to Vaccinia and Lassa Viruses in Unvaccinated Guinea Pigs and Guinea Pigs Vaccinated with NYBH or Recombinant (V-LSGPC and V-LSN) Vaccinia Viruses before and after Challenge with Lassa Virus[a]

	Vaccinia		Lassa					
Vac.	Prevac.	Postvac.[b]	Prevac.	Postvac.	Days 4–5	Days 7–8	Days 9–10	Days 11–13
None (14)	ND (14)	ND (14)	ND (14)	ND (14)	ND (14)	ND (14)	ND (1) 8 (1) 32 (2) 64 (1) 128 (3) 256 (1) 512 (4)	32 (1) 64 (1) 128 (3) 256 (3) 512 (2) 1024 (1) 2048 (1)
NYBH (23)	ND (23)	16 (3) 32 (6) 64 (7) 128 (4) 256 (3)	ND (23)	ND (23)	ND (23)	ND (23)	ND (1) 8 (2) 16 (2) 32 (5) 64 (10) 128 (3)	32 (2) 64 (9) 128 (6) 256 (3) 512 (1) 2048 (2)
V-LSGPC (19)	ND (19)	32 (6) 64 (3) 128 (9) 256 (1)	ND (19)	ND (19)	ND (19)	ND (8) 8 (3) 128 (1) 256 (1) 512 (6)	ND (1) 64 (5) 128 (2) 256 (3) 512 (3) 1024 (5)	32 (1) 64 (2) 128 (6) 256 (2) 1024 (3) 2048 (5)
V-LSN (17)	ND (17)	8 (1) 16 (2) 32 (5) 64 (3) 128 (5) 256 (1)	ND (17)	ND (17)	ND (9) 8 (7) 64 (1)	8 (6) 32 (1) 64 (3) 128 (1) 256 (3) 512 (2) 1024 (1)	16 (3) 32 (1) 64 (1) 256 (2) 512 (4) 1024 (5) 2048 (1)	64 (1) 128 (3) 256 (2) 512 (5) 1024 (3) 2048 (2) 4096 (1)
V-LSGPC + V-LSN (24)	ND (24)	32 (1) 64 (3) 128 (9) 256 (8) 512 (3)	ND (24)	ND (8) 8 (2) 16 (5) 32 (7) 64 (1) 128 (1)	ND (14) 8 (8) 32 (2)	ND (7) 16 (3) 32 (4) 64 (4) 128 (3) 256 (2) 512 (1)	64 (1) 128 (4) 256 (10) 512 (6) 1024 (1)	128 (3) 256 (5) 512 (4) 1024 (3) 2048 (3)

[a] Figures in parentheses indicate the number of individual animals in each test group with specific antibody titers. ND, none detected.
[b] Animals were bled 28 days following vaccination.

to plaque virus from liver, spleen, and kidney samples taken at necropsy from several of the animals were unsuccessful.

The antibody response to vaccination and challenge infection is shown in Table II. IFAs to vaccinia virus were detected 28 days postvaccination at titers ranging from 8 to 512 in all animals receiving a vaccinia virus inoculation. IFAs to Lassa virus were detected in only 16 of the 24 animals receiving the double vaccination and in none of the animals receiving single vaccinations. Following infection with Lassa virus, a clear difference was noted between the animals that had received a recombinant virus vaccine and the controls. Forty-five of sixty vaccinated animals developed IFAs to Lassa virus 7–8 days postinfection; the titers ranged from 8 to 1024. None of the 37 control animals had antibody at that time. Antibodies to Lassa virus developed in the control animals 2–3 days later, as expected for a primary response.

The complete level of protection reported by Clegg and Lloyd (1987) after vaccination with their nucleoprotein-expressing recombinant virus is probably the result of multiple injections at vaccination. This also may account for their ability to detect Lassa virus antibodies after vaccination. Apparently a single vaccination with a TKNYBH recombinant vaccinia virus produces a very moderate antibody response to the foreign protein. Nevertheless, it is apparent that both the nucleoprotein and envelope glycoproteins of Lassa virus can elicit a protective immune response in guinea pigs.

B. Monkeys

Rhesus monkeys infected with Lassa virus provide the best animal model for the human disease. Jahrling *et al.* (1980) were first to describe the pathogenesis of Lassa virus infection of rhesus monkeys in detail. The clinical signs of illness include anorexia, normal to slightly elevated temperatures, which fall precipitously just before death, severe petechial rash that is most apparent on the face, and occasional mucosal bleeding from the gums and the nares. Serum glutamic oxalacetic transaminase (SGOT) and serum glutamic pyruvic transaminase (SGPT) activities were elevated 7–12 days postinfection; the peaks occurred on day 10 (mean SGOT 400 IU/liter and mean SGPT 123 IU/liter). A transient and modest leukopenia (minimal mean count; 3100 WBC/mm^3) was evident by day 7; platelet counts remained within normal limits. Virus-specific antibodies detectable by indirect immunofluorescence appeared on schedule, 7–10 days postinfection, and high titers of antibody were present in the serum when viremias reached maximum levels of 10^5 to 10^6 pfu/ml from day 9 onward.

Substantial quantities of virus were maintained in the serum of infected monkeys for a week or more despite the presence of high-titer Lassa antibodies. Lassa virus was found in all visceral organs examined, including liver, lung, adrenal gland, pancreas, spleen, kidney, and lymph node. Mean titers ranged from $10^{6.2}$ pfu/ml in lymph node to $10^{7.6}$ pfu/ml in liver and in all cases exceeded the titer of virus in serum. A more detailed description of the pathophysiology of Lassa fever is presented in the chapter by Dr. Susan P. Fisher-Hoch.

The systemic Lassa virus infection involving nearly all the visceral organs is quite comparable to observations made in humans. As in humans, the development of serum antibody does not result in rapid clearance of the virus or ensure clinical recovery. Damage to certain tissues varies from animal to animal (Walker *et al.*, 1982). The many similarities of the clinical course and the associated pathology of Lassa fever in monkeys and humans make monkeys an ideal model for a critical evaluation of prototype vaccines. The evaluation of recombinant virus vaccines for Lassa fever in monkeys is in its infancy. Fisher-Hoch and colleagues at the CDC have reported preliminary success with a recombinant vaccinia virus expressing the Lassa virus glycoprotein gene (Fisher-Hoch *et al.*, 1989). Table III summarizes the immunization schedules and outcome after challenge of rhesus monkeys given various vaccine regimens. The two NYBH-vaccinated rhesus monkeys died of Lassa fever 12 and 15 days postinfection. Their hematological and clinical chemistry profiles were consistent with those previously reported for Lassa fever in rhesus monkeys and severely ill human patients (Fisher-Hoch *et al.*, 1987). The two monkeys immunized with Mopeia virus survived with little or no signs of constitutional illness. All four V-LSGPC-vaccinated monkeys survived the lethal Lassa virus infection; mild, transient illness developed in two of the four. The interval between vaccination and challenge ranged from 37 to 284 days, which suggests that a single vaccination with V-LSGPC provides long-term immunity.

TABLE III. Monkey Immunization Schedules and Outcome after Lassa Virus Challenge

Monkey[a]	Immunizing[b] virus	Interval (days) vacc./chall.	Illness	Challenge outcome
Rh1	NYBH vaccinia	37	Yes	Died
Rh3	NYBH vaccinia	37	Yes	Died
Rh2	Mopeia virus	37	No	Survived
Rh7	Mopeia virus	37	No	Survived
Rh6	V-LSGPC	37	Mild	Survived
Rh8	V-LSGPC	37	Mild	Survived
Rh4	V-LSGPC	284	No	Survived
Rh5	V-LSGPC	284	No	Survived

[a] Rh = rhesus.
[b] Vaccinations consisted of either a total of 10^9 pfu of respective vaccinia virus administered intradermally at four sites on the upper arms and lateral aspect of the thighs or 10^4 pfu of Mopeia virus by subcutaneous injection.

Vaccination with Mopeia virus or V-LSGPC, however, did not completely protect the animals from infection with Lassa virus and all symptoms of disease. One Mopeia-vaccinated animal (Rh7) was febrile intermittently from day 9 through day 25 postinfection with Lassa virus and continued to have episodes of low-grade fever for the next 3 months. All four V-LSGPC-vaccinated monkeys became febrile 6–7 days after infection with Lassa virus; three became afebrile by day 10. Rh6 continued to have intermittent episodes of low-grade fever for 3 months. Low levels of Lassa virus, 10^2 to 10^3 pfu/ml, were detected in the serum of the two Mopeia and the four V-LSGPC vaccinated animals 7–9 days postinfection, but not at any other times. No abnormal platelet functions were detected in the Mopeia-vaccinated monkeys after infection with Lassa virus. A transient mild depression of platelet aggregation responses was observed in the V-LSGPC-vaccinated animals while they were viremic.

Antibodies capable of immunoprecipitating Lassa virus proteins were detected in monkeys vaccinated with Mopeia virus or V-LSGPC after vaccination. Cross-reactive antibodies in the sera of Mopeia-vaccinated monkeys immunoprecipitated the N, GPC, G1, and G2 proteins of Lassa virus. The antibodies elicited in response to vaccination with V-LSGPC immunoprecipitated only the glycoproteins of Lassa virus. An increase in glycoprotein-specific antibodies was detected in the sera of V-LSGPC-vaccinated animals after infection with Lassa virus but not in the Mopeia vaccine recipients. Repeated attempts failed to detect neutralizing antibodies until 3 months after the Lassa virus challenge. Nevertheless, virus was cleared rapidly by all surviving monkeys during the acute infection, which supports previous observations that neutralizing antibodies do not have a role in this process.

Unlike the results in guinea pigs, vaccination with V-LSN fails to protect monkeys from a lethal Lassa virus infection (S. P. Fisher-Hoch and J. C. S. Clegg, personal communications). Individual animal species seem to differ significantly in the ability to respond immunologically to a recombinant virus vaccine. The reasons for this are unknown.

IV. CONCLUDING REMARKS

A recombinant vaccinia virus vaccine for Lassa fever appears promising. The initial results with primates indicate that the Lassa virus glycoproteins elicit a protective immune response when expressed from a live virus such as vaccinia and that the nucleoprotein does not. The reasons for the gross differences exhibited by guinea pigs and monkeys vaccinated with V-LSN and challenged with Lassa virus are unknown. One assumes that humans will more closely resemble primates in their response to vaccination and that the envelope glycoproteins of Lassa virus will provide protective immunity. The fact that species-specific responses to these recombinant vaccinia viruses have been observed means that caution must be maintained in making such extrapolations. The best vaccine for humans may be one that presents the envelope glycoproteins and the nucleoprotein together or includes the proteins encoded on the Lassa virus L RNA segment. Recently, a recombinant vaccinia virus that simultaneously expresses the Lassa virus N and GPC genes has been constructed (Morrison *et al.*, 1990) and is undergoing evaluation in animal models. Clearly more research is required to ascertain what viral proteins, and in what context they should be presented to elicit the most effective immune response.

Lassa virus is endemic in widespread areas of western sub-Saharan Africa, and there is certain to be genetic variability between various isolates. The amino acid sequences of the nucleoproteins of a Nigerian and a Sierra Leoneian isolate have been compared and were found to be 90% homologous (Auperin and McCormick, 1989). Less homology would be expected for the G1 glycoproteins. The extent to which a recombinant virus vaccine constructed from the cloned genes of one isolate will effectively protect against exposure to other strains of Lassa virus has not yet been examined. Clearly this is an important consideration in the development of an effective vaccine.

The prototype vaccines constructed thus far have failed to prevent Lassa virus infection, the early rounds of virus replication, and the development of mild symptoms of Lassa fever. Protection is believed to result from the priming of a cell-mediated immune response that would not be expected to prevent the initial infection but would respond to it by amplifying viral-specific cytotoxic T cells, which ultimately contain and clear the infection. We do not know whether a more complete level of protection (e.g., prevention of a detectable viremia and elimination of all symptoms of disease after Lassa virus infection) can be obtained through increased levels of protein expression by the recombinant vaccinia virus or the use of an alternate virus vector that produces a systemic rather than a local infection (i.e., adenovirus or varicella zoster). Increased levels of protein expression by recombinant viruses can presumably be obtained through the engineering of promoter sequences or the use of multiple expression vectors to insert two or more copies of a given gene into a single virus.

The future for vaccine development looks promising in light of the recombinant virus gene expression technology developed during the last 10 years. Certainly for diseases such as Lassa fever, in which a killed antigen vaccine has proven ineffective and an attenuated virus vaccine is not available, a live recombinant virus vaccine provides a very attractive alternative. A major consideration these days of any live virus vaccine is its safety in a population consisting of significant numbers of immunocompromised people. This is particularly evident in Africa where human immunodeficiency virus (HIV) infection is more prevalent than anywhere else in the world. Although complications arising from vaccination with vaccinia virus are extremely rare in healthy populations and recombinant vaccinia viruses with a TK⁻ phenotype are less virulent than wild-type virus (Buller *et al.*, 1985), the safety of such a vaccine in today's population is largely unknown. Ultimately the risks associated with recombinant virus vaccination will need to be determined and weighed against those associated with the disease.

REFERENCES

Arita, I., and Fenner, F., 1985, Complications of smallpox vaccination, in: *Vaccinia Viruses as Vectors for Vaccine Antigens* (G. V. Quinnan, Jr., ed.), pp. 49–60, Elsevier, New York.

Auperin, D. D., and McCormick, J. B. M., 1989, Nucleotide sequence of the Lassa virus (Josiah strain) S genome RNA and amino acid sequence comparison of the N and GPC proteins to other Arenaviruses, *Virology* **168**:421.

Auperin, D. D., Sasso, D. R., and McCormick, J. B., 1986, Nucleotide sequence of the glycoprotein gene and intergenic region of the Lassa virus S genome RNA, *Virology* **154**:155.

Auperin, D. D., Esposito, J. J., Lange, J. V., Bauer, S. P., Knight, J., Sasso, D. R., and McCormick, J. B., 1988, Construction of a recombinant vaccinia virus expressing the Lassa virus glycoprotein gene and protection of guinea pigs from a lethal Lassa virus infection, *Virus Res.* **9**:233.

Bangham, C. R. M., Openshaw, P. J. M., Ball, L. A., King, A. M. Q., Wertz, G. W., and Askonas, B. A., 1986, Human and murine cytotoxic T cells specific to respiratory syncytial virus recognize the viral nucleocapsid (N), but not the major glycoprotein (G), expressed by vaccinia virus recombinants, *J. Immunol.* **137**:3973.

Bennink, J. R., Yewdell, J. W., Smith, G. L., Moller, C., and Moss, B., 1984, Recombinant vaccinia virus primes and stimulates influenza haemagglutinin-specific cytotoxic T cells, *Nature* **311**:578.

Bennink, J. R., Yewdell, J. W., Smith, G. L., and Moss, B., 1987, Anti-influenza virus cytotoxic T lymphocytes recognize the three viral polymerases and a nonstructural protein: Responsiveness to individual viral antigens is major histocompatibility complex controlled, *J. Virol.* **61**:1098.

Bishop, D. H. L., and Auperin, D. D., 1987, Arenavirus gene structure and organization, in: *Arenaviruses: Biology and Immunotherapy* (M. B. A. Oldstone, ed.), Curr. Top. in Microb. and Immunol., Vol. 133, pp. 5–17, Springer-Verlag, Berlin.

Buckley, S. M., and Casals, J., 1970, Lassa fever, a new virus disease of man from West Africa. III. Isolation and characterization of the virus, *Am. J. Trop. Med. Hyg.* **19**:680.

Buller, R. M. L., Smith, G. L., Cremer, K., Notkins, A. L., and Moss, B., 1985, Decreased virulence of recombinant vaccinia virus expression vectors is associated with a thymidine kinase-negative phenotype, *Nature* **317**:813.

Byrne, J. A., and Oldstone, M. B. A., 1984, Biology of cloned cytotoxic T lymphocytes specific for lymphocytic choriomeningits virus: Clearance of virus *in vivo, J. Virol.* **51**:682.

Carey, D. E., Kemp, G. E., White, H. A., Pinneo, L., Addy, R,F., Fom, A. L. M. D., Stroh, G., Casals, J., and Henderson, B. E., 1972, Lassa fever: Epidemiological aspects of the 1970 epidemic, Jos, Nigeria, *Trans. Roy. Soc. Trop. Med. Hyg.* **66**:402.

Chakrabarti, S., Brechling, K., and Moss, B., 1985, Vaccinia virus expression vector: coexpression of β-galactosidase provides visual screening of recombinant virus plaques, *Mol. Cell. Biol.* **5**:3403.

Clayton, A. J., 1977, Lassa immune serum, *Bull. WHO* **5**:435.

Clegg, J. C. S., and Lloyd, G., 1983, Structural and cell-associated proteins of Lassa virus, *J. Gen. Virol.* **64**:1127.

Clegg, J. C. S., and Lloyd, G., 1987, Vaccinia recombinant expressing Lassa virus internal nucleocapsid protein protects guinea pigs against Lassa fever, *Lancet* **2**:186.

Clegg, J. C. S., and Oram, J. D., 1985, Molecular cloning of Lassa virus RNA: Nucleotide sequence and expression of the nucleocapsid protein gene, *Virology* **144**:363.

Clegg, J. C. S., Wilson, S. M., and Oram, J. D., 1990, Nucleotide sequence of the S RNA of Lassa virus (Nigerian strain) and comparative analysis of arenavirus gene products, *Virus Res.* **18**:151.

Cole, G. A., Nathanson, N., and Pendergast, R. A., 1972, Requirement for theta-bearing cells in lymphocytic choriomeningitis virus-induced central nervous system disease, *Nature (Lond.)* **238**:325.

Cole, G. A., Pendergast, R. A., and Henney, C. S., 1973, *In vitro* correlates of LCM virus-induced immune response, in: *Lymphocytic Choriomeningitis Virus and Other Arenaviruses* (F. Lehman-Grube, ed.), pp. 61–71, Springer-Verlag, New York.

Cremer, K. J., Mackett, M., Wohlenberg, C., Notkins, A. L., and Moss, B., 1985, Vaccinia virus recombinant expressing herpes simplex type 1 glycoprotein D prevents latent herpes in mice, *Science* **228**:737.

Doherty, P. C., and Zinkernagel, R. M., 1975, Capacity of sensitized thymus-derived lymphocytes to induce fatal lymphocytic choriomeningitis is restricted by the H-2 gene complex, *J. Immunol.* **114**:30.

Earl, P. L., Moss, B., Morrison, R. P., Wehrly, K., Nishio, J., and Cheseboro, B., 1986, T lymphocyte priming and protection against Friend leukemia by vaccinia-retrovirus *env* gene recombinant, *Science* **234**:728.

Enders, J. F., Katz, S. L., Milovanovic, M. V., and Holloway, A., 1960, Studies on an attenuated measles-virus vaccine. 1. Development and preparation of the vaccine: Techniques for assay of effects of vaccination, *N. Engl. J. Med.* **263**:153.

Fisher-Hoch, S. P., Mitchell, S. W., Sasso, D. R., Lange, J. V., Ramsey, R., and McCormick, J. B., 1987, Physiological and immunologic disturbance associated with shock in a primate model of Lassa fever, *J. Infect. Dis.* **155**:465.

Fisher-Hoch, S. P., McCormick, J. B., Auperin, D. D., Brown, B. G., Castor, M., Perez, G., Ruo, S., Conaty, A., Brammer, L., and Bauer, S. P., 1989, Protection of rhesus monkeys from fatal Lassa fever by vaccination with a recombinant vaccinia virus containing the Lassa virus glycoprotein gene, *Proc. Natl. Acad. Sci. USA* **86**:317.

Frame, J. D., Baldwin J. M.,Jr., Gocke, D. J., and Troup, J. M., 1970, Lassa fever, a new virus disease of man from West Africa. 1. Clinical description and pathological findings, *Am. J. Trop. Med. Hyg.* **19**:670.

Fraser, D. W., Campbell, C. C., Monath, T. P., Goff, P. A., and Gregg, M. B., 1974, Lassa fever in the eastern province of Sierra Leone, 1970–1972. I. Epidemiologic studies, *Am. J. Trop. Med. Hyg.* **23**:1131.

Jahrling, P. B., 1983, Protection of Lassa virus–infected guinea pigs with Lassa-immune plasma of guinea pig, primate, and human origin, *J. Med. Virol.* **12**:93.

Jahrling, P. B., and Peters, C. J., 1984, Passive antibody therapy of Lassa fever in Cynomolgus monkeys: Importance of neutralizing antibody and Lassa virus strain, *Infect. Immun.* **44**:528.

Jahrling, P. B., and Peters, C. J., 1986, Serology and virulence diversity among Old World arenaviruses, and the relevance to vaccine development, *Med. Microbial Immunol.* **175**:165.

Jahrling, P. B., Hesse, R. A., Eddy, G. A., Johnson, K. M., Callis, R. T., and Stephen, E. I., 1980, Lassa virus infection of rhesus monkeys: Pathogenesis and treatment with ribavirin, *J. Infect. Dis.* **141**:580.

Jahrling, P. B., Smith, S., Hesse, R. A., and Rhoderick, J. B., 1982, Pathogenesis of Lassa virus infection in guinea pigs, *Infect. Immun.* **37**:771.

Jahrling, P. B., Peters, C. J., and Stephen, E. L., 1984, Enhanced treatment of Lassa fever by immune plasma combined with ribavirin in cynomolgus monkeys, *J. Infect. Dis.* **149**:420.

Johnson, K. M., McCormick, J. B., Webb, P. A., Smith, E. S., Elliott, L. H., and King, I. J., 1987, Clinical virology of Lassa fever in hospitalized patients, *J. Infect. Dis.* **155**:456.

Keane, E., and Gilles, H. M., 1977, Lassa fever in Panguma hospital, Sierra Leone, 1973–6, *Br. Med. J.* **1**:1399.

Kendal, A. P., Maassab, H. F., Alexandrova, G. I., and Ghendon, Y. Z., 1981, Development of cold-adapted recombinant live, attenuated influenza A vaccines in the U.S.A. and U.S.S.R., *Antiviral Res.* **1**:339.

Kew, O. M., Nottay, B. K., Hatch, M. H., Nakano, J. H., and Obijeski, J. F., 1981, Multiple genetic changes can occur in the oral poliovaccines upon replication in humans, *J. Gen. Virol.* **56**:337.

Kieny, M. P., Lathe, R., Drillien, R., Spehmer, D., Skory, S., Schmitt, D., Wiktor, T., Koprowski, H., and Lecocq, J. P., 1984, Expression of rabies virus glycoprotein from a recombinant vaccinia virus, *Nature* **312**:163.

Lange, J. V., Mitchell, S. W., McCormick, J. B., Walker, D. H., Evatt, B. L., and Ramsey, R., 1985, Kinetic study of platelets and fibrinogen in Lassa virus–infected monkeys and early pathologic events in Mopeia virus-infected monkeys, *Am. J. Trop. Med. Hyg.* **34**:999.

Leifer, E., Gocke, D. J., and Bourne, H., 1970, Lassa fever, a new virus disease of man from West Africa. II. Report of a laboratory-acquired infection treated with plasma from a person recently recovered from the disease, *Am. J. Trop. Med. Hyg.* **19**:677.

Maassab, H. F., 1967, Adaptation and growth characteristics of influenza virus at 25° C, *Nature* **213**:612.

Mackett, M., Smith, G. L., and Moss, B., 1982, Vaccinia virus: A selectable eukaryotic cloning and expression vector, *Proc. Natl. Acad. Sci. USA* **79**:7415.

Mackett, M., Yilma, T., Rose, J. K., and Moss, B., 1985, Vaccinia virus recombinants: expression of VSV genes and protective immunization of mice and cattle, *Science* **227**:433.

Marker, O., and Volkert, M., 1973, Studies on cell-mediated immunity on lymphocytic choriomeningitis virus in mice, *J. Exp. Med.* **137**:1511.

McCormick, J. B., King, I. J., Webb, P. A., Scribner, C. L., Craven, R. B., Johnson, K. M., Elliott, L. H., and Belmount-Williams, R., 1986, Lassa fever effective therapy with ribavirin, *N. Engl. J. Med.* **314**:20.

McCormick, J. B., Webb, P. A., Krebbs, J. W., Johnson, K. M., and Smith, E. S., 1987a, A prospective study of the epidemiology and ecology of Lassa fever, *J. Infect. Dis.* **155**:437.

McCormick, J. B., King, I. J., Webb, P. A., Johnson, K. M., O'Sullivan, R., Smith, E. S., Tripper, S., and Tong, T. C., 1987b, A case control study of the clinical diagnosis and course of Lassa fever, *J. Infect. Dis.* **155**:445.

Mims, C. A., and Blanden, R. V., 1972, Antiviral action of immune lymphocytes in mice infected with lymphocytic choriomeningitis virus, *Infect. Immun.* **6**:695.

Monath, T. P., Mertens, P. E., Patton, R., Mosur, G. R., Baum, J. J., Pinneo, L., Gary, G. W., Jr., and Kissling, R. E., 1973, A hospital epidemic of Lassa fever in Zorzor, Liberia, March–April, 1972, *Am. J. Trop. Med. Hyg.* **22**:773.

Monath, T. P., Maher, M., Casals, J., Kissling R. E., and Cacciapuoti, A., 1974a, Lassa fever in the eastern province of Sierra Leone, 1970–1972. II. Clinical observations and virological studies on selected hospital cases, *Am. J. Trop. Med. Hyg.* **23**:1140.

Monath, T. P., Newhouse, V. F., Kemp, G. E., Setzer, H. W., and Cacciapuoti, A., 1974b, Lassa virus isolation from *Mastomys natalensis* rodents during an epidemic in Sierra Leone, *Science* **183**:263.

Morrison, H. G., Bauer, S. P., Lange, J. V., Esposito, J. J., McCormick, J. B., and Auperin, D. D., 1989, Protection of guinea pigs from Lassa fever by vaccinia virus recombinants expressing the nucleoprotein or the envelope glycoproteins of Lassa virus, *Virology* **171**:179.

Morrison, H. G., Goldsmith, C. S., Regnery, H. L., and Auperin, D. D., 1990, Simultaneous expression of the Lassa virus N and GPC genes from a single recombinant vaccinia virus, *Virus Res.* **18**:231.

Murphy, F. A., and Walker, D. H., 1978, Arenaviruses: persistent infection and viral survival in reservoir hosts, in: *Viruses and Environment* (E. Kurstak and K. Maramorosch, eds.), pp. 155–180, Academic Press, New York.

Nkowane, B. M., Wassilak, S. G. F., Orenstein, W. A., Bart, K. J., Schonberger, L. B., Hinman, A. R., and Kew, O. M., 1987, Vaccine-associated paralytic poliomyelitis, United States: 1973 through 1984, *JAMA* **257**:1335.

Oldstone, M. B. A., Blount, P., and Southern, P. J., 1986, Cytoimmunotherapy for persistent virus infection reveals a unique clearance pattern from the central nervous system, *Nature* **321**:239.

Olmstead, R. A., Elango, N., Prince, G. A., Murphy, B. R., Johnson, P. R., Moss, B., Chanock, R. M., and Collins, P. L., 1986, Expression of the F glycoprotein of respiratory syncytial virus by a recombinant vaccinia virus: comparison of the individual contribution of the F and G glycoproteins to host immunity, *Proc. Natl. Acad. Sci. USA* **83**:7462.

Panicali, D., and Paoletti, E., 1982, Construction of poxviruses as cloning vectors: Insertion of the thymidine kinase gene from herpes simplex virus into the DNA of infectious vaccinia virus, *Proc. Natl. Acad. Sci. USA* **79**:4927.

Paoletti, E., Lipinskas, B. R., Samsonoff, C., Mercer, S., and Panicali, D., 1984, Construction of live vaccines using genetically engineered poxviruses: Biological activity of vaccinia virus recombinants expressing the hepatitis B virus surface antigen and the herpes simplex virus glycoprotein D, *Proc. Natl. Acad. Sci. USA* **81**:193.

Peters, C. J., Jahrling, P. B., Liu, C. T., Kenyon, R. H., McKee, K. T., Jr., and Barrera Oro, J. G., 1987, Experimental studies of arenaviral hemorrhagic fevers, in: *Arenaviruses: Biology and Immunotherapy* (M. B. A. Oldstone, ed.), Curr. Top. in Microb. and Immunol., Vol. 134, pp. 5–68, Springer-Verlag, Berlin.

Puddington, L., Bevan, M. J., Rose, J. K., and Lefroncois, L., 1986, N protein is the predominant antigen recognized by vesicular stomatitis virus-specific cytotoxic T cells, *J. Virol.* **60**:708.

Sabin, A. B., 1955, Characteristics and genetic potentialities of experimentally produced and naturally occurring variants of poliomyelitis virus, *Ann. NY Acad. Sci.* **61**:924.

Smith, G. L., Mackett, M., and Moss, B., 1983, Infectious vaccinia virus recombinants that express hepatitis B virus surface antigen, *Nature* **302**:490.

Speir, R. W., Wood, O., Liebhaber, H., and Buckley, S. M., 1970, Lassa fever, a new virus disease of man from West Africa. IV. Electron microscopy of vero cell cultures infected with Lassa virus, *Am. J. Trop. Med. Hyg.* **19**:692.

Takahashi, M., Otsuka, T., Okuno, Y., Asano, Y., Yazaki, T., and lsomura, S., 1974, Live vaccine used to prevent the spread of varicella in children in hospital, *Lancet* **2**:1288.

Theiler, M., and Smith, H. H., 1937, The use of yellow fever virus modified by *in vitro* cultivation for human immunization, *J. Exp. Med.* **65**:787.

Townsend, A. R. M., McMichael, A. J., Carter, N. P., Huddleston, J. A., and Brownlee, G. G., 1984, Cytotoxic T cell recognition of the influenza virus nucleoprotein and hemagglutinin expressed in transfected mouse L cells, *Cell* **39**:13.

Troup, J. M., White, H. A., A. L. M. D. Fom, and Carey, D. E., 1970, An outbreak of Lassa fever on the Jos plateau, Nigeria, in January–February 1970: A preliminary report, *Am. J. Trop. Med. Hyg.* **19**:695.

Walker, D. H., Wulff, H., Lange, J. V., and Murphy, F. A., 1975, Comparative pathology of Lassa virus infection in monkeys, guinea pigs, and *Mastomys natalensis, Bull. WHO* **52**:523.

Walker, D. H., Johnson, K. M., Lange, J. V., Gardner, J. J., Kiley, M. P., and McCormick, J. B., 1982, Experimental infection of rhesus monkeys with Lassa virus and a closely related arenavirus, Mozambique virus, *J. Infect. Dis.* **146**:360.

Whitton, J. L., Gebhard, J. R., Lewicki, H., Tishon, A., and Oldstone, M. B. A., 1988,

Molecular definition of a major cytotoxic T-lymphocyte epitope in the glycoprotein of lymphocytic choriomeningitis virus, *J. Virol.* **62**:687.

Wiktor, T. J., Macfarlan, R. I., Reagan, K. J., Dietzschold, B., Curtis, P. J., Wunner, W. H., Kieny, M. P., Lathe, R., Lecocq, J. P., Mackett, M., Moss, B., and Koprowski, H., 1984, Protection from rabies by a vaccinia virus recombinant containing the rabies virus glycoprotein gene, *Proc. Natl. Acad. Sci. USA* **81**:7194.

Yewdell, J. W., Bennink, J. R., Smith, G. L., and Moss, B., 1985, Influenza A virus nucleoprotein is a major target antigen for cross-reactive anti-influenza A virus cytotoxic T lymphocytes, *Proc. Natl. Acad. Sci. USA* **82**:1785.

Zarling, J. M., Morton, W., Moran, P. A., McClure, J., Kosowski, S. G., and Hu, S-L., 1986, T-cell responses to human AIDS virus in Macaques immunized with recombinant vaccinia viruses, *Nature* **323**:344.

Zeller, W., Bruns, M., and Lehmann-Grube, F., 1988, Lymphocytic choriomeningitis virus X. Demonstration of nucleoprotein on the surface of infected cells, *Virology* **162**:90.

Zinkernagel, R. M., and Doherty, P. C., 1974, Restriction of *in vitro* T cell–mediated cytotoxicity in lymphocytic choriomeningitis within a syngeneic or semiallogeneic system, *Nature (Lond.)* **248**:701.

Zinkernagel, R. M., and Doherty, P. C., 1979, MHC-restricted cytotoxic T cells: Studies on the biological role of polymorphic major transplantation antigens determining T cell restriction specificity, function, and responsiveness, *Adv. Immunol.* **7**:151.

Zinkernagel, R. M., and Welsh, R. M., 1976, H-2 compatibility requirement for virusspecific T cell mediated effector functions *in vivo:* 1. Specificity of T cells conferring anti-viral protection against lymphocytic choriomeningitis virus is associated with H-2K and H-2D, *J. Immunol.* **117**:1495.

CHAPTER 16

The Tacaribe Complex

The Close Relationship between a Pathogenic (Junin) and a Nonpathogenic (Tacaribe) Arenavirus

LILIANA A. MARTINEZ PERALTA,
CELIA E. COTO, AND
MERCEDES C. WEISSENBACHER

I. INTRODUCTION

The Tacaribe complex of Arenaviridae is comprised of a group of 10 serologically cross-related viruses that are transmitted by rodents living in different geographic locations of the Western hemisphere from Florida to Argentina. So far, members of the Tacaribe group, also referred to as the New World arenaviruses, include Junin (the agent of Argentine hemorrhagic fever), Machupo (the agent of Bolivian hemorrhagic fever), Guanarito (the agent of Venezuelan hemorrhagic fever), and Pichinde, Tacaribe, Tamiami, Latino, Amapari, Flexal, and Parana viruses, none of which are currently known to be naturally infectious or pathogenic for humans (Howard, 1986).

All agents of the Tacaribe complex are strongly rodent-associated

LILIANA A. MARTINEZ PERALTA AND MERCEDES C. WEISSENBACHER • Catedra de Microbiologia, Parasitologia e Immunologia, Facultad de Medicina de la Universidad de Buenos Aires, 1121 Buenos Aires, Argentina. CELIA E. COTO • Laboratorio de Virologia, Departamento de Quimica Biologica, Facultad de Ciencias Exactas y Naturales, Ciudad Universitaria, Pabellón 2, Piso 4, 1428 Buenos Aires, Argentina.

The Arenaviridae, edited by Maria S. Salvato. Plenum Press, New York, 1993.

except Tacaribe virus, which was isolated from the tissue of fructivorous bats (Downs *et al.*, 1963). A clinically silent, chronic infection of one or two rodent species with both horizontal and vertical intraspecific virus transmission represents the fundamental mechanism for the natural maintenance of these agents (Johnson *et al.*, 1973; Vitullo and Merani, 1988). A full listing of the Tacaribe group, together with the geographic location, year of isolation, and the main reservoir in nature, is presented in Table 1. For more detailed information on Tacaribe virus group ecology, the reader is referred to Howard (1986).

II. HISTORICAL BACKGROUND OF TACARIBE COMPLEX

Recognition of Junin and Machupo viruses as human pathogens dates back to no more than 30 years ago. However, as Johnson *et al.* pointed out (1973), the restricted natural host range and highly evolved state of parasitism in specific indigenous rodents strongly suggest that Junin and Machupo viruses have lived in South America for millennia. Since most members of the Tacaribe virus group are nonpathogenic in humans, they remained unknown until their rodent hosts were captured in the course of virus survey programs (Pinheiro *et al.*, 1966; Trapido and Sanmartin, 1971). Thus it is conceivable that there exist many more arenaviruses, which will remain silent in chronically infected rodents, until a virologist brings them to light or continuous mutations in their RNA genomes originates some new virus strain pathogenic to humans.

Junin virus was first isolated during a serious outbreak of hemorrhagic fever in Argentina in 1958 (Parodi *et al.*, 1958). Five years later it

TABLE I. Tacaribe Complex: Isolation Data and Reservoirs

Name of virus	Year of isolation	Location	Reservoir
Tacaribe	1956	Trinidad	*Artibeus lituratus* *Artibeus jamaicensis*
Junin	1957	Argentina	*Calomys musculinus* *Calomys laucha* *Akodon azarae*
Machupo	1962	Bolivia	*Calomys callosus*
Tamiami	1963	U.S.A.	*Sigmodon hispidus*
Amapari	1964	Brazil	*Oryzomys goeldii* *Neacomys guianae*
Latino	1965	Bolivia	*Calomys callosus*
Parana	1965	Paraguay	*Oryzomys buccinatus*
Pichinde	1970	Colombia	*Oryzomys albigularis* *Thomasomys fuscatus*
Flexal	1975	Brazil	*Oryzomys* spp.
Guanarito	1990	Venezuela	*Oryzomys* spp.

was reported that a new virus named Tacaribe had been isolated from bats living in Trinidad during a study of rabies infection in the bat population (Downs *et al.*, 1963). The agent was found not to be related to any of the then-known "arborviruses" and was named Tacaribe after a pre-Columbian tribe of Trinidadian Indians. It was tentatively assigned at first to the arthropod-borne virus group on the basis of its inactivation by deoxycholate. Soon a serological relationship was found between Junin and Tacaribe viruses isolated some 3000 miles apart from each other.

Machupo virus, discovered in 1963 (Johnson *et al.*, 1965), was also found to be related to Junin and Tacaribe viruses, leading to the foundation of the "Tacaribe complex." From then on, any new virus isolated in South and North American territory presenting a serological relationship with Tacaribe virus was included in the complex. Among viruses of the Tacaribe group extensive studies were performed with Pichinde virus, selected as a model for biochemical studies of Arenaviridae, and with Junin and Tacaribe viruses, the former for its high pathogenicity in humans and the latter because of its potential use as a vaccine against Argentine hemorrhagic fever (AHF).

III. ANTIGENIC RELATIONSHIP OF TACARIBE COMPLEX VIRUSES

A. Old World Arenaviruses

By the time arenaviruses were proposed as a new taxonomic group, Tacaribe virus was related to the prototypic arenavirus, lymphocytic choriomeningitis virus (LCMV), by a similar ultrastructure and a weak, one-way, antigenic cross-relationship (Rowe *et al.*, 1970).

Immunological techniques such as complement fixation, indirect immunofluorescence, and immune precipitation using hyperimmune sera allowed the formation of the Tacaribe complex but showed less relationship with Old World arenaviruses (Casals *et al.*, 1975; Wulff *et al.*, 1978). This situation promptly changed with the availability of monoclonal antibodies reacting with several arenaviruses (Buchmeier *et al.*, 1988) and the analysis of their molecular structure, which revealed a closer relationship between New and Old World arenaviruses than previously thought. Five of six monoclonal antibodies against Pichinde virus were not cross-reactive with LCMV by indirect immunofluorescence, but one (P3.B-3) cross-reacted with Amapari, Junin, Parana, Tamiami, LCM, and Lassa but not Machupo and Tacaribe. Since P3.B-3 antibody reacted with the NP of both Pichinde and LCM by immunoprecipitation, it was concluded that it defines an antigenic determinant shared by both New and Old World arenaviruses.

Arenaviruses are readily distinguished in neutralization tests using hyperimmune polyclonal sera. Neutralizing antibodies to LCMV have been shown to be directed against GP-1 (Parekh and Buchmeier, 1986). However, a monoclonal antibody that recognized GP-2 glycopeptide on LCM virus-infected cells reacted with nine Mozambique strains but not with Lassa virus (Buchmeier *et al.*, 1988), showing that conserved antigens are present not only in the nucleocapsid proteins but also in viral glycoproteins. More recently, a monoclonal antibody (33.6) that reacts with GP-2 glycoprotein of LCMV and cross-reacts with both New World (Pichinde, Junin, Tacaribe, Amapari, Parana, and Machupo) and Old World (LCM and Lassa, Mopeia, and Mobala) viruses was described (Buchmeier *et al.*, 1988). A series of synthetic peptides derived from the deduced amino acid sequence of the LCMV ARM GP-2 were generated and assayed (Weber and Buchmeier, 1988) for reactivity against monoclonal antibody 33.6. It was found that it reacted only with one peptide corresponding to GPC (GP-2) residues 370–382. Further analyses demonstrated that the 33.6 epitope mapped to a nine-amino-acid segment (370–378) that contains five amino acids conserved among Lassa, LCMV, and Pichinde viruses. Thus a single peptide (Lys-Phe-Trp-Tyr-Leu) contains a universal antigenic determinant of arenaviruses. The presence of a common epitope in all arenaviruses justifies the previous inclusion of Tacaribe complex viruses with LCMV in the same taxon, which was based mainly on similar structure and biological properties.

B. The Tacaribe Complex

Tacaribe virus contains a major internal nucleocapsid protein (NP) and a single glycoprotein (GP) species in its outer envelope (Gard *et al.*, 1977). This is in contrast to other members of the Tacaribe group such as Pichinde and Junin viruses, both of which contain two glycoprotein moieties in the viral envelope (Howard, 1986; Mersich *et al.*, 1988). (See the Chapter by Burns and Buchmeier for recent evidence that Tacaribe retains two envelope glycoproteins like other arenaviruses when it is purified in low-salt conditions.)

It was demonstrated that a soluble antigen extractable from infected cells is the complement-fixing antigen. In fact, antisera directed against partially purified complement-fixing antigen from cells infected by Pichinde virus reacted against the internal nucleocapsid protein (NP) of the virus but not with the surface antigens of infected cells. The antigenic cross-reactivity observed by complement fixation among the New World Tacaribe complex viruses was shown to be due to conservation of NP-related antigens (Rawls and Leung, 1979). Complement-fixation serological cross-reactivity showed that Junin, Machupo, Amapari, and Tacaribe viruses appear to be particularly closely related (Casals *et al.*, 1975; Wulff *et al.*, 1978). As described later, a similar grouping of Tacaribe viruses was derived from cross-protection studies performed in guinea pigs. Furthermore, five monoclonal antibodies prepared against Tacaribe virus nucleoprotein cross-reacted with Junin virus antigen and four of them with Amapari virus–infected cells (Howard, 1986). Additionally, seven monoclonal antibodies specific for Junin virus reacted with Tacaribe virus but also with Machupo virus antigen.

Interestingly, by a completely different approach, resistance to the superinfection of persistently infected cultures against homotypic and heterotypic viruses, it was suggested that Junin, Tacaribe, and Amapari viruses constituted a group closely related, with Tamiami and Pichinde in another subgroup (Damonte *et al.*, 1983). A taxonomic relationship between Pichinde and LCM virus was also established by these studies (Weissenbacher *et al.*, 1987).

C. Close Relationship between Junin and Tacaribe Viruses

Members of the Tacaribe virus group, like all arenaviruses, are readily distinguished by neutralization tests using convalescent or hyperimmune antisera (Johnson *et al.*, 1973). Sera from patients recovered from AHF disease contain high titers of anti-Junin neutralizing antibodies (Weissenbacher and Damonte, 1983) and are routinely employed to identify new Junin virus isolates.

Using human convalescent serum in the virus dilution-constant serum neutralization test, Junin virus strains have been considered a homogeneous serotype. However, it is possible to distinguish between Junin virus strains by cross-neutralization assays with hyperimmune antisera raised in rabbits against each strain (Damonte *et al.*, 1986) and establish quantitatively their relatedness through a dendrogram based on taxonomic distance coefficients (Candurra *et al.*, 1989). Similarly, by kinetic neutralization assays it is possible to differentiate between Junin virus wild-type and antigenic mutants isolated from *Calomys musculinus* persistently infected with Junin virus (Alche and Coto, 1986; 1988).

The envelope of Junin virus particles contains two glycoproteins GP-1 (MW 45–48 kDa) and GP-2 (MW 36–38 kDa). Both glycoproteins arise by proteolytic cleavage of a glycoprotein precursor (GPC, MW 63–65 kDa) present in infected cells (De Mitri and Martinez Segovia, 1985). Immunization of rabbits and guinea pigs with GP-2 emulsified with complete Freund's adjuvant elicits neutralizing antibodies against infectious virus (Weissenbacher *et al.*, 1987), suggesting that this glycoprotein is exposed in the outer envelope. Recently, by surface iodination of Junin virus (Mersich *et al.*, 1988) it was shown that GP-2 is the predominant virus glycoprotein accessible to lactoperoxidase-catalyzed iodination on the envelope.

Tacaribe virus structural antigenicity was investigated by Howard (1986); working with five monoclonals prepared against GP 40–44 kDa it was found that all of them neutralized virus infectivity and immunoprecipitated viral glycoprotein from radiolabeled cell extracts.

As early as 1965, Henderson and Downs presented evidence that Junin virus was weakly neutralized by hyperimmune serum against Tacaribe virus. Tauraso and Shelokov (1965) found that sera from guinea pigs immunized with Tacaribe virus contained negligible amounts of neutralizing antibodies against heterologous Junin on day 49 p.i. Later studies (Weissenbacher *et al.*, 1976) demonstrated that neutralizing antibodies rose steadily from day 49 onwards, reaching the same levels of homologous neutralizing antibodies against Tacaribe. It was further reported that human convalescent serum against Junin also reacted by neutralization tests with Tacaribe virus, a fact also observed with two different hyperimmune sera prepared in rabbits against Junin virus that reduced Tacaribe virus infectity by 99–99.9% (Weissenbacher *et al.*, 1976). More recently, cross-neutralization reactivity was demonstrated between Tacaribe virus and four strains of Junin virus (Damonte *et al.*, 1986). Hyperimmune polyclonal antiserum was raised in rabbits by intravenous inoculation of purified virus. Tacaribe virus antisera neutralized all strains of Junin virus, but reciprocally only one Junin virus strain antisera cross-reacted with Tacaribe virus by the plaque reduction test. In the same study the cross-reactivity between Junin and Tacaribe viruses was also investigated by immune precipitation of [^{35}S]methionine-labeled Junin or Tacaribe virus-infected cell lysates. Junin virus–specific polypeptides, NP, GPC (glycopeptide precursor), GP-1, and GP-2, were immunoprecipitated by either Junin or Tacaribe antisera. Similarly, P77, GPC, and GP of Tacaribe virus were immune and cross-immune precipitated. Identical results were obtained using cytosol radiolabeled with [^{3}H]glucosamine hydrochloride, confirming the cross-reactivities of glycoproteins.

These findings strongly suggest that Junin and Tacaribe viruses share at least one or more antigenic determinants in their surface glycoproteins. The molecular analyses of the S-RNA segment of Junin and Tacaribe viruses show 77% homology for the NP and 44% for GP-1, which is much higher than the homology found between Junin virus and other arenaviruses (see Chapter 4). The high homology of the NP would explain why Junin and Tacaribe viruses cannot be distinguished by complement fixation and cross-immunofluorescence, whereas the high homology of GP supports the assumption that these two viruses share antigenic determinants in their glycopeptides.

IV. PATHOGENESIS OF HUMAN INFECTION

A. Argentine Hemorrhagic Fever

AHF is an endemoepidemic disease, geographically restricted to the central areas of the pampas. Although the epidemic zone is small, the population density and economic significance make AHF an important health problem. During recent years the steady expansion of the endemic area has raised the number of exposed people to 1.5 million while a seroepidemiological survey in AHF endemic areas indicated that about 90% of the population in this zone is still susceptible (Weissenbacher *et al.*, 1987). On the other hand, studies carried out in nonendemic zones have shown positive results in humans as well as in wild rodents (Weissenbacher *et al.*, 1985b).

The possible routes of viral entrance in human beings are cutaneous and mucosal, as confirmed by accidental laboratory infections as well as by experimental oral or nasal infection of guinea pigs and marmosets (Weissenbacher *et al.*, 1987). The initial symptoms are quite nonspecific and at 6–10 days after onset in most patients cardiovascular, digestive, renal, or neurological involvement becomes more severe, together with hematological and clotting alterations. Blood and urine tests show thrombocytopenia, leukopenia, albuminuria, and cylindruria. In each individual case, findings are mainly hemorrhagic or neurological, and at 10–15 days, over 80% of the patients improve while the remainder are likely to worsen. Total mortality reaches 16% but drops to 1–2% when immune plasma is administered early. Convalescence is quite lengthy, but total recovery takes place without aftereffects, except for the so-called neurological syndrome, which appears during full convalescence in 8–10% of plasma-treated patients (Weissenbacher and Damonte, 1983).

Pathological studies in fatal cases show that most lesions are nonspecific and consist of a generalized vasocongestion with multiple hemorrhages in different organs and subcutaneous tissue. The only pathological lesions consistently found in fatal cases are widespread necrosis in the lymphatic tissue and a global cell depression in bone marrow, confirmed by *in vivo* studies. Correspondingly, the highest virus titers at postmortem are found in the spleen, lymph nodes, and lungs, occasionally in bone marrow and liver (Weissenbacher and Damonte, 1983).

Immunofluorescence demonstrates Junin virus antigens in cells presumably belonging to the mononuclear–phagocytic system in the lymphatic tissue, lung, and liver and occasionally in renal and hepatic parenchyma, while in the CNS the capillary endothelium shows viral antigens apparently complexed with autologous antibodies (Weissenbacher *et al.*, 1987).

There are several questions to be answered on the pathogenesis of AHF, considering that unusually high levels of alpha-interferon are found in cases with a fatal outcome (Levis *et al.*, 1985). Further, it has recently been found that human PMN can be damaged *in vitro* by Junin virus, producing enzyme elimination and death (Laguens *et al.*, 1986b) and that the immunodepression already observed in AHF is selective, since there is a transient decrease in the number of OKT4 peripheral blood lymphocytes (Vallejos *et al.*, 1985).

B. Is Tacaribe Virus a Potential Hazard in Humans?

The only reported human infection with Tacaribe virus was laboratory acquired. A laboratory investigator handling Tacaribe virus became ill in the latter part of 1976 with severe headaches, fever, and muscle pains. Symptoms occurred over a 4-day period. Examination of serial serum samples by immunofluorescence and complement fixation showed a seroconversion to Tacaribe virus. It was presumed that the infection was acquired by aerosol. None of the other co-workers showed any evidence of infection with the virus (Bishop and Casals, personal communication, 1979).

In Argentina we have worked for a long time with Tacaribe virus, inoculating animals or handling large amounts of virus for purification procedures, and no seroconversion was observed in frequent serological trials of laboratory workers.

V. PATHOGENESIS OF JUNIN AND TACARIBE VIRUS IN ANIMALS

The animal models developed for the study of Junin experimental infection have formed the basis for the knowledge of the mechanisms of human infection, thus providing tools for practical control measures against Junin and other pathogenic arenaviruses.

There are two animal models in which Junin virus mimics the human illness in its physiopathogenic and even clinical aspects: guinea pigs and primates. In these models, where the viral mechanism of damage appears due to virus cytolytic activity, Tacaribe virus infects the animals with no apparent damage and with development of an immune response. In contrast, in rats and mice where the host's immune response is responsible for the CNS damage, both Tacaribe virus and Junin virus produce lethal disease.

A. Guinea Pigs

Depending on the strain, adult guinea pigs inoculated with Junin virus will present a mortality rate from 16 to almost 100% with different clinical and pathological pictures. The more-pathogenic strains, such as XJ, are viscerotropic and lymphotropic, while the more-attenuated ones, such as XJ clone 3 (XJCl3), are mainly neurotropic, with intermediate pathogenic strains, such as MC2, with 70% lethality and showing both lympho- and neutropism (Weissenbacher *et al.*, 1987; Candurra *et al.*, 1989).

In guinea pigs infected with the XJ strain the most frequent findings are massive bone marrow and lymphatic tissue necrosis and interstitial pneumonia as well as multiple hemorrhages. The highest virus titers are obtained from bone marrow, lymphatic tissue, and lungs, but even if most of the organs replicate the virus, it is not recovered from the CNS when animals are infected by peripheral routes (Weissenbacher *et al.*, 1987).

The high virus titers in bone marrow as well as the visualization of virus particles in megakaryocyte channels suggest that these cells may be a primary site for virus replication, consistent with the severe thrombocytopenia induced. The thrombocytopenia leads to a prolonged serum clotting time, and the clotting factors are reduced to approximately 10% of the normal values. Sequential studies have shown that the interstitial pneumonia is similar to the human adult pulmonary distress syndrome and that in the early stages a massive polymorphonuclear leukocyte infiltrate is present in the alveolar capillary lumina. Furthermore, it has been shown that the severe neutropenia induced by anti-PMN serum prevented lesions in the lungs of treated guinea pigs (Gonzalez *et al.*, 1987). Recently the presence of interferon has been detected in some infected guinea pigs from day 2 postinfection (p.i.) and was generalized when the animals were in the premortem stage (Dejean *et al.*, 1988). The titers were much higher than those attained with attenuated and intermediate virulent strains. Junin virus also induces immunosuppression against nonspecific and specific viral antigens, and the number of T lymphocytes in the spleen, lymph nodes, and blood is diminished.

Inoculation of guinea pigs with the attenuated XJC13 strain results in the development of a mild or inapparent infection in 80% of the animals, evidence of virus replication being restricted to the spleen, lymph nodes, lungs, pancreas, and salivary glands, although attenuated virus has been demonstrated by cocultivation methods in bone marrow and brain (Guerrero *et al.*, 1985). Attenuated strains show a marked neurotropic potential, although repeated passage of the virulent XJ strain beyond the 30th pass also leads to a marked increase in neurovirulence (Oubina and Carballal, 1985). It has been suggested that the strain variations are related to their respective tropism for different cells of the lymphoid system. Laguens *et al.* (1986a) have shown that macrophages are the main targets for virus replication in the spleens of animals infected with the virulent XJ strain, whereas in animals infected with the attenuated strain, attempts to detect either viral antigens or infectious virus from macrophages were unsuccessful and virus could only be recovered from dendritic cells.

In contrast to Junin virus, it was demonstrated as early as 1964 (Parodi and Coto) that Tacaribe virus is not pathogenic for guinea pigs, by any route of inoculation. In spite of the replication of Tacaribe virus in the lymph nodes, spleen, and lungs and short-lived viremia, no significant variation in the weight and temperature of the infected animals was observed during a period of 30 days. Tacaribe virus–infected guinea pigs did not show detectable immunosuppression in either the humoral or cell-mediated immunity (Weissenbacher *et al.*, 1987). Moreover, a long-term study has shown that Tacaribe virus titers and antigen become undetectable after 30 days p.i. and by coculture as long as 540 days p.i. Besides, no immunoglobulins or C3 deposits were detected by immunofluorescence in the brain or kidney at any stage, and histological lesions were absent throughout (Carballal *et al.*, 1987). When cell affinities in the spleen of Tacaribe virus–infected guinea pigs were studied, it was observed that, as in the case of the attenuated strain of Junin virus, Tacaribe virus could only be recovered from dendritic cells (Laguens *et al.*, 1986a).

B. Primates

Although a number of nonhuman primates have been studied for their sensitivity to Junin virus, only two species reproduce a severe disease when infected with pathogenic Junin virus strains: the South American marmoset *Callithrix jacchus* and the rhesus macaque *Macaca mulatta.*

Marmosets infected with the XJ pathogenic strain show 100% mortality, with hemorrhagic rash, tremors, and sometimes convulsions. Virus is readily recovered from blood and organs, especially nervous and hemopoietic tissue. The histopathological lesions most frequently found are multiple hemorrhages and severe meningoencephalitis with extensive brain necrosis and demyelinization. Immunofluorescence shows Junin virus antigenic determinants in neurons and blood vessels and in cells resembling macrophages in the lymphatic tissue and lungs (Weissenbacher and Damonte, 1983).

On the other hand, no deaths, clinical signs, or hematological alterations were observed in animals inoculated with the attenuated XJ clone 3 strain. Viremia was transient and lower than for the pathogenic strains, and viral spread was limited to a few organs, with the marmosets promptly developing humoral antibodies (Avila *et al.*, 1985).

C. jacchus inoculated with Tacaribe virus did not show any clinical signs, hematological alterations, or mortality whatsoever. Both viremia and viral spread were lacking, but serum antibodies appeared after the second week p.i. (Weissenbacher *et al.*, 1982, 1985a). In a long-term study there were no histopathological changes due to Tacaribe virus or virus recovery after primary culture or cocultures with susceptible cells in primates sacrificed up to 480 days p.i. (Samoilovich *et al.*, 1988).

M. mulatta infected with Junin virus have demonstrated clinical, histopathological and laboratory findings very closely approximating those of humans with AHF. Monkeys infected with the pathogenic strains of Junin virus show 75–100% mortality, with loss of weight, hemorrhagic rash, and diathesis. Histopathological findings support clinical impressions of virus-specific disease patterns, while hematological findings include lympho- and neutropenia, with thrombocytopenia in some cases.

Interestingly, when low-passage isolates of Junin virus are obtained from humans, each of the clinically recognized variants of AHF shows a correspondingly similar clinical pattern upon infecting the monkeys (Peters *et al.*, 1987). In Rhesus macaques infected with the attenuated strains of Junin virus (Candid, XJC13) neither clinical symptoms nor hematological disturbances are observed.

C. Mice

The newborn mouse is highly susceptible to any strain of Junin virus administered by any route. Inoculated animals develop an immune cell-mediated lethal meningoencephalitis. It has been shown that immunosuppression produced by thymectomy, antithymocyte serum, drugs, and the congenital lack of thymus prevents the development of acute disease in mice (Weissenbacher and Damonte, 1983), leading to a persistent infection (Weissenbacher *et al.*, 1983). The reconstitution of persistently infected thymectomized as well as athymic mice leads to a neurological lethal disease (Oubina *et al.*, 1988), confirming the immune pathogenesis of tissue damage.

Immune depression renders adult mice susceptible to Junin virus, whereas adult nu/nu athymic mice show a persistent infection with high virus yield, no antibodies, and no pathological alterations (Weissenbacher *et al.*, 1986a). On the other hand, the splenocytes of persistently infected nu/+ athymic mice are incapable of inducing death or viral clearance in athymic recipients (Rabinovich *et al.*, 1988). This leads to the conclusion that the T-cell compartment is the main factor determining development of tissue damage or the establishment and/or maintenance of a persistent infection in mice.

In support, it was observed that more mice survived with persistent infection when inoculated within 24 hr of birth, and that Junin virus replicated earlier and in higher titers in the thymus and bone marrow than when inoculation was carried out at 2–3 days of age (Calello *et al.*, 1986).

Laboratory studies have shown that Tacaribe virus is uniformly lethal for neonatal mice by the intracerebral route, but that adults survive (Downs *et al.*, 1963). Moribund newborn mice show pronounced paralysis and ataxia (Downs *et al.*, 1963), and the histological lesions observed in Tacaribe virus–infected animals are similar to those observed with Junin virus. This pathology is absent in thymectomized or immunosuppressed mice (Borden and Nathanson, 1974).

Coto and Leon (1978) have examined the relationship between age of Tacaribe virus–infected mice and mortality. They have shown that the pathogenicity of the Tacaribe virus decreases with increasing mouse age, as previously found for Junin virus, thereby showing another similarity between the two viruses in this model.

D. Rats

The adult rat is resistant to Junin virus, but newborn and suckling animals present different responses according to the viral strain and route of inoculation. We are presented with the enigma that although 10-day-old rats are equally susceptible to XJ and XJC13 strains by the intracerebral route, in newborn rats inoculated with XJC13 a fatal neurological disease invariably follows, while if the latter are inoculated with XJ, the disease is milder and most animals recover, although some present arrested growth and neurological symptoms (Weissenbacher *et al.*, 1987). These chronically infected animals show histopathological lesions with viral antigen in CNS, and virus can be isolated only by cocultivation with permissive cells (Weissenbacher *et al.*, 1986b). To our knowledge, this is the first time that a severe chronic neurological illness was found in a persistently Junin virus–infected animal.

E. Reservoirs

In the endemic area of AHF, Junin virus has been isolated from several wild rodents, such as *Calomys musculinus*, *Calomys laucha*, and *Akodon azarae*. However, the species that presented the highest number of infected animals by virus isolation or antibody detection was *C. musculinus*, which is therefore considered the main Junin virus reservoir. The pioneer laboratory studies performed demonstrated that infection of *C. musculinus* was mainly horizontally by means of close contact between those animals, and that persistent viremia invariably followed infection by the intranasal route (Sabattini *et al.*, 1977). In animals infected by the intranasal route, the highest titer of virus was recovered from the salivary glands. The presence of viral antigen in these organs with no histological changes, as well as the finding of virus particles in the ducts, confirmed the method of transmission already observed in the laboratory. In recent studies performed by Vitullo and co-workers (Vitullo *et al.*, 1987; Vitullo and Merani, 1988), it was elegantly shown that vertical transmission is insufficient to overcome the effect of highly depressed survival and fertility of the infected host. This supports the need for horizontal transmission for viral maintenance in nature. On the other hand, adults that are horizontally infected may secure intergeneration transmission by both vertical and horizontal means, since they do not show altered reproduction and survival. Under these circumstances, as the mentioned authors speculate, Junin virus maintenance may arise from an equilibrium between both modes of transmission, with the horizontal representing the main route for viral persistence in nature and the vertical transmission being an added option for intergeneration transfer when population numbers are reduced and horizontal transmission is precluded.

The mechanisms leading to viral persistence in *C. musculinus* are still unexplored. Up to now persistent infection has been correlated with high titers of anti-Junin virus neutralizing antibodies, but on the other hand, serological variants were observed in the blood of chronically infected rodents, suggesting that persistence may be due to escape from neutralization by genetic variation (Alche and Coto, 1986, 1988).

Tacaribe virus is the only member of the arenavirus family to be isolated from a nonrodent source. The first strain was obtained from the brain of a bat caught in Port of Spain, Trinidad, and subsequent isolations were all made from either brain or salivary gland homogenates of these animals, although one strain was obtained from a pool of mosquitoes collected in the forested region to the east of Trinidad (Downs *et al.*, 1963). Besides, no evidence of Tacaribe virus infection was found in any of 2000 indigenous mammals. As the species of bats that carried Tacaribe virus are not carnivores, the possibility that virus has been acquired by the hunting of small mammals is unlikely, although the possibility that virus spread resulted from contact with rodents sharing a nomadic habit cannot be excluded (Howard, 1986).

Although signs of persistent Junin virus infection have been reported in mice (Coto and Leon, 1978), it is surprising that no other chronic experimental infection in any other system has been found, in spite of intensive long-term studies in guinea pigs and marmosets (Carballal *et al.*, 1987; Samoilovich *et al.*, 1988). On the other hand, extensive attempts to cause persistent infection in mice with Tacaribe virus under the same conditions that rendered mice persistently infected with Junin virus were all unsuccessful (Calello *et al.*, 1987).

The apparent failure of Tacaribe virus to induce chronic infections could be due in part to the lack of virus tropism for lymphoid cells, which could prevent the occurrence of a transient immunosuppression during the early stages of maturation of the lymphoid system (Borden and Nathanson, 1974). This mechanism has been postulated for the classical persistent infection of the mouse with LCM virus (Lehmann-Grube *et al.*, 1983).

VI. CROSS-PROTECTION STUDIES

Soon after the Junin and Tacaribe viruses were found to be antigenically related, it was soundly demonstrated that guinea pigs inoculated only once with Tacaribe virus were fully protected against challenge with otherwise lethal doses of Junin virus (Parodi and Coto, 1964; Tauraso and Shelokov, 1965). The basis for the use of a live heterologous nondangerous vaccine against AHF was provided.

Cross-protection studies were extended to other members of the Tacaribe complex by taking advantage of the fact that only Junin virus is 100% lethal for guinea pigs. Animals inoculated intramuscularly with a massive dose (10^6 LD_{50}) of Tacaribe or Machupo viruses were resistant to challenge with 10^3 LD_{50} of Junin virus (Weissenbacher *et al.*, 1976). By contrast, inoculation with Tamiami or Pichinde virus only produced a delay in the mortality rate. Only one dose of Amapari virus allowed survival of some animals, but administration of three doses induced protection in nearly 60% of inoculated guinea pigs. Cross-protection studies in guinea pigs advanced a grouping of Tacaribe complex viruses similar to that derived from serological cross-reactivity.

Our finding that Machupo virus protected against Junin virus prompted us to suggest that a vaccine prepared against Junin virus could protect against other viruses of the Tacaribe group as pathogenic in humans as Machupo. About 10 years later this hypothesis was confirmed by investigators of the USAMRI and the Salk Institute. Rhesus monkeys were vaccinated with different doses of an attenuated strain of Junin virus, Candid #1, presently tried as a vaccine on human volunteers. The animals were challenged 11 months after vaccination with Machupo virus and survived with good health up to 6 months after challenge, whereas, the control animals died (Barrera Oro *et al.*, 1988; Peters *et al.*, 1987). Low heterologous neutralizing antibody titers against Machupo were present at the moment of challenge with Machupo virus in monkeys inoculated with high doses of virus but not with low doses.

Heterologous vaccination against Junin by inoculation with Tacaribe virus was analyzed in guinea pigs and monkeys. A single dose of Tacaribe virus, intramuscularly inoculated at 10^5 LD_{50}, protected guinea pigs against challenge with 1000 LD_{50} of Junin virus up to 6 months post-Tacaribe infection (Weissenbacher *et al.*, 1976). Heterologous neutralizing antibodies against Junin virus appeared at 45 days p.i., before challenge, and increased steadily until day 190. The protective effect of Tacaribe virus in guinea pigs may be demonstrated as early as 3 days after immunization before the appearance of cross-reactive neutralizing antibodies in the sera. Percentage survival increases with the extension of the interval between Tacaribe administration and Junin virus challenge. From 30 days onward, animals are all protected against Junin virus. Unexpectedly, between days 12 and 13 post-Tacaribe vaccination, animal resistance against Junin is lost. One day later this situation is reverted, when a 60% survival rate appears. There is not yet a clear explanation of this phenomenon.

The protection afforded by Tacaribe virus to animals challenged with Junin 3 days later correlates with a complete reduction in the extent of Junin virus replication in liver, lung, lymph nodes, and spleen analyzed on days 9 and 11 post-Junin infection. No detectable viremia was found in vaccinated animals. The protective effect from challenge after 15 days post-Tacaribe inoculation seems to be the result of priming the immune system with Tacaribe viral antigens similar to Junin virus antigenic determinants. In fact, control animals inoculated with Tacaribe virus only developed cross-reactive neutralizing antibodies to Junin virus 2 months later, but the challenge with Junin virus stimulates the appearance of Junin antibodies in a manner typical of a secondary response (Coto *et al.*, 1980). A recent study (Carballal *et al.*, 1987) showed that a single dose of Tacaribe virus fully protected animals against Junin virus challenge up to 550 days postvaccination. By day 660, protection decreased to 33%. Neutralizing homologous and heterologous antibodies persist in surviving animals at least until 660 days p.i.

Protection studies were also performed on the marmoset *Callithrix jacchus*, susceptible to Junin virus. Marmosets vaccinated with Tacaribe virus administered intramuscularly (Weissenbacher *et al.*, 1982) or by nasal route (Samoilovich *et al.*, 1984) become resistant to Junin virus challenge given on day 53–60 post-Tacaribe infection. Protection against pathogenic Junin virus was detected up to 480 days after a single dose of Tacaribe virus in a long-term study (Samoilovich *et al.*, 1988).

Tacaribe-infected marmosets were protected by mechanisms that inhibited viral replication, viremia, and spread of Junin virus challenge. Humoral neutralizing heterologous (anti-Junin) antibodies were detected in marmosets at 14 days after Junin virus infection. Prechallenge antibody levels to Junin virus were not correlated with levels of protection. The absence of anti-Junin antibodies in Tacaribe-infected monkeys up to the time of challenge contrasts with the results just described for guinea pigs. It is known that passive immunilation performed in guinea pigs and humans infected with Junin virus has demonstrated the important role of neutralizing antibodies in the protection against disease and death induced by the virus (Weissenbacher and Damonte, 1983). However, the protection observed in marmosets, even in the absence of neutralizing anti-Junin antibodies, could be due to a cell-mediated immune response.

All the studies discussed in this chapter suggest Tacaribe virus as an alternative candidate for a vaccine to immunize against AHF, and better than attenuated strains of Junin virus. There are two main reasons for this; one is the almost discountable risk of a persistent infection with Tacaribe virus, which also rules out the fear of pathology related to persistent infections whether direct or immunopathological. As a second reason, since Tacaribe virus is a naturally attenuated strain, not a strain obtained through laboratory manipulations, the risk of regression to the original pathogenicity is almost nil.

It has also been shown that it is possible to immunize with Tacaribe virus through the mucosal method as well as with the attenuated strains of Junin virus. It would be interesting, therefore, to take advantage of a virus that has been naturally provided as an immunizing agent against Junin virus.

REFERENCES

Alche, L. E., and Coto, C. E., 1986, Antigenic variants of Junin virus isolated from infected *Calomys musculinus*, *Arch.Virol.* **90**:343.

Alche, L. E., and Coto, C. E., 1988, Differentiation of Junin virus and antigenic variants isolated *in vivo* by kinetic neutralization assays, *J. Gen.Virol.* **69**:2123.

Arribalzaga, R. A., 1955, Una nueva enfermedad epidemica a germen desconocido: Hipertermia nefrotoxica, leucopenica y enantematica, *Dia Med.* **27**:1204.

Avila, M. M., Frigerio, M. J., Weber, E. L., Rondinone, B., Samoilovich, S. R., Laguens, R. P., Guerrero, L. B. de, and Weissenbacher, M. C., 1985, Attenuated Junin virus infection in *Callithrix jacchus*, *J. Med.Virol.* **15**:93.

Borden, E. C., and Nathanson, N., 1974, Tacaribe virus Infection of the mouse: An immunopathologic disease model, *Lab. Invest.* **30**:465.

Buchmeier, M. J,. Wright, K. E., Weber, E. L., and Parekh, B. S., 1988, Structure and expression of arenavirus proteins, *in: "Immunobiology and Pathogenesis of Persistent* Virus Infections" (Carlos Lopez, ed.), pp. 91–104, American Society for Microbiology, Washington, D.C.

Calello, M. A., Rabinovich, R. D., Boxaca, M. C., and Weissenbacher, M. C., 1986, Relationship between Junin virus infection of thymus and the establishment of persistence in rodents, *Med. Microbiol. Immunol.* **175**:109.

Calello, M. A., Rabinovich, R. D., Gomez Carrillo, M., Quintans, C. J., and Boxaca, M. C., 1987, Evolucion de la infeccion con virus Tacaribe en el modelo raton desarrollado para estudiar la persistencia del virus Junin. *Medicina (Buenos Aires)* **47**:568.

Candurra, N. A., Damonte, E. B., and Coto, C. E., 1989, Antigenic relationships among attenuated and pathogenic strains of Junin virus, *J. Med.Virol.* **27**:145.

Carballal, G., Calello, M. A., Laguens, R. P., and Weissenbacher, M. C., 1987, Tacaribe virus: A new alternative for Argentine hemorrhagic fever vaccine, *J. Med. Virol.* **23**:257.

Casals, J., Buckley, S. M., and Cedeno, R., 1975, Antigenic properties of arenaviruses, *Bull. WHO* **52**:421.

Coto, C. E., and Leon, M. E., 1978, Susceptibilidad exacerbada del raton de 10 dias a la infeccion con arenavirus, *Medicina (Buenos Aires)* **38**:281.

Coto, C. E., Damonte, E. B., Calello, M. A., and Weissenbacher, M. C., 1980, Protection of guinea pigs inoculated with Tacaribe virus against lethal doses of Junin virus, *J. Infect. Dis.* **141**(3):389.

Damonte, E. B., Mersich, S. E., and Coto, C. E., 1983, Response of cells persistently infected with arenaviruses to superinfection with homotypic and heterotypic viruses, *Virology* **129**:474.

Damonte, E. B., Mersich, S. E., Candurra, N. A., and Coto, C. E., 1986, Cross-reactivity between Junin and Tacaribe viruses as determined by neutralization test and immunoprecipitation, *Med. Microbiol. Immunol.* **175**:85.

Dejean, C. B., Oubina, J. R., Carballal, G., and Teyssie, A. R., 1988, Circulating interferon in the guinea pig infected with the XJ prototype Junin virus strain, *J. Med. Virol.* **24**:97.

De Mitri, M. I., and Martinez Segovia, Z. M. de, 1985, Polypeptide synthesis in Junin virus infected BHK-21 cells, *Acta Virol.* **29**:97.

Downs, W. G., Anderson, C. R., Spence, L., Aitken, T. H. G., and Greenhall, A. G., 1963, Tacaribe virus, a new agent isolated from Artibeus bats and mosquitoes in Trinidad, West Indies, *Am. J. Trop. Med. Hyg.* **12**:640.

Gard, G. P., Vezza, A. C., Bishop, D. H. L., and Compans, R. W., 1977, Structural proteins of Tacaribe and Tamiami virions, *Virology* **83**:84.

Gonzalez, P. H., Ponzinibbio, C., and Laguens, R. P., 1987, Effect of polymorphonuclear depletion on experimental Argentine hemorrhagic fever in guinea-pigs, *J. Med. Virol.*, **22**(3):289.

Guerrero, L. B. de, Boxaca, M. C., Malumbres, E., Dejean, C., and Caruso, E., 1985, Early protection to Junin virus of guinea pigs with an attenuated Junin virus strain, *Acta Virol.* **29**:334.

Howard, C. R., 1986, Arenaviruses, in: *Perspectives in Medical Virology* Vol. 12 (A. Zuckerman, ed.) Elsevier, Amsterdam.

Johnson, K. M., Wiebenga, N. H., Mackenzie, R. B., Kuns, M. L., Tauraso, N. M., Shelokov, A., Webb, P. A., Justines, G., and Beye, H. K., 1965, Virus isolations from human cases of hemorrhagic fever in Bolivia, *Proc. Soc. Exp. Biol. Med.* **118**:113.

Johnson, K. M., Webb, P. A., and Justines, G., 1973, Biology of Tacaribe-complex viruses, in: *Lymphocytic Choriomeningitis Virus and Other Arenaviruses* (F. Lehmann-Grube, ed.), pp. 241–258, Springer-Verlag, Vienna.

Laguens, R. M., Chambo, J. G., and Laguens, R. P., 1986a, Splenic dendritic cells and Junin virus, *Med. Microbiol. Immunol.* **175**:187.

Laguens, R. P., Gonzalez, P. H., Ponzinibio, C., and Chambo, J., 1986b, Damage of human polymorphonuclear leucocytes by Junin virus, *Med. Microbiol. Immunol.* **175**:177.

Lehmann-Grube, F., Martinez Peralta, L. A., Bruns, M., and Lohler, J., 1983, Persistent infection of mice with the lymphocytic choriomeningitis virus, in: *Comprehensive Virology*, Vol. 18 (H. Fraenkel-Conrat and R. P. Wagner, eds.), pp. 43–103, Plenum Press, New York.

Levis, S. C., Saavedra, M.C., Ceccoli, C., Feuillade, M. R., Enria, D. A., Maiztegui, J. I., and Falcoff, R., 1985, Correlation between endogenous interferon and the clinical evolution of patients with Argentine hemorrhagic fever, *J. Interferon Res.* **5**:383.

Mersich, S. E., Castilla, V., and Damonte, E. B., 1988, Lectin affinity of Junin virus glycoproteins, *Ann. Inst. Pasteur Virol.* **139**:277.

Oubina, J. R., and Carballal, G., 1985, Neurotropism of a high-passage XJ strain of Junin virus, *J. Med. Virol.* **15**:157.

Oubina, J. R., Carballal, B., Laguens, R. P., Quintans, C., Merani, S., and Weissenbacher, M. C., 1988, Mortality induced by adoptive immunity in Junin virus–infected athymic mice, *Intervirology* **29**:61.

Parekh, B. S., and Buchmeier, M. J., 1986, Proteins of lymphocytic choriomeningitis virus: Antigenic topography of the viral glycoproteins, *Virology* **153**:168.

Parodi, A. S., and Coto, C. E., 1964, Inmunizacion de cobayos contra el virus Junin por inoculacion de virus Tacaribe, *Medicina (Buenos Aires) 24:*151.

Parodi, A. S., Greenway, D.J., Rugiero, H. R., Rivero, S., Frigerio, M. J., Mettler, N. E., Garzon, F., Boxaca, M., Guerrero, L. B. de, and Nota, N. R., 1958, Sobre la etiologia del brote epidemico de Junin, *Dia Med.* **30**:2300.

Peters, C. J., Jahrling, P. B., Liu, C. T., Kenyon, R. H., McKee, K. T., Jr., and Barrera Oro, J. G., 1987, Experimental studies of arenaviral hemorrhagic fevers, in: *Curr. Topics in Microbiol. and Immunol.*, Vol.134 (M. Oldstone, ed.), pp. 5–68, Springer-Verlag, Berlin.

Pinheiro, F. P., Shope, R. E., de Andrade, A. H. P., Ben Sabeth, G., Cacios, G. V., and Casals, J., 1966, Amapari, a new virus of the Tacaribe group from rodents and mites of Amapa territory, Brazil, *Proc. Soc. Exp. Biol. Med.* **122**:531.

Rabinovich, R. D., Calello, M. A., Boxaca, M. C., Quintans, C. J., and Weissenbacher, M. C., 1988, Mouse splenocyte transfer effect depends on donor's Junin virus infection stage, *Intervirology* **29**:21.

Rawls, W. E., and Leung, W. C., 1979, Arenaviruses, in: *Comparative Virology*, Vol. 14 (H. Fraenkel-Conrat and R. Wagner, eds.), pp. 157–192, Plenum Press, New York.

Rowe, W. P., Murphy, F. A., Bergold, G. H., Casals, J., Hotchin, J., Johnson, K. M.,

Lehmann-Grube, F., Mims, C. A., Traub, E., and Webb., P.A., 1970, Arenaviruses: A proposed name for a newly defined virus group, *J. Virol.* **5**:651.

Sabattini, M. S., Rios, L. E. G. de, Diaz, G., and Vega V. R., 1977, Natural and experimental infection of rodents with Junin virus, *Medicina (Buenos Aires)* **37**(3):149.

Samoilovich, S. R., Pecci Saavedra, J., Frigerio, M. J., and Weissenbacher, M. C., 1984, Nasal and intrathalamic inoculations of primates with Tacaribe virus: Protection against Argentine hemorrhagic fever and absence of neurovirulence, *Acta Virol.* **28**:1277.

Samoilovich, S. R., Calello, M. A., Laguens, R. P., and Weissenbacher, M. C., 1988, Long-term protection against Argentine hemorrhagic fever in Tacaribe virus infected marmosets: Virologic and histopathologic findings, *J. Med. Virol.* **24**:236.

Tauraso, N. M., and Shelokov, A., 1965, Protection against Junin virus by immunization with live Tacaribe virus, *Proc. Soc. Exp. Biol. Med.* **119**:608.

Trapido, H., and Sanmartin, C., 1971, Pichinde virus: a new virus of Tacaribe group from Colombia, *Am. J. Trop. Med. Hyg.* **20**:621.

Vallejos, D. A., Ambrosio, A. M., Gamboa, G., Briggiler, A. N., and Maiztegui, J. I., 1985, Alteraciones de las subpoblaciones linfocitarias en la fiebre hemorragica Argentina, *Medicina (Buenos Aires)* **45**:407.

Vitullo, A. D., and Merani, M. B., 1988, Is vertical transmission sufficient to maintain Junin virus in nature? *J. Gen. Virol.* **69**:1437.

Vitullo, A. D., Hodara, V. L., and Merani, M. S., 1987, Effect of persistent infection with Junin virus on growth and reproduction of its natural reservoir, *Calomys musculinus*, *Am. J. Trop. Med. Hyg.* **37**(3):663.

Weber, E. L., and Buchmeier, M. J., 1988, Fine mapping of a peptide sequence containing an antigenic site conserved among arenaviruses, *Virology* **164**:30.

Weissenbacher, M. C., and Damonte, E. B., 1983, Fiebre hemorragica Argentina, *Adel. Microbiol. Enf. Infecc.* **2**:119.

Weissenbacher, M. C., Coto, C. E., and Calello, M. A., 1976. Cross-protection between Tacaribe complex viruses. Presence of neutralizing antibodies against Junin virus (Argentine hemorrhagic fever) in guinea pigs infected with Tacaribe virus, *Intervirology* **6**:42.

Weissenbacher, M. C., Coto, C. E., Calello, M. A., Rondinone, S. N., Damonte E. B., and Frigerio, M. J., 1982, Cross-protection in nonhuman primates against Argentine hemorrhagic fever, *Infect. Immun.* **35**:425.

Weissenbacher, M. C., Laguens, R. P., Quintans, C. J., Calello, M. A., Montoro, L., Woykowsky, N. M., and Zannoli, V. H., 1983, Persistencia viral y ausencia de lesiones en el encefalo de ratones congenitamente atimicos infectados con virus Junin, *Medicina (Buenos Aires)* **43**:403.

Weissenbacher, M. C., Avila, M. M., Calello, M. A., Frigerio, M. J., and Guerrero, L. B. de, 1985a, The marmoset *Callithrix Jacchus* as an attenuation marker for arenaviruses, *Comuni. Biol.* **3**:375.

Weissenbacher, M. C., Calello, M. A., Carballal, B., Planes, N., de la Vega, M. T., and Kravetz, F., 1985b, Actividad del virus Junin en humanos y roedores de areas no endemicas de la provincia de Buenos Aires, *Medicina (Buenos Aires)* **45**:263.

Weissenbacher, M. C., Calello, M. A., Merani, M. S., Oubina, J. R., Laguens, R. P., Montoro, L., and Carballal, G., 1986a, Induction of Junin virus persistence in adult athymic mice, *Intervirology* **25**:210.

Weissenbacher, M. C., Lascano, E. F., Avila, M. M., and Berria, M, I., 1986b, Chronic neurologic disease in Junin virus infected rats, *J. Med. Virol.* **20**:57.

Weissenbacher, M. C., Laguens, R. P., and Coto, C. E., 1987, Argentine hemorrhagic fever, in: *Curr. Topics Microbiol. Immunol.*, Vol. 134 (M. Oldstone, ed.), pp. 80–116, Springer-Verlag, Berlin.

Wulff, H., Lange, J. V., and Webb, P., 1978, Interrelationships among arenaviruses measured by indirect immunofluorescence, *Intervirology* **9**:344.

CHAPTER 17

Arenavirus Pathophysiology

SUSAN P. FISHER-HOCH

I. INTRODUCTION

Arenavirus infections are virtually without pathophysiology in the naturally infected rodent host. The animal normally experiences silent but persistent and lifelong infection, the viruses being adapted to this highly successful ecological niche by asymptomatic horizontal or vertical transmission at or near birth (Traub, 1935; McCormick, 1990). In adult rodents, however, infection does induce pathology and an acute self-limiting disease, ending in death or recovery, with virus clearance depending on the dose and route of inoculation and the genetic background of the rodent. Arenavirus infections in humans takes an altogether different course. Infections range from a febrile disease with aseptic meningitis with lymphocytic choriomeningitis virus (LCMV) (Armstrong and Sweet, 1939), to total collapse and death with circulatory and respiratory failure with the hemorrhagic fever viruses: Lassa fever (Rose, 1956; Buckley *et al.*, 1970; Frame *et al.*, 1970), Argentine hemorrhagic fever (AHF) (Ruggiero *et al.*, 1964a; Maiztegui, 1975), and Bolivian hemorrhagic fever (BHF) (Aribalzaga, 1955; MacKenzie *et al.*, 1964; Johnson *et al.*, 1965). A meningoencephalitis is characteristic of LCMV in humans, but most infections are apparently mild and self-limiting. In contrast, the arenavirus hemorrhagic fevers are often severe, generalized febrile diseases with multiorgan involvement and case/fatality rates of about 16% in untreated hospitalized patients. The hemo-

SUSAN P. FISHER-HOCH • Special Pathogens Branch, Division of Viral and Rickettsial Diseases, Centers for Disease Control, Atlanta, Georgia 30333. *Present address:* Mycotic Diseases Branch, Division of Bacterial and Mycotic Diseases, Centers for Disease Control, Atlanta, Georgia 30333.

The Arenaviridae, edited by Maria S. Salvato. Plenum Press, New York, 1993.

static defect, which is their hallmark, is characterized by platelet dysfunction rather than coagulation cascade activation, with little evidence of major hepatorenal failure or organ destruction by direct viral replication. Tissue and pulmonary edema with hypovolemic shock and adult respiratory distress syndrome (ARDS) are prominent in severe disease, and nervous system involvement is a feature of both acute and convalescent stages of these infections.

Infection of humans by these viruses is an accidental event, reflecting minimal host/parasite adaptation and bearing little resemblance to the course of infection in rodents. However, except for LCMV, which may be acquired from house mice or pet hamsters and laboratory hamsters or nude mice almost anywhere in the world, human arenavirus infections are rural diseases. Since they usually occur in areas with limited facilities for medical care, details of the clinical and pathogenic features of the diseases are scarce. Human-to-human spread has been reported for Lassa fever in the community and in hospital settings (McCormick *et al.*, 1987a), whereas only a few cases of nosocomial transmission have been reported for Bolivian hemorrhagic fever (Peters *et al.*, 1974), and none for Argentine hemorrhagic fever.

II. INFECTIONS IN THE NATURAL RODENT HOST

All arenaviruses establish persistent infection in the natural rodent host following virus acquisition *in utero* or within a few days of birth. Most of the persistently infected animals have viremia and viruria throughout life (Traub, 1935; Webb *et al.*, 1975). This persistence is not only a highly efficient means of virus perpetuation in most of the rodent offspring, but it is also the primary source of contamination of the environment leading to human infections. These mice are relatively deficient in virus-specific immune lymphoctcytes, but are able to mount high titers of antiviral antibodies (Oldstone, 1987). Intracerebral inoculation of Lassa virus may be lethal to mice (Lukashevich, 1985).

A. LCMV in Mice

When immunocompetent adult mice are inoculated with LCMV by the intracerebral route, they develop an acute inflammatory leptomeningitis, choroiditis, and ventriculitis, leading to death within 6–10 days (Oldstone, 1987). High white blood cell counts are found in the CSF, with only a few red blood cells, although the the blood–brain barrier has been violated. The animals develop tremors with characteristic extensor spasms of the legs and finally convulse and die. The white cell infiltrate of the CSF comprises T cells (Thy-1^+ and Lyt-2^+), NK cells, and macrophages. Inflammation is mediated by class I MHC-restricted cytotoxic T cells specific for LCMV antigens, apparently recognizing viral epitopes different from those which induce humoral antibodies, and which may vary depending on the genotype of the mouse (Allan *et al.*, 1987). This model can be reproduced in persistently infected immunosuppressed mice by reconstitution with virus-specific Lyt-2^+ cells from immunized animals.

When adult animals are inoculated peripherally, the outcome is variable. Viremia peaks about 4–5 days postinoculation and then rapidly declines. Antibodies are detectable by 4–5 days postinfection, and virus-specific cytotoxic T cells (CTL) appear about day 5, with highly activated macrophages, and reach peak activity around days 7–9 (Cole *et al.*, 1972; Marker and Volkert, 1973).

Classically the disease results from widespread immune-mediated damage to the meninges, choroid plexus, and ependyma, which are heavily infected with the virus and which become targets for effector T lymphocytes (Cole and Johnson, 1975). The processes of cellular immune responses are discussed in detail elsewhere (see Chapters 11 and 12).

In newborn mice less than 24 hr old, LCMV inoculation by any route, including intracerebral, results in silent but persistent infection (Oldstone, 1987). Virus replication in the brain soon weakens the ventricles and virus rapidly enters the circulation, reaching tissues, serum, and urine. In the central nervous system in these animals, virus is primarily expressed in neuronal cells, in contrast to the acutely infected adults, in which the virus replicates mainly in the leptomeninges and lining of the ventricles (Cole *et al.*, 1971; Southern *et al.*, 1984). Ultrastructural studies of persistently infected neural tissue show localization of antigen to neurons in the cortex, limbic system, and hypothalmus and the anterior horn of the spinal cord and Purkinje cells in the cerebellum (Rodriguez *et al.*, 1983; Monjan *et al.*, 1975). The infection is not associated with any morphological changes, and expression is mainly of the nucleocapsid protein. The viral antigens appear to be mostly associated with ribosomes (Rodriguez *et al.*, 1983). The persistent infectious state can also be achieved in adult mice if they are immunosuppressed by thymectomy, irradiation, antilymphocytic serum, or immunosuppressive drugs. Transfer of virus-specific immune lymphocytes to either neonatally infected or adult immunosuppressed, persistently infected mice results in clearance of the virus (Oldstone *et al.*, 1986). The ultimate site of persistence appears to be renal tissue (Ahmed *et al.*, 1987). In a recent outbreak in laboratory personnel, inoculation of persistently LCMV-infected tumor cell lines into nude mice resulted in a number of human infections (Dykewicz *et al.*, 1992). The nude mouse provides an immunological host comparable to the neonatal mouse, and virus persistence may be expected (Frei *et al.*, 1988).

There is now considerable evidence from studies in LCMV-infected mice that replication of the virus in a specialized cell can result in selective disruption of cell function while incurring no structural injury (Oldstone *et al.*, 1984a,b). Growth hormone deficiency, caused by reduction in transcription initiation for the growth hormone gene, results in low serum glucose levels and down-regulation of the gene for thyroid-stimulating hormone, but not "housekeeping genes" (Klavinskis and Oldstone, 1989). These effects are apparently dependent on the strain of infecting virus and mapped to the small (S) segment of the viral RNA (Riviere *et al.*, 1985b). The virus has also been show to be diabetogenic, and viral persistence in this model has been demonstrated in the beta cells of the islets of Langerhans without obvious cytolysis (Oldstone *et al.*, 1984b). In the nonobese diabetic (NOD) mouse model, virus infection averts insulin-dependent diabetes, and this maps to the small (S) segment of LCMV (Oldstone *et al.*, 1991).

B. Lassa Virus in *Mastomys*

The only known reservoir of Lassa virus in West Africa is *Mastomys natalensis*, one of the most commonly occurring rodents in Africa (Wulff *et al.*, 1977; McCormick *et al.*, 1987a). At least two species of *Mastomys* (diploid types with 32 and 38 chromosomes) inhabit West Africa, and both have been found to harbor the virus. All species are equally susceptible to silent persistent infection in the same way that LCMV infects mice. Experimental inoculation of laboratory-reared neonatal *Mastomys* leads to persistent excretion of Lassa virus in urine, but adult animals clear virus from serum and urine within three–weeks. Studies of wild-caught *Mastomys* show that over half the captured animals in some foci may be chronically infected, and antibody and virus may be present at the same time in about one-third of these (McCormick, personal observations).

C. Junin and Machupo Viruses in *Calomys*

The major rodent hosts for Junin virus are *Calomys* species. Both of the South American viruses may cause illness and death in newborn mice or may induce persistence (Sabattini *et al.*, 1977; Webb *et al.*, 1975). The rodents are affected by the virus, with up to 50% fatality among infected suckling animals and stunted growth in many others. Machupo virus renders its major natural host *Calomys callosus* essentially sterile, with the young dying *in utero*. Machupo virus also induces a hemolytic anemia in its rodent host with significant splenomegaly, often an important marker of infected rodents in the field. Transmission from rodent to rodent is horizontal, not vertical, and is believed to occur through contaminated saliva and urine. Both viruses induce a humoral immune response, which may include neutralizing antibody, in the face of persistent infection.

III. INFECTIONS IN LABORATORY RODENTS

A. LCMV in Hamsters

In adult hamsters LCMV behaves much as in adult mice, with variable outcome depending on the virus strain used, the WE strain giving the highest mortality (Hotchin *et al.*, 1975). Some hamster strains were more susceptible than others to developing high serum virus titers, severe disease, and death. Virus was cleared by the fourth week postinoculation in survivors. There were no differences in outcome when the virus was inoculated intraperitoneally or into the footpad. Immunosuppressed hamsters had a higher frequency of severe disease and death when inoculated with LCMV (WE or Armstrong strains) and, unlike mice, did not develop persistent infection (Peters *et al.*, 1987). Histopathological studies show that the virus is pantropic in lethally infected animals, though it appears to be concentrated in or around blood vessels, with little necrosis or inflammatory infiltrate or significant organ damage. There is marked wasting and profuse diarrhea, with up to 40% loss of body weight. This model has interesting similarities with the processes observed in humans and primates infected with African and South American arenaviruses. In another study, neonatally infected hamsters and cyclophosphamide-treated adult animals inoculated with the less pathogenic LM_4 strain of LCMV appeared to develop subclinical persistent infection, as do mice (Hotchin *et al.*, 1975). This study also reported persistence in immunocompenent "golden" hamsters inoculated at 30 days of age. From these limited data it seems both hamster genotype and virus strain are important in determining outcome.

B. LCMV in Guinea Pigs

The lethal dose of LCMV is also highly dependent on virus strain, ranging from less than 1 plaque-forming unit (pfu) for the WE strain to more than 10^6 pfu for the Armstrong strain. Reassortants of the WE and Armstrong strains of LCMV have been used to map the lethality of WE to the S RNA segment (Riviere *et al.*, 1985a). As in hamsters, the virus is pantropic, with high titers of virus and little evidence of histopathological damage. Endothelial cells and mesangial cells in the kidneys are sites of abundant replication, and large amounts of virus in the transitional epithelium of the bladder provide an obvious source for viruria (Parker *et al.*, 1976; Peters *et al.*, 1987).

C. Lassa Virus in Guinea Pigs

Guinea pigs have been used as an experimental laboratory model for Lassa fever with variable results. Outbread guinea pigs exhibit a range of responses with 20–30% survival following a mild or moderate febrile episode. Those that die have pathological evidence of myocarditis, pulmonary edema, and hepatocellular damage (Walker *et al.*, 1975; Callis *et al.*, 1982; Jahrling *et al.*, 1982). Lethality for inbred guinea pig strains may vary with virus strains. However, some strains that do not kill adult inbred guinea pigs are uniformly lethal for 3- to 5-day-old animals and for pregnant guinea pigs, which abort infected fetuses. Animals destined to die develop earlier viremia and reach higher persisting viral titers.

D. Junin Virus in Guinea Pigs

Outcome with Junin infection is again related to high viremia and depends on the infecting strain (Kenyon *et al.*, 1988). In the guinea pigs, virus is mainly viscerotropic, with no evidence of an immunopathological mechanism, so this laboratory animal is a reasonable model for human disease (Oubina *et al.*, 1984). These animals develop florid hemorrhagic disease with extensive necrosis of lymphatic tissue (Weissenbacher *et al.*, 1975; Molinas *et al.*, 1978). There are occasional reports of neurological disease as a later development in animals with lower viremia, but with high-titer virus replication in the brain (Contigiani and Sabattini, 1977). As in primates, a late neurological syndrome is also reported associated with effective administration of immune serum (Kenyon *et al.*, 1986). The authors of this study suggest that the brain provides a site where virus may evade the antibody and subsequently replicate to generate symptoms typified by prominent rear-limb paralyses. Although this type of lesion suggests ischemic or thrombotic events in the spinal cord, apparently neurotropism is frequent in any animals surviving Junin virus infection long enough to allow it to reach the CNS, and indeed, guinea pigs develop encephalitis with recoverable virus from the brain if death is delayed by treatment with immune plasma. Transplacental transmission of Junin virus in guinea pigs has also been demonstrated, with variable outcome depending on the stage of pregnancy (Sangiorgio and Weissenbacher, 1983). Some fetuses died and were aborted with hemorrhagic manifestations and recoverable virus, and a few survived. Only in early preganancy was the outcome favorable for both mother and fetus, as has been observed in human Lassa virus infections (Price *et al.*, 1988).

In experimental Junin virus infections of guinea pigs depleted of polymorphonuclear monocytes (PMN), the disease was apparently more severe, with higher virus titers and lung pathology suggesting pulmonary distress syndrome (Gonzalez *et al.*, 1987). The implications of these data are as yet obscure, but they do suggest a role for PMN in the pathology of arenavirus infections. It is unclear whether they are responsible for generation of or protection from disease processes.

IV. ARENAVIRUSES IN NONHUMAN PRIMATES

A. LCMV in Monkeys

The disease produced by the WE strain of LCMV in rhesus and cynomolgus monkeys is fatal within about 2 weeks, with a course similar to that described in monkeys and humans for Lassa fever (Peters *et al.*, 1987). The course is relentless fever and viremia reaching 10^7 to 10^8 pfu/ml by death. In a report of six cynomolgus monkeys infected with this strain of LCMV, an initial leukopenia was followed by a leukocytosis, primarily neutrophilia as previously observed in Lassa fever–infected primates and humans. All monkeys had intradermal hemorrhage (petechiae and ecchymoses) and *epistaxis.* At autopsy, large effusions were found, as well as high titers of virus in all tissues, including vascular components in the brain. Serum aspartate amino transferase (AST) levels were elevated as in Lassa fever. Monkeys similarly infected with the Armstrong strain of LCMV failed to show illness, despite a seroconversion to the virus. Development of antibodies with ability to neutralize LCMV was markedly delayed in the Armstrong infection.

B. Lassa Virus in Monkeys

Infection of rhesus and cynomolgus monkeys by Lassa virus causes fever after 5 days, significant anorexia, and progressive wasting (Jahrling *et al.*, 1980). This laboratory model has proved extremely useful in studying the pathogenesis of fulminating viral infections and in testing potential vaccines for Lassa fever. In rhesus and cynomolgus monkeys the infection almost invariably ends after 10–15 days in death from vascular collapse and shock, with mild to moderate hemorrhage affecting primarily mucosal surfaces (Fisher-Hoch *et al.*, 1987).

Although Lassa virus is pantropic, pathological findings are limited to mild hepatic focal necrosis without significant inflammatory response, some evidence of pulmonary interstitial pneumonitis, chiefly interstitial edema, and occasional focal adrenal cortical necrosis (Walker *et al.*, 1982a). Although the liver is the most affected (Edington and White, 1972; Winn *et al.*, 1975; McCormick *et al.*, 1986b), biochemical measures of liver function and the extent of tissue necrosis are inadequate to account for death due to hepatic failure. A marked discrepancy in AST:ALT ratios, as in Marburg and Ebola infections, is found. Since elevation of alanine amino transferase (ALT) is the closer marker of hepatocyte failure, it is possible that the AST levels are not due solely to damage to hepatocytes. This conclusion is further supported by the observation that prothrombin coagulation times are only marginally prolonged, and blood glucose is within the normal range (Fisher-Hoch *et al.*, 1987).

Absence of significant disturbances in coagulation, low titers of fibronogen degradation products (FDPS), and absence of evidence for increased platelet and fibrinogen consumption make disseminated intravascular coagulation (DIC) unlikely as a primary pathological process (Fisher-Hoch *et al.*, 1987; Lange *et al.*, 1985). In primates thrombocytopaenia is rarely seen, though petechiae have been observed in two animals challenged during vaccine studies (Fisher-Hoch, personal observation). Platelet function, on the other hand, is markedly depressed, and as in humans, a circulating inhibitor of platelet function has been observed (Cummins *et al.*, 1989a; Fisher-Hoch, personal observation). There is also evidence for disturbance of endothelial function in that prostacylin production in *postmortem* vascular samples is depressed compared with material from normal control animals (Fisher-Hoch *et al.*, 1987). Subtle changes in vascular function obviously occur, and appear to be rapidly reversible. Presumably these are sufficient to account for the failure of integrity of the intravascular compartment, leading to edma of the face, shock, and the effusions regularly observed at autopsy.

Early lymphopenia followed by rising relative and absolute neuthrophilia has been reported in primates (Fisher-Hoch *et al.*, 1987). *In vitro* lymphocyte proliferation tests during the acute phase of the illness show impaired responses to nonspecific mitogens, suggesting the function of lymphocytes is also inhibited, a fact that may be associated with the observation that no inflammatory infiltration of tissues is observed. The viral glycoprotein G2 has been observed to be associated with circulating neutrophils, but not with lymphocytes. The significance of this finding is uncertain, but in view of the capacity of the platelet inhibitor of Lassa fever to affect neutrophil function (Roberts *et al.*, 1989), some consideration needs to be given to the role of neutrophils in the pathogenesis of severe Lassa fever.

Serum antibodies to Lassa virus do not usually neutralize Lassa virus *in vitro* in a classical replication inhibition, serum dilution neutralization assay (Jahrling *et al.*, 1982; Wulff and Lange, 1975; Fisher-Hoch *et al.*, 1989). In one study of primates vaccinated with a vaccinia recombinant vaccine expressing the Lassa glycoproteins that survived challenge with Lassa virus, neutralizing antibody by a fixed-serum, varying-virus-dilution technique could only be demonstrated in a few samples between 21 and 97 days postchallenge, at a time when the IFA antibody was as much as 1:250,000, and when all animals had long cleared virus from their serum. Antibodies to viral proteins G1, G2, and N can be detected in dying unprotected animals following challenge. None of the prechallenge specimens from vaccinated, protected animals had any measurable neutralizing activity *in vitro* (Fisher-Hoch *et al.*, 1989). Neutralizing antibody does not play any role in clearance of virus from serum in acute infection, and it seems unlikely to play a major role in protection from challenge.

C. Junin and Machupo Viruses in Monkeys

Junin and Machupo virus infection of primates, from rhesus monkeys to marmosets, closely approximates the disease in humans, with fever, anorexia, weight loss, and gastrointestinal symptoms (Eddy *et al.*, 1975a; Kastello *et al.*, 1976; Avila *et al.*, 1987; Weissenbacher *et al.*, 1979). As the disease progresses, the animals develop a rash, flushing, thrombocytopenia, and petechiae with bleeding, especially in the mucous membranes. The animals die with cachexia and severe dehydration and at postmortem have hemorrhages and lymphocyte depletion from nodes, spleen, and bone marrow. In addition, the animals appear to reflect the same biological response as that of humans to virus-determined factors (McKee *et al.*, 1985). Thus a strain that produces primarily hemorrhagic disease in humans elicits the same in monkeys. Likewise, a primarily neurotropic strain in humans also produces a similar disease in rhesus monkeys. Studies in neotropical primates, such as *Callithrix jacchus*, also reveal an acute hemorrhagic disease with early, severe thrombocytopenia (Molinas *et al.*, 1983). Both the coagulation and complement cascades were activated.

Machupo virus infection has been studied in a variety of Old World monkeys (Eddy *et al.*, 1975a,b; Scott *et al.*, 1978; Kastello *et al.*, 1976; Terrell *et al.*, 1973; Peters *et al.*, 1987; McLeod *et al.*, 1976). The same variation in pathology observed with Junin is observed with Machupo. Some of the monkeys develop progressive fever, weight loss, hemorrhage, and eventually shock and death after 3–4 weeks. These animals demonstrate progressive thrombocytopenia, anemia, lymphocytopenia, and neutropenia. Inconstant and highly variable degrees of pathological change were also noted in the liver, myocardium, bowel, adrenals, and lymphoid tissues. A second group of animals survived the initial infection only to develop a late neurological disease manifested by tremors, ataxia, nystagmus, and paresis with rare survival. At this stage there is no viremia, and the brain shows lymphocytic infiltration and vasculitis. The acute hemorrhagic disease in monkeys is more like that observed in humans, whereas the neurological disease is much less frequently observed in human infections.

Treatment of Junin virus–infected primates with immune serum, as in guinea pigs and humans, even though successful in treating the acute hemorrhagic disease, predisposes to a late neurological syndrome involving hind limb paralyses (Avila *et al.*, 1987). Similar observations have been made in Machupo-infected monkeys (Eddy *et al.*, 1975b). Acute-phase phenomena may be due to direct viral replication in the brain (encephalitis or meningitis), but there are no data to suggest this is the case (Weissenbacher *et al.*, 1987). The etiology of these events is obscure, though obviously the apparent induction of lesions by use of immune serum indicates that some immune-mediated phenomenon is involved, strongly supported by the observation of lymphocytic infiltration with vasculitis in Junin-infected monkeys. Whether the treatment induces the disease or whether successful salvage of an otherwise fatal infection uncovers the neurological involvement is unclear. The cerebellar syndrome described in Machupo infections resembles that observed in patients with a variety of viral infections, including, in particular, varicella zoster, which is known to be neurotropic. On the other hand, localization to hind limb paralyses suggests spinal cord injury, which could be vascular. The only coherent unifying hypothesis would be that focal vasculitis, possibly mediated by immune complexes, may be involved in both syndromes.

V. ARENAVIRUS INFECTIONS IN HUMANS

A. Lymphocytic Choriomeningitis

LCMV illness in humans follows an incubation period of 1–3 weeks (Armstrong and Sweet, 1939; Hinman *et al.*, 1975). There has never been a reported case of person-to-person transmission, however. In a recent outbreak, inoculation of persistently LCMV-infected tumor cell lines into nude mice led to a number of infections in laboratory workers. In this study there is very close statistical association with regular contact with the nude mice and their bedding (Dykewicz *et al.*, 1992). Data from this and other studies show that close exposure to aerosol of animal excreta may be a route of infection.

LCM virus infection in humans may be asymptomatic, mild, or moderately severe with central nervous system (CNS) manifestation requiring hospitalization. In the largest studies of human infections reported, 33 of 94 (35%) infections were asymptomatic, 47 (50%) were mild to moderate febrile illnesses without significant CNS manifestations, and 14 (15%) had typical LCM (Deibel *et al.*, 1975; Hinman *et al.*, 1975.). Although rarely fatal (Smadel *et al.*, 1942; Warkel *et al.*, 1973), the disease can be severe with a prolonged convalescence. Severity of illness may depend on both dose and route of infection as well as the host immunogenetic background.

Typical lymphocytic choriomeningitis, from which the virus derives its name, begins with fever, malaise, weakness, myalgia, and headache, which is often severe, retro-orbital, and associated with photophobia. Myalgia is marked in the lumbar region. Anorexia, nausea, and dizziness are common. As many as 50% of patients may have combination of sore throat, vomiting, and arthralgias, with chest pain and pneumonitis occurring less frequently (Biquard *et al.*, 1977; Farmer and Janeway, 1942). Alopecia, orchitis, and transient arthritis of the hands have also been reported. The white blood cell count is often $3000/mm^3$ or less with a mild thrombocytopenia.

Physical examination shows pharyngeal inflammation, usually without exudate, and in more severely ill patients, meningeal signs including nuchal rigidity. About one-third of patients with CNS manifestations will develop encephalopathy, while the rest exhibit primarily aseptic meningitis (Meyer *et al.*, 1960). An interstitial pneumonia has also been described in two atypically fatal human cases and in nonhuman primate postmortem studies (Smadel *et al.*, 1942). Convalescence is prolonged, with persistent fatigue, somnolence, and dizziness.

In LCMV infections, cerebrospinal fluid (CSF) from patients with meningeal signs contains several hundred white cells per cubic centimeter, predominantly lymphocytes (>80%), with mildly increased protein and occasionally low sugar levels. Virus is often found in spinal fluids taken during acute disease (Vanzee *et al.*, 1975). In mice experimentally infected with LCMV, intrathecal levels of B-cell-stimulating factor 2 (BSF-2) and interferon-gamma have been correlated with the development of meningitis, and there is evidence that intrathecal BSF-2 also rises in CSF from patients with acute viral meningitis. This suggests that one component of the pathology of meningitis in mice and humans is invasion of the CNS by lymphocytes and plasma cells and their attendant cytokines (Frei *et al.*, 1988).

There are few published descriptions of the pathology of LCMV infection in humans. In one report of a fatal case with primarily neurological manifestations, there was evidence of perivascular infiltration of macrophages in multiple areas of the brain (Warkel *et al.*, 1973). Antigen was observed in the meninges and cortical cells by IFA, consistent with viral replication in the CNS. Neurological sequelae to LCMV infection are unusual but have been reported (Meyer *et al.*, 1960). Some experimental and indirect epidemiological evidence of hydrocephalus in newborns following maternal LCMV infection has been reported (Casals, 1977; Farmer and Janeway, 1942; Sheinbergas, 1975).

Less well-substantiated neurological associations with LCMV infection in humans are reported in a study from the U.S.S.R. of 12 patients with amyotrophic lateral sclerosis (Tkachenko *et al.*, 1984). The long incubation period of this disease is markedly at variance with the acute nature of the disease in all the animal models currently known. Three had antibody to LCM virus, and antigen was detetected in serum from two patients and tissue from one at postmortem by ELISA and solid-phase radioimmunoassay techniques. If the virus is indeed responsible for the pathology, this would represent a further instance of fatal outcome from LCMV infection. Confirmation of these observations in independent studies would obviously throw new light on the potential of LCMV to induce a wide range of human pathology, including chronic neurological disease. Also of interest is a single report of unilateral deafness caused by LCMV infection (Ormay and Kovacs, 1989). Since this is a well documented late complication of Lassa virus infections in humans, it may be that arenaviruses have tropism in humans for the auditory nerve or organ of Corti, or that some other processes resulting in local ischemia lead to localized nerve injury.

B. Lassa Fever in Humans

Lassa fever virus normally infects humans via mucosa or cuts or abrasions. The incubation period of 1–3 weeks that follows infection suggests a silent primary replication site as yet unknown, though in all probability this is within the reticuloendothelial system. Subtle onset with generalized symptoms including high fever, joint pain, back pain, and severe headache does not distinguish this infection from many other viral syndromes (McCormick *et al.*, 1987b). The characteristic dry cough and exudative pharyngitis suggest some upper and lower respiratory tract involvement early in the disease, but though virus may be isolated from the pharynx, this is variable and at low titer (Johnson *et al.*, 1987) and viral pneumonia is not a feature of Lassa fever. It does not appear that the upper respiratory tract is a major site of viral replication or excretion despite the severe sore throat.

Route and titer of infecting dose may be important determinants of outcome. Case fatality in hospitalized patients with Lassa fever is about 16% (McCormick *et al.*, 1987b). However, in recent outbreaks in Nigeria, much higher death rates have been observed (Tomori *et al.*, personal communication). Whether this is due to variation in virulence with different virus strains or the high dose and route of inoculation in the Nigerian outbreak (by sharing of needles and syringes in hospitals administering parenteral drugs) remains to be seen.

The degree of organ damage in fatal human infections is mild; sharply at variance with the clinical course and collapse of the patient (Walker *et al.*, 1982b). Liver damage is variable, with concomitant cellular injury, necrosis, and regeneration (McCormick *et al.*, 1986a). Nevertheless serum AST levels over 150 IU/liter are correlated with poor outcome, and an ever-increasing level is also associated with increased risk of death (McCormick *et al.*, 1986a,b). ALT is only marginally raised, and the ratio of AST:ALT in natural infections and in experimentally infected primates is as high as 11:1 (Fisher-Hoch *et al.*, 1987). Furthermore, prothrombin coagulation times and glucose and bilirubin levels are near-normal, excluding biochemical hepatic failure and suggesting that some of the AST may be nonhepatic in origin.

It is clear that the outcome in Lassa fever is associated with the degree of virus replication (McCormick *et al.*, 1987b). An increasing viremia is associated with an increasing case fatality. In addition to the liver, high virus titers occur in brain, ovary, pancreas, uterus, and placenta, but no significant pathological or functional lesions are observed. Since these patients also have high viremia, the titer in these organs may reflect their blood content rather than specific parenchymal replication of the virus. Immunofluorescence studies suggest that though parenchymal replication occurs, it is limited. Individual hepatocytes or groups of cells contain viral antigen, as may a few endothelial cells, but there is little evidence for extensive replication in other parenchyma such as neuronal tissue or alveolar cells. Electron microscopy does not show extensive cellular damage. Indeed, there are few clues to the pathogenesis of Lassa fever in standard pathological studies.

On admission to the hospital the hematrocrit of Lassa fever patients is often elevated (mean of 50.6/100 ml), presumably owing to dehydration (McCormick *et al.*, 1987a). Severe cases progress with vomiting and diarrhea. Some patients develop severe pulmonary edema and adult respiratory distress syndrome, gross head and neck edema, pharyngeal stridor, and hypovolemic shock (Fisher-Hoch *et al.*, 1985). This pattern is consistent with edema due to capillary leakage rather than cardiac failure and impaired venous return. Endothelial cell dysfunction has been demonstrated in experimentally infected primates dying of Lassa fever, in that there is apparently a marked decrease in prostacyclin production by endothelial cells (Fisher-Hoch *et al.*, 1987). Loss of integrity of the capillary bed presumably causes the leakage of fluids and macromolecules into the extravascular spaces and the subsequent hemoconcentration, hypoalbuminemia, and hypovolemic shock. Proteinuria is common, occurring in two-thirds of patients. The blood urea nitrogen (BUN) may be moderately elevated, probably owing to dehydration.

Edema and bleeding may occur together or independently. There is no characteristic skin rash in Lassa fever and petechiae and ecchymoses are not seen. In only about 15–20% of patients is there frank bleeding, manifest as oozing gums, epistaxis, gastrointestinal or vaginal bleeding, and conjunctival hemorrhages. The case fatality of patients with hemorrhage is 50%. Since there is minimal disturbance of the intrinsic and almost none of the extrinsic coagulation system, and there is no increase in fibrinogen breakdown products, disseminated intravascular coagulation (DIC) is excluded (Fisher-Hoch *et al.*, 1988). Furthermore, platelet and fibrinogen turnover in experimental primate infections are normal (Lange *et al.*, 1985; Fisher-Hoch *et al.*, 1987).

Though platelet numbers are only moderately depressed, in severe disease their function is almost completely abolished by a circulating inhibitor of platelet function (Cummins *et al.*, 1989a). The origin of this inhibitor is not known; however, it cannot be reproduced with viral material nor can it be blocked by antibodies to Lassa virus. In the platelet it blocks dense granule and ATP release and thus abolishes the secondary wave of *in vitro* aggregation while sparing the arachidonic acid metabolite–dependent primary wave. Its inhibition is probably by interference with calcium channel mechanisms and phosphatidyl inositol secondary messenger pathway systems. In a few Lassa fever cases there is later relative or absolute neutrophilia, (Fisher-Hoch *et al.*, 1988). Polymorphonuclear leukocyte (PMN) counts as high as $30 \times 10^3/mm^3$ have been recorded in very sick or terminally ill patients. The inhibitor of platelet function also interferes with the generation of the FMLP-induced superoxide generation in PMN (Roberts *et al.*, 1989).

Acute neurological manifestations are common in Lassa fever. These range from isolated unilateral or bilateral deafness, with or without tinnitus, to moderate or severe diffuse encephalopathy with or without general seizures (McCormick *et al.*, 1987a). The encephalopathic complications generally carry a poor prognosis, while the deafness usually occurs just as recovery is underway. Manifestations during the acute phase range from mild confusion and tremors to grand mal seizures and decerebrate coma. Focal fits are not seen. CSF examination usually shows a few lymphocytes, but otherwise is normal, with low virus titers where blood-free samples have been reliably drawn. Furthermore, brain pathology is minimal both in humans and in primates infected with Lassa virus, so it is unclear whether cerebral involvement is a result of direct infection. Observations of cardiac pathology have been limited to hemorrhage and a lymphocytic infiltrate in the pericardium, and occasional interstitial myocarditis. The severe retrosternal or epigastric pain seen in many patients may be due to pleural or pericardial involvement, and late in the disease about 20% of patients have pleural or pericardial "rubs" (grating noises heard as the heart beats) (McCormick *et al.*, 1987a). Up to 70% of electrocardiogram (ECG) observations made on 32 patients in a recent study showed a range of abnormalities. These do not suggest a consistent cardiac pathology associated with Lassa virus infection, such as myocarditis or pericarditis (Cummins *et al.*, 1989b). The changes included nonspecific abnormalities in the ECG wave patterns (ST-segment and T-wave abnormalities, ST-segment elevation, generalized low-voltage complexes); all changes reflected electrolyte disturbance, but none correlated with clinical severity of infection, serum transaminase levels, or eventual outcome. Thus ECG changes were common in this limited study, but usually unassociated with clinical manifestations of myocarditis.

Lassa fever is particularly severe during the third trimester of pregnancy (Price *et al.*, 1988). Studies have shown that the overall case fatality in pregnant women infected by Lassa virus is about 20%, and very high levels of virus replication have been found in placental tissue in third trimester patients. A fourfold reduction was noted in case fatality among women in all trimesters who were spontaneously or therapeutically aborted compared to those who were not (odds ratio for fatality with pregnancy intact is 5.47 compared to those with uterine evacuation). The excess mortality in the third trimester may be due to the relative immunosuppression of pregnancy, which peaks at that time. Lassa fever is also devastating to the fetus.

Fetal/neonatal loss is 87%. Lassa virus is known to be present in the breast milk of infected mothers, and neonates are therefore at risk of congential, intrapartum, and postpartum infection with Lassa virus. Lassa fever also occurs in children, in whom the disease is similar to that in adults. However, a "swollen baby syndrome," consisting of widespread edema, abdominal distention, and bleeding, has been associated with serological evidence of Lassa virus infection of young children in a report from Liberia (Monson *et al.*, 1987). This syndrome has not been seen in adjacent areas of Sierra Leone. It would be of interest to compare viruses isolated from such cases to confirm the etiology and possible involvement of particular strains of virus.

Rare complications of Lassa fever include uveitis and orchitis. Virus replication occurs extensively in the adrenal glands, but no functional studies of the adrenal system have been done, though adrenal insufficiency during convalescence has been seen (McCormick, unpublished observation). The most significant sequel of Lassa fever is acute VIIIth nerve deafness, (Cummins *et al.*, 1990a). The onset is invariably during the convalescent phase of illness, and its development and severity are unrelated to severity of the acute disease. Nearly 30% of patients with Lassa fever infection suffer an acute loss of hearing in one or both ears. The mean auditory threshold of these patients is 55 dB (normal < 25 dB), and the mean disability is over 20%. About half the patients show a near or complete recovery over the 3–4 months after onset, but the other half continue with permanent, significant sensorineural deafness. Many patients also exhibit cerebellar signs during convalescence from severe disease, particularly tremors and ataxia, but this usually resolves with time. As with the late neurological syndrome of the South American hemorrhagic fevers, it is unclear whether this deafness is due to damage by neurotropic viruses, thrombosis, vasculitis, or focal hemorrhage. A late event observed in a few patients has been polyserositis (Hirabayashi *et al.*, 1988), but its pathology is obscure. A single case report describes an interesting complex of hemorrhagic pericarditis and cardiac tamponade with pleural effusions and ascites 6 months after acute Lassa fever. Repeated cultures failed to isolate virus from effusion fluids, but these specimens contained high titers of Lassa-specific IgG and numerous lymphocytes, suggesting an immune-mediated mechanism.

The immunological response to Lassa virus infection is complex. There is a substantial macrophage response, with little if any lymphocytic infiltrate. There appears to be a brisk B-cell response with a classic primary IgG and IgM antibody response to Lassa virus early in the illness. This event does not, however, coincide with virus clearance, and high viremia and high IgG and IgM titers often coexist in both humans and primates (Johnson *et al.*, 1987; Fisher-Hoch *et al.*, 1987). Indeed, virus may persist in the serum and urine of humans for several months after infection, and possibly in occult sites, such as renal tissue, for years. Acute lesions in Lassa virus infections are minimal, and without significant lymphocyte infiltration. There may be some impairment in the T-cell arm of the immune response during the acute infection, as has been seen in nonhuman primates. *In vitro* studies have shown that Lassa-virus is able to replicate in a continuous monocytic cell line, and this capacity is apparently enhanced by Lassa-specific antibody (Lewis *et al.*, 1988).

Neutralizing antibodies to Lassa virus are absent in the serum of patients at the beginning of convalescence, and in most people they are never detectable. In a minority of patients some low-titer serum-neutralizing activity may be observed several months after resolution of the disease and clearance of the virus (Jahrling *et al.*, 1980). The biological significance of this observation is not understood. Passive protection with antibody to Lassa virus has been demonstrated in animals given selected antiserum at the time of, or soon after, inoculation with virus, but clinical trials of human plasma have shown no protective effect (McCormick *et al.*, 1986a). Thus the clearance of Lassa virus appears to be independent of antibody formation and presumably depends on the cell-mediated immune (CMI) response. That the major immune response in human Lassa virus infection may be CMI-dependent is supported by recent experience with experimental Lassa vaccines in primates (Fisher-Hoch *et al.*, 1989; see also Chapter 15). A vaccinia virus recombinant vaccine expressing the surface glycoproteins of Lassa virus confers partial protection from Lassa challenge without eliciting neutralizing antibodies.

On the other hand, a second recombinant virus expressing the Lassa virus nucleoprotein produced a good IgG response by immunofluorescence. However, this vaccine is ineffective in protecting primates from lethal challenge with Lassa virus and may even accelerate the disease course (Fisher-Hoch, unpublished observations). During natural virus replication, nucleoprotein is produced in excess of glycoprotein, and antibody measured against this product *in vitro* is dominant. The role of this antibody in human infection could be important in the generation of disease.

These studies led to the conclusion that the cellular immune response must be critical in virus clearance and protection against disease possibly by limiting the extent of viral replication. Reinfection following natural Lassa infection does occur in humans, but it does seem that clinical disease does not ensue (McCormick, unpublished observation). It is possible that the apparent paralysis of the CMI component of the immune system during acute disease could also be associated with the host derived platelet inhibitory factor mentioned earlier. Perhaps this factor is itself an aberrant product of the acute immune response with very broad inhibitory effects.

C. Argentine and Bolivian Hemorrhagic Fevers

Argentine hemorrhagic fever (AHF) and Bolivian hemorrhagic fevers (BHF) are clinically similar diseases caused by arenaviruses related to Lassa virus (Aribalzaga, 1955; Rugiero *et al.*, 1964a,b,c; Maiztegui, 1975; MacKenzie *et al.*, 1964; Weissenbacher *et al.*, 1987; Johnson *et al.*, 1965; Peters *et al.*, 1974). After an incubation period of about 12 days, both AHF and BHF have insidious onset of a nonspecific illness consisting of malaise, high fever, severe myalgia, anorexia, lumbar pain, epigastric pain and abdominal tenderness, conjunctivitis, and retro-orbital pain, often with photophobia. Reports of subclinical or asymptomatic infections are rare. There is no lymphadenopathy or splenomegaly. Although there may be a pharyngeal enanthem (mucous membrane erruption), there is no sore throat or cough. These viruses presumably involve the respiratory tract to a lesser degree than Lassa virus. Another important difference is that there is marked erythema of the face, neck, and thorax, and petechiae may be observed.

In severe cases involvement of the gastrointestinal and nervous systems is manifest with nausea, vomiting, tremors, and convulsions. Intense proteinuria, microscopic hematuria with subsequent oliguria, and uremia are frequent. Localization of renal damage is shown by severe structural damage in the distal tubular cells and collecting ducts with relative sparing of the glomeruli and proximal tubules (Cossio *et al.*, 1975), confirmed by clinical observations of normal glomerular filtration rates, renal plasma flow, and creatinine clearance values (Maiztegui, 1975). Renal failure has been reported (Agrest *et al.*, 1969). Fatal cases demonstrate vascular collapse with hypotensive shock, hypothermia, and pulmonary edema. There is some electrocardiographic evidence of myocarditis (Ruggiero *et al.*, 1964b). Case fatality may be as high as 30% in clinically diagnosed BHF, and 16% in laboratory confirmed hospitalized patients with untreated AHF. In contrast to Lassa fever, bleeding is frequent, especially in AHF, and may be manifest as gingival hemorrhages, epistaxis, metrorrhagia (inappropriate menstrual bleeding), petechiae, ecchymoses, purpura, melena (stool darkened with altered blood), and hematuria (Maiztegui, 1975). Histological observations include large areas of intra-alveolar or bronchial hemorrhage. Gross examination of organs at necropsy shows petechiae on the organ surfaces, and ulcerations of the digestive tract have been described. Nearly half the patients with South American hemorrhagic fevers have hemorrhagic manifestation, most commonly epistaxis and/or hematemesis (Melcon and Herskovits, 1981). Platelet counts under 100,000 are invariable, and bleeding and clot retraction times are concomitantly prolonged. Though reductions of levels of factors II, V, VII, VIII, and X and of fibrinogen are observed, alterations in clotting functions are minor. Despite some reports of the presence of fibrinogen degradation products and absence of fibrinolysis, DIC, though it has been occasionally reported, is apparently not a significant feature (Agrest *et al.*, 1969; Molinas and Maiztegui, 1981; Weissenbacher *et al.*, 1987). Recently a circulating inhibitor of platelet aggregation has been described, as in Lassa fever (Cummins *et al.*, 1990b).

Nevertheless, bleeding is not the cause of shock and death. As in Lassa fever, pulmonary edema is common in severely ill patients, and intractable shock accounts for the majority of deaths. Persistent hypovolemic shock in the face of intravascular volume expanders suggests that this is due to the loss of endothelial function and leakage of fluid into extravascular spaces. Clinical observations led to the conclusion that vascular endothelial dysfunction and subsequent circulatory failure are also important in AHF and BHF (Rugiero *et al.*, 1964a). Microscopic examination shows a general alteration in endothelial cells and mild edema of the vascular walls, with capillary swelling and perivascular hemorrhage.

Fifty percent of AHF and BHF patients have acute neurological symptoms, such as tremors of the hands and tongue, progressing in some patients to delirium, oculogyrus, and strabismus. Meningeal signs and cerebrospinal fluid abnormalities are rare. As in Lassa fever, the pathology of central nervous system involvement is obscure, and there is, again, no evidence for direct infection. A late neurological syndrome has also been described, consisting mainly of cerebellar signs (Melcon and Herskovits, 1981; Maiztegui *et al.*, 1979; Enria *et al.*, 1986).

Virus titers in serum are not as high as in Lassa fever, but the infection is also apparently pantropic (Weissenbacher *et al.*, 1975). Outcome may, again, be related to virus titer in blood or tissues, though definitive studies demonstrating this have not been published. Electron microscopy studies have shown intracytoplasmic and intranuclear inclusions and marked nonspecific cellular damage in all organs examined. Immunofluorescence studies suggest that viral antigen but no immunoglobulins or C3 are associated with these damaged cells. Clinical studies reveal activation of the complement system, but no evidence of immune complex formation (de Bracco *et al.*, 1978). Cellular damage is probably mainly due to direct viral replication, rather than immune processes (Maiztegui, 1975; Maiztegui *et al.*, 1975).

The role of antibody in Junin and Lassa virus infections appears to be different (Cossio *et al.*, 1975; de Bracco *et al.*, 1978). There may be leukopenia with the thrombocytopenia (Rugiero *et al.*, 1964b). The antibody response to Junin virus may be very effective in clearing virus during acute infection and may also be sufficient to protect against future infections. Antibody, especially neutralizing antibody, is detectable at the time the patient begins to recover from the acute illness, and the therapeutic efficacy of immune plasma in patients with Junin infection is directly associated with the titer of neutralizing antibody in the plasma given (Enria *et al.*, 1984). Although elements of a CMI response to Junin virus have been shown, its importance in virus clearance and subsequent protection is not known. Interferon levels in patients with AHF may be very high (up to 64,000 IU/ml), and these high levels correlate with severity of disease and with outcome (Levis *et al.*, 1984, 1985).

VI. SUMMARY

The pathophysiology of arenavirus infections in rodents, nonhuman primates, and humans is discussed. Arenaviruses naturally infect rodent hosts in which the viruses are normally persistent but silent, with minimal histopathology. Transmission is *in utero* or at birth. In the rodent infected as an adult, acute, self-limiting disease may occur. Nonhuman primates develop severe, hemorrhagic disease and death, with pathology resembling that in humans when experimentally inoculated with AHF, BHF, or Lassa virus, and sometimes also with LCMV.

In humans, LCMV infection produces a lymphocytic choriomeningitis of varying severity, but usually mild and, though convalescence is prolonged, self-limiting and without sequelae. The virus replicates in the CNS where there are dense infiltrations of lymphocytic cells. Lassa fever and AHF and BHF have case fatality of about 16% in hospitalized patients, characterized by tissue and pulmonary edema with prominent hypovolemic shock and adult respiratory distress syndrome (ARDS). The hemostatic defect is characterized by platelet dysfunction rather than coagulation cascade activation, with little evidence for major hepatorenal failure or organ destruction by direct viral replication.

Bleeding with characteristic severe thrombocytopenia is more common in AHF and BHF than Lassa fever. In Lassa fever, despite normal platelet counts, an inhibitor of platelet function has been demonstrated that may affect the function of other cells, such as lymphocytes and endothelial cells. DIC appears to be confined to the terminal phases of all these diseases. Acute encephalopathy is seen in severe Lassa fever without evidence of direct viral invasion of the CNS, and in all three hemorrhagic fevers, late neurological syndromes are observed. In Lassa fever, as many as one-third of patients may develop a sensorineural hearing deficit in the early phase of convalescence, which is some cases may be total and permanent.

Lassa fever is particularly severe in the third trimester of pregnancy, possibly owing to the immunomodulation of pregnancy. Neutralizing antibodies are produced in AHF, and convalescent serum is effective treatment, whereas in Lassa fever neutralizing antibodies are, if present at all, low titer and evolve late in convalescence. Cell mediated immunity is probably critical in virus clearance and resistance to reinfection.

REFERENCES

Agrest, A., Sanchez Avalos, J. C., Arce, M., and Slepoy, A., 1969, Fiebre Hemorragica Argentine y coagulpatia por consume, *Medicina (Buenos Aires)* **29**:194.

Ahmed, R., Jamieson, B. D., and Porter, D. D., 1987, Immune therapy of a persistent and disseminated viral infection, *J. Virol.* **61**:3920.

Allan, J. E., Dixon, J. E., and Doherty, P. C., 1987, Arenaviruses: Biology and immunotherapy, *Curr. Top. Microbiol. Immunol.* **134**:131.

Aribalzaga, R. A., 1955, Una nueva enfermedad epidemica a germen desconocido: Hipertermia nefrotoxica, leucopenica y enantematica, *Dia. Med.* **27**:1204.

Armstrong, C., and Sweet, L. K., 1939, Lymphocytic choriomeningitis, *Public Health Rep.* **54**:673.

Avila, M. M., Samoilovich, S. R., Laguens, R. P., Merani, M. S., and Weissenbacher, M. C., 1987, Protection of Junin virus infected marmosets by passive administration of immune serum: Association with late neurologic signs, *J. Med. Virol.* **21**:67.

Biquard, H. A., Figini, D. A., Monteverde, M. J., Somoza, S., and Alvarez, F., 1977, Manifestaciones neurologies de la fiebre hemorragica Argentine, *Medicina (Buenos Aires)* **37**:193.

Buckley, S. M., Casals, J., and Downs, W. G., 1970, Isolation and antigenic characterization of Lassa virus, *Nature* **227**:174.

Callis, R. T., Jahrling, P. B., and DePaoli, A., 1982, Pathology of Lassa virus infection in the rhesus monkey, *Am. J. Trop. Med. Hyg.* **31**:1038.

Casals, J., 1977, Serological reactions with arenaviruses, *Medicina (Buenas Aires)* **37**:59.

Cole, G. A., and Johnson, E. D., 1975, Immune responses to LCM virus infection *in vivo* and *in vitro*, *Bull. WHO* **52**:465.

Cole, G. A., Gilden, D. H., Monjan, A. A., and Nathansen, N., 1971, Lymphocytic choriomeningitis virus. Pathogenesis of acute central nervous system disease, *Fed. Proc.* **30**:1831.

Cole, G. A., Nathanson, N., and Prendergast, R. A., 1972, Requirement for theta-bearing cells in lymphocytic choriomeningitis virus-induced central nervous system disease, *Nature* **238**:335.

Contigiani, M. S., and Sabattini, M. S., 1977, Virulencia diferencial de cepas de virus Junin por marcadores biologicos en ratones y cobayos, *Medicine (Buenos Aires)* **37**:244.

Cossio, P. M., Laguens, R. P., Arana, R. M., Segal, A., and Maiztegui, J. I., 1975, Ultrastructural and immunohistochemical study of the human kidney in Argentine hemorrhagic fever, *Virchows. Arch.* **368**:1.

Cummins, D., Fisher-Hoch, S. P., Walshe, K. J., Mackie, I. J., McCormick, J. B., Bennett, D., Perez, G., Farrar, B., and Machin, S. J., 1989a, A plasma inhibitor of platelet aggregation in patients with Lassa fever, *Br. J. Hematol.* **72**:543.

Cummins, D., Bennett, D., Fisher-Hoch, S. P., Farrar, B., and McCormick, J. B., 1989b, Electrocardiographic abnormalities in patients with Lassa fever, *Am. J. Trop. Med. Hyg.* **92**:1.

Cummins, D., McCormick, J. B., Bennett, D., Samba, J. A., Farrar, B., Machin, S. J., and Fisher-Hoch, S. P., 1990a, Acute sensory neural deafness in Lassa fever, *JAMA* **264**: 2093.

Cummins, D., Molinas, F. C., Lerer, G., Maiztegui, J. I., Faint, R., and Machin, S. J., 1990, A plasma inhibitor of platelet aggregation in patients with Argentine hemorrhagic fever, *Am. J. Trop. Med. Hyg.* **42**:470–475.

de Bracco, M. M., Rimoldi, M. T., Cossio, P. M., Rabinovich, A., Maiztegui, J. I., Carballal, G., and Arana, R. M., 1978, Argentine hemorrhagic fever: Alterations of the complement system and anti-Junin virus humoral response, *New Engl. J. Med.* **299**:216.

Deibel, R., Woodall, J. P., Decher, W. J., and Schryver, G. D., 1975, Lymphocytic choriomeningitis virus in man. Serologic evidence of association with pet hamsters, *JAMA* **232**:501.

Dykewicz, C. A., Dato, V. M., Fisher-Hoch, S. P., Howarth, M. P., Perez-Oronoz, G. I., Ostroff, S. M., Gary, H. Schonberger, L. B., McCormick, J. B., 1992, Lymphocytic choriomeningitis outbreak associated with nude mice in a cancer research institute, *JAMA* **267**:1349–1353.

Eddy, G. A., Scott, S. K., Wagner, F. S., and Brand, O. M., 1975a, Pathogenesis of Machupo virus infection in primates, *Bull. WHO* **52**:517.

Eddy, G. A., Wagner, F. S., Scott, S. K., and Mahlaudt, B. J., 1975b, Protection of monkeys against Machupo virus by the passive administration of Bolivian hemorrhagic fever immunoglobulin (human origin), *Bull. WHO* **52**:723.

Edington, G. M., and White, H. A., 1972, The pathology of Lassa fever, *Trans. R. Soc. Trop. Med. Hyg.* **66**:381.

Enria, D., Briggiler, A. M., Fernandez, J. H., Levis, S. C., and Maiztegui, J. I., 1984, Importance of dose of neutralizing antibodies in treatment of Argentine haemorrhagic fever with immune plasma, *Lancet* **2**:255.

Enria, D. A., de Damilano, A. J., Briggiler, A. M., Ambrosio, A. M., Fernandez, N. J., Feuillade, M. R., and Maiztegui, J. I., 1986, Sindrome neurologico tardio en enfermos de fiebre hemorragica Argentina tratados con plasma immune, *Medicine (Buenos Aires)* **45**:615.

Farmer, T. W., and Janeway, C. A., 1942, Infections with the virus of lymphocytic choriomeningitis, *Medicine* **21**:1.

Fisher-Hoch, S. P., Price, M. J., Craven, R. B., Price, F., Forthal, D. N. F., Sasso, D., Scott, S. M., Elliott, L. E., and McCormick, J. B., 1985, Safe intensive care management of a severe case of Lassa fever using simple barrier nursing techniques, *Lancet* **2**:1227.

Fisher-Hoch, S. P., Mitchell, S. W., Sasso, D. R., Lange, J. V., Ramsey, R., and McCormick, J. B., 1987, Physiologic and immunologic disturbances associated with shock in Lassa fever in a primate model, *J. Infect. Dis.* **155**:465.

Fisher-Hoch, S. P., McCormick, J. B., Sasso, D., and Craven, R. B., 1988, Hematologic dysfunction in Lassa fever, *J. Med Virol.* **26**:127.

Fisher-Hoch, S. P., McCormick, J. B., Auperin, D., Brown, B. G., Castor, M., Perez, G., Ruo, S., Conaty, A., Brammer, L., and Bauer, S., 1989, Protection of rhesus monkeys from fatal Lassa fever by vaccination with a recombinant vaccinia virus containing the Lassa virus glycoprotein gene, *Proc. Natl. Acad. Sci. USA* **85**:317.

Frame, J. D., Baldwin, M. N., Gocke, D. J., and Troup, J. M., 1970, Lassa fever: A new virus disease of man from West Africa. 1. Clinical description and pathological findings, *Am. J. Trop. Med. Hyg.* **19**:670.

Frei, K., Leist, T. P., Meager, A., Gallo, P., Leppert, D., Zinkernagel, R. M., and Fontana, A., 1988, Production of a B cell stimulatory factor-2 and interferon gamma in the central nervous system during viral meningitis and encephalitis. Evaluation of a murine model infection and in patients, *J. Exp. Med.* **168**:449.

Gonzalez, P. H., Ponzinibbio, C., and Laguens, R. P., 1987, Effect of polymorphonuclear depletion on experimental Argentine hemorrhagic fever in guinea pigs, *J. Med. Virol.* **22**:289.

Hinman, A. R., Fraser, D. W., Douglas, R. G., Bowen, G. S., Kraus, A. L., Winkler, W. G., and Rhodes, W. W., 1975, Outbreak of lymphocytic choriomeningitis virus infections in medical center personnel, *Am. J. Epidemiol.* **101**:103.

Hirabayashi, Y., Oka, S., Goto, H., Shimada, K., Kurata, T., Fisher-Hoch, S. P., and McCormick, J. B., 1988, An imported case of Lassa fever with late appearance of polyserositis, *J. Infect. Dis.* **158**:872.

Hotchin, J., Kinch, W., and Sikora, E., 1975, Some observations on hamster-derived human infection with lymphocytic choriomeningitis virus, *Bull. WHO* **52**:561.

Jahrling, P. B., Hesse, R. A., Eddy, G. A., Johnson, K. M., Callis, R. T., and Stephen, E. L., 1980, Lassa virus infection of rhesus monkeys: Pathogenesis and treatment with ribavirin, *J. Infect. Dis.* **141**:580.

Jahrling, P. B., Smith, S., Hesse, R. A., and Rhoderick, J. B., 1982, Pathogenesis of Lassa virus infection in guinea pigs, *Infect. Immun.* **37**:771.

Johnson, K. M., Wiebenga, N. H., Mackenzie, R. B., Juns, M. L., Tauraso, N. M., Shelokov, A., Webb, P. A., Justines, G., and Beye, H. K., 1965, Virus isolations from human cases of hemorrhagic fever in Bolivia, *Proc. Soc. Exp. Biol. Med.* **118**:113.

Johnson, K. M., McCormick, J. B., Webb, P. A., Smith, E., Elliott, L. H., and King, I. J.,

1987, Lassa fever in Sierra Leone: Clinical virology in hospitalized patients, *J. Infect. Dis.* **155**:456.

Kastello, M. D., Eddy, G. A., and Kuehne, R. W., 1976, A rhesus monkey model for the study of Bolivian hemorrhagic fever, *J. Infect. Dis.* **133**:57.

Kenyon, R. H., Green, D. E., Eddy, G. A., and Peters, C. J., 1986, Treatment of Junin virus-infected guinea pigs with immune serum: Development of late neurological disease, *J. Med. Virol.* **20**:207.

Kenyon, R. H., Green, D. E., Maiztegui, J. I., and Peters, C. J., 1988, Viral strain dependent differences in experimental Argentine hemorrhagic fever (Junin virus) infection of guinea pigs, *Intervirology* **29**:133.

Klavinskis, L. S., and Oldstone, M. B. A., 1989, Lymphocytic choriomeningitis virus selectively alters differentiated but not housekeeping function: Block in expression of growth hormone gene is at the level of transcriptional initiation, *Virology* **168**:232.

Lange, J. V., Mitchell, S. W., McCormick, J. B., Walker, D. H., Evatt, B. L., and Ramsey, R. R., 1985, Kinetic study of platelets and fibrinogen in Lassa virus–infected monkeys and early pathologic events in Mopeia virus–infected monkeys, *Am. J. Trop. Med. Hyg.* **34**:999.

Levis, S. C., Saavedra, M. C., Ceccoli, C., Falcoff, E., Feullade, M. R., Enria, D. A. M., Maiztegui, J. I., and Falcoff, R., 1984, Endogenous interferon in Argentine hemorrhagic fever, *J. Infect. Dis.* **149**:428.

Levis, S. C., Saavedra, M. C., Ceccoli, C., Feuillade, M. R., Enria, D. A., Maiztegui, J. I., and Falcoff, R., 1985, Correlation between endogenous interferon and the clinical evolution of patients with Argentine hemorrhagic fever, *J. Interferon Res.* **5**:383.

Lewis, R. M., Cosgriff, T. M., Griffin, B. Y., Rhoderick, J., and Jahrling, P. B., 1988, Immune serum increases arenavirus replication in monocytes, *J. Gen. Virol.* 69:1375.

Lukashevich, I. S., 1985, Lassa virus lethality for inbred mice, *Ann. Soc. Belg. Med. Trop.* **65**:207.

MacKenzie, R. B., Beye, H. K., Valverde, C. L., and Garron, H., 1964. Epidemic hemorrhagic fever in Bolivia. 1. A preliminary report of the epidemiologic and clinical findings in a new epidemic area in South America, *Am. J. Trop. Med. Hyg.* **13**:620.

Maiztegui, J. I., 1975, Clinical and epidemiological patterns of Argentine hemorrhagic fever, *Bull. WHO* **52**:567.

Maiztegui, J. I., Laguens, R. P., Cossio, P. M., Casanova, M. B., de la Vega, M. T., Ritacco, V., Segal, A., Fernandez, M. J., and Arana, R. M., 1975, Ultrastructural and immunohistochemical studies in five cases of Argentine hemorrhagic fever, *J. Infect. Dis.* **132**:35.

Maiztegui, J. I., Fernandez, N. H., and Damilano, A. J., 1979, Efficacy of immune plasma in treatment of Argentine hemorrhagic fever and association between treatment and a late neurological syndrome, *Lancet* **8**:1216.

Marker, O., and Volkert, M., 1973, Studies on cell-mediated immunity to lymphocytic choriomeningitis virus in mice, *J. Exp. Med.* **137**:1511.

McCormick, J. B., 1990, Arenaviridae, in: *Virology* (B. N. Fields and D. M. Knipe, eds.), pp. 1245–1267, Raven Press, New York.

McCormick, J. B., King, I. J., Webb, P. A., Schribner, C. L., Craven, R. B., Johnson, K. M., Elliott, L. H., and Williams, B. 1986a, Lassa fever: Effective therapy with ribavirin, *N. Engl. J. Med.* **314**:20.

McCormick, J. B., Walker, D., King, I. J., Webb, P. A., Elliott, L. H., Whitfield, S. G., and Johnson, K. M., 1986b, Lassa virus hepatitis: A study of fatal Lassa fever in humans, *Am. J. Trop. Med. Hyg.* 35:401.

McCormick, J. B., Webb, P. A., Krebbs, J. W., Johnson, K. M., and Smith, E. S., 1987a, A prospective study of the epidemiology and ecology of Lassa fever, *J. Infect. Dis.* **155**:437.

McCormick, J. B., King, I. J., Webb, P. A., Johnson, K. M., O'Sullivan, R., Smith, E. S., and Trippel, S., 1987b, Lassa fever: A case–control study of the clinical diagnosis and course, *J. Infect. Dis.* **155**:445.

McKee, K. T., Mahlandt, B. G. Maiztegui, J. E., Eddy, G. A., and Peters, C. J., 1985, Experimental Argentine hemorrhagic fever in rhesus macaques: Virus strain-dependent clinical response, *J. Infect. Dis.* **152**:218.

McLeod, C. G., Stookey, J. C., Eddy, G. A, and Scott, S. K., 1976, Pathology of chronic Bolivian hemorrhagic fever in the rhesus monkey, *Am. J. Pathol.* **84**:211.

Melcon, M. O., and Herskovits, E., 1981, Late neurological complications of Argentine hemorrhagic fever, *Medicina—Buenos Aires* **41**:137.

Meyer, Jr., H. M., Johnson, R. T., Crawford, I. P., Dascomb, H. E., and Rogers, N. G., 1960, Central nervous system syndromes of "viral" etiology, *Am. J. Med.* **29**:334.

Molinas, F. C., and Maiztegui, J. I., 1981, Factor VIII:C and factor VIII R:Ag in Argentine hemorrhagic fever, *Thromb. Haemost.* **46**:525.

Molinas, F. C., Paz, R. A., Rimoldi, M. T., and de Bracco, M. M. E., 1978, Studies of blood coagulation and pathology in experimental infection of guinea pigs with Junin virus, *J. Infect. Dis.* **137**:740.

Molinas, F. C., Giavedoni, E., Frigerio, M. J., Calello, M. A., Barcat, J. A., and Weissenbacher, M. C., 1983, Alteration of blood coagulation and complement system in neotropical primates infected with Junin virus, *J. Med Virol.* **12**:281.

Monjan, A. A., Bohl, L. S., and Hudgens, G. A., 1975, Neurobiology of LCM virus infection in rodents, *Bull. WHO* **52**:487.

Monson, M. H., Cole, A. D., Frame, J. D., Serwint, J. R., Alexander, S., and Jahrling, P. B., 1987, Pediatric Lassa fever: A review of 33 Liberian cases, *Am. J. Trop. Med. Hyg.* **36**:408.

Oldstone, M. B. A., 1987, Immunotherapy for virus infection, *Curr. Top. Microbiol. Immunol.* **134**:212.

Oldstone, M. B., Rodriguez, M., Daughaday, W. H., and Lampert, P. W., 1984a, Viral perturbation of endocrine function: Disordered cell function leads to disturbed homeostasis and disease, *Nature* **307**:278.

Oldstone, M. B., Southern, P., Rodriguez, M., and Lampert, P., 1984b, Virus persists in beta cells of islets of Langerhans and is associated with chemical manifestations of diabetes, *Science* **224**:1440.

Oldstone, M. B. A., Blount, P., Southern, P. J., and Lampert, P. W., 1986, Cytoimmunotherapy for persistent virus infection: Unique clearance pattern from the central nervous system, *Nature* **321**:239.

Oldstone, M. B. A., Salvato, M. S., and Ahmed, R., 1991, Viruses as therapeutic agents. II. Viral reassortants map prevention of insulin-dependent diabetes mellitus to the small RNA of lymphocytic choriomeningitis virus, *J. Exp. Med.* **171**:2091.

Ormay, I., and Kovacs, P., 1989, Lymphocytic choriomeningitis causing unilateral deafness [in Hungarian], *Or. Hetil.* **130**:789.

Oubina, J. R., Carballal, G., Videla, C. M., and Cossio, P. M., 1984, The guinea pig model for Argentine hemorrhagic fever, *Am. J. Trop. Med. Hyg.* **33**:1251.

Parker, J. C., Igel, H. J., Reynolds, R. K., Lewis, A. M., and Rowe, W. P., 1976, Lymphocytic choriomeningitis virus infection in fetal, newborn and young adult syrian hamsters (*Mesocricetus auratus*), *Infect. Immun.* **13**:967.

Peters, C. J., Kuehne, R. W., Mercado, R. R., Le Bow, R. H., Spertzel, R. O., and Webb, P. A., 1974, Hemorrhagic fever in Cochabamba, Bolivia, 1971, *Am. J. Epidemiol.* **99**:425.

Peters, C. J., Jahrling, P. B., Liu, C. T., Kenyon, R. H., McKee, K. T., and Barrera Oro, J. G., 1987, Experimental studies of arenaviral hemorrhagic fevers, *Curr. Top. Microbiol. Immunol.* **134**:5.

Price, M. E., Fisher-Hoch, S. P., Craven, R. B., and McCormick, J. B., 1988, Lassa fever in pregnancy, *Br. Med. J.* **297**:584.

Riviere, Y., Ahmed, R., Southern, P., Buchmeier, M. J., and Oldstone, M. B. A., 1985a, Genetic mapping of lymphocytic choriomeningitis virus pathogenicity: Virulence in guinea pigs is associated with the L RNA segment, *J. Virol.* **55**:704.

Riviere, Y., Ahmed, R., Southern, P., and Oldstone, M. B. A., 1985b, Perturbation of differentiated functions during viral infection *in vivo.* II. Viral reassortants map growth hormone defect to the S RNA of the lymphocytic choriomeningitis virus genome, *Virology* **142:**175.

Roberts, P. J., Cummins, D., Bainton, A. L., Walshe, K., Fisher-Hoch, S. P., McCormick, J. B., Gribben, J. G., Machin, S. J., and Linch, D. C., 1989, Plasma from patients with severe Lassa fever profoundly modulates f-Met-Leu-Phe–induced superoxide generation in neutrophils, *Br. J. Haematol.* **73:**152.

Rodriguez, M., Buchmeier, M. H., Oldstone, M. B. A., and Lampert, P. W., 1983, Ultrastructural localization of viral antigens in the CNS of mice persistently infected with lymphocytic choriomeningitis virus (LCMV), *Am. J. Pathol.* **110:**95.

Rose, J. R., 1956, A new clinical entity? *Lancet* **2:**197.

Ruggiero, H. R., Cintora, A., Libonatti, E., Magnoni, C., Castiglione, E., and Locisera, R., 1960, Formas nerviosas de la fiebre hemorragica epidemica, *Prensa Med. Argent.* **47:**1845.

Ruggiero, H. R., Parodi, A. S., Ruggiero, H. G., Mettler, N., Boxaca, M., de Guerrero, A. L., Cintora, A., Magnoni, C., Milani, H., Maglio, F., Gonzalez-Cambaceres, C., Astarloa, L., Squassi, G., Fernandez, D., and Giacosa, A., 1964a, Fiebre hemorragica Argentina. I. Periodo de incubacion e invasion, *Rev. Asoc. Med. Argent.* **78:**221.

Ruggiero, H., Cintora, F. A., Magnoni, C., Ruggiero, H., Gonzalez-Cambaceres, C., Maglio, F., Astarloa, L., Squassi, A., and Fernandez, D., 1964b, Fiebre hemorragica Argentina. III. Aparat cardiovascular, *Rev. Asoc. Med. Argent.* **78:**360–371.

Ruggiero, H. R., Cintora, F. A., Magnoni, C., Ruggiero, H., Gonzalez-Cambaceres, C., Maglio, F., Astarloa, L., Squassi, A., and Fernandez, D., 1964c, Fiebre hemorragica Argentina. IV. Formas clinicas, *Rev. Asoc. Med. Argent.* **78:**500.

Sabattini, M. S., de Rios, L. E. G., Diaz, G., and Vega, V. R., 1977, Natural and experimental infection of rodents with Junin virus, *Medicina (Buenos Aires)* **37:**149.

Sangiorgio, P., and Weissenbacher, M. C., 1983, Congenital and perinatal infection with Junin virus in guinea pigs, *J. Med Virol.* **11:**161.

Scott, S. K., Hockman, R. L., Lang, C. M., Eddy, G. A., Hilmas, D. E., and Spertzel, R. W., 1978, Studies of the coagulation system and blood pressure during experimental Bolivian hemorrhagic fever in rhesus monkeys, *Am. J. Trop. Med. Hyg.* **27:**1232.

Sheinbergas, M. M., 1975, Antibody to lymphocytic choriomeningitis virus in children with congenital hydrocephalus, *Acta Virol.* **19:**165.

Smadel, J. E., Green, R. H., Paltauf, R. M., and Gonzales, T. A., 1942, Lymphocytic choriomeningitis: Two human fatalities following an unusual febrile illness, *Proc. Soc. Exp. Biol. Med.* **49:**683.

Southern, P. J., Blount, P., and Oldstone, M. B. A., 1984, Analysis of persistent virus infections by *in situ* hybridization to whole-mouse sections, *Nature* **312:**555.

Terrell, T. G., Stookey, J. L., Eddy, G. A., and Kastello, M. D., 1973, Pathology of Bolivian hemorrhagic fever in the rhesus monkey, *Am. J. Pathol.* **73:**477.

Tkachenko, E. A., Dekonenko, E. P., Ivanov, A. P., Korolev, M. B., and Rezapkin, G. V., 1984, Results of a serological examination of amyotropic lateral sclerosis patients, *Vopr. Virusol.* **29:**191.

Traub, E., 1935, A filterable virus recovered from white mice, *Science* **81:**298.

Vanzee, B. E., Douglas Jr., R. G., Betts, R. F., Bauman, A. W., Frazer, D. W., and Hinman, A. R., 1975, Lymphocytic choriomeningitis in university hospital personnel: Clinical features, *Am. J. Med.* **58:**803.

Walker, D. H., Wolff, H., Lange, J. V., and Murphy, F. A., 1975, Comparative pathology of Lassa virus infection in monkeys, guinea pigs and Mastomys natalensis, *Bull. WHO* **52:**523.

Walker, D. H., Johnson, K. M., Lange, J. V., Gardner, J. J., Kiley, M. P., and McCormick, J. B., 1982a, Experimental infection of rhesus monkeys with Lassa virus and a closely related arenavirus, Mozambique virus, *J. Infect. Dis.* **146:**360.

Walker, D. H., McCormick, J. B., Johnson, K. M., Webb, P. A., Komba-Kono, G., Elliott,

L. H., and Gardner, J. J., 1982b, Pathologic and virologic study of fatal Lassa fever in man, *Am. J. Pathol.* **107**:349.

Warkel, R. L., Rinaldi, D. F., Bancroft, W. H., Cardiff, R. D., Holmes, G. E., and Wilsnack, R. E., 1973, Fatal acute meningoencephalitis due to lymphocytic choriomeningitis virus, *Neurology* **23**:198.

Webb, P. A., Justines, G., and Johnson, K. M., 1975, Infection of wild and laboratory animals with Machupo and Latino viruses, *Bull. WHO* **52**:493.

Weissenbacher, M. C., de Guerrero, L. B., and Boxaca, M. C., 1975, Experimental biology and pathogenesis of Junin virus infection in animals and man, *Bull. WHO* **52**:507–515.

Weissenbacher, M. C., Calello, M. A., Collas, D. J., Golfera, H., and Frigerio, M. H., 1979, Argentine hemorrhagic fever: A primate model, *Intervirology* **11**:363.

Weissenbacher, M. C., Laguens, R. P., and Coto, C. E., 1987, Argentine hemorrhagic fever, *Curr. Top. Microbiol. Immunol.* **133**:79.

Winn, W. C., Monath, T. P., Murphy, F. A., and Whitfield, S. G., 1975, Lassa virus hepatitis: Observations on a fatal case from the 1972 Sierra Leone epidemic, *Arch. Pathol.* **99**:599.

Wulff, H., and Lange, J. V., 1975, Indirect immunofluoresence for the diagnosis of Lassa fever infection, *Bull. WHO* **52**:429.

Wulff, H., McIntosh, B. M., Hamner, D. B., and Johnson, K. M., 1977, Isolation of an arenavirus closely related to Lassa virus from *Mastomys natalensis* in south-east Africa, *Bull. WHO* **55**:441.

CHAPTER 18

Lassa Virus and Central Nervous System Diseases

MARYLOU V. SOLBRIG

I. INTRODUCTION

Lassa fever, famous and infamous for its hemorrhagic shock syndrome, hospital epidemics, and high mortality in untreated cases, can now also be linked to neurological and psychiatric syndromes. Infrequent but important manifestations of disease in hospitalized Lassa fever patients are early acute encephalitis, late or convalescent ataxia, and subacute or chronic neuropsychiatric syndromes, including mania, depression, asthenia, sleep disorders, dementia, and psychosis. Readers are referred to Solbrig and McCormick (1991) for details of the clinical cases. The natural history of disease in Lassa patients includes neurological disease; neurological syndromes were recognized in 41% of hospitalized Lassa patients.

Entry of Lassa virus and immune system cells into the central nervous system (CNS) poses no great conceptual problem. The blood–brain barrier that isolates the CNS from infection and immune reactivity has limitations. White blood cells will cross inflamed meninges and the ependymal layer lining cerebral ventricles. Microglia begin life in the periphery and enter the nervous system as perivascular mesenchymal cells and may become macrophages for the CNS (Truex and Carpenter, 1969a). Blood crosses into the cerebrospinal fluid (CSF) compartments if a subarachnoid hemorrhage occurs, as seen in one of the clinical cases

MARYLOU V. SOLBRIG • Lassa Fever Research Project, Centers for Disease Control, Atlanta, Georgia 30333. *Present address:* Department of Neurology, University of California, Irvine, Irvine, California 92717.

The Arenaviridae, edited by Maria S. Salvato. Plenum Press, New York, 1993.

(Solbrig and McCormick, 1991). Also, there are a few areas where the blood–brain barrier is absent or incomplete: the pituitary, pineal, area postrema, supraoptic crest, and subfornical organ (Truex and Carpenter, 1969b). The area postrema is the floor of the fourth ventricle, and the cerebellum its roof; the supraoptic crest and subfornical organ are midline hypothalamic and limbic structures, respectively.

Some of these sites match suspected sites of neuropathological effect following Lassa infection. Ataxia localizes to the cerebellum or its outflow tracts; some of the neuropsychiatric syndromes may relate to the hypothalamic and limbic systems, and complaints of amenorrhea and fatigue may pertain to the pituitary.

Clinical observations have yielded reports of distinct neurological syndromes associated with Lassa fever. Therefore , Lassa virus may do harm in the CNS in several different ways. Taking the view that Lassa viral infection is a neural disorder for some patients, in addition to a hemorrhagic and hepatic one, investigators may choose to explore how Lassa may be involved in CNS diseases. Lassa virus may emerge as another arenavirus that has a contribution to make to understanding the relation of viral infection to neurochemical or neurotransmitter function, the immune system, and CNS diseases.

II. ENCEPHALITIS

Lassa infection causes encephalitic illnesses. A febrile, systemic illness typically progresses in 4 days to generalized seizures, dystonia, and memory difficulties; or seizures, agitation, and personality and cognitive changes. CSF inflammatory changes may be limited, and the neurological illness may be short, with recovery in days to weeks (Solbrig and McCormick, 1991).

Examinations of the brains from necropsy of nonhuman primates following experimental infection with Lassa virus, as well as studies of human brain specimens at postmortem examinations, have failed to detect any specific pathological changes or processes by light microscopy. In addition, virus isolations from brain specimens have generally been low titer relative to titers in blood and other organs (Johnson *et al.*, 1987; Walker *et al.*, 1982). These findings suggest that some of the neurological manifestations of Lassa fever may not result from a direct viral effect, but may instead have a more indirect cause.

The pathogenesis of CNS disease in Lassa fever remains poorly understood. Seizures, delirium, memory difficulties, and abnormal movements are the final common pathways of many metabolic changes. Blood chemistry facilities have not been available in rural West Africa and would be required for meaningful inquiry in this direction.

III. NEUROCHEMICAL HYPOTHESES

A. Excitotoxic Effects

Convulsions, short-term memory loss, and difficulty in learning extrapyramidal motor programs, all may relate to pathological increased activity at excitatory amino acid synapses (Choi, 1988; McGeer and McGeer, 1989). Seizures, agitation, memory changes, tremor and other movement disorders are also the clinical features observed in some acute Lassa patients. If Lassa infection alters the formation or turnover of neurotransmitter amino acids, then the clinical symptoms could relate to agonist effects on excitatory amino acid or glutamate class receptors or to the acute neurotoxic effects of these amino acids and their agonists.

To test for metabolic effect by excitatory transmitters, endogenous glutamate receptor agonists, such as quinolinic acid, can be assayed in the CSF. Quinolinic acid levels are increased in mouse brains after systemic administration of endotoxin (Heyes *et al.*, 1988), and the relationship of glutamate receptor agonists and their metabolites in CSF to viral diseases of humans has already been under investigation for HIV infections (Heyes *et al.*, 1989).

B. Peptides

It has been shown that platelet function is inhibited during severe Lassa fever (Fisher-Hoch *et al.* , 1988), and that dense granule release from platelets (Cummins *et al.*, 1989) and neutrophils (Roberts *et al.*, 1989) is inhibited. Neutrophils, in the presence of Lassa patients' serum, have blunted secretory responses to chemotactic peptides. A similar mechanism may play a role in the CNS manifestations, modulating neurotransmitter release or modifying signal transduction.

Much of the understanding of the organization of peptidergic neuronal systems comes from studies of the hypothalamic–pituitary and opiate systems (Brownstein, 1989). Although neuroendocrine disorders have not traditionally been associated with viral illnesses, interestingly, for months following Lassa infection, women are amenorrheic, lethargic, and asthenic (Samba, personal communication; Solbrig, unpublished observations), symptoms characteristic of panhypopituitarism.

IV. POSTVIRAL FATIGUE

Fatigue, without clear neuroendocrine disturbance, is a problem for another group of patients who recover from Lassa fever. Lassa convalescence is a 3- to 6-month period of disabling fatigue, motivation loss, and poor concentration, for many. There is loss of productivity as individuals are unable to resume normal work, daily activities, and household chores for 3–6 months after their illnesses.

In the postinfectious fatigue syndrome of Lassa, opportunity exists for assessing the role of persistent infection and defining patterns of host reactivity that predispose to chronic symptoms. Finding laboratory markers, excessive levels of lymphokines (Straus *et al.*, 1989) or cytokines (Chao *et al.*, 1990) in serum or CSF, for example, may inform our thinking on other prolonged fatigue states, such as the emotive chronic fatigue syndrome, or have direct bearing on other subacute or long-term neurological sequelae of Lassa fever.

As in asthenic or fatigue states, the contribution of acute or persistent infection, acute or prolonged neurotransmitter disturbance, loss of cellular homeostatic or "luxury" functions (Oldstone *et al.*, 1976; Lipkin *et al.*, 1988), CNS immune injury, and stress of overcoming a severe illness to other psychological and neuropsychiatric disturbances with Lassa, is unknown at this time.

V. CONVALESCENT CEREBELLAR SYNDROME

One syndrome likely to be related to idiosyncratic differences in immune responsiveness to Lassa virus is the convalescent cerebellar syndrome. Ataxia was exclusively a convalescent phenomenon, occurring 2 weeks after illness in two patients observed (Solbrig and McCormick, 1991). The ataxic illness clinically resembled the pure cerebellar ataxias that complicate or rapidly follow chicken pox and, less commonly, measles and mumps (Walton, 1977). Such syndromes have been attributed to an autoimmune or allergic demyelinative process triggered by viral illness. The delay in onset of neurological findings in the Lassa-associated ataxia cases, along with the frequent absence or low titer of virus in CSF and brain, raises the possibility that Lassa virus may induce a demyelinative illness with ataxic features by immunopathological mechanisms in a portion of infected persons who recover.

A similar convalescent-stage ataxic syndrome has been described in patients with Argentine hemorrhagic fever (Enria *et al.*, 1985; Maiztegui *et al.*, 1979). Nystagmus, ataxia, cerebellar tremor, and lateralizing gait occurred in the convalescent period or after a latent interval 2 or 3 weeks following discharge in approximately 10% of patients treated with Junin immune plasma. Brainstem-evoked potentials were abnormal in some AHF cases (Cristiano *et al.*, 1985), localizing portions of the pathology to brainstem myelinated tracts.

Immune response, to the detriment of the CNS of the host, may be a feature shared by several arenaviruses. The mechanism that produces classic CNS disease in lymphocytic choriomeningitis, a cytotoxic CD8+ T-cell response (Doherty *et al.*, 1990), may be a candidate mechanism to explain the postrecovery neurological syndromes of Lassa and Junin. Again, study of patient CSF offers possibilities for analyzing virus-induced CNS pathology. CSF cellular elements could be identified, and antibody and cytokine levels assessed. For the evaluation of cytokines, the longer biological half-life in CSF relative to serum is advantageous (Moskophidis *et al.*, 1991).

VI. CONCLUSION

There is, therefore, considerable scope for interference by Lassa virus in the normal functioning of the brain and neuroendocrine systems. Many people in West Africa develop acute Lassa fever each year. Lassa virus causes certain neurological disorders in certain patients. Clinical and laboratory observations in the population of patients who acquire and survive particular Lassa viruses may hold answers to the pathogenesis of encephalitic syndromes without cell destruction, postviral ataxic and other demyelinating conditions, postviral fatigue syndromes, and virus-related neuropsychiatric and neuroendocrine conditions.

REFERENCES

Brownstein, M. J., 1989, Neuropeptides, in: *Basic Neurochemistry,* 4th ed. (G. Siegel, B. Agranoff, R. W. Albers, and P. Molinoff, eds.), pp. 287–309, Raven Press, New York.

Chao, C. C., Gallagher, M., Phair, J., and Peterson, K., 1990, Serum neopterin and interleukin-6 levels in chronic fatigue syndrome, *J. Infect. Dis.* **162**:1412.

Choi, D. W., 1988, Glutamate neurotoxicity and diseases of the nervous system, *Neuron* **1**:623.

Cristiano, F., Huerta, M., D'Avino, P., Enria, D., Briggiler, A., and Maiztegui, J., 1985, Potenciales evocados auditivos de tronco cerebral en periodos alejados de la fiebre hemorragica argentina (FHA), *Medicina (Buenos Aires)* **45**:409.

Cummins, D., Fisher-Hoch, S. P., Walshe, K., Mackie, I., Bennett, D., Perez, C., Farrar, B., Machin S. J., and McCormick, J. B., 1989, A plasma inhibitor of platelet aggregation in patients with Lassa fever, *Br. J. Hematol.* **72**:543.

Doherty, P. C., Allan, J. E., Lynch, F., and Ceredig, R., 1990, Dissection of an inflammatory process induced by CD8+ T cells, *Immunol. Today.* **11**:55.

Enria, D. A., deDamilano, A. J., Briggiler, A. M., Ambrosia A. M., Fernandez, N. J., Feuillade, M. R., and Maiztegui, J. I., 1985, Sindrome neurologico tardio en enfermos de fiebre hemorragica argentina tratados con plasma immune, *Medicina (Buenos Aires).* **45**:615.

Fisher-Hoch, S. P., McCormick, J.B., Sassa, D., and Craven, R. B., 1988, Hematologic dysfunction in Lassa fever, *J. Med. Virol.* **26**:127–135.

Heyes, M. P., Kim, P., and Markey, S. P., 1988, Systemic lipopolysaccharide and pokeweed mitogen increase in quinolinic acid concentrations in cerebral cortex, *J. Neurochem.* **51**:1946.

Heyes, M. P., Rubinow, D., Lane, C., and Markey, S. P., 1989, Cerebrospinal fluid quinolinic acid concentrations are increased in acquired immune deficiency syndrome, *Ann. Neurol.* **26**:275.

Johnson, K. M., McCormick, J. B., Webb, P. A., Smith, E.S., Elliott, L. H., and King, I. J., 1987, Clinical virology of Lassa fever in hospitalized patients, *J. Infect. Dis.* **155**:456.

Lipkin, W. I., Battenberg, E. L. F, Bloom, F. E. , and Oldstone, M. B. A. 1988. Viral infection of neurons can depress neurotransmitter mRNA levels without histologic injury, *Brain Res.* **451**:333.

Maiztegui, J. H., Fernandez, N. J., and deDamilano, A. J., 1979, Efficiency of immune plasma in treatment of Argentine hemorrhagic fever and association between treatment and a late neurological syndrome, *Lancet* **2**:1216.

McGeer, P. L., and McGeer, E. G., 1989, Amino acid neurotransmitters, in: *Basic Neurochemistry,* 4 ed. (G. Siegel, B. Agranoff, R. W. Albers, and P. Molinoff, eds.), pp. 323–328, Raven Press, New York.

Moskophidis, D., Frei, K., Lohler. J., Fontana, A., and Zinkernagel, R. M., 1991, Production of random classes of immunoglobulins in brain tissue during persistent viral infection paralleled by secretion of interleukin-6 (IL-6) but not IL-4, IL-5, and gamma interferon, *J. Virol.* **65**:1364.

Oldstone, M. B. A., Holmstoen, J., and Welsh, R. M., 1976, Alterations of acetylcholine enzymes in neuroblastoma cells persistently infected with lymphocytic choriomeningitis virus, *J. Cell. Physiol.* **91**:459.

Roberts, P. J., Cummins, D., Bainton, A. L., Walshe, K., Fisher-Hoch, S. P., McCormick, J. B., Machin, S. J., and Linch, D. C., 1989, Plasma from patients with severe Lassa fever profoundly modulates f-Met-Leu-Phe~–induced superoxide generation in neutrophils, *Br. J. Haematol.* **73**:152.

Solbrig, M. V., and McCormick, J. B., 1991, Lassa fever: central nervous system manifestations, *J. Trop. Geograph. Neurol.* **1**:23.

Straus, S., Dale, J. K., Peter, J. B., and Dinarello, C. A., 1989, Circulating lymphokine levels in chronic fatigue syndrome, *J. Infect. Dis.* **160**:1085.

Truex, R. C., and Carpenter, M. B., 1969a, *Human Neuroanatomy,* 6th ed., pp. 155–159, Williams & Wilkins, Baltimore.

Truex, R. C., and Carpenter, M. B., 1969b, *Human Neuroanatomy,* 6th ed., pp. 12–25, Williams & Wilkins, Baltimore.

Walker, D. H., McCormick, J. B., Johnson, K. M., Webb, P. A., Kombo-Kono, G., and Gardner, J. J., 1982, Pathologic and virologic study of fatal Lassa fever in man, *Am. J. Pathol.* **107**:349.

Walton, J. N., 1977, *Brain's Diseases of the Nervous System,* 8th ed., pp. 514–541, Oxford University Press, Oxford.

CHAPTER 19

Ecology and Epidemiology of Arenaviruses and Their Hosts

JAMES E. CHILDS AND CLARENCE J. PETERS

I. INTRODUCTION

Arenavirus epidemiology must be understood on three levels: (1) The most fundamental and scientifically interesting concerns the distribution of virus in rodent populations. The dominant characteristic of the known, well-characterized arenaviruses is their ability to establish chronic viremic infections in specific rodent hosts. At the population level, this complex problem depends on the interaction of virus, rodent, and ecological variables that mutually determine the abundance of infected rodents, which may in turn contaminate humans with virus. (2) The variables that bring humans in contact with these rodents and their excreta in such a way as to lead to human infection. This process depends on human ecology and habits, as well as the dynamics of infected rodents. (3) The situations in which infected humans may be responsible for secondary infections of humans.

Several recent reviews of the specifics of arenavirus epidemiology are available (Peters, 1991), and other chapters in this volume will touch on those areas. This chapter focuses on selected issues of general importance, using the three categories described above.

JAMES E. CHILDS • Department of Immunology and Infectious Diseases, The Johns Hopkins University School of Hygiene and Public Health, Baltimore, Maryland 21205. *Present address:* Viral and Rickettsial Zoonoses Branch, Centers for Disease Control and Prevention, Atlanta, Georgia 30333. CLARENCE J. PETERS • Disease Assessment Division, U.S. Army Medical Research Institute of Infectious Diseases, Fort Detrick, Frederick, Maryland 21701. *Present address:* Special Pathogens Branch, Centers for Disease Control and Prevention, Atlanta, Georgia 30333.

The Arenaviridae, edited by Maria S. Salvato. Plenum Press, New York, 1993.

TABLE I. Species and Characteristics of Mammalian Hosts of Arenaviruses[a]

Virus	Species	Location	Habitat	Human contact[b]
Old World				
LCM	*Mus musculus*	Europe, Asia and the Americas	Peridomestic, grasslands	Primarily within houses
Lassa	*Mastomys natalensis*	West Africa	Savannah, forest clearings	Primarily within houses
Mopeia	*Mastomys natalensis*	Southern Africa	Savannah	
Mobala	*Praomys* spp.	Central African Republic	Savannah	
Ippy	*Arvicanthus* spp.	Central African Republic	Grassland/savannah	
New World				
Junin	*Calomys musculinus* *Calomys laucha* *Akodon azarae* *Bolomys obscurus*	Argentina	Grassland, cultivated fields, hedgerows	Primarily occupational, in cultivated fields
Machupo	*Calomys callosus*	Bolivia	Grasslands, peridomestic	Primarily within houses, associated with seasonal factors or severe weather
Guanarito	*Sigmodon alstoni* *Oryzomys* spp.?	Venezuela	Grasslands, brush	Primarily within houses?
Tacaribe	*Artibeus* spp.	Trinidad	Tropical forest	
Amapari	*Oryzomys goeldii* *Neacomys guianae*	Brazil	Tropical forest	
Flexal	day	*Oryzomys* spp.	Brazil	Tropical forest
Pichinde	*Oryzomys albigularis* *Thomasomys fuscatus*	Columbia	Tropical forest, riparian	
Latino	*Calomys callosus*	Bolivia, Brazil	Grasslands, peridomestic	
Parana	*Oryzomys buccinatus*	Paraquay	Tropical forest, savannah	
Tamiami	*Sigmodon hispidus*	Florida, U.S.A.	Grasslands, marsh	

[a] After Howard, 1986; Arata and Grantz, 1975; Johnson, 1981; Gonzalez *et al.*, 1983; Salas *et al.*, 1991; R. Tesh and M. Wilson, personal communication.
[b] Shown for human pathogens only.

II. GEOGRAPHIC DISTRIBUTION OF ARENAVIRUSES AND HOSTS

Arenaviruses are zoonotic agents maintained exclusively within rodent hosts, with the single exception of Tacaribe virus isolated from *Artibeus* spp. of the Phyllostomatid family of bats (Table I; Downs *et al.*, 1963). The various arenaviruses have been identified from areas subsuming either the entire or partial geographic ranges of the primary host species. Within those ranges the prevalence of each virus may be spatially or temporally patchy.

The distribution of lymphocytic choriomeningitis virus (LCMV) within house mouse populations has been most intensively studied. LCMV has the largest potential geographic range, and may occur wherever the house mouse, *Mus musculus* (*domesticus*), is native or has been introduced, probably including all continents with the exception of Antarctica (Sage, 1981). The most detailed analyses of LCMV distribution within a specific geographic region are from Germany and indicate an overall low prevalence of LCMV in feral *Mus* (3.6%; $n = 1795$ mice) and an uneven spatial distribution, with the majority of positive mice coming from northern locations (Ackermann *et al.*, 1964; Ackermann, 1973). In the United States the distribution of LCMV has been shown to be highly variable within populations of *Mus* sampled from the same geographic region. Infection with LCMV, as determined by presence of complement fixating (CF) antibody, was focal in squab farms surveyed in southern California, with prevalences of 0% (0/753 mice), 6.6% (44/667), 19.8% (552/2768), and 26.9% (214/594) found on four separate farms in two counties (Gardner *et al.*, 1973). Emmons *et al.* (1978) also found clustering of LCMV infections in California, where 5/63 mice from an apartment complex were antibody positive compared with a total of 3/296 mice from other sites. In Baltimore, LCMV antibody prevalences, measured by enzyme-linked immunosorbent assay (ELISA), varied from 3.9 to 13.4% in mice from inner-city locations with evidence that infections were clustered within residential blocks and even within households (Fig. 1; Childs *et al.*, 1991a). Clustering of infected mice may also have been present in Washington, D.C., where LCMV was isolated from >20% (65/307) of *Mus* captured from 35/78 households sampled (Armstrong, 1940). High prevalences of LCMV have been reported in some locations, including 67% of 76 wild mice trapped in or near a research institute in Surrey, England (Skinner *et al.*, 1977).

Human infections with LCMV appear to reflect its zoonotic distribution in *Mus* in Europe (Blumenthal *et al.*, 1968), England (Smithard and Macrae, 1951), and the United States (Armstrong and Sweet, 1939). In Baltimore, house mouse infestations are common in the inner city (Childs *et al.*, 1991a,b), and residents had a 4.7% ($n = 1149$) antibody prevalence to LCMV (Childs *et al.*, 1991c). LCMV in humans and *Mus* is also reported from South America, where LCMV circulates in the same

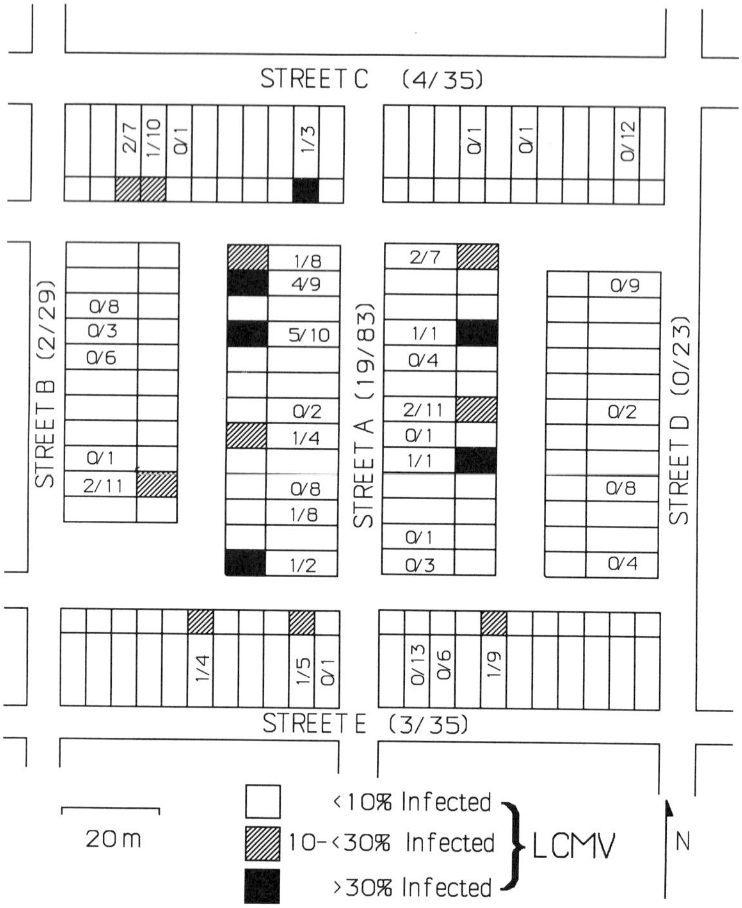

FIGURE 1. LCMV is abundant and focal in the urban United States and other areas where house mice occur. Street map of a five block area in residential Baltimore showing the prevalence of antibody to LCMV in house mice (number of mice infected/number of mice tested) within individual residences. Prevalence is indicated by the shading of a residence, and prevalence summaries for each block are indicated. Antibody was significantly higher in houses on either side of street A, and clustering of infections in some households was significantly greater than could be expected from chance. (From Childs *et al.*, 1991a.)

region as Junin virus in Argentina (Sabattini *et al.*, 1970; Sabattini and Contigiani, 1982). LCMV may be absent from Australia and Scandanavia (Pevear and Pfau, 1989).

By comparison, Lassa virus *sensu strictu* has been demonstrated by isolation from humans or *Mastomys natalensis* only in Nigeria, Guinea, Liberia and Sierra Leone (Fig. 2; Frame *et al.*, 1970; Fraser *et al.*, 1974; Monath *et al.*, 1973, 1974; Wulff *et al.*, 1975), a small portion of the entire range of this rodent which occurs throughout most of sub-Saharan Africa (Coetzee, 1975). In Sierra Leone, *M. natalensis* were found infected with Lassa virus, by viral isolation, at prevalences vary-

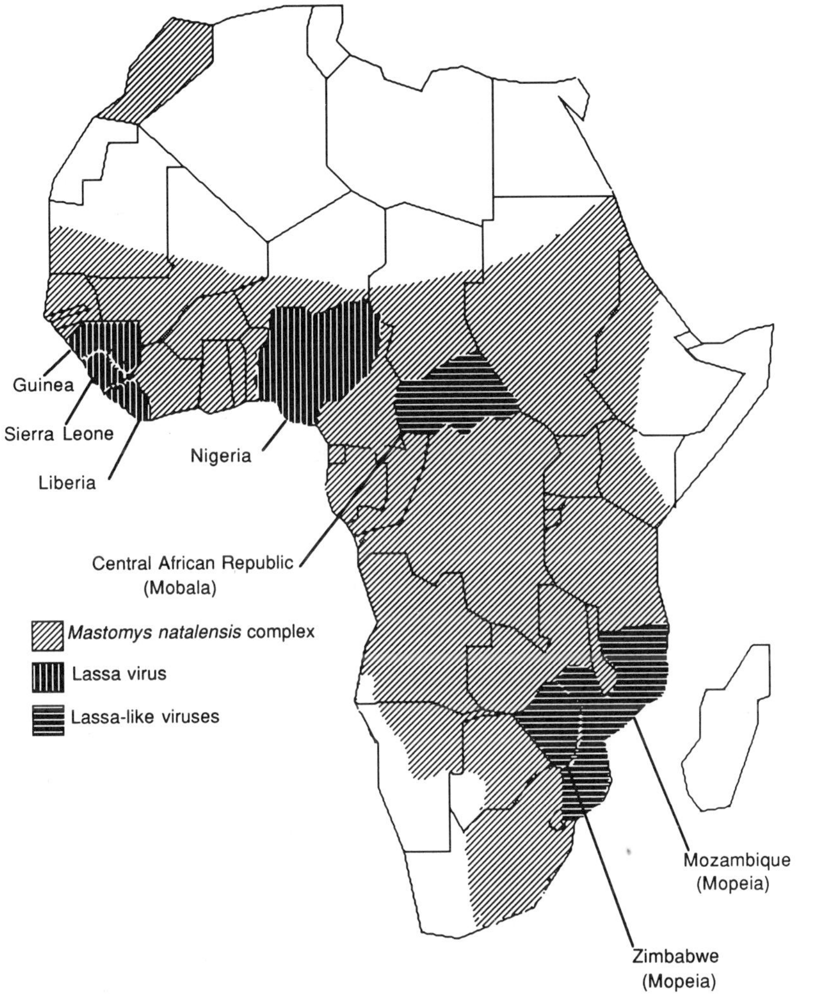

FIGURE 2. Map of Africa showing the geographic distribution of Lassa and related viruses by country. Lassa fever is known to be a major health problem in Sierra Leone and Liberia and suspected to be so in Guinea and Nigeria. Other West African countries need more epidemiological work. The distribution of the species complex of the major rodent reservoir, *Mastomys natalensis*, is also indicated. See text for references and details.

ing from 0% ($n = 8$) to >80% (26/32) from 14 villages, where it was also the numerically dominant species of rodent in houses (McCormick *et al.*, 1987). In comparison, an isolation rate of 4% ($n = 350$) was found in *M. natalensis* during an epidemic in Sierra Leone in 1972 (Monath *et al.*, 1974), while Keenlyside *et al.*, (1983) found 39% ($n = 23$) of *M. natalensis* trapped from the houses of human Lassa cases to be viremic, compared to 3.7% of those from control houses ($n = 27$).

Serological evidence indicating some degree of human infection with Lassa or Lassa-like viruses has been obtained from other West African countries, including Senegal, Gambia, Guinea, Ghana, Upper Volta, Mali, and Ivory Coast (Fig. 2; Frame, 1975; McCormick and Johnson, 1978; Monath, 1975). Lassa-like viruses, such as Mopeia, Mobala, Ippy, and others, have been obtained from *Mastomys* or *Praomys* from several different geographic regions in Africa: Mozambique (Wulff *et al.*, 1977); Central African Republic (Gonzalez *et al.*, 1983; Swanepoel *et al.*, 1985); and Zimbabwe (Johnson *et al.*, 1981). These viruses are less pathogenic for experimental animals and may be less virulent for humans than Lassa; they are serologically distinct from Lassa, but cross-protect experimental animals by cellular immune mechanisms (Jahrling and Peters, 1986). In Africa there are taxonomic difficulties in differentiating *Mastomys* species within what is now regarded as a species complex (Green *et al.*, 1980; Gordon, 1984; Happold, 1987). Thus, the actual distribution of Lassa and Lassa-like viruses may turn out to be highly localized and associated with a particular *Mastomys* sibling species, although markers such as chromosome number [$n = 32$ or $n = 36$ in South Africa (Green *et al.*, 1980); $n = 32$ or $n = 38$ in West Africa (Bellier, 1975)] do not appear to segregate with Lassa virus infection in this taxon, on the basis of limited field data (McCormick, 1987; McCormick *et al.*, 1987).

The Lassa–*Mastomys* pattern of incomplete virus overlap of a host species range is most representative of the other pathogenic arenaviruses. It should be noted, however, that intensive virus surveys outside of endemic foci are rarely undertaken to provide a definitive delineation of the arenavirus–host distribution. Descriptions of the geographic limits of arenaviruses are, therefore, tentative and in most instances defined by the occurrence of human disease, not by the true distribution of the virus. In Argentina, high prevalences (5–57%) of Junin virus isolation are reported from tissues or secretions obtained from *Calomys musculinus* in epidemic areas of Argentine hemorrhagic fever (AHF) (Sabattini and Contigiani, 1982; Sabattini *et al.*, 1977), but no infection (Sabattini and Contigiani, 1982) or lower levels of infection are reported from this species outside these locations (Mills *et al.*, 1991a). However, the recovery of Junin virus from other rodents, such as *C. laucha, Akodon azarae, Oxymycterus rufus, Bolomys (Akodon) obscurus,* and *M. musculus* in endemic foci (Mills *et al.*, 1991a; Sabbatini and Contigiani, 1982), and from *A. azarae* outside of recognized endemic AHF foci (Weissenbacher *et al.*, 1985) indicates our incomplete understanding of factors that determine current Junin virus distribution and the potential for the spread of AHF.

Recently, a new arenaviral disease of humans, clinically similar to AHF, has been described in rural areas in the municipality of Guanarito, Venezuela (Salas *et al.*, 1991). Although epizootiological investigations of the etiological agent, named Guanarito virus, are preliminary, the virus has been isolated from the spleen of a cotton rat (*Sigmodon alstoni*), and antibodies were identified in serum of a rice rat (*Oryzomys* spp.) (Salas *et al.*, 1991; R. Tesh and M. Wilson, personal communication). Subspecies of *S. hispidus* occur in northern South America, and have extensive geographic ranges in the United States (Hall and Kelson, 1959; Cameron and Spencer, 1981; Cockrum, 1948). The presence of a new, highly pathogenic arenavirus within this genus is of great interest given its large numbers and extensive distribution. A second nonpathogenic arenavirus (Tamiami) has been isolated from S. *hispidus* in Florida (Jennings *et al.*, 1970).

Reasons for a patchy spatial distribution of arenaviral infections within the geographic range of their mammalian reservoirs are unclear. Immunological mechanisms may play a role, but viral species, although cross-reactive in fluorescent antibody tests, generally do not share significant neutralizing antigens (Rowe *et al.*, 1970). Some degree of cross-protection has been seen between Tacaribe and Junin viruses (Tauraso and Shelokov, 1965), and this may be due to the late appearance of Junin neutralizing antibodies in Tacaribe-inoculated animals (Coulombie *et al.*, 1984). In the case of old World arenaviruses, there is cross-protection in guinea pigs and primates mediated by cellular immune responses in spite of lack of cross neutralization (Jahrling and Peters, 1986; Peters *et al.*, 1987). In addition, a cell infected with one arenavirus is refractory to infection with a heterologous arenavirus. Highly specific virus–rodent pairing is another possible mechanism which may explain the distribution of Lassa and the Lassa-like viruses isolated from *Mastomys* or *Praomys* in different regions of Africa (Johnson, 1985). Genetic variation in host susceptibility has not yet been shown to play a significant role in the distribution of any arenavirus, although host genetics are crucial for defining the outcome of individual infections (see below). The potential for viral exclusion between Machupo and Latino viruses was tested directly by cross-infections of *Calomys callosus* colonized from the two nearby areas where the viruses were isolated (Webb *et al.*, 1973, 1975). There was no evidence that prior infection with one virus (even viremic infection) prevented infection with the other nor that geographic host–virus matching was more efficient in producing persistent infections.

Junin virus in Argentina may be spreading within the reservoir population, as a five- to sixfold expansion in the size of the AHF endemic area has occurred since recognition of the disease in 1958 (Maiztegui, 1975; Maiztegui *et al.*, 1986). A gradient of infection is also found in Junin virus surveys of *Calomys* across the boundaries of AHF endemic–epidemic regions (Mills *et al.*, 1991a; Sabattini and Contigiani, 1982). The prevalence of Junin virus infection in *Calomys* is highest in endemic regions and is reported to be nonexistent or low outside the endemic zone (Carballal *et al.*, 1988, and Mills *et al.*, 1991a, respectively).

Accurate descriptions of arenaviral distributions remain problematic. Most infections within natural host populations are too scantily studied over the time periods necessary to differentiate geographic from seasonal effects or to identify processes that may be driven by fluctuations of rodent populations. As an example, Machupo virus has been hypothesized to drive cyclical population changes in *C. callosus* (Johnson, 1985); if this is true, the distribution and prevalence of this virus may vary in time in a manner related to the period of the cycle. Current assessments of virus–host ranges must then be viewed within the framework of dynamic processes that are poorly understood and have been inadequately defined because of the time scales and geographic areas involved.

III. INDIVIDUAL PROCESSES IN ARENAVIRUS MAINTENANCE

In natural primary hosts, arenaviruses can establish at least two types of chronic infections that result in persistent and copious viral shedding into the environment. LCMV, Junin virus, Machupo virus, and Lassa virus can establish what has been called a "persistent tolerant infection." This type of infection results in long-term, or lifelong, viremias with few signs of disease in carriers, although a "late disease" may occur. In the LCMV–*Mus* pairing, the term "tolerant" is now recognized as a misnomer (Oldstone and Dixon, 1967). Antibody is produced by these animals but does not result in a clearing of viremia; rather, it is bound in circulating immune complexes. Antigen–antibody complexes deposited on the basement membranes of the kidney fix complement and result in glomerulonephritis and death as the animals age (Hotchin and Weigand, 1961; Hotchin, 1962; Hotchin and Collins, 1964; Oldstone and Dixon, 1967). Other arenaviruses cause chronic infections without persistent viremia. Some *C. callosus* inoculated i.p. with Latino virus develop antibody, slowly clear their viremias, but retain virus in a number of tissues (Webb *et al.*, 1973). *S. hispidus* infected with Tamiami virus show long-term presence of virus in kidneys and urine, but only brief, transient viremia (Jennings *et al.*, 1970). Hamsters (*Mesocricetus auratus*) can also respond to LCMV infection in a manner similar to that of *Sigmodon* infected with Tamiami (Parker *et al.*, 1976).

Although the chronic carrier state in rodents has been extensively studied, its exact nature remains controversial and is only selectively reviewed here in the context of naturally acquired infections. Five factors can be identified as contributing to the development of different types of infections within the host; age of host at infection; genetic susceptibility (or strain) of the host; route of exposure; dose of virus; and the strain or passage history of infecting virus. Within this complex matrix of variables, manipulations that cause immunosuppression of the host, such as lymphocytotoxic drugs, X-ray-irradiation, T-cell-specific antiserum, or neonatal thymectomy, can determine the course of disease or infection (Lehmann-Grube, 1984).

A. Age at Infection

1. *In Utero* and Neonatal Infection

The chronic viremic carrier state typically results from arenaviral infection early in ontogeny. With LCMV, infection can occur transovarially (Mims, 1966) and then usually involves the entire litter (Traub, 1938, 1973). In this manner LCMV could presumably infect entire lineages of *Mus, in perpetuo,* without reverting to horizontal transmission (Mims, 1975). In most instances, experimental evidence indicates that neonatal inoculation with LCMV, i.c., within 2–3 days of birth (Hotchin, 1962; Weigard and Hotchin, 1961), or within 10 days by the i.p. route (Volkert *et al.*, 1975), also results in the chronic carrier state. Viremic titers may be lower and of shorter duration as a result of natural, early contact transmission of LCMV, compared to *in utero* or transplacental infection (Traub, 1938).

Machupo virus infections may also occur transovarially, infecting entire litters (Johnson *et al.*, 1973) or, through infectious semen or maternal blood, resulting in partial infection of a litter (Webb *et al.*, 1975). These authors speculate that either random infection of ova or local infection of implantation sites on the uterine epithelium causes the sometimes haphazard infection pattern. A similar pattern of partial infection in litters has also been reported for LCMV when uninfected females are mated with persistently infected males (Traub, 1973). Skinner and Knight (1974) demonstrated that adult female mice inoculated i.p. earlier than 9 days prior to mating were not infective to the fetus, but from this time through 11 days after mating a high percentage of litters were infected. These data define a brief window of opportunity by which acute infection of an adult rodent could result in congenital infection of a litter and potentially initiate the carrier state in subsequent generations. Mims (1969) showed that viral replication in the placenta begins within 24 hr of i.v. exposure, and Skinner and Knight (1974) demonstrated that placental infection followed local viral replication at the site of intramuscular injection.

Recovery from arenavirus infection, particularly LCMV and Lassa virus, is believed to be mediated significantly or entirely by cellular immunity, probably by virus-specific cytolytic T cells. Most of the Tacaribe complex or New World arenaviruses (Pichinde and Latino viruses being exceptions) elicit a more significant neutralizing antibody response, and antibody or antibody-dependent cellular cytotoxicity may be correspondingly more important (Peters *et al.*, 1987; Cerny *et al.*, 1988; Kenyon. *et al.*, 1990; Trapido and Sanmartin, 1971). When neutralizing antibodies are not readily detectable by standard serum dilution neutralization tests, the indirect fluorescent antibody, CF, or ELISA tests are often used to measure the humoral response (Peters *et al.*, 1973). Rodents that develop the chronic carrier state usually have only low levels of detectable CF or immunofluorescent (IF) antibodies and little or no detectable neutralizing antibody; this is described for the IFA test in Machupo, Junin, Latino, and LCMV viruses (Johnson *et al.*, 1965; Sabattini and Contigiani, 1982). Noncarriers will have higher levels of IF antibodies, and viruses such as Machupo and Junin will induce neutralizing antibodies. Natural infection of *Oryzomys capita* with Amapari virus, for example, results in a long-lived viremia concurrent with circulating CF antibody, but no neutralizing (N) antibody, whereas typical Amapari infections lead to clearance of viremia and a readily measurable serum neutralizing antibody response (Pinheiro *et al.*, 1977).

2. Adult Infection

As rodents age, they develop an immune competence that may result in the ability to effectively clear arenaviruses when challenged. Older *Mus* inoculated i.p., s.c., intranasally (i.n.), or in the footpad with LCMV usually develop transient viremias, followed by CF antibody, and permanent immunity, even when challenged by the i.c. route (Lehmann-Grube, 1964). Vertical transmission is, therefore, a particularly efficient mechanism for maximizing duration of infectiousness and the viral load contaminating the environment. Lassa virus seems to follow a similar pattern in *M. natalensis* as LCMV in *M. musculus.* Neonatal inoculation leads to persistent viremia and high tissue, throat swab, and urine virus titers. Inoculation of adults results in transient appearance of virus in blood with gradual clearing of virus from the host; only one of two animals tested 103 days postinfection had virus recovered from lymph nodes and brain and showed low titers in urine (Walker *et al.*, 1975; Murphy and Walker, 1978). Breeding studies in the Lassa virus system have shown vertical transmission of virus to offspring of carriers (K. M. Johnson, personal communication).

Adult animals may also develop chronic viremic infections. For example, inoculation of *C. callosus* less than 9 days of age with Machupo virus always results in lifelong viremia and its accompanying phenomena. About half of colonized animals older than 9 days will respond in a neotenic fashion to i.m. Machupo challenge and develop the chronic carrier state (called immunotolerant or type A animals; Justines and Johnson, 1969; Webb *et al.*, 1975), while others (immunocompetent or type B) develop a sterilizing and protective immunity. Type B animals show N, CF, and IF antibody, minimal or no viruria, and none of the anemia or splenomegaly found in type A animals (Webb *et al.*, 1975). This "split" response of *C. callosus* has been shown to be under complex polygenic control and is hypothesized to mediate cyclical population fluctuations via a differential impact on type A and B rodent reproduction and survival (Fig. 3; and see below). The ratio of type A or B responders can also be influenced by the dose of inoculated virus, with

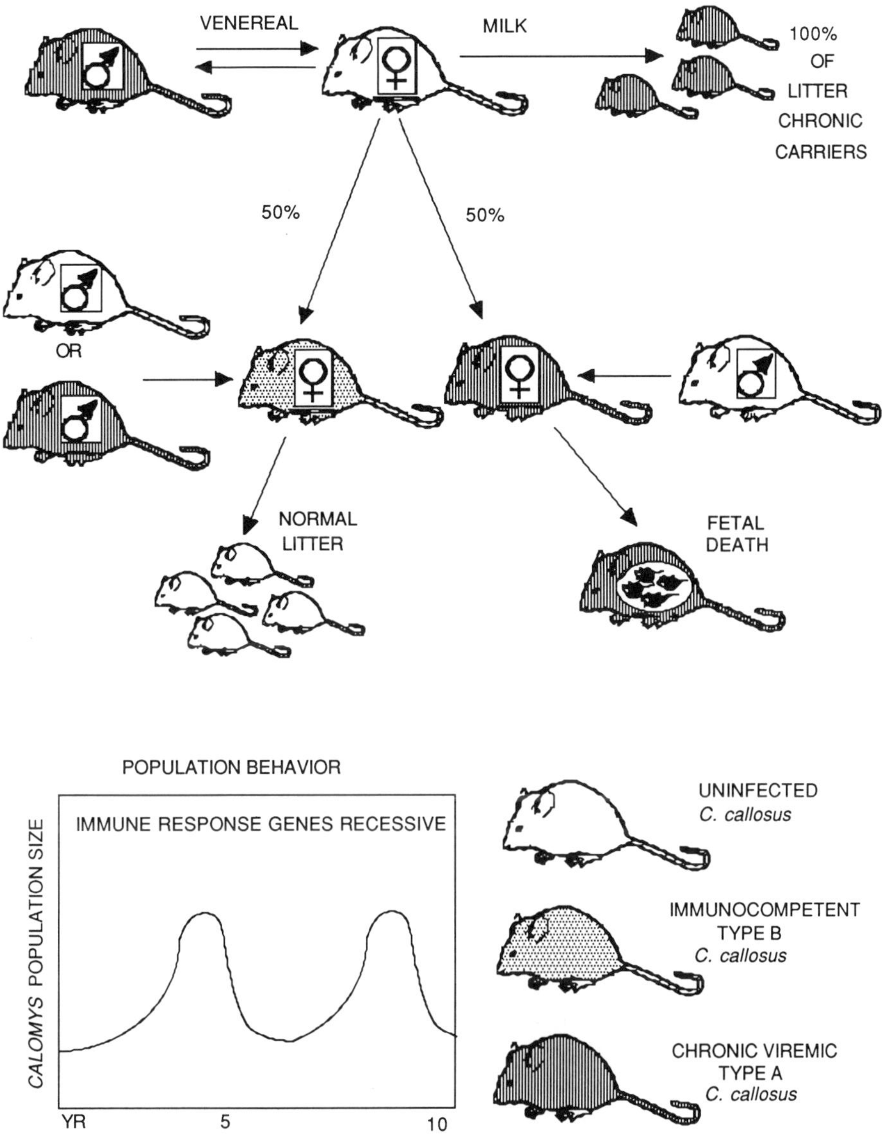

(after Johnson, 1985)

FIGURE 3. Modes of transmission and outcome of infection for Machupo virus in the rodent *Calomys callosus*. Two different genotypic responses to infection are shown: type A, chronically viremic but subfertile, and type B, which makes an effective immune response and has normal fertility. *Calomys* populations are shown to fluctuate cyclically as virus spreads within the rodent population generating type A and B mice. Recently infected mothers produce neonatally infected 100% type A litters. Adult mice will be about half type A and half type B. The infertility of type A infections will result in a fall in total rodent numbers and a decrease of the genotype leading to this response to infection. As numbers of type A mice dwindle, the two genetically distinct populations are either isolated or infection decreases until populations grow and virus again initiates the cycle. Presumably in the absence of Machupo virus, type A mice have a competitive advantage over type B. See text for additional details. (Adapted from Johnson, 1985.)

more animals becoming immunotolerant with increasing virus dose (Webb *et al.*, 1975).

Infection of adult *C. musculinus* with Junin virus following exposure to infected cage mates can also result in the development of the carrier state (Peralta *et al.*, 1979b). Clearly, *in utero* or neonatal exposures are not required for *Calomys* species to develop persistent viremic infections with South American arenaviruses. Indeed, the Junin virus–*C. musculinus* system resembles the Machupo virus–*C. callosus* interaction in many ways. Infection of colonized neonates by the i.n. route leads to persistent viremia in all animals, whereas infections at 3 weeks of age or later led to about 50% chronic viremia and 50% effective clearance of virus (Vitullo *et al.*, 1987; Vitullo and Merani, 1990).

3. Age-Specific Infection Rates in Natural Populations

Limited data exist on patterns of arenaviral infection in rodents of different age or size classes in natural populations. Antibody prevalence to LCMV has been shown to increase from a range of 0–16.4% in mice weighing <10 g (presumably juveniles) to 10–68.8% in mice > 25 g in the same populations in California (Gardner *et al.*, 1973). In Baltimore, there was also a trend for increasing LCMV antibody prevalence with mouse size, but rates varied only from 7.7% in animals < 10 g to 9.7% in animals > 15 g (Childs *et al.*, 1991a). In Argentina, *C. musculinus* were found to be antigen positive at all ages, but prevalence increased with age and size (J. N. Mills and J. E. Childs, unpublished data).

B. Genetic Determinants of Host Susceptibility

Most of our knowledge of the genetics of the host's immune response and susceptibility to infection comes from the LCMV-*Mus* model, as this is the prototype virus, and the primary host is a species for which genetically well-defined inbred strains are commonly available. Much of the difficulty in duplicating results from studies of LCMV across laboratories has undoubtedly been due to the great variety of outcomes possible when various permutations of mouse strain and virus strains of differing origin and passage history are considered (Traub, 1973).

Although our understanding of the genetics mediating the individual rodent's response to arenaviral infection is rapidly progressing, we still know little about the population genetics that influence the potential or actual geographic and host range of these viruses. Population genetic studies for hosts of arenaviruses are primarily based on electrophoretic analysis of various proteins in the hope of correlating allelic patterns with disease occurrence or behavior of the reservoir (see below). However, ecological studies on host genetics are usually divorced from simultaneous virological assessments of host susceptibility, so progress is slow.

1. Strains of Mice and LCMV Infection

Typically, inoculation of new-born laboratory mice with commonly used strains of LCMV leads to chronic viremia regardless of the route; there is usually no disease, although death and severe runting can occur within days of inoculation in some virus/mouse strain combinations. Some of the carrier mice will develop late immune complex disease. Adult mice inoculated by a peripheral route, such as i.n., s.c., or i.p., usually have a mild, immunizing infection. However, when adult mice are inoculated i.c., fatal T cell-mediated immunopathological meningitis occurs (Hotchin, 1962; Oldstone and Dixon, 1967; Cole and Nathanson, 1974; Oldstone *et al.*, 1985). The i.c. route is obviously unnatural, but the research value of this system for immunology has been enormous.

The variation in susceptibility of different strains of mice to LCMV (and the variation of different LCMV viral strains) is well documented, although there is a less clear consensus about the exact nature of the variation. In general, inbred or outbred strains of mice inoculated i.c. with a particular strain of LCMV will show differences in mortality rate, LD_{50}, and time to death. Tosolini and Mims (1971) showed that pathological changes in the viscera and pleural and peritoneal effusions were greatest in outbred Walter and Eliza Hall Institute (WEHI) mice, which also showed the highest mortality. These researchers reported that viral titers in the brains and spleens of WEHI mice infected as adults did not exceed those in three less susceptible inbred strains. WEHI mice also showed the greatest delayed hypersensitivity reaction, measured by footpad inoculation, and the conclusion was that "the more exuberant immune response" could explain the spectrum of results. Other studies have documented that mouse strains that are more susceptible as adults to LCMV have significantly higher titers of LCMV in tissues when neonates are inoculated i.c. to establish the carrier state (Oldstone and Dixon, 1968). In this latter study the LD_{50}, for the most resistant mouse strain (C_3H) was >10,000-fold greater than that of the least resistant (SWR/J), while SWR/J mice had >1,000-fold greater infectious virus titers in their tissues than 3-month-old carriers of the C_3H strain.

2. Major Histocompatibility Complex Genes and LCMV

Work with LCMV in mice clearly demonstrated the absolute requirement for MHC class I compatibility between the cytotoxic T lymphocyte (CTL) and the infected cell for killing to occur (Zinkernagel and Doherty, 1974). Indeed, the role of the class I molecule in presenting target epitopes leads to a selection of specific molecular sequences as immunogenic, depending on the MHC type. Both MHC and non-MHC genes affect the magnitude of the response (Allan and Doherty, 1989).

As a consequence of the selection of epitopes by class I molecules, different mouse strains may develop T cells reacting with different epitopes; for example, H-2^d and H-2^k mice each produce a CTL response predominantly directed against a different single epitope on the N protein of LCMV. H-2^b mice, however, develop CTL specific for a single epitope on Gl and a lesser response to an epitope on N (Whitton *et al.*, 1989; Klavinskis *et al.*, 1990). This limited number of epitopes presented by a single MHC haplotype has led to concern about the possibility of subversion of the immune response and interesting experiments demonstrating enhancement of disease by T-cell mechanisms under certain laboratory conditions (Oehen *et al.*, 1991). Another consequence of this mechanism of T-cell-antigen recognition is that the cross-reactivity pattern of LCMV-virus-specific CTLs generated in mice of different H-2 haplotypes is completely different for different strains of LCMV. CTL clones from H-2^b, mice (C57Bl/6) inoculated with the Arm strain of LCMV lysed targets infected with five different strains of LCMV (Arm, Traub, WE, Pasteur, and UBC), while a clone restricted to the H-2^d haplotype lysed only cells infected with three strains of LCMV (Arm, Traub, and WE), but not the other two (Byrne *et al.*, 1984). One suggestion is that common determinants of LCMV are seen in the H-2^b context, whereas regions of viral proteins recognized in the context of H-2^d molecules differed between viral strains. In related work, these same researchers suggest that the specificity of the CTL response could be used to dissect differences among various LCMV strains (Ahmed *et al.*, 1984a). The field significance of these findings is not clear, given the extensive polymorphisms of wild mouse MHC. One might imagine that superinfection and segment reassortment between strains could be facilitated.

Strains of mice can also be separated into categories of high or low responder with regard to the levels of circulating anti-LCMV antibodies and Clq binding material (Oldstone *et al.*, 1983). Levels of Clq-binding and anti-LCMV antibody were closely correlated, although levels of anti-LCMV antibody were not correlated to total IgG. F_1 hybrids between high and low responders more closely resembled high responders, although at intermediate levels, and backcrossing clearly indicated that Clq binding was influenced by gene(s) in the H-2 complex (Oldstone *et al.*, 1983; 1985). Immune response (Ir) genes in the MHC complex are probably influencing Clq binding through their influence on levels of antibody response to LCMV infection, but the variation in these responses indicates that other genes within or outside the Ir region may also play a role (Oldstone *et al.*, 1983). These findings may help unravel the variation in an individual host's response to virus infections.

3. Strains of Mice and Other Arenaviruses

Although the host specificity of arenaviruses has been emphasized, there may be chronic infections of other rodent species, albeit not typical of the fully developed carrier state and probably not adequate to maintain the virus in nature. The laboratory mouse provides a good example. Either Machupo or Junin viruses will infect mice, and in the case of Machupo, chronic virus shedding in throat swabs has been demonstrated (Johnson *et al.*, 1973). Chronic viremia does not occur. Intracranial inoculation of mice often leads to fatal immunopathological encephalitis resembling that seen in the adult mouse inoculated i.c. with LCMV unless, as has been demonstrated for Junin virus, T-cell function is ablated (Weissenbacher *et al.*, 1975).

Lassa virus inoculated i.c. into inbred mice produces a variety of possible outcomes, including viremia persisting at least 4 weeks, depending on the mouse strain and age at inoculation (Peters, 1991; Lukashevich, 1985). For example, Lukashevich (1985) studied four strains of inbred mice; Lassa virus caused mortality in nearly 100% of C3H/Sn mice, when challenged as newborns or young adults, whereas adult C3H/Sn mice and BALB/c mice of all ages were almost totally resistant to lethal effects. C57Bl/6 and AKR mice were intermediate in their susceptibility to lethal challenge. Lukashevich (1985) speculates that the increasing susceptibility of different strains of mice to Lassa virus challenge correlates with the amounts of endogenous interferon generated following challenge, as has been shown for LCMV and Mus (Riviere *et al.*, 1980).

4. Population Genetics

Population-based studies on the reservoir species for arenaviruses have focused on sampling animals from regions within and outside the known distribution of human disease. Genetic studies aimed at documenting the cytogenetic relationship of different sympatric *Calomys* species and the allelic frequencies and heterozygosity of *Akodon* and *Calomys* have focused primarily on Cordoba province where AHF is endemic in the south (Gardenal *et al.*, 1977a,b; Apfelbaum and Blanco, 1985a,b). Earlier work established diploid complements for three species of *Calomys*; *C. musculinus*$^{2n=38}$, *C. laucha*$^{2n=64}$, and *C. callidus*$^{2n=56}$ (Gardenal *et al.*, 1977a). *C. musculinus* is recognized as an ecological generalist as it is found in a large variety of habitats throughout its range, and most studies have as a final goal the identification of genetic markers that could segregate with habitat selection or Junin virus susceptibility. Initial surveys of genetic variation, based on examination of 15 loci of *C. musculinus* laboratory and field populations, showed high variation, with mean heterozygosity per locus of 20% (Gardenal *et al.*,

1980). Further efforts confirmed this observation, and on examination of 22 loci, heterozygosity was found to be 19% with 72.7% of the loci polymorphic (Gardenal and Blanco, 1985). Gardenal *et al.* (1986) sampled *C. musculinus* collected from six locations along a north-south transect that followed the progression of the AHF endemic area in Cordoba. The six populations showed remarkable homogeneity at three of the polymorphic loci previously identified. The authors conclude that either there is significant gene exchange between these populations or they are experiencing similar selection pressures (see Section III.G). The data indicate no obvious genetic barrier with regard to the competency of populations of *C. musculinus* to act as Junin virus reservoirs. The future spread of AHF into new regions is anticipated.

The population genetics of *M. musculus* has received considerable attention (see Berry, 1977, 1979, for review), but never in the context of LCMV susceptibility or distribution. This could provide some insights on the causes of virus focality, as has been shown for other infections in Mus. For example, Gardner *et al.* (1980) found a gene (*Akvr-1*) in wild *M. musculus* (*domesticus*) captured from California that restricted viremia by AKR endogenous retrovirus and Molony virus. Gene complexes in some of the various species of *Mus* may be similarly coadapted to LCMV. The taxonomic status of this species in various geographic locations is being debated, and Marshall (1981) splits the composite species *M. musculus* into seven different species. Up to four of these species may coexist in eastern Europe, and, although the worldwide distribution of the genus is extensive, the distribution of species or subspecies may be shown to be limited (Sage, 1981).

House mice are among the most variable of mammalian species with a mean heterozygosity ranging from 0 to 17% in different populations (Berry, 1981) and an overall average of 7% compared to 5.6% for other rodents (Selander and Kaufman, 1973). Because the house mouse establishes highly segregated demes, an excellent opportunity exists to integrate current laboratory work on MHC complex genes and LCMV infection with population-based studies. *Mus* populations are strictly divided into territorial subunits to the point where some researchers believe the effective population (breeding) size in natural populations is less than four individuals (DeFries and McClearn, 1972). The potential for inbreeding in physically unrestricted populations of *Mus* has been demonstrated by experimental introduction of male mice bearing *t*-alleles into island populations, where the allele was expected to spread rapidly until the segregation advantage in male heterozygotes was balanced by the frequency of lethal homozygotes (Anderson *et al.*, 1964). The *t*-allele spread far more slowly than if the island population had behaved as a breeding unit, and eventually disappeared, as might be expected by chance in small breeding populations (Myers, 1974). Studies from Baltimore suggest variation in LCMV infection on the microscale (perhaps household to household), which in part could be driven by the nature of demes (Childs *et al.*, 1991a).

Detailed population studies have not been conducted on other reservoirs of arenaviruses, but analyses of hemoglobin (Hb) electrophoretic patterns in *M. natalensis* collected in Sierra Leone provided a reliable marker for distinguishing the two sympatric species of *Mastomys* having 32 and 38 chromosomes (Robbins *et al.*, 1983). The 32 chromosome species (tentatively named *M. huberti*; Happold, 1987) is more frequently found in houses, while the 38 chromosome species (*M. erythroleucus*; Happold, 1987) is more frequently found in bush habitats (Robbins *et al.*, 1983).

Genetic factors determining the split response of *C. callosus* to Machupo challenge were investigated by Webb *et al.*, (1975), by inbreeding 52 families and their descendants. A 3:1 ratio of offspring phenotype in accord with parental phenotype was found in the F_1 generation from homotypic parents; however, analysis of six generations indicated that type A offspring were more easily obtained than type B (Webb *et al.*, 1975). The data suggest that genetic factors are significant in determining *C. callosus* response to Machupo virus, but indicate that more than a single locus is involved.

C. Routes of Exposure and Viral Entry

The outcome of arenavirus challenge is highly dependent on the site at which virus is introduced. The variation in response between i.c. inoculation (usually fatal in previously unexposed adults, tolerated in many neonates) and other routes of exposure with LCMV has been described in other contexts and will not be detailed here. Other more natural routes of exposure that result in significant horizontal transmission are via the nasal epithelium (Traub, 1936b) and introduction i.m. or s.c. into the dermis via contrived rodent bite (Skinner *et al.*, 1977). The i.m. or s.c. route may effectively mimic the introduction of virus by bite, and arenaviruses, such as Junin virus, have been shown to replicate to high titer in the salivary glands of animals following infection by natural routes involving contact with infected cagemates (Peralta *et al.*, 1979a,b). Contact transmission via aerosols or direct exposure of mucous membranes during allogrooming could also be a significant mechanism for transmission, although in a situation where LCMV was transmitted from wild to laboratory mice, bite appeared a more likely exposure route (Skinner *et al.*, 1977). The gastrointestinal tract of mice does not appear to be an important portal of entry for LCMV (Traub, 1936b), although in zoo primates transmission of LCMV or a LCMV-related virus is believed to have occurred by this route (R. J. Montali, personal cormnunication; see below). *Per os* transmission could be significant for rodents that engage in cannibalism, but this habit is apparently rare in *Mus* (Sage, 1981). Obviously, the transovarial, or *in utero*, infection of fetal tissues can result in transmission of LCMV, Machupo virus, and Junin virus, but the costs to the host of the latter two viruses may preclude total reliance on this mode of transmission (Vitullo *et al.*, 1987).

D. Dose of Viral Inoculum

The dose of arenavirus in conjunction with the route of exposure and age of the host can have a critical effect on the type of infection that results. Lehmann-Grube (1963) demonstrated that 100 LD_{50} of LCMV killed most mice when inoculated i.c. at 1 day of age, while 10^7 or 10^9 LD_{50} of Arm or WE LCMV viral strains, respectively, were tolerated when delivered i.p., and the higher doses initiated long-term viremias. Decreasing the challenge dose of Machupo virus increases the proportion of *C. callosus* that respond as type A (immunotolerant) animals (Webb *et al.*, 1975). Hotchin (1962) noted that "Swiss" mice survived very high doses (10^6 LD_{50}) of i.c.-inoculated "docile" LCMV, although a lower dose of the same virus strain proved fatal.

E. Strain and Passage History of Virus

Strain and passage history of arenaviruses inoculated into experimental animals have a marked effect on lethality, tissue tropism, and development of persistent infection (Dutko and Oldstone, 1983; Traub, 1938). Differences in the virulence of different LCMV isolates (Hotchin, 1962; Tosolini and Mims, 1971) and those with varied passage histories are well documented (Hotchin, 1962; Shwartzman, 1946). Hotchin (1962) designated the passage history of his experimental LCMV strains and documented that early mouse brain passages of LCMV resulted in lower mortality and tolerance in newborns ("docile" virus), whereas higher passage virus killed all newborns ("aggressive" virus). Viral stocks passed sequentially by i.p. inoculation and liver harvest retained their docility while brain passage resulted in aggressive characteristics.

Ahmed at al. (1984b) demonstrated that LCMV variants emerge during the course of individual mouse infections, and spleen cells carrying these variants can cause LCMV-specific CTL suppression when adoptively transferred from carrier mice. These strains, which differ only in a single amino acid in a viral glycoprotein, can initiate persistent viremic infections even in adult, immunocompetent mice (Ahmed *et al.*, 1991; Salvato *et al.*, 1988). Thus LCMV variants that arise naturally during the course of infection may play a crucial role in the maintenance of viral persistence.

The special biological relationship of Machupo and Junin viruses to the *Calomys* host is lost after adaptation through passage. For example, Machupo virus passed 20 times in suckling mice lost virulence for guinea pigs and an experimental primate host, but actually acquired virulence for newborn *C. callosus* (Johnson, 1975). As noted above, peripheral inoculation of wild strains of Junin virus results in the persistent carrier state in *C. musculinus* characterized by long-term presence of virus in brain, saliva, urine, and blood (Sabattini *et al.*, 1977; Peralta *et al.*, 1979b). Even direct i.c. inoculation of adult *C. musculinus* resulted in no disease and a carrier state. However, use of the attenuated XJCL3 virus resulted in undetectable viremia, neurovirulence, and some acute lethality for the rodents (Medeot *et al.*, 1987, 1990). Similar inverse results were seen in younger *Calomys* when virulent and derived strains were tested. There is also considerable variation when different wild-type Junin strains are studied in rhesus macaques; virus strains produced a spectrum of tissue tropisms and clinical disease that correlated well with those of AHF patients from which the strains were isolated (McKee *et al.*, 1985). A guinea pig model for Junin virus also produces a range in mortality, LD_{50}, and tissue distribution depending on viral strain, but disease in this animal is not as similar to human AHF as in the monkey model (Kenyon *et al.*, 1988). The results from arenavirus challenge of a number of experimental hosts with a wide variety of viral strains have been reviewed recently (Peters *et al.*, 1987).

Strains can be differentiated on a biochemical and molecular level by oligonucleotide mapping of L and S RNA (Dutko and Oldstone, 1983) by direct sequence analysis (Salvato and Shimomaye, 1989) or by tryptic mapping and monoclonal antibody analysis of GP1 (Buchmeier, 1983; Buchmeier *et al.*, 1981).

F. Viral Shedding and Transmission

1. Routes of Natural Transmission

The natural routes of arenavirus transmission from rodent to rodent or from rodent to human are a direct result of tissues infected during the chronic carrier state in the natural host (see below). Arenaviruses can be isolated with varying degrees of success from blood, urine, feces, and oropharyngeal secretions of experimental and naturally infected rodent hosts. Throat, oral cavity, or nasal swabs have been shown to be equal or superior to urine for detection of Machupo virus in experimental or field-collected *C. callosus* and experimentally infected hamsters (Justines and Johnson, 1968; Johnson *et al.*, 1973), and for culturing of Junin virus in four species of field-collected rodents in Argentina (Sabattini and Contigiani, 1982). LCMV was also isolated by oropharyngeal swabs during the Junin virus survey cited above, and bite was suggested as the most efficient and, indeed, the principal route of transmission for this virus from wild to laboratory mice (Skinner *et al.*, 1976, 1977). Skinner and Knight (1973) demonstrated that even very short contact (30 min) was sufficient to result in transmission of LCMV between cagemates, and transmission between pairs of mice that fought was substantially greater than for pairs that did not (48% and 9%, respectively). Traub (1936b) demonstrated the efficacy of nasal secretions in transmitting LCMV by experimental exposure, but his attempts to transmit virus by feeding infected brains to mice failed. Ingestion of milk from persistently infected dams is an efficient means of LCMV transmission when mice are young (78% of those 0–1 days of age), but is less certain as young mice age (39% in mice 2–7 days of age; Skinner and Knight, 1973). Milk is also an effective vehicle for Junin virus transmission in *C. musculinus* (Sabattini *et al.*, 1977; Weissenbacher *et al.*, 1987).

Various arenaviruses also have been isolated from arthropods, although the epidemiological significance of these findings is unclear. Isolates of Junin virus from ticks in 1958 initially confounded the recognition of the central role of rodents as reservoirs of the virus (Pirosky *et al.*, 1959). Pichinde virus was isolated from mites (*Gicrantolaelaps* sp.) and ticks (*Ixodes tropicalis*) obtained from viremic *Oligoryzomys albigularis* (Trapido and Sanmartin, 1971). Amapari virus has been isolated from mites (*G. oudemansi*) removed from *Oryzomys capito.* Quaranfil virus, a newly defined arenavirus, has been isolated from *Argas arboreus* ticks from Egypt (Zeller *et al.*, 1989). However, extensive efforts to isolate Machupo virus from 366 pools of >28,000 arthropods were unsuccessful (Kuns, 1965).

2. The Natural and Experimental Carrier State

The carrier state in rodent reservoirs, as documented by a high-titered viremia or viruria, is a natural phenomenon that has been shown to occur in field collections of *Mus* infected with LCMV (Skinner *et al.*, 1977), *C. callosus* infected with Machupo (Johnson *et al.*, 1965), *C. musculinus* infected with Junin virus (Sabattini *et al.*, 1977), *O. albigularus* infected with Pichinde (Trapido and Sanmartin, 1971), and *O. capito* and *Neacomys quianae* infected with Amapari (Pinheiro *et al.*, 1977). The carrier state has also been reproduced successfully in the laboratory for the species mentioned above, and for *M. natalensis* and Lassa virus (Walker *et al.*, 1975; Murphy and Walker, 1978), *C. callosus* and Latino virus (Webb *et al.*, 1973, 1975), *S. hispidus* with Tamiami virus (Murphy *et al.*, 1976), *Akodon molinae* and Junin virus (Carballal *et al.*, 1986), *Calomys callidus* and Junin virus (Videla *et al.*, 1989), and for the hamster and LCMV (Parker *et al.*, 1976). The hamster–LCMV relationship is artifactual in one sense because this species is not a natural reservoir of LCMV, but the accidental infection of hamster colonies is epidemiologically important since several documented outbreaks of human LCMV have resulted from exposure to this species (Ackermann *et al.*, 1972; Biggar *et al.*, 1975; Deibel *et al.*, 1975; Hirsch *et al.*, 1974). The hamster is a permissive host for LCMV when challenged as a weanling, and viremias usually decline after 2 months, although kidney infection with persistent high-titered viruria can last for at least 6 months (Parker *et al.*,, 1976). Neonatal hamsters may clear their viremias by 8 months, although others will respond like carrier *Mus* and continue to be viremic and viruric and go on to develop fatal progressive glomerulonephritis at about 1 year of age (Parker *et al.*, 1976; Skinner *et al.*, 1976).

Other natural reservoirs for arenaviruses may not develop the chronic viremic carrier state, as appears to be the situation with *S. hispidus* and Tamiami virus (Winn and Murphy, 1975) or *C. callosus* and Latino virus. These rodents may eventually clear their viremia after a variable period of infection, but can develop chronic infections of the kidney with persistent viruria (Jennings *et al.*, 1970). Although other virus–host pairings need to be examined, and many of the examples detailed above require additional experimental and field documentation, the carrier state in rodents has become identified as the hallmark of arenaviral infection in natural hosts.

G. Costs of Infection to Individual Fitness

Many arenaviral infections in their natural hosts cause, or can be manipulated to cause, measurable pathological effects that can influence individual fitness (Table II). Fitness in this sense can be measured by the number of one's offspring that survive to pass their genes to the next generation (Horn, 1981). Any process that has an impact on survival or age-specific fecundity will influence the reproductive rate of an individual and its relative fitness. If susceptibility to arenaviral infection has a genetic component, then the particular phenotype (such as type A versus type B *Calomys* with regard to Machupo virus; Webb *et al.*, 1975) can be the target of selection. In most instances these effects may be exerted directly on individual survival and number of offspring, but subtle effects on growth rates and age at sexual maturation could also have a significant impact on short lived rodents (cf. Childs *et al.*, 1989). Mass-based demographic modeling of rodent populations has as a fundamental assumption that reproductive effort is stage (size) specific (Sauer and Slade, 1986, 1987), so that runting may extend the prereproductive period and decrease the effective reproductive lifespan of an individual.

In the laboratory LCMV, Machupo virus and Junin virus can be shown to have significant actual or potential effects on reproductive

TABLE II. Examples of Some Experimental Effects of Selected Arenaviruses on Their Rodent Hosts

Rodent	Virus/strain	Exposure route/ age	Mortality	Weight	Reproduction	Ref.[a]
Calomys musculinus	Junin/Cba An9446	i.n./1–2 days	67% vs. 35% of controls at day 40	65–75% of controls at weaning	Decreased litter size	1
				Normal as adults[b]	No successive pregnancies	
Calomys callosus	Machupo/Carvallo	i.n., i.p./1–3 days	65% vs. 31% of controls at 1 year	80–90% of controls at 4 months	Decreased litter size[c]	2, 3
			Still low at 9 months			
Calomys callosus	Latino	i.p./1–4 days	Not tested	No effect	No effect?	3
Mus musculus	LCMV/WE	i.p./1 day	Normal at 1 years, > than controls after	70% of controls at weaning	Decreased litter size	4
WEHI[d]				Lower as adults	Neonatal loss[e]	

[a] 1 = Vitullo *et al.*, 1987; 2 = Johnson *et al.*, 1973; 3 = Webb *et al.*, 1975; 4 = Mims, 1970.
[b] Most animals with retarded growth were dead by 60 days of age.
[c] Decreased fertility of infected males and smaller litter sizes from infected females.
[d] Walter and Eliza Hall outbred mice.
[e] Described from carrier colony infected with WE, not from neonatally exposed mice.

effort by: (1) decreasing the survival of hosts infected *in utero* or as neonates; (2) retarding the growth rate of animals that do survive; and (3) decreasing the number of offspring that subsequent generations produce (Table II). In some instances distinctions between these various effects are unclear, and in all cases their roles in influencing natural population processes are speculative.

Traub (1936a,b, 1973) and others (Hotchin and Weigand, 1961; Hotchin, 1962; Lehmann-Grube, 1963; Mims, 1970; Seamer, 1964) have shown that certain strains of LCMV in some strains of laboratory mice result in a decreased life expectancy. This effect is present in chronic carrier mice, and only a portion of this effect is ascribed to chronic glomerulonephritis (Mims, 1973). Mice surviving their infections frequently show profound runting that can result in adult weights 70% of uninfected controls (Mims, 1970). Typically, weight differences on the order of 5–10% at weaning (21 days) are described (Hotchin and Weigand, 1961; Seamer, 1964). First ovulation and reproductive development in mice are highly correlated with body mass, as well as with other factors (Perrigo and Bronson, 1983; Hamilton and Bronson, 1985), so any delay in reaching the threshold mass necessary to begin reproductive life will result in a significant decrease in reproductive effort. Recently, LCMV has been shown to infect the pituitary and decrease levels of growth hormone in mice (Oldstone *et al.*, 1984; Klavinskis and Oldstone, 1986), providing one proposed mechanism for runting.

Potential costs to a host's fitness resulting from increased mortality and fetal wastage, with decreasing growth, have been well documented in *C. callosus* infected with Machupo virus and have recently been reported for *C. musculinus* infected with Junin virus. In the Machupo model, Carvallo strain virus obtained from a human spleen and passed twice in hamster brain was inoculated i.p. into *Calomys* only recently adapted to the laboratory (<21 generations; Johnson *et al.*, 1973). The infected young suffered increased mortality and decreased growth rates. At 4 months of age mean weights of infected and uninfected males were 55 and 62 g, respectively, and female weights were 40 and 52 g, respectively (Johnson *et al.*, 1973). Male fertility was estimated to be decreased by 50%, and normal females exposed to chronically infected carrier males gave birth to litters 60% that of controls, whereas chronically infected females exposed to normal males produced litters only 5% of controls (Johnson *et al.*, 1973).

Vitullo *et al.*, (1987) showed similar effects with Junin virus whereby *C. musculinus* infected i.n. at birth had significantly greater mortality, lower weight gain, and decreased reproductive efficiency compared to controls. Based on these findings, the reality of maintaining Junin virus in rodent populations solely through vertical transmission was questioned (Vitullo *et al.*, 1987; Vitullo and Merani, 1988). Interestingly, infection of adult *C. musculinus* resulted in some viremic carrier mice and some antibody-positive animals without viremia. In contrast to the neonatally infected animals, these persistently viremic rodents have normal fertility and produce about 50% infected offspring.

Obviously these effects are significant and suggest that a susceptible population of mice exposed to the selective pressures of a rodent pathogenic arenavirus would form an unstable balance with elimination of the rodent or the virus (cf. Vitullo and Merani, 1988). However, the importance of these processes in natural populations under field conditions has never been shown. In the case of LCMV, the suggestion has been made that many effects are epiphenomena resulting from highly selected laboratory *Mus* stocks exposed to manipulated LCMV strains with a variety of passage histories. Other researchers have documented few or transient effects in their mouse colonies as a result of LCMV introduction from a presumably feral *Mus* source (Skinner *et al.*, 1977). Traub (1939, 1973) documented a diminution of LCMV's deleterious effects in a chronically infected colony followed for 4 years and noted that congenital infection, with the resultant persistent carrier state and no early disease, became the only mode of transmission within 2 years.

Direct field evidence to support such dramatic effects of an arenaviral agent on a rodent's survival and reproduction is either entirely lacking or circumstantial. In Sierra Leone, wild-caught infected *M. natalensis* weighed less (mean = 33 g) than uninfected animals (54 g) captured from the same location, suggesting that runting may occur in natural populations (DeMartini *et al.*, 1975). However, as the authors acknowledge, rodent weight varies with age and the pattern could indicate an age-specific infection pattern as has been described for other rodent-borne viruses (Childs *et al.*, 1985, 1987). The variation in LCMV antibody prevalence with *Mus* body weight has already been described (Childs *et al.*, 1991a; Gardner *et al.*, 1973). Johnson *et al.*, (1973) report that in Bolivia, the prevalence of Machupo infection in rodents is fivefold higher in locations where Bolivian hemorrhagic fever (BHF) has occurred in the last year than in places where no disease has occurred for 3 or more years. The inference is that hemorrhagic fever epidemics in humans precede or are coincident with population declines in the reservoir caused by autosterilizing mechanisms described above. However, without data on absolute numbers of *Calomys* before and after BHF epidemics or surveys of relative numbers (or relative phenotypic frequency of type A and B mice) in the two regions of different BHF incidence, these interpretations remain speculative.

In Argentina, peak annual incidence of AHF and maximal populations of *Calomys* have been reported to cycle with a period of approximately 4 years (Carballal *et al.*, 1988), which is also compatible with an autosterilizing hypothesis. However, simple epidemiological models often assume that the rate of disease transmission is proportional to the number of random encounters between infective and susceptible individuals in the population (Anderson and May, 1979). Given a constant fraction of infective *Calomys*, epidemics of AHF in humans could be driven by "normal" population cycles whereby an unusually large number of young (susceptibles) rodents are produced, or survive, during a particularly auspicious year (see below). Such a phenomenon would not necessarily result in any variation in the prevalence of rodent infections, although it could lead to a greater absolute number of infected rodents. Without concomitant long-term studies of rodent populations and the incidence and prevalence of arenaviral infection, it will be impossible to differentiate viral effects on host fitness, and the resulting changes of host and virus gene frequencies, from population effects driven by unrelated processes.

IV. POPULATION PROCESSES IN VIRUS MAINTENANCE AND TRANSMISSION

A. Community Structure of Arenaviral Hosts

Although arenaviral infections in natural rodent hosts are usually discussed within the context of one virus–one host, there is ample evidence that more than one species or genus may be infected in a given locale (Arata and Gratz, 1975). This finding is usually dismissed as epidemiologically irrelevant, as one host usually dominates in absolute numbers (e.g., *Mastomys* in Sierra Leone, McCormick *et al.*, 1987) or frequency of infection (e.g., *C. musculinus* in Argentina; Mills *et al.*, 1991a; Sabattini and Contigiani, 1982), or develops the carrier state while other species show transient infection and antibody development (e.g., *A. azarae* infected with Junin virus (CbaAN 9446); Sabattini *et al.*, 1977; Sabattini and Contigiani, 1982).

In Santa Fe province, Argentina, at least 10 species of sympatric rodents can be trapped from cultivated fields and bordering weedy fence rows (Mills *et al.*, 1991b). Evidence of Junin virus infection can be found in six species: *C. musculinus, C. laucha, A. azarae, Bolomys obscurus, Oligoryzomys flavescens*, and M. *musculus* (Mills *et al.*, 1991a; Sabattini and Contigiani, 1982). *C. musculinus* is the most commonly infected and is most frequently found excreting or circulating virus, whereas the other species are commonly antibody positive (Sabattini and Contigiani, 1982; J. Mills, personal communication). In contrast, Lassa virus *sensu strictu* is usually recovered only from *M. natalensis*, though *Praomys jacksoni*, a closely related species, harbors Lassa-like viruses (Johnson *et al.*, 1981; Gonzalez *et al.*, 1983), and Lassa-like isolates have been identified by CF tests from *R. rattus* and *Mus minutoides* in Nigeria (Wulff *et al.*, 1975) and *Mastomys* and *Arvicanthus* spp. in Tanzania. (Leirs *et al.*, submitted). LCMV also has a highly restricted host range and is usually only associated with *M. musculus* in natural situations, although *Apodemus sylvaticus* and A. *flavicolis* have been reported as infected in Europe (Lehmann-Grube, 1971), and the hamster has become a linkage host for transmission to humans. Pichinde virus is similar to Lassa virus and LCMV in its host restriction in Columbia, where 54 of 55 isolates were obtained from *O. albiqularis* and a single isolate from *Thomasomys fuscatus*, although 23 species of small mammal were captured (Trapido and Sanmartin, 1971). The host range of Guanarito virus is as yet undefined, although from 11 rodents tested, virus was isolated from one *S. alstoni* and antibodies were identified in an *Oryzomys* sp., raising the possibility that more than a single reservoir species is involved in viral maintenance (Salas *et al.*, 1991; R. Tesh and M. Wilson, personal communication).

B. Habitat Requirements

The majority of the rodents involved in arenaviral maintenance are either grassland/savannah species (e.g., *Calomys, Mus,* and *Mastomys;* Arata and Gratz, 1975; O'Connell, 1982) or dwell primarily in ecotones such as forest edges (e.g., *Oryzomys;* Trapido and Sanmartin, 1971). Most are primarily terrestrial in their habits, although members of both *Calomys* and *Oligoryzomys* genera can be arboreal/scansorial (O'Connell, 1982). Nesting habits can be variable even within a genus; *C. musculinus* uses only ground nests while *C. laucha* also utilizes underground burrows (Bush *et al.*, 1984; Yunes *et al.*, 1991). This latter habit may offer a degree of protection from cattle grazing and farming equipment and may explain the ability of *C. laucha* populations to grow more rapidly than *C. musculinus* when grazing pressures are reduced (Bush *et al.*, 1984).

Many of the rodents associated with arenaviruses can be found as commensals or semicommensals, living within human dwellings [*Mus* (Laurie, 1946), *Mastomys* (Coetzee, 1975), and *C. callosus* (Mercado, 1975)]. Others are primarily found in cultivated fields or weedy borders (*C. musculinus, C. laucha, B. obscurus,* and *A. azarae*), although the species composition and rodent population dynamics of cultivars and border areas vary (Fig. 4, 5, 6; Crespo, 1966; Mills *et al.*, 1991b). The habitat preferences of each species dictate the epidemiological patterns of the diseases they cause in humans with regard to occupational, sexual, or seasonal biases in occurrence (see below).

The occurrence of many of these species in periodically disturbed habitats, such as cultivated fields, has led to detailed analyses and modeling of anthropogenic fluctuations in rodent distributions and population sizes (De Villafañe *et al.*, 1977; De Villafañe and Bonaventure, 1987; Kravetz and De Villafañe, 1981). In the case of Junin virus, several studies have examined the mammalian fauna associated with agricultural regions in Argentina and shown that far higher population densities of *C. musculinus*, and other rodents, are found in corn fields than in soybean fields (De Villafañe *et al.*, 1977; Kravetz *et al.*, 1986; Mills *et al.*,

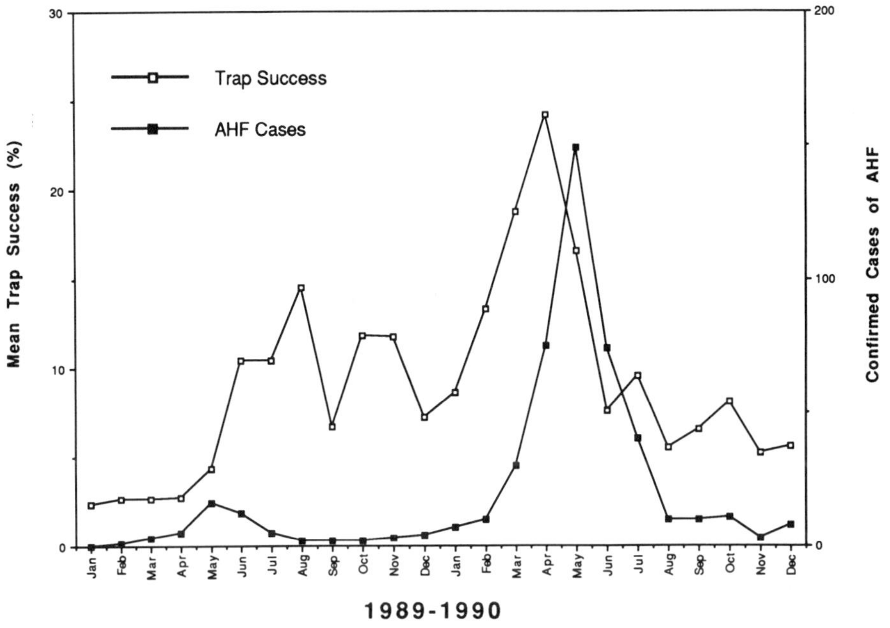

FIGURE 4. Incidence of confirmed cases of AHF in humans and population densities of rodents (six species), as estimated by trap success, followed longitudinally over 2 years in Santa Fe and Buenos Aires provinces, Argentina. Human cases of AHF increased from 181 in 1989 to 689 during 1990. Rodent populations of all species, including *C. musculinus* (see Fig. 5), increased more than fivefold during this period (J. N. Mills, J. E. Childs, and J. I. Maiztegui, unpublished data).

1991b). The gradual shift in Argentina from the cultivation of corn to that of soybeans has been suggested as a cause of the decreasing incidence of AHF in once epidemic foci, although this mechanism is not established (Maiztegui *et al.*, 1996). However, based on this apparent relationship tween AHF and corn, some ecologists have suggested a switch to soybean or sunflower cultivation, with intensive postharvest grazing, as a control for AHF spread (Bush *et al.*, 1984; Percich, 1988).

C. Rodent Competition, Exclusion, and Predation

Although an extensive body of literature exists on this subject and its importance in structuring rodent communities (for review, see Grant, 1972), little is known of the impact species interactions have on arenavirus distribution. In West Africa, Lassa fever is predominantly a rural disease, because *Mastomys* is typically a rural or small-village rodent (Rosevear, 1969; Happold, 1987). However, Rosevear (1969) suggests that this species' current distribution in villages is the result of displacement following competition with black and Norway rats (*Rat-*

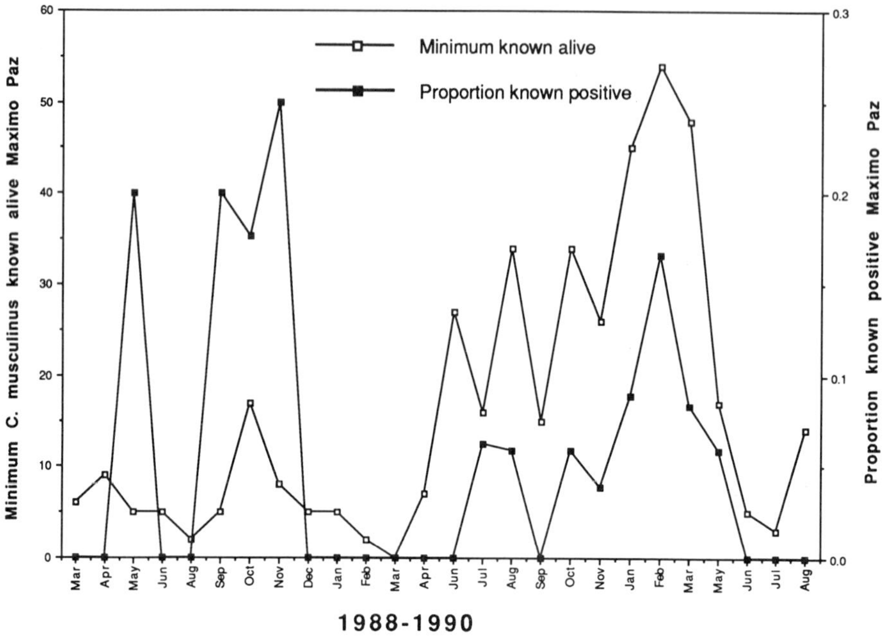

FIGURE 5. Population numbers of *C. musculinus* (minimum number known to be alive) and prevalence of Junin virus (minimum number known to be infected) as determined by antigen ELISA, at an AHF endemic site in Sante Fe Province, Argentina. Although *C. musculinus* densities on the two mark-release study grids increased more than fivefold over the 2 years, and absolute numbers of Junin virus–infected animals also increased, the fraction of infected rodents did not increase (J. N. Mills and J. E. Childs, unpublished data).

tus rattus and *R. norvegicus*, respectively) and cites trapping records from Liberia and Sierra Leone to support this contention.

In Argentina, *Calomys* spp. have been shown to be behaviorally subordinate to *Akodon, Oligorvzomys*, and *Bolomys* (Kravetz *et al.*, 1986). Competition between these species may explain the relative exclusion of *Calomys* from the more stable habitats bordering cultivated fields (Crespo, 1966; De Villafañe *et al.*, 1977; Mills *et al.*, 1991b) and, in part, their drastic population crashes following harvesting of cultivars (Fig. 6). *Calomys* spp. are also regarded as a better colonizing species than other rodents in similar habitats and have a shorter gestation period, larger litter size, and a higher frequency of postpartum estrus than other rodents, such as *A. azarae*, captured from identical locales (De Villafañe *et al.*, 1977, 1988; Mills *et al.*, 1991c).

The influence of predation in controlling rodents involved in arenavirus outbreaks is rarely mentioned. Johnson (1965) notes that cats were extremely rare in San Joaquin, Bolivia, in 1963 and had been dying for several years owing to an unexplained neurological disease (unrelated to Machupo virus infection). It is conceivable that the release of *Calomys*

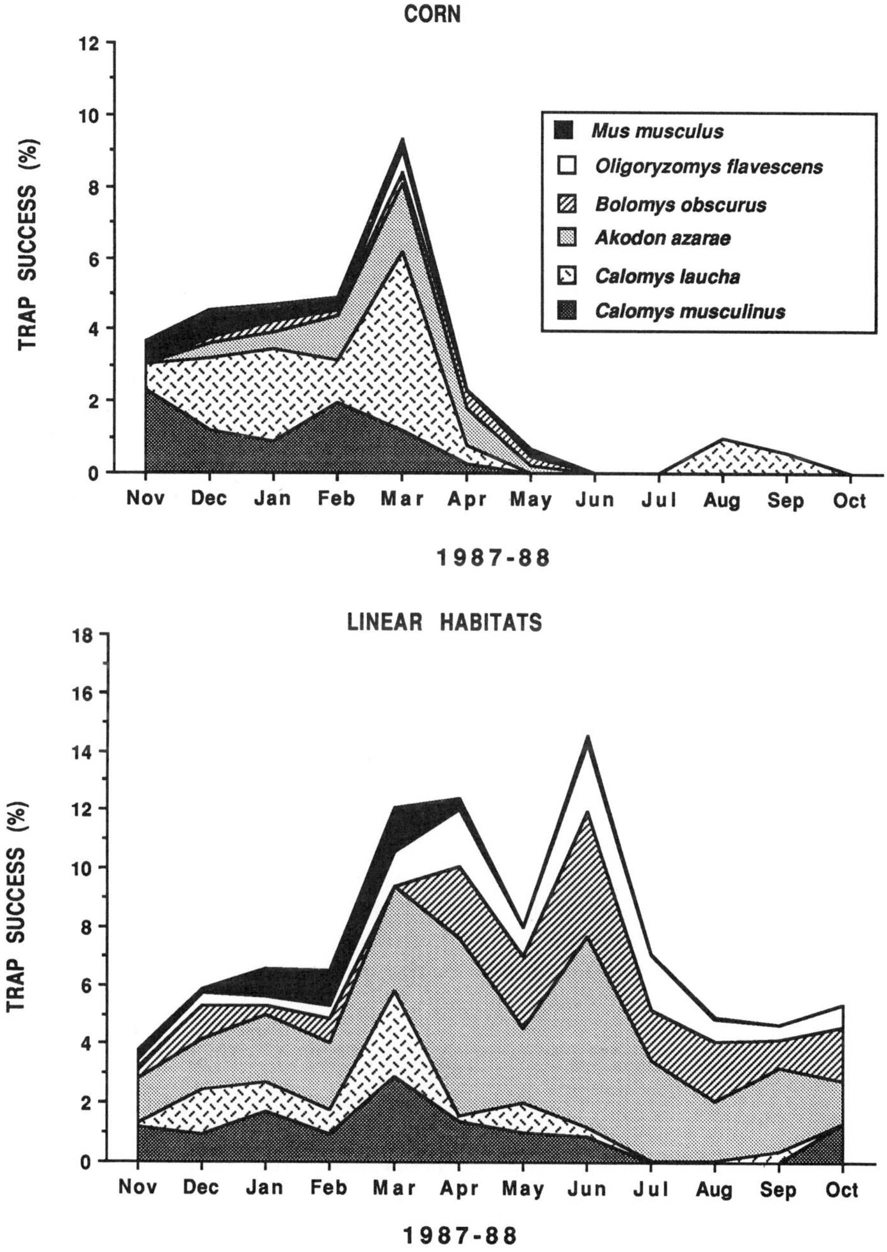

FIGURE 6. Percent trap success (number of rodents caught/number of traps set × 100) for two major habitats in Argentina. (Top) Trap success in corn. (Bottom) Trap success in linear habitats (fence rows, railroad beds, and roadside). Note the differences in species dominance between the two habitats and the population crash in corn after harvesting in March–April. The persistence of *C. laucha* in corn fields after harvest may be a consequence of their burrowing habits. (Adapted from Mills *et al.*, 1991b.)

from predation pressure fueled their population growth and contributed to the subsequent epidemic of BHF. In any case, BHF-free areas had no *Calomys*, but rather *Mus* and *Rattus* (Johnson *et al.*, 1967).

D. Population Dynamics

During the course of investigating South American arenaviral hemorrhagic fevers, several authors have commented on the relationship of annual outbreaks of human disease to levels of rodent infestation (Johnson *et al.*, 1982; Maiztegui *et al.*, 1986; J. N. Mills and J. E. Childs, unpublished data). In Argentina, rodent population densities were monitored from 1989 to 1990, during which time the numbers of human cases of AHF varied from 181 to 689. The population densities of all rodents, including *C. musculinus*, were severalfold higher the year of greater AHF incidence (Fig. 4). Junin viral antigen was detected in a much greater number of samples during years of greater rodent numbers (Fig. 5), but not in a significantly greater proportion of animals (J. N. Mills and J. E. Childs, unpublished data). In contrast, prevalence of LCMV antibody in house mice collected from Baltimore was greater in blocks with greater mouse densities (Childs *et al.*, 1991a). The relationship between arenaviral prevalence and host densities requires additional attention through long-term monitoring of rodent populations with active surveillance of human disease.

Studies of the abundance and diversity of rodents in various habitats are common in northern temperate regions but rare in tropical areas. Field studies of arenaviral infections in their rodent hosts conducted over ecologically meaningful periods (e.g., >1 year) are correspondingly much rarer. The following discussion will therefore highlight what has been studied and review a few pertinent aspects of reservoir population biology as they might relate to the maintenance and transmission of arenaviruses within and among species.

Seasonal and Cyclical Factors

In temperate woodlands and grasslands, many rodents are seasonal breeders, although length of breeding season often varies with latitude. Widely distributed species, such as *Mus*, may be seasonal breeders in northern latitudes but breed continuously in tropical climates (Table III). Genera such as *Calomys* and *Mus* reach sexual maturity at a young age, and animals born early in the year can reproduce during the season of their birth (Dalby, 1975). In some years this ability for rapid population growth can result in spectacular plagues of murid and cricetid rodents (Elton, 1942) that could result in increased prevalence of viral infections in rodents and increased human–rodent contact. In temperate and tropical grasslands these population eruptions are generally

TABLE III. Ecological Characteristics of Some Rodent Reservoirs of Pathogenic Arenavirus Based on Selected Studies and Reviews

Species	Locality studied	Timing of breeding	Age at maturity (months)	Litter size (mean)	Ref.[a]
Calomys musculinus	Argentina[b]	Seasonal	1.2[c]	5.4	1, 2
Calomys laucha	Uruguay	Seasonal	—	6.0	3
	Argentina			5.3	2
Calomys callosus	Brazil[b]	Seasonal	1.3[c]	4.5	4
	Bolivia[b]	Seasonal	1.4[c]	6.0–6.5	5
Mus musculus	Worldwide	Variable	1.5–2.0[d]	2.2–8.4	6–8
Mastomys natalensis	West Africa	Year-round (> in late rainy–early dry)	3.5[d]	11–12	9
	South Africa	Seasonal	3.7[d]	10–11	10, 11
Sigmodon alstoni	South America	Seasonal	—	5	12

[a] 1 = De Villafañe, 1981; 2 = Mills *et al.*, 1991c; 3 = Barlow, 1969; 4 = Mello, 1978; 5 = Justines and Johnson, 1970; 6 = Southwick, 1966; 7 = Laurie, 1946; 8 = Sage, 1981; 9 = Rosevear, 1969; 10 = Hanney, 1965; 11 = Coetzee, 1975; 12 = O'Connell, 1982.
[b] Laboratory colony.
[c] Based on vaginal patency.
[d] Based on age at first reproduction.

correlated with climatic conditions, such as rainfall or food resources (O'Connell, 1982), and are not necessarily true cyclical fluctuations such as the 3 to 4-year cycle of tundra zone microtine rodents (Southern, 1979).

To date, no definitive evidence exists that true population cycles occur in any South American grassland rodents (O'Connell, 1982; Dalby, 1975). However, long-term population fluctuations of *Calomys* have been reported from Argentina with approximately a 4-year period. These surges in 1964, 1967, 1974, and 1977 coincided with greater numbers of AHF cases in Buenos Aires province (Maiztegui *et al.*, 1986; see also Fig. 4).

Typical seasonal changes in density are the norm for temperate rodents, with populations peaking in fall and early winter and reaching their nadir in spring and early summer (French *et al.*, 1975). Although peak seasonal populations of rodents may coincide with peak incidence of arenaviral disease, as is the case with AHF (Maiztegui, 1975), this could also result from increased seasonally dependent occupational contact with rodents rather than from changes in virus prevalence or shedding.

Some evidence exists that within a given community arenaviral infections may fluctuate on both a seasonal and annual basis. In the longest continuous study of rodents and arenaviruses (1964–1971), Pinheiro *et al.* (1977) documented a 9-month period when Amapari virus infection could not be found in *Neacomys,* which had been preceded by a period of high infection. This study combined both removal trapping and mark–release–recapture methods, so it is difficult to interpret the effects of rodent sampling on the virological results. Although no quantitative data on population density fluctuations are presented, the authors indicate that declines in Amapari infection rate preceded a decline in rodent captures by 5 months (Pinheiro *et al.*, 1977).

Trapido and Sanmartin (1971) isolated Pichinde from 8–32% of *O. albigularis* captured during all months of the year, except May, when no isolates were obtained from 12 animals. Sabattini *et al.* (1977) found *C. musculinus* infected with Junin virus in each season over a 3-year field study and, by summing the results of nonconsecutive surveys, could show that virus had been present in the same locale for 11 years. Mills *et al.* (1991a) found Junin virus antigen–positive rodents during every month of the year when 2 years of surveillance data were pooled (Mills *et al.*, 1991a). Johnson (1985) reports that viremic–viruric *M. natalensis* were found year-round in Sierra Leone, and population levels were stable within individual villages monitored over a 1- to 2-year period (McCormick, 1987). This led to endemic Lassa fever in humans, with year-round transmission, as *Mastomys* is a household pest with continuous presence in dwellings (Johnson, 1985).

Persistent infection and long-term shedding of viruses may be a mechanism of viral maintenance during adverse seasonal conditions when population numbers are low and rodent-to-rodent contact is rare. In Brazil, chronic excretion or viremia of Amapari virus was documented for 10 months in wild-captured and, subsequently, laboratory-maintained *O. capito* and for 6 months in *N. quianae* released and recaptured in the field (Pinheiro *et al.*, 1977). In a similar study, Pichinde was recovered from the blood of four wild-caught *O. albigularis*, maintained for up to 455 days in the laboratory after their initial capture (Trapido and Sanmartin, 1971). In Argentina, Sabattini *et al.*, (1977) documented chronic viremias of Junin virus for up to 56 days in field-caught *C. musculinus* by capture-and-release studies. Transmission was believed to be primarily vertical, with newborn *Calomys* becoming infected when nursing from infected dams (Sabattini *et al.*, 1977). These studies emphasize that the carrier state in rodents, so well described from laboratory studies, plays a critical role in ensuring arenaviral maintenance in a manner analogous to overwintering of arboviruses. Thus, as populations dwindle in the winter, a few long-lived, persistently infected adults survive to infect a new cohort of susceptibles or give birth to chronically infected young in the spring.

E. Population Reduction and Control

Rodent control aimed at reducing the incidence of arenaviral disease in humans has been reported as successful only in the isolated instance of Machupo virus in San Joaquin, Bolivia (Kuns, 1965; Johnson *et al.*, 1973; Mercado, 1975). In this instance intensive house-to-house trapping of *C. callosus* was shown to coincide with a dramatic decrease in BHF incidence. The effort required was considerable, as 318,726 trap-nights were registered between May 1 and July 26, 1964, with a total catch of 2896 *C. callosus*, 70 *Oryzomys bicolor*, 32 *O. subflavus*, three *Zygodontomys lasiurus*, three *Holochilus brasiliensis*, and 16 *Mus musculus* (Kuns, 1965). The body count is given here to stress the fact that BHF had only a single reservoir, and this was the most common rodent present. In Sierra Leone, rodent trapping in individual houses failed to reduce the rate of human seroconversion to Lassa virus, although rodent reductions of two- to threefold were achieved (Keenlyside *et al.*, 1983). McCormick (1987) reports that Lassa virus transmission was decreased in West African villages following >90% reduction of *Mastomys* within houses. In certain situations such house-by-house control may be warranted and effective.

Control of sylvatic rodents has been successfully applied over a long time period in specific locales (Richards, 1988) but should probably be regarded as a temporary measure for suppressing outbreaks of grassland- or forest-dwelling species (Arata and Gratz, 1975). The specific problems of rodent pests and their control in the tropics (Gratz and Arata, 1975; Elias and Fall, 1988; Fiedler, 1988) and temperate areas

(Lund, 1988; Marsh, 1988) have recently been reviewed. The potential for habitat modification, through crop replacement, as a measure for reducing *Calomys* spp. populations has been suggested in Argentina (Bush *et al.*, 1984). but the practicality of such measures is uncertain.

F. Evolution

The evolution of arenaviruses must be closely linked to their respective hosts and their own evolution. The rodent families Cricetidae and Muridae are within a single superfamily (Muroidea) and are the two largest families of rodents (Eisenberg, 1981). The largest diversity of arenaviruses has been described from South America, where they infect rodents of a single subfamily (either the Cricetinae or Sigmodontinae depending on the authority; Webb and Marshall, 1982, and Reig, 1987, respectively). The only exceptions are Tacaribe, isolated from *Artibeus* spp., a genus of Phyllostomatid bats (Downs *et al.*, 1963), and LCMV found in the introduced murid, *M. musculus.* The cricetid rodents are new arrivals to South America, having immigrated from North America within the last 7 million years during the great American faunal exchange (Webb and Marshall, 1982). The Cricetinae have radiated dramatically in temperate and tropical South America, where 47 (or 53–54 of the Sigmodontinae; Reig, 1987) genera occur, compared to 23 in North America (Webb and Marshall, 1982). The only North American arenavirus, Tamiami virus, probably immigrated back to Florida with the reinvasion of its host, *S. hispidus* (Arata and Gratz, 1975).

In Africa, Lassa or Lassa-like viruses have been isolated only from murid rodents, most commonly from *Mastomys* and *Praomys.* The Muridae probably originated in central Asia and invaded Africa sometime during the Pleistocene some 2 million years ago (Eisenberg, 1981). The Muridae are believed to have branched from an early cricetine-like form and radiated in southeast Asia (Eisenberg, 1981; Petter, 1966). Gonzalez *et al.* (1986) envision a three-way split in rodents and the prototypical arenavirus ancestor, beginning some 20 million years ago. At that time ancestral cricetid rodents with some LCMV-like predecessor crossed Beringea to establish themselves in North America. There they evolved with what became the Tacaribe complex of viruses prior to their invasion of South America (how an arenavirus ended up in a frugivorous bat is a mystery). Later, murids from central Asia spread to Europe with an evolving LCMV, and others colonized Africa and evolved with what became the Lassa complex of arenaviruses, (Gonzalez *et al.*, 1986).

The relationship of LCMV with wild *Mus* has been described as the perfect parasitism and the outcome of a long process of coevolution (Traub, 1939). Machupo virus and *C. callosus* have been seen to have an imperfect relationship because of the deleterious effects of infection on host fitness, and it has been suggested that this presumably newer evolutionary relationship will in time reach a more benign equilibrium (Mims, 1973). The traditional theory that a well-balanced host–parasite relationship is one in which the parasite does as little damage to the host as possible has recently been questioned from both a theoretical and empirical perspective (cf. Anderson and May, 1982). These authors contend that the outcome of any particular coevolutionary host–parasite relationship will depend on the way in which virulence and the production of transmission stages of the parasite are linked; a gradient from zero to very high virulence can be expected, and the example of myxoma virus in Australian rabbits is examined. Thus, if the exuberant production of virus in type A infections of *Calomys* is more important to the spread of Machupo in the population than the reproductive penalty, the apparently deleterious effects on the host will be overridden.

One possible direction for arenavirus evolution is away from reliance on horizontal transmission. Mims (1975) has hypothesized that highly vertical transmission of LCMV could be so efficient that ultimately there might be no need of infectious virus to perpetuate viral genes in the population and that this infectious trait might disappear. This is an appealing theory especially as various other viruses, such as the endogenous retroviruses, have opted for this alternative. As much as 0.04% of the DNA of mice (*M. cervicolor*) can be comprised of these viral genomes (Callahan and Todaro, 1978). However, if our current understanding of arenaviral relationships with their hosts is applicable to natural populations, then pairings such as Junin–*C. musculinus* cannot become dependent on a strictly vertical transmission mode, unless significant changes occur (Vitullo and Merani, 1988).

V. RODENT-TO-HUMAN INFECTION

A. Risk to Humans: Seasonal and Occupational Differences

The preceding section focused on the prevalence of viral infection in rodent communities. The risk to humans occurs when fluctuations in rodent populations and other factors result in potential exposure of humans to virus, principally in infectious excreta. The theme is the same for all known arenavirus diseases, although the details differ.

In the case of LCMV in Germany, England, Argentina, and the United States, where information is available, the infection is focal in murine demes and preferentially spreads to humans in a rural setting or when rodents invade dwellings either to escape winter conditions or because of suboptimal neighborhood hygiene. This results in the typical fall–winter preponderance of LCMV-induced aseptic meningitis (Adair *et al.*, 1953; Meyer *et al.*, 1960). At times LCMV may reach humans through infection of an intermediate host that is capable of excreting large quantities of virus over prolonged periods; the hamster-associated cases of the 1970s and a more recent report of human disease acquired from nu/nu laboratory mice are good examples (Gregg, 1975). LCMV may be introduced into laboratory rodents by contact with feral *Mus* or by contamination in the laboratory. The typical noncytopathic interaction of arenaviruses with cells readily results in chronic infections of transplanted tumors or tissue culture cell lines that are difficult to detect and readily survive liquid nitrogen preservation (Biggar *et al.*, 1976). These cells may directly infect laboratory workers, or transplantation into rodents can lead to infection of the recipient. Passage through infected animals may result in tumor lines acquiring LCMV. Indeed, chronic infections of laboratory mouse colonies have passed unnoticed on many occasions and may result in spurious research findings.

Extensive human infection and major disease impact from Lassa virus is documented in Sierra Leone and Liberia, and there are indications that Lassa fever may also be important in Nigeria (Carey *et al.*, 1972; Monath *et al.*, 1973; Fraser *et al.*, 1974; McCormick *et al.*, 1987; Frame, 1989). There is fragmentary information suggesting transmission in other western African nations, particularly Guinea. Cases of Lassa fever are more common in dry season, and the aerosol stability of Lassa fever virus in the laboratory is greater in conditions of low relative humidity (Stephenson, *et al.*, 1984). This type of correlation has also been invoked to explain seasonal trends in another highly aerosol-infectious disease, measles (Langmuir, 1980). Since *Mastomys* are abundant in the west African bush and also inhabit dwellings, it is not surprising that Lassa fever as a disease of males and females, adults and children. Modes of human infection other than aerosol may be particularly efficient in the African village. For example, disease is particularly common in villages that are engaged in diamond mining with abandonment of traditional life-styles, leaving abundant unprotected food supplies for potential rodent contamination. *Mastomys* may also serve as supplemental food sources, and their pursuit and processing is a risk factor for human infection (Johnson *et al.*, 1982; McCormick *et al.*, 1987).

In the case of Junin virus, the major rodent reservoir, *C. musculinus*, is in proximity to humans most often in cultivated fields and adjacent fence rows, resulting in a predominant impact on men in agriculture-related occupations. The infection of drivers and machinery operators working rural areas, but with urban residence, has demonstrated the major role for aerosol transmission in this disease (Maiztegui, 1975).

Machupo virus is transmitted by another *Calomys* species, *C. callosus*. This grassland rodent readily invades houses, particularly when the plains of the region are partly flooded and food is plentiful within dwellings. The resulting epidemics noted in small towns have some male preponderance, but affect both sexes in villages. There are also sporadic cases acquired in grassland or open secondary brush, predominantly occupationally acquired by men (Mackenzie, 1965; Johnson *et al.*, 1967).

B. Rodent Populations and Human Disease Incidence

The prevalence of infected rodents clearly is a potentially crucial factor in determining the risk to humans, but there is remarkably little data that bears directly on the issue. As noted above, the number (but not the proportion) of Junin virus–infected rodents was higher in a year with appreciable numbers of AHF cases than in the previous year when a small number of cases were observed. BHF cases seem to occur in a setting of high populations and high infection rates of *C. callosus*, although there are no quantitative studies (Johnson *et al.*, 1967). No correlation was noted between the percentage of infected *M. natalensis* and human antibody prevalence in Sierra Leone villages (McCormick *et al.*, 1987).

Both Junin and Lassa virus–infected rodent reservoirs have been found in zones where no human cases have been reported, although quantitative factors in the rodent populations, intensity of surveillance, and virus strain variation all need to be explored (McCormick *et al.*, 1987; Kravetz *et al.*, 1986; Mills *et al.*, 1991a).

C. Mechanism of Transmission to Humans

1. Aerosols

The exact mode of transmission of arenaviruses to humans is unknown. Aerosols from infected rodent urine are probably the most important mechanism, but it is difficult to prove this assertion. Infectious aerosols are readily suspected when there is no possibility of direct contact between the infected person and the source or when large numbers of persons in an enclosed space are infected. However, if the aerosol source strength is low, those in most direct contact with the offending source are at highest risk and attack rates may be low.

The large number of laboratory infections indirectly attests to the aerosol infectivity of arenaviruses, and this is further supported by direct studies of LCM, Junin, and Lassa virus infections induced by quantitative exposure of experimental animals to small particle aerosols (Danes *et al.*, 1963; Stephenson *et al.*, 1984; Lehman-Grubbe, 1984). Anecdotal evidence exists describing LCMV infection of people exposed only by entering a contaminated room or working in an area receiving potentially infective air currents (Hinman *et al.*, 1975; Biggar *et al.*, 1975). Junin virus infections in the endemic zone often occur in truck or farm machinery drivers with no direct rodent contact and are virtually certain to be aerosol-transmitted (J. I. Maiztegui, unpublished observations).

Aerosols arise when infected rodents shed virus-laden urine, a behavior that may be precipitated by fear or surprise. In addition, lung and salivary glands are commonly infected and throat swabs contain high-titered virus. The efficiency with which secondary aerosols are generated after urine and other secretions are in contact with earth or dried on surfaces is unknown, but is likely to be low.

2. Droplets and Fomites

It seems likely that arenaviruses in droplets or on surfaces coming in contact with mucous membranes are infectious, but there is little direct information. This is an important deficiency in our knowledge since contamination of food and subsequent ingestion with exposure of pharyngeal tissues is a particularly common possibility, especially in Lassa-endemic areas of west Africa.

3. Contact with Rodent Blood

Corn-harvesting machinery generates bloody aerosols when rodents are entrapped in Argentina. In west Africa, preparation of *Mastomys* for food provides intensive exposure to small particle aerosols, droplets, and fomites of blood, as well as other potentially infectious secretions.

D. Target Species Other than Humans

It is evident that one of the most important ways in which LCMV gains access to human subjects is through infection of mouse or hamster colonies. In other countries, the possibility of introduction of other arenaviruses into mouse or other rodent colonies (e.g., Lassa may cause chronic infections of some mouse strains; Peters, 1991) should be considered.

Recently a new target population for arenavirus disease has been identified. Oral transmission of an arenavirus has been implicated as the cause of a disease referred to as Callitrichid hepatitis (CH), affecting several species of tamarins and marmosets in zoos of the United States and England (Montali *et al.*, 1989; Ramsay *et al.*, 1989). The etiological agent was identified as an arenavirus, which appears to be identical to LCMV (Stephensen *et al.*, 1991). CH falls within the spectrum of other primate-infections with arenaviruses (Peters *et al.*, 1987; Peters, 1991). It is an acute, frequently fatal, disease with pathological changes including multifocal hepatocellular necrosis with the presence of characteristic acidophilic bodies and minor inflammation (Montali *et al.*, 1989). In addition to liver and spleen, other organs including the CNS may be affected. Hemorrhagic signs, such as occasional petechiae and sanguineous and nonsanguineous pleuro-pericardial effusions, also occur (Montali *et al.*, 1989; Ramsay *et al.*, 1989). The virus is believed to be transmitted to tamarins by feeding of infected suckling mice (R. J. Montali, personal communication), although outbreaks have occurred at zoos where mouse feeding has been precluded, suggesting transmission from wild mice.

VI. HUMAN-TO-HUMAN INFECTION

A. Nosocomial Outbreaks

There has never been any suggestion of person-to-person transmission of LCMV. Hundreds of cases of AHF and BHF have been cared for in the endemic areas without suspicion of secondary cases in medical personnel, despite the common occurrence of hemorrhage. However, there is a single example of explosive spread to family and hospital contacts with each of the South American hemorrhagic fevers (Peters *et al.*, 1974; Maiztegui, personal communication). Lassa fever initially came to medical attention in a setting of nosocomial spread (Frame *et al.*, 1970; Monath, 1975), and a hospital-based Nigerian epidemic claimed 15 lives among 28 cases in 1970 (Carey *et al.*, 1972). Nevertheless, subsequent experience has indicated that hospital spread of Lassa virus is distinctly uncommon in the endemic area and in the relatively small number of exported cases (McCormick, 1987). It must be borne in mind that invasive procedures and patient contact are vastly less than would be expected to occur in the United States.

All three of the nosocomial outbreaks referred to above had common characteristics: the index case was critically ill and died, aerosol spread was the most likely explanation for the route of infection of at least some of the secondary cases, lethality was high, and transmission stopped after the second or third generation. It is not known whether these episodes represented an unusual pattern of host pathogenesis (perhaps higher viremia as observed in Bolivia or prominent pulmonary involvement as seen in Nigeria), a variation in virus strain (some minor differences were detected between the Machupo strain isolated in Bolivia and the prototype), or some combination of these factors. The unusual patient who disseminates infection to large numbers of contacts by aerosol constitutes a well-known, if often ignored, phenomenon in several common virus diseases, and viral hemorrhagic fevers furnish another example (Langmuir, 1980; Hattis *et al.*, 1973).

Such outbreaks are sufficiently rare that the overall pattern of contagion in hospitals is low and the most dangerous route is parenteral exposure through improperly sterilized needles, autopsy incidents, or other breaks in technique (Peters *et al.*, 1991). This has led to the reasonable recommendation that ordinary mask, gown, and glove isolation procedures be used to protect medical staff. Additional respiratory protection would be a useful precaution in some settings.

Common clinical laboratory tests are performed on viremic blood samples from patients in Argentina and Africa with no apparent increased risk to workers. Once again, it must be remembered that the number of tests and workers exposed is low. Necropsy is extremely hazardous, and two scalpel accidents have led to a fatal outcome for the attending pathologists (Carey *et al.*, 1972; Peters *et al.*, 1974).

All arenaviruses that have been manipulated in virology laboratories have been infectious to humans under circumstances implicating aerosol spread. The hemorrhagic fever agents have induced serious disease and should be manipulated with caution and generally under BSL-4 containment.

B. Person-to-Person Transmission

Virological data explain why transmission to contacts outside the hospital is uncommon in Argentina and Bolivia. Virus has rarely been isolated from throat swab or urine. Once hemorrhagic manifestations are present the disease is recognized by local inhabitants, leading to precautions and/or hospitalization to minimize contact with viremic blood.

Low titers of Lassa virus have been detected inconstantly in throat swabs and, beginning later in disease, also in urine from some patients with continued excretion for weeks. Blood also contains higher titers of virus than in the South American hemorrhagic fevers, particularly in severely ill patients.

The field data on person-to-person transmission of Lassa fever virus are unclear for two reasons. The first, related to host factors, is the reliance on IF antibodies to assess exposure and immunity in humans. Antibodies as measured by this test decline after infection and may disappear in several years, leaving a seronegative host with previous experience with Lassa virus and probably some residual immunity. Conversely, febrile disease and concomitant Lassa IF antibody titer increases have been observed in seropositive subjects prospectively studied in Sierra Leone, suggesting that the presence of Lassa IF antibody does not guarantee immunity to reinfection (McCormick *et al.*, 1987). The ambiguity of the data on reinfection and subsequent disease spectrum, as well as the deficiencies of the IF antibody test in serosurveys, must be resolved before a clear picture of Lassa fever in such highly endemic zones can be fully developed. Any epidemiological study of "seronegative" patients must be interpreted in light of the possibility that an unknown number of those patients may have some degree of immunity from previous exposure; indeed, the presence of antibodies in younger age groups in serosurveys and the importance of pediatric Lassa fever suggest that this effect might be considerable (Webb *et al.*, 1986).

The second confounding factor is environmental. The presence of heavy infestation of houses with rodents makes it difficult to be certain that clustering of cases in houses may not simply be related to the intensity of Lassa-infected *Mastomys* exposure. For example, Keenlyside *et al.* (1983) found 20–30% of rodents taken in houses of Lassa cases or controls positive in Sierra Leone.

C. Sexual Transmission

Anecdotal reports have suggested that Lassa and Machupo viruses may have been transmitted from incubating or convalescing patients to their female consorts (Douglas *et al.*, 1965). The epidemiology of Junin virus lent itself to the systematic exploration of this issue since the main thrust of disease was in the occupationally exposed adult male population. Prospective and retrospective studies of conjugal partners of AHF cases convincingly demonstrated a small, but significant, incidence of clinical or subclinical cases occurring after the acute disease in the female partner and during the accepted incubation period for AHF (Briggiler *et al.*, 1987a,b).

D. Prenatal Transmission

In at least two instances, prenatal transmission has been documented for LCMV (Akermann *et al.*, 1974). Abortion, neonatal hydrocephalus, and teratogenic effects have also been described following LCMV infection (Johnson, 1985).

VII. SUMMARY

More than a dozen arenaviruses have been recognized with conventional methods of virus isolation, and it is certain that new ones will be encountered as humans enter the remaining undisturbed areas of our planet, as rodent habits change, and, speculatively, as arenaviruses evolve. It is not known whether application of newer virological methods will detect other arenaviruses, perhaps with very different properties [compare the morphologically identical agents from ticks (Zeller *et al.*, 1989)].

Nevertheless, the known agents are primarily viruses of rodents, and the biological constant is persistence in rodent populations through a specific virus–rodent interaction. In general, a single virus chronically infects a single rodent species, although there are circumstances in which there may be infection of other rodents, either closely or distantly related and with proven or possible epidemiological significance. The importance of LCMV infection of hamsters and the possibility of alternate rodent hosts for Junin virus in the pampas of Argentina are two examples where cross-species infection may be of considerable importance.

In spite of the specificity of the host–virus relationship, there appear to be significant variations in the strategies of viral persistence, and these strategies are reflected in the epidemiology of the virus in its reservoir. Unfortunately, the field and laboratory studies to fully explore this question have not been performed. Available evidence in the case of LCMV and *M. musculus* suggests that vertical infection, acquired *in utero*, is the major mechanism of viral maintenance at the population level. This involves a close relationship of the virus to the host, probably with little effective cost to the host in survival or rcproductive potential. This type of relationship would seem to be particularly suitable to the habits of the house mouse, with its limited range and small deme social structure, and is reflected in the extreme focality that can be seen in urban *Mus* infections. Vertical, congenital transmission of chronic viremic infection may also be accompanied by horizontal infection, which would lead, depending on host and viral variables, to either chronic viremia, or more often, transient infection with lesser degrees of virus shedding. Lassa virus and *M. natalensis* have the LCMV–*Mus* pattern (neonatal infection leads to chronic viremia, adult-infection results in immunizing infection), but the dynamics of transmission and congenital infections have not been studied.

The precision of the virus–host match in nature can be inferred from the many pathological effects seen in laboratory mouse strains infected with passaged LCMV or infected by a contrived route. These studies have, however, allowed the delineation of the dominant role of cytolytic T lymphocytes in viral clearance from the arenavirus-infected host, the importance of immune complex disease in many chronic viral infections, and the participation of chronic interferonemia in "runting" syndromes. Studies of LCMV itself have also shown the emergence in infected spleens of immunosuppressive viral variants capable of causing chronic viremic infection of mice through T-lymphocyte suppression.

Two related South American hemorrhagic fever viruses, Junin and Machupo, appear to have a different strategy from that of LCMV. Young animals develop chronic, viremic infections and excrete large quantities of virus, but the viral infection carries with it a decrease in fitness and a severe restriction in reproductive potential. Thus, horizontal infection from persistently viremic mice seems to play a more important role in viral persistence in the population. A significant proportion of rodents infected as adults will also develop chronic viremic infections; in the case of Machupo virus and *C. callosus*, there is a host genetic basis for the response. The significance of the development of chronic viremia after adult rodent infection to viral survival at the population level is unknown, but is a different pattern than seen with LCMV and suggests a different epidemiological mechanism. In the case of Junin virus and *C. musculinus*, these adult carriers seem not to suffer the decrement to survival and fertililty that neonatally infected animals bear, and they may serve as important reservoir amplifiers of virus by maternal vertical (but apparently not congenital) transmission to about half their offspring. Obviously, these virus–host infection patterns could lead to complicated cycles in which variations in the numbers and genetic constitution of the rodents interacted with viral transmission in a mutually regulatory fashion. Anecdotal field data have suggested increased Machupo transmission to humans in settings in which both the number and proportion of rodents infected were increased, but systematic 2-year studies of Junin virus showed a decreased proportion of infected rodents in the year in which total rodent numbers, total infected rodents, and increased human transmission occurred. Multiple mechanisms are probably operative, varying temporally and by location.

Most arenaviruses studied have produced chronic viremia in their natural hosts, either in laboratory studies-or by field observations of naturally infected hosts: LCMV, Lassa, Junin, Machupo, Pichinde, Amapari. Latino, however, produced a prolonged, but not a truly persistent viremia in *C. callosus* colonized from the same area in which the virus circulates. Tamiami virus has given only a transient viremia in laboratory studies using *S. hispidus*, and limited field observations on wild-caught infected animals have not yet yielded evidence for chronic viremia. The behavior of Latino and Tamiami suggests the possibility of arenavirus persistence in rodent populations by chronic infection and virus shedding without a true persistently viremic carrier state being present in its host. This would presumably require vertical transmission in the postnatal period and/or horizontal infection for virus maintenance.

Clearly, more work on the rodent–arenavirus dynamic would provide us with a better understanding of the biology and evolution of this interesting taxon, as well as approaches to the control of human infection. LCM remains a common but underdiagnosed disease in the Americas and Europe; AHF plagues the rich farmlands of the Pampas and is increasing its geographic range; Lassa fever is responsible for thousands of cases and many deaths every year in West Africa; and new agents such as Guanarito virus, recently isolated from Venezuelan hemorrhagic fever cases, await our attention.

REFERENCES

Ackermann, R., 1973, Epidemiological aspects of lymphocytic choriomeningitis in man, in: *Lymphocytic Choriomeningitis Virus and Other Arenaviruses* (F. Lehmann-Grube, ed.), pp. 234–237, Springer-Verlag, New York.

Ackermann, R., Bloedhorn, H., Küpper, B., Winkens, I., and Scheid, W., 1964, Ober die verbreitung des virus der lymphozytaren choriomeningitis unter den Mausen in Westdeutschland, *Zentralbl. Bakter. Parasit. Infekt. Hyg.* **194**:407.

Ackermann, R., Stille, W., Blumenthal, W., Helm, E. B., Keller, K., and Baldus, O., 1972, Syrische goldhamster als uberträger von lymphozytären choriomeningitis, *Deut. Med. Wochen.* **97**:1725.

Ackermann, R., Korver, G., Turss, R., Wonne, R., and Hochgesand, P., 1974, Pranatale infektion mit dem virus der lymphozytären choriomeningitis, *Dtsch. Med. Wochenschr.* **99**:629.

Adair, C. V., Gauld, R. L., and Smadel, J. E., 1953, Aseptic meningitis, a disease of diverse etiology: Clinical and etiologic studies on 854 cases, *Ann. Intern. Med.* **39**:675.

Ahmed, R., Byrne, J. A., and Oldstone, M. B. A., 1984a, Virus specificity of cytotoxic T lymphocytes during acute lymphocytic choriomeningitis virus infection; role of the H-2 region in determining cross reactivity for different lymphocytic choriomeningitis virus strains, *J. Virol.* **51**:34.

Ahmed, R., Salmi, A., Butler, L. D., Chiller, J.-M., and Oldstone, M. B. A., 1984b, Selection of genetic variants of lymphocytic choriomeningitis virus in spleens of persistently infected mice: Role in suppression of cytotoxic T lymphocyte response and viral persistence, *J. Exp. Med.* **60**:521.

Ahmed, R., Hahn, C. S., Somasundram, T., Villarete, L., Matloubian, M., and Strauss, J. H., 1991, Molecular basis or organ-specific selection of viral variants during chronic infection, *J. Virol.* **65**:4242.

Allan, W., and Doherty, P. C., 1989, Influence of non-major histocompatibility complex differences on the severity of lymphocytic choriomeningitis, *J. Neuroimmunol.* **24**:55.

Anderson, P. K., Dunn, L. C., and Beasley, A. B., 1964, Introduction of a lethal allele into a feral house mouse population, *Am. Nat.* **98**:57.

Anderson, R. M., and May, R. M., 1979, Population biology of infectious diseases: Part I, *Nature* **280**:361.

Anderson, R. M., and May, R. M., 1982, Coevolution of hosts and parasites, *Parasitol.* **85**:411.

Apfelbaum, L. I., and Blanco, A., 1985a, Genetic polymorphism in populations of *Akodon* rodents, *Can. J. Genet. Cytol.* **27**:12.

Apfelbaum, L. I., and Blanco, A., 1985b, Temporal variation of allele frequencies in populations of *Akodon dolores* (Rodentia, Cricetidae), *Theor. Appl. Genet.* **70**:569.

Arata, A. A., and Gratz, N. G., 1975, The structure of rodent faunas associated with arenaviral infections, *Bull. WHO* **52**:621.

Armstrong, C., 1940, Studies on choriomeningitis and poliomyelitis, *Harvey Lect.* **1940–41**:39.

Armstrong, C., and Sweet, L. K., 1939, Lymphocytic choriomeningitis, *Public Health Rep.* **54**:673.

Barlow, J. C., 1969, Observations on the biology of rodents in Uruguay, *Life Sci. Con. Roy. Ontario Mus.* **75**:1.

Bellier, L., 1975, The genus *Mastomys* in the Ivory Coast, *Bull. WHO* **52**:665.

Berry, R. J., 1977, The population genetics of the house mouse, *Sci. Progr. (Oxf.)* **64**:341.

Berry, R. J., 1979, Genetical factors in animal population dynamics, in: *Population Dynamics* (R. M. Anderson, L. R. Taylor, and B. D. Turner, eds.), pp. 53–80, Blackwell, Oxford.

Berry, R. J., 1981, Population dynamics of the house mouse, *Symp. Zool. Soc. Lond.* **47**:395.

Biggar, R. J., Woodall, J. P., Walter, P. D., and Haughie, G. E., 1975, Lymphocytic choriomeningitis outbreak associated with pet hamsters: Fifty-seven cases from New York state, *JAMA* **232**:494.

Biggar, R. J., Deibel, R., and Woodall, J. P., 1976, Implications, monitoring, and control of accidental transmission of lymphocytic choriomeningitis virus within hamster tumor cell lines, *Cancer Res.* **36**:537.

Blumenthal, W., Ackermann, R., and Scheid, W., 1968, Durchseuchung mit dem virus der lymphozytaren choriomeningitis in einem Endemiegebiet, *Dtsch. Med. Wochenschr.* **93**:944.

Briggiler, A. M., Enria, D. A., Feuillade, M. R., and Maiztegui, J. I., 1987a, I. Contagio interhumano e infeccion clinica con virus Junin en matrimonios residentes del area endemica de fiebre hemorragica Argentina, in: Resumenes de las comunicaciones de la XXXII Reunion Cientifica de la Sociedad Argentina de Investigacion Clinica, Nov. 15–19, Mar del Plata, Argentina (abstract Nx1), *Medicina (Buenos Aires)* **47**:565.

Briggiler, A. M., Enria, D. A., Feuillade, M. R., and Maiztegui, J. I., 1987b, 1. Contagio interhumano e infeccion inaparente por virus Junin en matrimonios residentes del area endemica de fiebre hemorragica Argentina, in: Resumenes de las comunicaciones de la XXXII Reunion Cientifica de la Sociedad Argentina de Investigacion Clinica, Nov. 15–19, Mar del Plata, Argentina (abstract Nx2), *Medicina (Buenos Aires)* **47**:565.

Buchmeier, M. J., 1983, Antigenic and structural studies on glycoproteins of lymphocytic choriomeningitis virus, in: *Negative Strand Viruses* (R. Compans and D. H. Bishop, eds.), Elsevier North-Holland, New York.

Buchmeier, M. J., Lewicki, H. A., Tomori, O., and Oldstone, M. B. A., 1981, Monoclonal antibodies to lymphocytic choriomeningitis and Pichinde viruses: Generation, characterization and cross-reactivity with other arenaviruses, *Virology* **113**:7.

Bush, M., Kravetz, F. O., Perich, R. E., and Zuleta, G. A., 1984, Propuesta para un control ecológico de la fiebre hemorrahica Argentina a traves del manejo del habitat, *Medicina (Buenos Aires)* **44**:34.

Byrne J. A., Ahmed, R., and Oldstone, M. B. A., 1984, Biology of cloned cytotoxic T lymphocytes for lymphocytic choriomeningitis virus. 1. Generation and recognition of virus strains and H-2^b mutants, *J. Immunol.* **133**:433.

Callahan, R., and Todaro, G., 1978, Four major endogenous retrovirus classes each genetically transmitted in various species of *Mus* in: *The Origins of Inbred Mice* (H. C. Morse, III, ed.), pp. 689–713, Academic Press, New York.

Cameron, G. N., and Spencer, S. R., 1981, *Sigmodon hispidus*, in: *Mammalian Species* (D. F. Williams and T. E. Lawlor, eds.), No. 158, pp. 1–9, American Society of Mammalogists, Lawrence, KS.

Carballal, G., Videla, C. M., Dulout, F., Cossio, P. M., Acuna, A. M., and Bianchi, N. O., 1986, Experimental infection of *Akodon molinae* (Rodentia, Cricetidae) with Junin virus, *J. Med. Virol.* **19**:47.

Carballal, G., Videla, C. M., and Merani, M. S., 1988, Epidemiology of Argentine hemorrhagic fever, *Eur. J. Epidemiol.* **4**:259.

Carey, D. E., Kemp, G. E., White, H. A., Pinneo, L., Addy, R. F., Fom, A. L. M. O., Stroh, G., Casals, J., and Henderson, B. E., 1972, Lassa fever: epidemiological aspects of the 1970 epidemic, Jos, Nigeria, *Trans. Roy. Soc. Trop. Med. Hyg.* **66**:402.

Cerny, A., Sutter, S., Bazin, H., Hengartner, H., and Zinkernagel, R. M., 1988, Clearance of lymphocytic choriomeningitis virus in antibody- and B-cell-deprived mice, *J. Virol.* **62**:1803.

Childs, J. E., Korch, G. W., Smith, G. A., Terry, A. D., and LeDuc, J. W., 1985, Geographical distribution and age related prevalence of antibody to Hantaan-like virus in rat populations of Baltimore, Maryland, USA, *Am. J. Trop. Med. Hyg.* **34**:385.

Childs J. E., Glass, G. E., Korch, G. W., and LeDuc, J. W., 1987, Prospective seroepidemiology of hantaviruses and population dynamics of small mammal communities of Baltimore, Maryland, U. S. A., *Am. J. Trop. Med. Hyg.* **37**:648.

Childs J. E., Glass, G. E., Korch, G. W., and LeDuc, J. W., 1989, Effects of hantaviral infection on survival, growth, and fertility in wild rat (*Rattus norvegicus*) populations of Baltimore, Maryland, *J. Wildlife Dis.* **25**:469.

Childs, J. E., Glass, G. E., Korch, G. W., Ksiazek, T. G., and Leduc, J. W., 1991a, Lymphocytic choriomeningitis virus infection and house mouse (*Mus musculus*) distribution in urban Baltimore, *Am. J. Trop. Med. Hyg.* **46**:390.

Childs, J. E., Glass, G. E., and LeDuc, J. W., 1991b, Rodent sightings and contacts in an inner-city population of Baltimore, Maryland, USA, *Bull. Soc. Vect. Ecol.* **16**:245.

Childs, J. E., Glass, G. E., Ksiazek, T. G., Rossi, C. A., Barrera Oro, J. G., and Leduc, J. W., 1991c, Human-rodent contact and infection with Lymphocytic choriomeningitis and Seoul viruses in an inner-city population, *Am. J. Trop. Med. Hyg.* **44**:117.

Cockrum, E. L., 1948, The distribution of the hispid cotton rat in Kansas, *Trans. Kansas Acad.* Sci. 51:306.

Coetzee, C. G., 1975, The biology, behaviour, and ecology of *Mastomys natalensis* in southern Africa, *Bull. WHO* **52**:637.

Cole, G. A., and Nathanson, N., 1974, Lymphocytic choriomeningitis, *Progr. Med. Virol.* **18**:94.

Coulombie, F. C., Damonte, E. G., and Coto, C. E., 1984, Influencia de la celula huesped en las reacciones de neutralizacion cruzada entre virus Junin y Tacaribe, *Rev. Argent. Microbiol.* **16**:159.

Crespo, J. A., 1966, Ecologia de una communidad de roedores silvestres en al Partido de Rojas, Provincia de Buenos Aires, *Rev. Mus. Argent. Cien. Nal. Inst. Nac. Invest. Cien. Nat. Ecol.* **1**:79.

Dalby, P. L., 1975, Biology of pampa rodents, Belcarce area, Argentina, *Pub. Mus. Mich. State Biol. Ser.* **5**:149.

Danes, L., Benda, R., and Fuchsova, M., 1963, Experimental inhalation infection with the lymphocytic choriomeningitis virus (WE strain) of the monkeys of the *Macacus cynomolgus* and *Macacus rhesus* species, *Bratisl. lek. listy* **43**:21.

DeFries, J. C., and McClearn, G. E., 1972, Behavioral genetics and the fine structure of mouse populations: A study in microevolution, in: *Evolutionary Biology*, Vol. 5 (T. Dobzhansky, M. K. Hecht, and W. C. Steere, eds.), pp. 279–291, New York.

Deibel, R., Woodall, J. P., Decher, W. J., and Schryver, G. D., 1975, Lymphocytic choriomeningitis virus in man: Serological evidence of association with pet hamsters, *JAMA* **232**:501.

De Martini, J. C., Green, D. E., and Monath, T. P., 1975, Lassa virus infection in *Mastomys natalensis* in Sierra Leone, *Bull. WHO* **52**:651.

De Villafañe, G., 1981, Reproduccion y crecimiento de *Calomys musculinus murillus* (Thomas, 1916), *Hist. Nat. (Argent.)* **1**:237.

De Villafañe, G., and Bonaventure, S. M., 1987, Ecological studies in crop fields of the endemic area of Argentine hemorrhagic fever: *Calomys musculinus* movements in relation to habitat and abundance, *Mammalia* **51**:233.

De Villafañe, G., Kravetz, F. O., Donadio, O., Pecich, R. E., Kencher, L., Torres, M. P., and Fernandez, N., 1977, Dinamica de las communidades de roedores en agro-ecosistemas Pampasicos, *Medicina (Buenos Aires)* **37**:128.

Douglas G. R., Wiebenga, N. H., and Couch, R. B., 1965, Bolivian hemorrhagic fever probably transmitted by personal contact, *Am. J. Epidemiol.* 83:85.

Downs, W. G., Anderson, C. R., Spence, L., Aitken, T. H. G., and Greenhall, A. H., 1963, Tacaribe virus, a new agent isolated from *Artibeus* bats and mosquitoes in Trinidad, West Indes, *Am. J. Trop. Med. Hyg.* **12**:640.

Dutko, F. J., and Oldstone, M. B. A., 1983, Genomic and biological variarion among commonly used lymphocytic choriomeningitis virus strains, *J. Gen. Virol.* **64**:1689.

Eisenberg, J. F., 1981, *The Mammalian Radiations: An Analysis in Trends in Evolution, Adaptation and Behavior*, University of Chicago Press, Chicago.

Ellas, D. J., and Fall, M. W., 1988, The rodent problem in latin America, in: *Rodent Pest Management* (I. Prakash, ed.), pp. 13–28, CRC Press, Boca Raton, FL.

Elton, C., 1942, "Voles, Mice and Lemmings," Oxford University Press, Oxford.

Emmons, R. W., Yescott, R. E., and Dondero, D. V., 1978, A survey for lymphocytic choriomeningitis virus in the San Francisco Bay area, *Calif. Vector Views* 25:21.

Fiedler, L. A., 1988, Rodent problems in Africa, in: *Rodent Pest Management* (I. Prakash, ed.), pp. 35–66, CRC Press, Boca Raton, FL.

Frame, J. D., 1975, Surveillance of Lassa fever in missionaries stationed in West Africa, *Bull. WHO* **52**:593.

Frame, J. D., 1989, Clinical features of Lassa fever in Liberia, *Rev. Infect. Dis.* **11**:S783.

Frame, J. D., Baldwin, Jr., J. M., Gocke, D. J., and Troup, J., 1970, Lassa fever: A new virus disease of man from west Africa. I. Clinical description and pathological findings, *Am. J. Trop. Med. Hyg.* **19**:630.

Fraser, D. W., Campbell, C. C., Monath, T. P., Goff, P. A., and Gregg, M. B., 1974, Lassa fever in the eastern province of Sierra Leone, 1970–1972. I. Epidemiological studies, *Am. J. Trop. Med. Hyg.* **23**:1131.

French, N. R., Stoddart, D. M., and Bobek, B., 1975, Patterns of demography in small mammals, in: *Small Mammals: Their Productivity and Population Dynamics* (F. B. Golley, K. Petrusewicz, and L. Ryszkowski, eds.), pp. 73–102, Cambridge University Press, Cambridge.

Gardenal, C. N., and Blanco, A., 1985, Polimorfismo enzimático en *Calomys muscilinus:* Nueva estimación, *Mendeliana* **7**:3.

Gardenal, C. N., dejuarez, N. T., Gutierrez, M., and Sabattini, M. S., 1977a, Contribucion al conocimiento de tres especies del genero *Calomys* (Rodentia, Cricetidae). I. Estudios citogenéticos, *Physis, Sec. C* **36**:169.

Gardenal, C. N., Blanco, A., and Sabattini, M. S., 1977b, Contribucion al conocimiento de tres especies del genero *Calomys* (Rodentia, Cricetidae). II. Analisis electroforetico de formas multiples de enzimas como criterio de diferenciacion taxonomica, *Physis Sec. C* **36**:179.

Gardenal, C. N., Sabattini, M. S., and Blanco, A., 1980, Enzyme polymorphism in a population of *Calomys musculinus* (Rodentia: Cricetidae), *Biochem. Genet.* **18**:563.

Gardenal, C. N., Sabattini, M. S., and Blanco, A., 1986, Geographic patterns of allele frequencies in *Calomys musculinus* reservoir-host of Junin virus, *Medicina (Buenos Aires)* **46**:73.

Gardner, M. B., Henderson, B. E., Estes, J. D., Menck, H., Parker, J. C., and Huebner, R. J., 1973, Unusually high incidence of spontaneous lymphomas in wild house mice, *J. Natl. Cancer Inst.* **50**:1571.

Gardner, M. B., Resheed, S., Pal, B. K., Estes, J. D., and O'Brien, S. J., 1980, *Akvr-I,* a dominant murine leukemia virus restriction gene, is polymorphic in leukemia-prone wild mice, *Proc. Natl. Acad. Sci. USA* **77**:531.

Gonzalez, J. P., McCormick, J. B., Saluzzo, J. F., Herve, J. P., Georges, A. J., and Johnson, K. M., 1983, An arenavirus isolated from wild-caught rodents (*Praomys* species) in the Central African Republic, *Intervirology* **19**:105.

Gonzalez, J. P., Georges, A. J., Kiley, M. P., Meunier, D. M. Y., Peters, C. J., and McCormick, J. B., 1986, Evolutionary biology of a Lassa virus complex, *Med. Microbiol. Immunol.* **175**:157.

Gordon, D. H., 1984, Evolutionary genetics of the *Praomys (mastomys) natalensis* species complex (Rodentia: Muridae), unpublished Ph.D. thesis, University of Witwaterstrand, 193 pp.

Grant, P. R., 1972, Interspecific competition among rodents, *Annu. Rev. Ecol. Syst.* **3**:79.

Gratz, N., and Arata, A., 1975, Problems associated with the control of rodents in tropical Africa, *Bull. WHO* **52**:697.

Green, C. A., Keogh, H., Gordon, D. H., Pinto, M., and Hartwig, E. K., 1980, The distribution, identification, and naming of the *Mastomys natalensis* species complex in southern Africa (Rodentia: Muridae), *J. Zool. (Lond.)* **192**:17.

Gregg, M. B., 1975, Recent outbreaks of lymphocytic choriomeningitis in the United States of America *Bull. WHO* **52**:549.

Hall, E. R., and Kelson, K. R., 1959, *The Mammals of North America,* p. 671, Ronald Press, New York.

Hamilton, G. D., and Bronson, F. H., 1985, Food restriction and reproductive development in wild house mice, *Biol. Reprod.* **32**:773.

Happold, D. C. D., 1987, *The Mammals of Nigeria*, Oxford Science Publications, Clarendon Press, Oxford.

Hattis, R. P., Halstead, S. B., Herrmann, K. L., and Witte, J. J., 1973, Rubella in an immunized island population, *JAMA* **223**:1019.

Hinman, A. R., Fraser, D. W., Douglas, R. G., *et al.* 1975, Outbreak of lymphocytic choriomeningitis virus infections in medical center personnel, *Am. J. Epidemiol.* **101**:103.

Hirsch, M. S., Moellering, Jr., R. C., Pope, H. G., and Poskanzer, D. C., 1974, Lymphocytic choriomeningitis virus infection traced to a pet hamster, *N. Engl. J. Med.* **291**:610.

Horn, H. S., 1981, Sociobiology, in: *Theoretical Ecology: Principles and Applications*, 2nd ed. (R. M. May, ed.), pp. 272–294, Blackwell Scientific Publications, Boston.

Hotchin, J., 1962, The biology of lymphocytic choriomeningitis infection: Virus-induced immune disease, *Cold Spring Harbor Symp. Quant. Biol.* **28**:479.

Hotchin, J., and Collins, D. N., 1964, Glomerulonephritis and late onset disease of mice following neonatal virus infection, *Nature* **203**:1357.

Hotchin, J., and Weigand, H., 1961, Studies of lymphocytic choriomeningitis in mice. 1. The relationship between age at inoculation and outcome of infection, *J. Immunol.* **86**:392.

Howard, C. R., 1986, Arenaviruses, *Perspect. Med. Virol.***2**:1.

Jahrling, P. B., and Peters, C. J., 1986, Serology and virulence diversity among Old-World arenaviruses, and the relevance to vaccine development, *Med. Microbiol. Immunol.* **175**:165.

Jennings, W. L., Lewis, A. L., Sethers, G. E., Pierce, L. V., and Bond, J. O., 1970, Tamiami virus in the Tampa Bay area, *Am. J. Trop. Med. Hyg.* **19**:527.

Johnson, K. M., 1965, Epidemiology of Machupo virus infection. III. Significance of virological observations in man and animals, *Am. J. Trop. Med. Hyg.* **14**:816.

Johnson, K. M., 1975, Status of arenavirus vaccines and their application, *Bull. WHO* **52**:729.

Johnson, K. M., 1981, Arenaviruses: Diagnosis of infection in wild rodents, in: *Comparative Diagnosis of Viral Diseases* (E. Kurstak and C. Kurstak, eds.), pp. 511, Academic Press, New York.

Johnson, K. M., 1985, Arenaviruses, in: *Virology* (B. N. Fields and D. M. Knipe, eds.), pp. 1033–1053, Raven Press, New York.

Johnson, K. M., Mackenzie, R. B., Webb, P. A., and Kuns, M. L., 1965, Chronic infection of rodents by Machupo virus, *Science*, **150**:1618.

Johnson, K. M., Halstead, S. B., and Cohen, S. N., 1967, Hemorrhagic fevers of Southeast Asia and South America: A comparative appraisal, *Progr. Med. Virol.* **9**:105.

Johnson, K. M., Webb, P. A., and Justines, G., 1973, Biology of Tacaribe-complex viruses, in: *Lymphocytic Choriomeningitis Virus and Other Arenaviruses* (F. Lehmann-Grube, ed.), pp. 241–258, Springer-Verlag, New York.

Johnson, K. M., Taylor, P., Elliott, L. H., and Tomori, O., 1981, Recovery of a Lassa-related arenavirus in Zimbabwe, *Am. J. Trop. Med. Hyg.* **30**:1291.

Johnson, K. M., McCormick, J. B., Webb, P. A., and Krebs, J. W., 1982, The comparative biology of Old World (Lassa) and New World (Junin-Machupo) arenaviruses, in: *International Symposium on Tropical Arboviruses and Haemorrhagic Fevers* (F. de. Pinheiro, ed.), pp. 287–294, Academia Brasileira de Ciencias, Rio de Janeiro.

Justines, G., and Johnson, K. M., 1968, Use of oral swabs for detection of Machupo virus infection in rodents, *Am. J. Trop. Med. Hyg.* **17**(5):788.

Justines, G., and Johnson, K. M., 1969, Immune tolerance in *Calomys callsous* infected with Machupo virus, *Nature* **222**:1090.

Justines, G., and Johnson, K. M., 1970, Observations on the laboratory breeding of the cricetine rodent *Calomys callosus*, *Lab. Anim. Care* **20**:57.

Keenlyside, R. A., McCormick, J. B., Webb, P. A., Smith, E., Elliott, L., and Johnson, K. M.,

1983, Case–control study of *Mastomys natalensis* and humans in Lassa virus-infected households in Sierra Leone, *Am. J. Trop. Med. Hyg.* **32**:829.

Kenyon, R. H., Green, D. E., Maiztegui, J. I., and Peters, C. J., 1988, Viral strain-dependent differences in experimental Argentine hemorrhagic fever (Junin virus) infection of guinea pigs, *Intervirology* **29**:133.

Kenyon, R. H., Condie, R. M., Jahrling, P. B., and Peters, C. J., 1990, Protection of guinea pigs against experimental Argentine hemorrhagic fever by purified human-IgG: Importance of elimination of infected cells, *Microb. Pathog.* **9**:219.

Klavinskis, L. S., and Oldstone, M. B. A., 1986, Perturbation of endocrine functions during persistent infection of mice with lymphocytic choriomeningitis virus, *Med. Microbial. Immun.* **175**:117.

Klavinskis, L. S., Whitton, J. L., Joly, E., and Oldstone, M. B. A., 1990, Vaccination and protection from a lethal viral infection: Identification, incorporation, and use of a cytotoxic T lymphocyte glycoprotein epitope, *Virology* **178**:393.

Kravetz, F. O., and De Villafañe, G., 1981, Poblaciones de roedores en cultivos de maiz durante las etapas de madurez y rastrojo, *Hist. Nat.* **1**:213.

Kravetz, F. O., Percich, R. E., Zuleta, G. A., Calello, M. A., and Weissenbacher, M. C., 1986, Distribution of Junin virus and its reservoirs: Argentine hemorrhagic fever risk, *Interciencia* **11**:185.

Kuns, M., 1965, Epidemiology of Machupo virus infeqtion. II. Ecological and control studies of hemorrhagic fever, *Am. J. Trop. Med. Hyg.* **14**:813.

Langmuir, A. D., 1980, Changing concepts of acute contagious diseases: A reconsideration of classic epidemiologic theories, *Ann. NY Acad. Sci.* **353**:35.

Laurie, E. M. O., 1946, The reproduction of the house mouse (*Mus musculus*) living in different environmental conditions, *Proc. Roy. Soc. (B)* **133**:248.

Lehmann-Grube, F., 1963, Lymphocytic choriomeningitis in the mouse. II. Establishment of carrier colonies, *Arch. Virusforsch.* 14:353.

Lehmann-Grube, F., 1964, Lymphocytic choriomeningitis in the mouse. I. Growth in the brain, *Arch. Ges. Virusforsch.* **14**:344.

Lehmann-Grube, F., 1971, *Lymphocytic Choriomeningitis Virus*, Virology Monograph 10, Springer-Verlag, Vienna.

Lehmann-Grube, F., 1984, Portrait of viruses: Arenaviruses, *Intervirology* **22**:121.

Leirs, H., Hoofd, G., Verhagen, W., Lloyd, G., Micheals, M., Sabuni, C., Verhagen, R., and van der Groen, G., Antibodies against lassa or a lassa-related virus in rodents in Tanzania (submitted).

Lukashevich, I. S., 1985, Lassa virus lethality for inbred mice, *Ann. Soc. Belge Med. Trop.* **65**:207.

Lund, M., 1988, Rodent problems in Europe, in: *Rodent Pest Management* (I. Prakash, ed.), pp. 29–34, CRC Press, Boca Raton, FL.

Mackenzie, R. B., 1965, Epidemiology of Machupo virus infection. I. Pattern of human infection, San Joaquin, Bolivia, 1962–1964, *Am. J. Trop. Med. Hyg.* **14**:808.

Maiztegui, J. I., 1975, Clinical and epidemiological patterns of Argentine haemorrhagic fever, *Bull. WHO* **52**:567.

Maiztegui, J. I., Briggiler, A., Enria, D., and Feuillade, M. R., 1986, Progressive extension of the endemic area and changing incidence of Argentine hemorrhagic fever, *Med. Microbiol. Immunol.* **175**:149.

Marsh, R. E., 1988, Rodent problems on the North American continent, in: *Rodent Pest Management* (I. Prakash, ed.), pp. 1–12, CRC Press, Boca Raton, FL.

Marshall, J. T., Jr., 1981, Taxonomy, in: *The Mouse in Biomedical Research.* Vol. 1. *History, Genetics, and Wild Mice* (H. L. Foster, J. D. Small, and J. G. Fox, eds.), pp. 17–27, Academic Press, New York.

McCormick, J. B., 1987, Epidemiology and control of lassa fever, in: *Arenaviruses: Biology and Immunotherapy* (M. B. A. Oldstone, ed.), pp. 69–78, Springer-Verlag, New York.

McCormick, J. B., and Johnson, K. M., 1978, Lassa fever: Historical review and contemporary investigation, in: *Ebola Virus Haemorrhagic Fever* (S. R. Pattyn, ed.), pp. 279–285, North Holland Biomedical Press, New York.

McCormick, J. B., Webb, P. A., Krebs, J. W., Johnson, K. M., and Smith, E. S., 1987, A prospective study of the epidemiology and ecology of Lassa fever, *J. Infect. Dis.* **155**:437.

McKee, K. T., Jr., Mahlandt, B. G., Maiztequi, J. I., Eddy, G. A., and Peters, C. J., 1985, Experimental Argentine hemorrhagic fever in rhesus macaques, virus-strain dependent clinical response, *J. Infect. Dis.* **152**:218.

Medeot, S. I., Contigiani, M. S., Diaz, F., and Sabattini, M. S., 1992, Experimental neuroinvasiveness of wild and laboratory Junin virus strains, *Res. Virol.* **143**:259.

Medeot, S. I., Contigiani, M. S., Brandan, E. R., and Sabattini, M. S., 1990, Neurovirulence of wild and laboratory Junin virus strains in animal hosts, *J. Med. Virol.* **32**:171–182.

Mello, D. A., 1978, Biology of *Calomys callosus* (Rengger, 1830) under laboratory conditions (Rodentia, Cricetinae), *Rev. Bras. Biol.* **38**:807.

Mercado, R. R., 1975, Rodent control programmes in areas affected by Bolivian haemorrhagic fever, *Bull. WHO* **52**:691.

Meyer, H. M., Jr., Johnson, R. T., Crawford, I. P., *et al.* 1960, Central nervous system syndromes of "viral" etiology, *Am. J. Med.* **29**:334.

Mills, T. N., Ellis, B. A., McKee, Jr., K. T., Ksiazek, T. G., Barrera Oro, J. G., Maiztegui, J. I., Calderon, G. E., Peters, C. J., and Childs, J. E., 1991a, Junin virus activity in rodents from endemic and nonendemic loci in central Argentina, *Am. J. Trop. Med. Hyg.* **44**:589.

Mills, J. N., Ellis, B. A., McKee, Jr., K., Maiztegui, J. I., and Childs, J. E., 1991b, Habitat associations and relative densities of rodent populations in cultivated areas of central Argentina, *J. Mammal.* **72**:470.

Mills, J. N., Ellis, B. A., McKee, Jr., K. T., Maiztegui, J. I., and Childs, J. E., 1992, Reproductive characteristics of the rodent cormnunity in cultivated areas of central Argentina, *J. Mammal.* **73**:515.

Mims, C. A., 1966, Immunofluorescence study of the carrier state and mechanism of vertical transmission in lymphocytic choriomeningitis virus infection in mice, *J. Pathol. Bacteriol.* **91**:395.

Mims, C. A., 1969, Effect on the fetus of maternal infection with lymphocytic choriomeningitis (LCM) virus, *J. Infect Dis.* **120**:582.

Mims, C. A., 1970, observations on mice infected congenitally or neonatally with lymphocytic choriomeningitis (LCM) virus, *Arch. ges. Virus* **30**:67.

Mims, C. A., 1973, Acute and chronic LCM disease, in: *Lymphocytic Choriomeningitis Virus and Other Arenaviruses* (F. Lehmann-Grube, ed.), pp. 167–173, Springer-Verlag, New York.

Mims, C. A., 1975, The meaning of persistent infections in nature, *Bull. WHO* **52**:747.

Monath, T. P., 1975, Lassa fever: Review of epidemiology and epizootiology, *Bull. WHO* **52**:577.

Monath, T. P., Mertens, P. E., Patton, R., Moser, C. R., Baum, J. J., Pinneo, L., Gary, G. W., and Kissling, R. E., 1973, A hospital epidemic of Lassa fever in Zorzor, Liberia, March–April, 1972, *Am. J. Trop. Med. Hyg.* 22:773.

Monath, T. P., Newhouse, V. F., Kemp, G. E., Setzer, H. W., and Cacciapuoti, A., 1974, Lassa virus isolations from *Mastomys natalensis* rodents during an epidemic in Sierra Leone, Science, 185:263.

Montali, R. J., Ramsay, E. C., Stephensen, C. B., Worley, M., Davis, J. A., and Holmes, K. V., 1989, A new transmissible viral hepatitis of marmosets and tamarins, *J. Infect. Dis.* **160**:759.

Murphy, F. A., and Walker, D. H., 1978, Arenaviruses: Persistent infection and viral survival in reservoir hosts, in: *Viruses and Environment* (E. Kurstak and K. Maramorosch, eds.), pp. 155–180, Academic Press, New York.

Murphy, F. A., Winn, W. C., Walker, D. H., Flemister, M. R., and Witfield, S. G., 1976,

Early lympho reticular viral tropism and antigen persistence. Tamiami virus infection in the cotton rat, *Lab. Invest.* **43**:125.

Myers, J. H., 1974, The absence of t alleles in feral populations of house mice, *Evolution* **27**:702.

O'Connell, M. A., 1982, Population biology of North and South American grassland rodents: A comparative review, in: *Mammalian Biology in South America* (M. A. Mares and H. H. Genoways, eds.), pp. 167–185, Pymamuning Laboratory, Linesville, PA.

Oehen, S., Hengartner, H., and Zinkernagel, R. M., 1991, Vaccination for disease, *Science* **251**:195.

Oldstone, M. B. A., and Dixon, F. J., 1967, Lymphocytic choriomeningitis: Production of antibody by "tolerant" infected mice, *Science* **158**:1193.

Oldstone, M. B. A., and Dixon, F. J., 1968, Susceptibility of different mouse strains to lymphocytic choriomeningitis virus, *J. Immunol.* **100**:355.

Oldstone, M. B. A., Tishon, T., and Buchmeier, M. J., 1983, Virus-immune complex disease: Genetic control of c1q binding complexes in the circulation of mice persistently infected with lymphocytic choriomeningitis virus, *J. Immunol.* **130**:912.

Oldstone, M. B. A., Rodriguez, M., Daughaday, W. H., and Lampert, P. W., 1984, Viral perturbation of endocrine function: Disordered cell function leads to disturbed homeostasis and disease, *Nature* **307**:278.

Oldstone, M. B. A., Ahmed, R., Byrne, J., Buchmeier, M. J., Riviera, Y., and Southern, P., 1985, Virus and immune responses: Lymphocytic choriomeningitis virus as a prototype model of viral pathogenesis, *Br. Med. Bull.* **44**:70.

Parker, J. C., Igel, H. J., Reynolds, R. K., Lewis, Jr., A. M., and Rowe, W. P., 1976, Lymphocytic choriomeningitis virus infection in fetal, newborn, and young adult syrian hamsters (*Mesocricetus auratus*), *Infect. Immun.* **13**:967.

Peralta, L. A. M., Cossio, P. M., Sabattini, M., Maiztegui, J. I., Arana, R. M., and Laguens, R. P., 1979a, Ultrastructural, immunohistochemical, and virological studies in organs of *Calomys musculinus* infected with Junin virus by natural routes, *Medicina (Buenos Aires)* **39**:213.

Peralta, L. A. M., Laguens, R. P., Cossio, P. M., Sabattini, M. S., Maiztegui, J. I., and Arana, R. M., 1979b, Presence of viral particles in the salivary gland of *Calomys musculinus* infected with Junin virus by a natural route, *Intervirology* **11**:111.

Percich, 1988, Habitat selection, social structure, density and predation in populations of *Calomys* rodents in the pampa region of Argentina and the effects of agricultural practices on them, *Mammalia* **52**:339.

Perrigo, G., and Bronson, F. H., 1983, Foraging effort, food intake, fat deposition and puberty in female mice, *Biol. Reprod.* **29**:455.

Peters, C. J., 1991, Arenaviruses, in: *Textbook of Human Virology,* 2nd ed. (R. Belshe, ed.), Chapter 20, pp. 541–570, Mosby Year Book, St. Louis.

Peters, C. J., Webb, P. A., and Johnson, K. M., 1973, Measurement of antibodies to Machupo virus by the indirect fluorescent technique (37060), *Proc. Soc. Exp. Biol. Med.* **142**:526.

Peters, C. J., Kuehne, R. W., Mercado, R. R., *et al.*, 1974, Hemorrhagic fever in Cochabamba, Bolivia, 1971, *Am. J. Epidemiol.* **99**:425.

Peters, C. J., Jahrling, P. B., Liu, C. T., Kenyon, R. H., McKee, Jr., K. T., and Oro, J. G. B., 1987, Experimental studies of arenaviral hemorrhagic fevers, *Curr. Top. Microbial. Immunol.* **134**:5.

Peters, C. J., Johnson, E. D., and McKee, K. T., 1991, Filoviruses and management of viral hemorrhagic fevers, in: *Textbook of Human Virology,* 2nd ed. (R. Belshe, ed.), Chapter 26, pp. 699–712, Mosby Year Book, St. Louis.

Petter, F., 1966, Origine des Murides Plan cricetin et murin, *Mammalia* **30**:202.

Pevear, D. C., and Pfau, C. J., 1989, Lymphocytic choriomeningitis virus, in: *Clinical and Molecular Aspects of Neurotropic Virus Infection* (D. H. Gilden and H. L. Lipton, eds.), pp. 141–172, Kluwer Academic Publishers, New York.

Pinheiro, F. P., Woodall, J. P., Travassos da Rosa, A. P. A., and Travassos da Rosa, J. F., 1977, Studies on arenaviruses in Brazil, *Medicina (Buenos Aires)* **37**:175.

Pirosky, I., Zuccarini, J., Molinelli, E. A., Di Pietro, A., Barrera Oro, J. G., Martini, P., and Copello, A. R., 1959, Virosis hemorragica del noroeste bonaerense, in: *Virosis Hemorragica del Noroeste Bonaerense*, Republica Argentina, Poder Ejecutivo Nacional, Ministerio de Asistencia Social y Salud Publica, Insituto Nacional de Microbiologia, Buenos Aires.

Ramsay, E. C., Montali, R. J., Worley, M., Stephensen, C. B., and Holmes, K. V., 1989, Callithrichid hepatitis: Epizootiology of a fatal hepatitis in zoo tamarins and marmosets, *J. Zoo Wildlife Med.* 20:178.

Reig, O. A., 1987, An assessment of the systematics and evolution of the Akodontini, with the description of new fossil species of *Akodon* (Cricetidae: Sigmodontinae), in: *Studies in Neotropical Mammalogy: Essays in Honor of Philip Hershkovitz* (B. D. Patterson and R. M. Timm, eds.), pp. 347–400, Field Museum of Natural History, Chicago.

Richards, C. G. J., 1988, Large-scale evaluation of rodent control technologies, in: *Rodent Pest Management* (I. Prakash, ed.), pp. 269–284, CRC Press, Boca Raton, FL.

Riviere, Y., Gresser, I., Guillon, J. C., Bandu, M. T., Ronco, P., Morel-Moroger, L., and Verroust, P., 1980, Severity of lymphocytic choriomeningitis virus disease in different strains of suckling mice correlates with increasing amounts of endogenous interferon, *J. Exp. Med.* **152**:633.

Robbins, C. B., Krebs, Jr., J. W., and Johnson, K. M., 1983, *Mastomys* (Rodentia: Muridae) species distinguished by hemoglobin pattern differences, *Am. J. Trop. Med. Hyg.* **32**:624.

Rosevear, D. R., 1969, *The Rodents of West Africa*, 604 pp, British Museum, London.

Rowe, W. P., Pugh, W. E., Webb, P. A., *et al.*, 1970, Serological relationship of the Tacaribe complex of viruses to lymphocytic choriomeningitis virus, *J. Virol.* **5**:289.

Sabattini, M. S., and Contigiani, M. S., 1982, Ecological and biological factors influencing the maintenance of arenaviruses in nature, with special reference to the agent of Argentinean haemorrhagic fever (AHF), in: *International Symposium on Tropical Arboviruses and Haemorrhagic Fevers* (F. deP. Pinheiro, ed.), pp. 251–262, Academia Brasileira de Ciencias, Rio de Janeiro.

Sabattini, M. S., Barrera Oro, J. G., Maiztegui, J. I., Fernandez, D., Contigiani, M. S., and Diaz, G. E., 1970, Aislamiento de un *Arenavirus* relacionado con el de la coriomeningitis linfocitica (LCM) a partir de un *Mus musculus* capturado en zona endemica de fiebre hemorragica Argentina (FHA), *Rev. Soc. Argent. Microbial.* **11**:182.

Sabattini, M. S., Gonzalez de Rios, L. E., Diaz, G., and Vega, V. R., 1977, Infeccion natural y experimental de roedores con virus Junin, *Medicina (Buenos Aires)* **37**:149.

Sage, R. D., 1981, Wild mice, in: *The Mouse in Biomedical Research.* Vol. 1. *History, Genetics, and Wild Mice* (H. L. Foster, J. D. Small, and J. G. Fox, eds.), pp. 40–90, Academic Press, New York.

Salas, R., de Manzione, N., Tesh, R. B., Rico-Hesse, R., Shope, R. E., Betancourt, A., Godoy, O., Bruzual, R., Pacheco, M. E., Ramos, B., Taibo, M. E., Tamayo, J. G., Jaimes, E., Vasquez, C., Araoz, F., and Querales, J., 1991, Venezuelan haemorrhagic fever—A severe multisystem illness caused by a newly recognized arenavirus, *Lancet* **338**:1033.

Salvato, M. S., and Shimomaye, E. M., 1989, The completed sequence of LCMV reveals a new gene encoding a zinc-finger protein, *Virology* **173**:1.

Salvato, M. S., Shimomaye, E. M., Southern, P. J., and Oldstone, M. B. A., 1988, Virus, lymphocyte interactions. IV. Molecular characteristics of LCMV Armstrong (CTL^+) small genomic segment and that of its variant, Clone 13 (CTL^-), *Virology* 164:517–522.

Sauer, J. R., and Slade, N. A., 1986, Size-dependent population dynamics of *Microtus ochrogaster*, *Am. Nat.* **127**:902.

Sauer, J. R., and Slade, N. A., 1987, Size-based demography of vertebrates, *Annu. Rev. Ecol. Syst.* **18**:71.

Seamer, J., 1964, The growth, reproduction and mortality of mice made immunologically tolerant to lymphocytic choriomeningitis virus by congenital infection, *Arch. ges. Virus.* **15**:169.

Selander, R. K., and Kaufman, D. W., 1973, Genic variabilty and strategies of adaptation in animals, *Proc. Natl. Acad. Sci. USA* **70**:1875.

Shwartzman, G., 1946, Alterations in pathogenesis of experimental lymphocytic choriomeningitis caused by prepassage of the virus through heterologous host, *J. Immunol.* **54**:293.

Skinner, H. H., and Knight, E. H., 1973, Natural routes for post-natal transmission of murine lymphocytic choriomeningitis, *Lab. Anim.* **7**:171.

Skinner, H. H., and Knight, E. H., 1974, Factors influencing prenatal infection of mice with lymphocytic choriomeningitis virus, *Arch. ges. Virus.* **46**:1.

Skinner, H. H., Knight, E. H., and Buckley, L. S., 1976, The hamster as a secondary reservoir host of lymphocytic choriomeningitis virus, *J. Hyg. (Camb.)* **76**:299.

Skinner, H. H., Knight, E. H., and Grove, R., 1977, Murine lymphocytic choriomeningitis: The history of a natural cross-infection from wild to laboratory mice, *Lab. Anim.* **11**:219.

Smithard, E. H. R., and Macrae, A. D., 1951, Lymphocytic choriomeningitis associated human and mouse infection, *Br. Med. J.* 51:1299.

Southern, H. N., 1979, The stability and instability of small mammal populations, in: *Ecology of Small Mammals* (D. M. Stoddart, ed.), pp. 103–134, Chapman and Hall, London.

Southwick, C. H., 1966, Reproduction, mortality and growth of murid rodent populations, in: *Indian Rodent Symposium* (D. W. Parrick, ed.), pp. 152–176, Johns Hopkins University and USAID, Calcutta.

Stephensen, C. B., Jacob, J. R., Montali, R. J., Holmes, K. V., Muchmore, E., Compans, R. W., Arms, E. D., Buchmeier, M. J., and Lanford, R. E., 1991, Isolation of an arenavirus from a marmoset with callitrichid hepatitis and its serologic association with disease, *J. Virol.* 65:3995.

Stephenson, E. H., Larson, E. W., and Dominik, J. W., 1984, Effect of environmental factors on aerosol-induced Lassa virus infection, *J. Med. Virol.* **14**:295.

Swanepoel, R., Leman, P. A., Shepherd, A. J., Shepherd, S. P., Kiley, M. P., and McCormick, J. B., 1985, Identification of Ippy virus as a Lassa-fever-related virus, *Lancet* **1**:639.

Tauraso, N., and Shelokov, A., 1965, Protection against Junin virus by immunization with live Tacaribe virus, *Proc. Soc. Exp. Biol. Med.* **119**:608.

Tosolini, F. A., and Mims, C. A., 1971, Effect of murine strain and viral strain on the pathogenesis of lymphocytic-choriomeningitis infection and a study of footpad responses, *J. Infect. Dis.* **123**:134.

Trapido, H., and Sanmartin, C., 1971, Pichindé virus: A new virus of the Tacaribe group from Columbia, *Am. J. Trop. Med. Hyg.* **20**:63.

Traub, E., 1936a, The epidemiology of lymphocytic choriomeningitis in white mice, *J. Exp. Med.* **64**:183.

Traub, E., 1936b, An epidemic in a mouse colony due to the virus of acute lymphocytic choriomeningitis, *J. Exp. Med.* **63**:533.

Traub, E., 1938, Factors influencing the persistence of choriomeningitis virus in the blood of mice after clinical recovery, *J. Exp. Med.* **68**:229.

Traub, E., 1939, Epidemiology of lymphocytic choriomeningitis in a mouse stock observed for four years, *J. Exp. Med.* **69**:801.

Traub, E., 1973, LCM virus research, retrospect and prospects, in: *Lymphocytic Choriomeningitis Virus and Other Arenaviruses* (F. Lehmann-Grube, ed.), pp. 3–10, Springer-Verlag, New York.

Videla, C., Kajon, A., Carballal, G., and Weissenbacher, M. C., 1989, *Calomys callidus* as a potential Junin virus reservoir, *J. Med. Virol.* 27:238.

Vitullo, A. D., and Merani, M. S., 1988, Is vertical transmission sufficient to maintain Junin virus in nature? *J. Gen Virol.* **69**:1437.
Vitullo, A. D., and Merani, M. S., 1990, Vertical transmission of Junin virus in experimentally infected adult *Calomys musculinus, Intervirology* **31**:339.
Vitullo, A. D., Hodera V. L., and Merani, M. S., 1987, Effect of persistent infection with Junin virus on growth and reproduction of its natural reservoir, *Callomys musculinus, Am. J. Trop. Med. Hyg.* **37**:663.
Volkert, M., Bro-Jorgensen, K., and Marker, O., 1975, Persistent LCM virus infection in the mouse: Immunity and tolerance, *Bull. WHO* **52**:471.
Walker, D. H., Wulff, H., Lange, J. V., and Murphy, F. A., 1975, Comparative pathology of Lassa virus infection in monkeys, guinea-pigs, and *Mastomys natalensis, Bull. WHO* **52**:523.
Webb, P. A., Johnson, K. M., Peters, C. J., and Justines, G., 1973, Behavior of machupo and latino viruses in *Calomys callosus* from two geographic areas of Bolivia, in: *Lymphocytic Choriomeningitis Virus and Other Arenaviruses* (F. Lehmann-Grube, ed.), pp. 314–322, Springer-Verlag, New York.
Webb, P. A., Justines, G., and Johnson, K. M., 1975, Infection of wild and laboratory animals with Machupo and Latino viruses, *Bull. WHO* **52**:493.
Webb, P. A., McCormick, J. B., King, I. J., *et al.* 1986, Lassa fever in children in Sierra Leone, West Africa, *Trans. Roy. Soc. Trop. Med. Hyg.* **80**:577.
Webb, S. D., and Marshall, L. G., 1982, Historical biogeography of recent South American land mammals, in: *Mammalian Biology in South America* (M. A. Mares and H. H. Genoways, eds.), University of Pittsburgh Press, Pittsburgh.
Weigand, H., and Hotchin, J., 1961, Studies of lymphocytic choriomeningitis in mice. 2. A comparison of the immune status of newborn and adult mice surviving inoculation, *J. Immunol.* **86**:401.
Weissenbacher, M. C., De Duerrero, L. B., and Boxaca, M. C., 1975, Experimental biology and pathogenesis of Junin virus infection in animals and man, *Bull. WHO* **52**:507.
Weissenbacher, M. C., Calello, M. A., Carballal, G., Planes, N., de la Vega, M. T., and Kravetz, F., 1985, Actividad del virus Junin en humanos y roedores de areas no endemical de la provincia de Buenos Aires, *Medicina (Buenos Aires)* **45**:263.
Weissenbacher, M. C., Laguens, R. P., and Coto, C. E., 1987, Argentine hemorrhagic fever, in: *Arenaviruses: Biology and Immunotherapy* (M. B. A. Oldstone, ed.), pp. 79–116, Springer-Verlag, New York.
Whitton, J. L., Tishon, A., Lewicki, H., Gebhard, J., Cook, T., Salvato, M., Joly, E., and Oldstone, M. B. A., 1989, Molecular analyses of a five-amino-acid cytotoxic T lymphocyte (CTL) epitope: An immunodominant region which induces nonreciprocal CTL cross-reactivity, *J. Virol.* **63**:4303.
Winn, W. C., and Murphy, F. A., 1975, Tamiami virus infection in mice and cotton rats, *Bull. WHO* **52**:501.
Wulff, H., Fabiyi, A., and Monath, T. P., 1975, Recent isolations of Lassa virus from Nigerian rodents, *Bull. WHO* **52**:609.
Wulff, H., McIntosh, B. M., Hammer, D. B., and Johnson, K. M., 1977, Isolation of an *Arenavirus* closely related to Lassa virus from *Mastomys natalensis* in south-east Africa, *Bull. WHO* **55**:441.
Yunes, R. M. F., Cutrera, R. A., and Castro-Vasquez, A., 1991, Nesting and digging behavior in 3 species of *Calomys* (Rodentia, Cricetidae), *Physiol. Behav.* **49**:489.
Zeller, H. G., Karabatsos, N., Calisher, C. H., Digoutte, J-P., Murphy, F. A., and Shope, R. E. 1989, Electron microscopy and antigen studies of uncharacterized viruses. I. Evidence suggesting the placement of viruses in family *Arenaviridae, Paramvxoviridae,* or *Poxiviridae, Arch. Virol.* **108**:191.
Zinkernagel R. M., and Doherty, P. C., 1974, Restriction of *in vitro* T cell-mediated cytotoxicity in lymphocytic choriomeningitis within a syngeneic or semiallogeneic system, *Nature* **248**:701.

Index